Complex Fluids in Biological Systems

BIOLOGICAL AND MEDICAL PHYSICS, BIOMEDICAL ENGINEERING

The fields of biological and medical physics and biomedical engineering are broad, multidisciplinary and dynamic. They lie at the crossroads of frontier research in physics, biology, chemistry, and medicine. The Biological and Medical Physics, Biomedical Engineering Series is intended to be comprehensive, covering a broad range of topics important to the study of the physical, chemical and biological sciences. Its goal is to provide scientists and engineers with textbooks, monographs, and reference works to address the growing need for information.

Books in the series emphasize established and emergent areas of science including molecular, membrane, and mathematical biophysics; photosynthetic energy harvesting and conversion; information processing; physical principles of genetics; sensory communications; automata networks, neural networks, and cellular automata. Equally important will be coverage of applied aspects of biological and medical physics and biomedical engineering such as molecular electronic components and devices, biosensors, medicine, imaging, physical principles of renewable energy production, advanced prostheses, and environmental control and engineering.

More information about this series at http://www.springer.com/series/3740

Saverio E. Spagnolie
Editor

Complex Fluids in Biological Systems

Experiment, Theory, and Computation

Springer

Editor
Saverio E. Spagnolie
Department of Mathematics
University of Wisconsin - Madison
Madison, WI, USA

ISSN 1618-7210 ISSN 2197-5647 (electronic)
ISBN 978-1-4939-2064-8 ISBN 978-1-4939-2065-5 (eBook)
DOI 10.1007/978-1-4939-2065-5
Springer New York Heidelberg Dordrecht London

Library of Congress Control Number: 2014953309

Printed on acid-free paper

Springer is part of Springer Science+Business Media (www.springer.com)

To life, in all its wondrous
and stupefying complexity

Preface

The complexity of biological systems, even on the smallest length scales, is staggering. Biological systems are replete with active functionality, heterogeneity, memory, and interconnectedness on a vast spectrum of length and time scales. With our ever-advancing abilities to observe nature in vivo at the microscale, and with continuing developments of mathematical and numerical machinery for understanding multiscale physical systems, the fields of complex fluids and biological systems are ripe for fruitful cross-pollination. There have already been many successful scientific advances along these lines, as will be made clear in the chapters that follow. The aim of this book is to introduce the reader to many of the exciting directions that this research is taking and to provide a valuable reference on fundamental phenomena, models, and analysis of complex fluids in a variety of biological systems.

The book is organized into four parts. In Part I, Newtonian and complex fluids are introduced, along with the terminology and models that will appear frequently throughout the book. The first chapter provides the mathematical framework of continuum mechanics and presents common constitutive laws used to describe fluids with such properties as shear-dependent viscosity and viscoelasticity. Classical rheological flows frequently used in experiments are introduced. The second chapter lays the foundations for the topics to be covered in the book and explores critical functional roles played by complex fluids in a familiar biological system, the human body. Using mucus as an illustrative example, a multidisciplinary approach to studying and modeling soft, complex biological matter is emphasized.

In Part II, the measurement of biological material properties, or *rheology*, takes center stage. The first chapter is devoted to microrheology, wherein the behavior of small immersed particles is used to infer material properties of the surrounding environment. Both passive microrheology and active microrheology are discussed, beginning with the famed Stokes-Einstein relation and marching through a history of the field towards a "nonequilibrium equation of state." The following two chapters return to specific biological structures, namely the cell membrane and cell cytoskeleton. Microrheology is revisited as a means of studying the viscoelastic properties of molecularly thin shells, and the intricate biopolymer network internal to individual cells is introduced. A final chapter in this section explores a variety of challenges faced by experimentalists in the study of complex biological fluids and shows how a misinterpretation of data can suggest complex fluid properties when there are none and vice versa.

Part III focuses on the locomotion of microorganisms through complex biological fluids, as described from experimental, analytical, and numerical perspectives. The first chapter reviews the recent experimental studies of biolocomotion in viscous and viscoelastic fluids and then turns to intriguing experimental results on the propulsion of a model organism, the roundworm *C. elegans*. This sets the stage for the following chapter, which covers a detailed mathematical theory of locomotion in complex fluids, and connections between microrheology and biolocomotion are described. The interaction of swimming organisms in complex fluids is also discussed, which leads naturally into the final chapter of the part. In the last chapter, the focus turns to a model of large collections of such swimming organisms, or an *active suspension*, which can exhibit large-scale correlated motions, pattern formation, and complex fluid properties including normal stress differences. The model is extended to the study of other systems, including the interaction of microtubules and translocating motor proteins as found in individual cells.

Finally, Part IV covers methods for computing fluid flows with intricate immersed boundaries. Common numerical approaches are made considerably more challenging when the fluid is highly elastic. The first chapter describes many of these challenges, including the catastrophic high-Weissenberg number problem, and offers solutions. The immersed boundary method is introduced, and the locomotion of *C. elegans* in viscoelastic fluids is revisited as a test problem from a numerical perspective. The final chapter of the book presents a cell-level numerical study of blood flow, where the shapes and dynamics of individual cells and their interactions are captured in a boundary integral formulation of the problem. The numerical method is used to understand physical effects well known to physiologists such as the Fåhræus effect, Fåhræus-Lindqvist effect, and the margination of leukocytes and platelets.

The chapters contained herein will provide the reader with an overview as well as a detailed inspection of the challenges and opportunities that await us in the coming decades of research in complex biological flows, and the observations, methods, and tools available for their study. Active areas of exploration are presented by many of the world's foremost experts in their respective fields. Consequently, each chapter both provides a substantial review of the literature and delivers the very cutting edge of our current knowledge. The book was developed with advanced undergraduate and early graduate students in the engineering, biological, and mathematical sciences in mind, but it will appeal to anyone interested in the intricate and beautiful nature of complex fluids in the context of living systems.

Numerous acknowledgements are in order. It has been a great pleasure to work with the many authors of this book, who continue to forge new paths in their respective fields and to inspire with their creativity and remarkably hard work. It is immensely gratifying to toil as a member of an extended scientific family that knows no geographical borders. I am particularly indebted to Harvey Segur, Michael

Shelley, Eric Lauga, Thomas Powers, and Jean-Luc Thiffeault, and I would like to thank Gwynn Elfring for dependable consultation on this project. Finally, I am forever grateful to my wife Elena for her love and support, and to my daughter Carina for joining the adventure.

Madison, WI, USA
July, 2014

Saverio E. Spagnolie

Contents

Part IV Computational Methods

Contributors

Paulo E. Arratia Department of Mechanical Engineering and Applied Mechanics, University of Pennsylvania, Philadelphia, PA, USA

John F. Brady Division of Chemistry and Chemical Engineering, California Institute of Technology, Pasadena, CA, USA

Lucas M. Caretta Department of Materials Science and Engineering, Massachusetts Institute of Technology, Cambridge, MA, USA

Gwynn J. Elfring Department of Mechanical Engineering, University of British Columbia, Vancouver, BC, Canada

Arthur A. Evans Department of Physics, University of Massachusetts Amherst, Amherst, MA, USA

Randy H. Ewoldt Department of Mechanical Science and Engineering, University of Illinois Urbana-Champaign, Urbana, IL, USA

M. Gregory Forest Department of Mathematics, Institute for Advanced Materials, University of North Carolina at Chapel Hill, Chapel Hill, NC, USA

Michael D. Graham Department of Chemical and Biological Engineering, University of Wisconsin-Madison, Madison, WI, USA

Robert D. Guy Department of Mathematics, University of California, Davis, 1 Shields Ave., Davis, CA 95616, USA

Hamed Hatami-Marbini School of Mechanical and Aerospace Engineering, Oklahoma State University, Stillwater, OK, USA

Michael T. Johnston Department of Mechanical Science and Engineering, University of Illinois Urbana-Champaign, Urbana, IL, USA

Amit Kumar Department of Chemical and Biological Engineering, University of Wisconsin-Madison, Madison, WI, USA

Eric Lauga Department of Applied Mathematics and Theoretical Physics, University of Cambridge, Cambridge, UK

Alex J. Levine Department of Chemistry and Biochemistry, The California Nanosystems Institute, UCLA, Los Angeles, CA, USA

Mohammad R. K. Mofrad Department of Bioengineering, University of California, Berkeley, CA, USA

Alexander Morozov SUPA, School of Physics and Astronomy, University of Edinburgh, JCMB, Edinburgh, UK

David Saintillan Department of Mechanical and Aerospace Engineering, University of California San Diego, La Jolla, CA, USA

Michael J. Shelley Courant Institute of Mathematical Sciences, New York University, New York, NY, USA

Saverio E. Spagnolie Department of Mathematics, University of Wisconsin-Madison, Madison, WI, USA

Josué Sznitman Department of Biomedical Engineering, Technion-Israel Institute of Technology, Technion City, Haifa, Israel

Becca Thomases Department of Mathematics, University of California, Davis, CA, USA

Paula A. Vasquez Department of Mathematics, University of South Carolina, Columbia, SC, USA

Roseanna N. Zia School of Chemical and Biomolecular Engineering, Cornell University, Ithaca, NY, USA

List of Symbols

Fluid Properties

ρ	Density
μ	Viscosity of a Newtonian fluid

Fluid Dynamics

$\mathbf{u}$	Fluid velocity
p	Pressure
$\boldsymbol{\sigma}$	Total stress tensor $(= -p\mathbf{I} + \boldsymbol{\tau})$
$\boldsymbol{\tau}$	Deviatoric stress tensor
D/Dt	Substantial/material time derivative $(= \partial/\partial_t + \mathbf{u} \cdot \nabla)$
$\overset{\nabla}{\boldsymbol{\tau}}$	Upper convected time derivative of $\boldsymbol{\tau}$ $\left(\equiv (D/Dt)\boldsymbol{\tau} - \left[(\nabla\mathbf{u})^T \cdot \boldsymbol{\tau} + \boldsymbol{\tau} \cdot \nabla\mathbf{u}\right]\right)$

Kinematic Tensors

$\nabla\mathbf{u}$	Velocity gradient tensor $[\nabla\mathbf{u}]_{ij} = \partial u_j/\partial x_i$
$\dot{\boldsymbol{\gamma}}$	Rate-of-strain tensor $\left(= \nabla\mathbf{u} + (\nabla\mathbf{u})^T\right)$
$\boldsymbol{\omega}$	Vorticity tensor $\left(= \nabla\mathbf{u} - (\nabla\mathbf{u})^T\right)$

Quantities in Constitutive Equations

η_0	Zero-shear-rate viscosity
η_∞	Infinite-shear-rate viscosity
λ	Relaxation time
$G(t-t')$	Relaxation modulus
$M(t-t')$	Memory function

Material Functions

$\eta(\lvert\dot{\boldsymbol{\gamma}}\rvert)$	Non-Newtonian viscosity $\left(= \eta\left(\sqrt{\mathrm{tr}(\dot{\boldsymbol{\gamma}}^2)/2}\right)\right)$
N_1	First normal stress difference
N_2	Second normal stress difference
Ψ_1	First normal stress difference coefficient

Ψ_2	Second normal stress difference coefficient
$\eta^*(\omega)$	Complex viscosity ($\eta^* = \eta' - i\eta''$)
$G^*(\omega)$	Complex modulus ($G^* = G' + iG''$)

Dimensionless Numbers

De	Deborah number
Re	Reynolds number
Wi	Weissenberg number

General

k_B	Boltzmann's constant
T	Absolute temperature
N_A	Avogadro's number

Mathematical Symbols

$\Re(z)$	Real part of complex number z
$\Im(z)$	Imaginary part of complex number z

Part I
Introduction to Complex Fluids

Chapter 1
Introduction to Complex Fluids

Alexander Morozov and Saverio E. Spagnolie

Abstract In this chapter we introduce the fundamental concepts in Newtonian and complex fluid mechanics, beginning with the basic underlying assumptions in continuum mechanical modeling. The equations of mass and momentum conservation are derived, and the Cauchy stress tensor makes its first of many appearances. The Navier–Stokes equations are derived, along with their inertialess limit, the Stokes equations. Models used to describe complex fluid phenomena such as shear-dependent viscosity and viscoelasticity are then discussed, beginning with generalized Newtonian fluids. The Carreau–Yasuda and power-law fluid models receive special attention, and a mechanical instability is shown to exist for highly shear-thinning fluids. Differential constitutive models of viscoelastic flows are then described, beginning with the Maxwell fluid and Kelvin–Voigt solid models. After providing the foundations for objective (frame-invariant) derivatives, the linear models are extended to mathematically sound nonlinear models including the upper-convected Maxwell and Oldroyd-B models and others. A derivation of the upper-convected Maxwell model from the kinetic theory perspective is also provided. Finally, normal stress differences are discussed, and the reader is warned about common pitfalls in the mathematical modeling of complex fluids.

1 Introduction

The complexity of biological systems is extraordinary and, from a mathematical modeling point of view, daunting. Even the continuum approximations that give rise to the classical equations of fluid and solid mechanics do not survive the intricacy

A. Morozov
SUPA, School of Physics & Astronomy, University of Edinburgh, JCMB,
King's Buildings, Mayfield Road, Edinburgh EH9 3JZ, UK
e-mail: Alexander.Morozov@ed.ac.uk

S.E. Spagnolie (✉)
Department of Mathematics, University of Wisconsin-Madison,
480 Lincoln Dr., Madison, WI 53706, USA
e-mail: spagnolie@math.wisc.edu

S.E. Spagnolie (ed.), *Complex Fluids in Biological Systems*, Biological and Medical Physics, Biomedical Engineering,
DOI 10.1007/978-1-4939-2065-5_1

of biological matter, and the systems of equations describing the relevant flows, deformations, and stresses are coupled and nonlinear. This book will be concerned with the dynamics of these complex fluid flows in relation to a number of important biological systems. Many of the biological fluids to be discussed are far from homogeneous. Highly heterogeneous biological materials include mucus, which forms a three-dimensional network with a potentially fractal length-scale distribution, and the cytoskeleton, which is an active structure that undergoes continuous remodeling in response to external and internal stimulation. Nevertheless, with the introduction of a more involved microstructure, such as the inclusion of long chain molecules (e.g., DNA, proteins, microtubules, etc.), continuum assumptions are commonly made to make mathematical modeling and analysis possible. Continuum modeling is even possible when each fundamental parcel includes numerous active particles such as swimming microorganisms, which allows for the derivation of partial differential equations describing *active suspensions*. As will be seen in the chapters to come, the continuum approach has already been an enormously successful method for modeling and understanding real biological systems.

In this first chapter we will lay out the mathematical framework of continuum mechanics and present common constitutive laws used to describe fluids with such properties as shear-dependent viscosity and viscoelasticity. The chapter is organized as follows. We begin with an introduction to the classical equations of Newtonian fluid mechanics in Sect. 2, covering material and spatial descriptions of variables, the mathematization of physical conservation laws, stress, the Navier–Stokes equations, and dimensional analysis. In Sect. 3 we take a first step away from the classical Newtonian constitutive law into elementary models of complex fluids where the viscosity depends on the local flow rate, so-called generalized Newtonian fluids, which include the power-law and Carreau–Yasuda models. More advanced differential constitutive models are the topic of Sect. 4, beginning with the linearly viscoelastic Maxwell fluid and Kelvin–Voigt solid models. After a discussion about objectivity (frame-invariance), the upper-convected Maxwell (UCM), Oldroyd-B, and many other models of nonlinear viscoelastic flow are introduced. A derivation of the UCM model from the perspective of kinetic theory is also provided. In Sect. 5, the material properties of viscoelastic fluids are discussed, and in particular we introduce normal stress differences and describe some of the classical rheological flows that are used to measure the various complex responses to deformation in real fluids. We conclude with a few words of caution about common but ill-advised choices made in the mathematical modeling of complex fluids in Sect. 6, and closing remarks in Sect. 7.

2 Newtonian Fluid Mechanics

The problems to be described in the chapters to come are extraordinarily involved when viewed at the molecular level. The basic mathematical idealization of a homogeneous liquid such as pure water assumes that the fundamental elements describing the material are *parcels* only somewhat larger than the molecular mean free path. Quantities such as density and pressure in the fluid are assumed to be constant throughout each such volume, but the parcel size is small enough so that the variations in such variables in neighboring parcels are effectively continuous. Partial differential equations modeling the response of a solid or fluid (or a material with both solid-like and fluid-like responses to deformations) are made possible by this *continuum approximation*.

The analysis of classical fluid flows has been one of the great successes of applied mathematics since the time of the Bernoullis, and there are countless excellent presentations of the subject. The reader is referred to the comprehensive texts by Batchelor [1], Landau and Lifshitz [2], Leal [3], and Pozrikidis [4] and the more concise introductions by Acheson [5] and Childress [6]. Here we introduce the basic concepts of mathematical fluid mechanics which, upon the application of mass and momentum conservation laws and specification of a particular constitutive law, result in the Navier–Stokes equations of classical Newtonian flow.

2.1 *Material (Lagrangian) and Spatial (Eulerian) Variables*

We begin by introducing two important descriptions of independent variables such as density, velocity, and pressure. The first is the material description, also known as the *Lagrangian* description, in which parcels of fluid (or other material) are associated with a "label" $\mathbf{a}$, commonly chosen to be the initial position of the parcel in space. The pressure, for instance, measured at a parcel of material as it moves through space may be written at time t as $P(\mathbf{a},t)$. The second is the spatial description, also known as the *Eulerian* description, in which the same variables are described in terms of a fixed position $\mathbf{x}$ in space, e.g., $P(\mathbf{a},t) = p(\mathbf{x}(\mathbf{a},t),t)$.

The relationship between the two descriptions is a map χ of each label $\mathbf{a}$ in the reference configuration to its current position, $\mathbf{x} = \chi(\mathbf{a},t)$, as illustrated in Fig. 1.1.

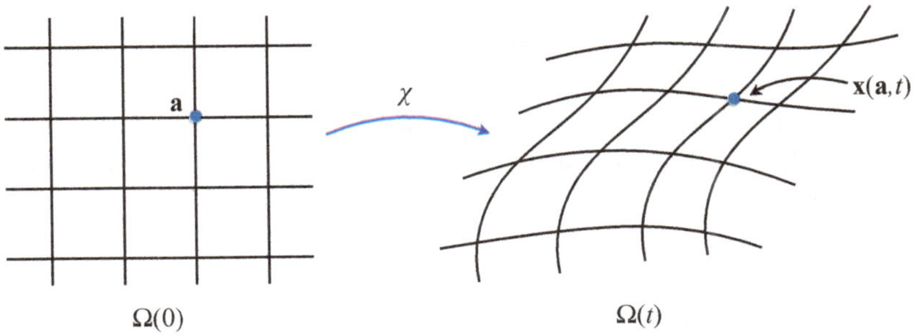

Fig. 1.1 The reference (*Lagrangian*) configuration is deformed to the current (*Eulerian*) configuration at a time t. The material "label" $\mathbf{a}$ maps to a new spatial position $\mathbf{x} = \chi(\mathbf{a},t)$, where $\chi(\mathbf{a},0) = \mathbf{a}$

A control volume in the reference domain $\Omega(0)$ maps to a volume in the current configuration, $\Omega(t)$. Defining the velocity of material at a point $\mathbf{x}$ in space as $\mathbf{u}(\mathbf{x},t)$, then the velocity of a material point labeled by $\mathbf{a}$ is given by

$$\left.\frac{d\mathbf{x}}{dt}\right|_{\mathbf{a}} = \frac{d}{dt}\chi(\mathbf{a},t) = \mathbf{u}(\mathbf{x}(\mathbf{a},t),t), \tag{1.1}$$

where $\mathbf{x}(\mathbf{a},0) = \mathbf{a}$. Similarly, the time rate of change of a scalar variable F in the material description is given simply as $\partial_t F(\mathbf{a},t)$. However, if F is described instead in terms of the spatial variables, $F(\mathbf{a},t) = f(\mathbf{x}(\mathbf{a},t),t)$, then the derivative must correct for the change of frame, and instead (using the chain rule) we find

$$\left.\frac{d}{dt}f(\mathbf{x}(\mathbf{a},t),t)\right|_{\mathbf{a}} = \frac{\partial f}{\partial t} + \frac{\partial f}{\partial x_i}\frac{\partial x_i}{\partial t} = (\partial_t + \mathbf{u}\cdot\nabla)f = \frac{Df}{Dt}. \tag{1.2}$$

We have used the Einstein summation notation where a summation is implied over the repeated index i. The operator $D/Dt = \partial_t + \mathbf{u}\cdot\nabla$ is the *material derivative*, which is a time derivative that follows the material as it deforms. The acceleration of a fluid particle, written in terms of the spatial representation of the velocity field $\mathbf{u}(\mathbf{x},t)$, is then given by

$$\frac{D\mathbf{u}}{Dt} = \frac{\partial \mathbf{u}}{\partial t} + \mathbf{u}\cdot\nabla\mathbf{u}. \tag{1.3}$$

Finally, an important measure of the fluid deformation is the Jacobian matrix of the map $\mathbf{x} = \chi(\mathbf{a},t)$, also known as the *deformation gradient tensor*, given by $\mathbf{F} = \partial\mathbf{x}/\partial\mathbf{a}$ (where $F_{ij} = \partial x_i/\partial a^j$), and $\mathbf{F}(\mathbf{x},0) = \mathbf{I}$, the identity operator. For instance, a line element $d\mathbf{a}$ from one material point to another in the reference configuration transforms to a new line element $d\mathbf{x}$ in the current configuration as $d\mathbf{x} = \mathbf{F}\cdot d\mathbf{a}$. As we proceed to consider conservation laws of mass and momentum we will require the time derivative of the determinant of the deformation gradient tensor $J = \det \mathbf{F}$, which is given by

$$\frac{dJ}{dt} = (\nabla\cdot\mathbf{u})J, \tag{1.4}$$

(see [6]). A volume-preserving or *incompressible* material is one for which $J(\mathbf{x},t) = 1$ and hence $\nabla\cdot\mathbf{u} = 0$ for all $\mathbf{x}$ and t. All of the fluids considered in this book are treated as incompressible.

2.2 Conservation of Mass

The density of a fluid, $\rho(\mathbf{x},t)$, is defined as the mass per volume in an infinitesimal fluid parcel centered at $\mathbf{x}$. Consider a material volume $\Omega(t)$ in the current configuration. As the material volume moves and deforms under the flow, the mass of fluid in $\Omega(t)$ is determined by integrating the density throughout the volume, $M(t) = \int_{\Omega(t)} \rho(\mathbf{x},t)\,dV$. Absent the creation or destruction of mass, we will then have that the mass in the material volume at any time is equal to its initial value, $M(t) = M(0)$ or $dM/dt = 0$. Mass conservation may then be written in the Eulerian form as

$$\frac{d}{dt}\int_{\Omega(t)} \rho(\mathbf{x},t)\,dV_x = 0, \tag{1.5}$$

with $dV_x = dx_1\,dx_2\,dx_3$. A differential form of mass conservation is achieved by first representing the density in the material coordinates,

$$\begin{aligned} 0 = \frac{d}{dt}\int_{\Omega(t)} \rho(\mathbf{x},t)\,dV_x &= \frac{d}{dt}\int_{\Omega(0)} \rho(\mathbf{x}(\mathbf{a},t),t)J\,dV_a \\ &= \int_{\Omega(0)} \left(\frac{D\rho}{Dt}(\mathbf{x}(\mathbf{a},t),t) + \rho(\mathbf{x}(\mathbf{a},t),t)(\nabla\cdot\mathbf{u})\right)J\,dV_a \\ &= \int_{\Omega(t)} \left(\frac{D\rho}{Dt}(\mathbf{x},t) + \rho(\mathbf{x},t)(\nabla\cdot\mathbf{u})\right)dV_x. \end{aligned} \tag{1.6}$$

This relation is a special case of the Reynolds transport theorem, or convection theorem, as applied to the scalar function $\rho(\mathbf{x},t)$. Since the above holds for all material volumes, we arrive at a differential form of mass conservation:

$$\frac{D\rho}{Dt} + \rho(\nabla\cdot\mathbf{u}) = 0. \tag{1.7}$$

In the event that the fluid is incompressible, $\nabla\cdot\mathbf{u} = 0$, so that $D\rho/Dt = 0$. In other words, in an incompressible flow, the density associated with any material point remains constant as it moves with the fluid.

2.3 *Conservation of Momentum*

While the mass in a control volume is given by the integrated fluid density, the fluid momentum contained in a volume $\Omega(t)$ may be written in terms of Eulerian variables as $\mathbf{p}(t) = \int_{\Omega(t)} \rho(\mathbf{x},t)\mathbf{u}(\mathbf{x},t)\,dV_x$. In a similar calculation as in the previous section and using Eq. (1.7), we have the following identity:

$$\begin{aligned} \frac{d}{dt}\mathbf{p}(t) = \frac{d}{dt}\int_{\Omega(t)} (\rho\mathbf{u})(\mathbf{x},t)\,dV_x &= \frac{d}{dt}\int_{\Omega(0)} (\rho\mathbf{u})(\mathbf{x}(\mathbf{a},t),t)J\,dV_a \\ &= \int_{\Omega(0)} \left(\frac{D\rho}{Dt}\mathbf{u} + \rho\frac{D\mathbf{u}}{Dt} + \rho\mathbf{u}(\nabla\cdot\mathbf{u})\right)J\,dV_a \end{aligned}$$

$$= \int_{\Omega(0)} \rho \frac{D\mathbf{u}}{Dt} J \, dV_a = \int_{\Omega(t)} \rho \frac{D\mathbf{u}}{Dt}(\mathbf{x},t) \, dV_x. \tag{1.8}$$

According to Newton's second law, the rate of change of the momentum in the material volume $\Omega(t)$ must balance with any forces acting on the contained fluid. The forces on the fluid come in two varieties: external body forces such as gravity, which we denote as a force per unit volume by **f**, and surface forces such as viscous or elastic stresses, which we denote as a force per unit area by **t**, the traction. The surface of the fluid volume is described locally by the outward-pointing unit normal vector, denoted by **n**, and the surface traction may be represented generally as $\mathbf{t} = \mathbf{n} \cdot \boldsymbol{\sigma}$, where $\boldsymbol{\sigma}$ is the Cauchy stress tensor (see Sect. 2.4). The standard proof of this representation is achieved by applying Newton's second law to a tetrahedron of shrinking volume (see [3]). Balancing the forces, another application of Newton's second law, now to an arbitrary material volume, provides the following integral form of momentum conservation,

$$\frac{d}{dt}\mathbf{p}(t) = \int_{\Omega(t)} \rho \frac{D\mathbf{u}}{Dt} \, dV_x = \int_{\Omega(t)} \mathbf{f} \, dV_x + \int_{\partial\Omega(t)} \mathbf{t} \, dS_x, \tag{1.9}$$

where $\partial\Omega(t)$ is the boundary of $\Omega(t)$ and dS_x is an infinitesimal surface area element. The last integral can be converted to a volume integral using the divergence theorem, so that

$$\int_{\Omega(t)} \rho \frac{D\mathbf{u}}{Dt} \, dV_x = \int_{\Omega(t)} \mathbf{f} \, dV_x + \int_{\Omega(t)} \nabla \cdot \boldsymbol{\sigma} \, dV_x. \tag{1.10}$$

Since the above holds for all material volumes, we arrive at the general differential form of momentum conservation:

$$\rho \frac{D\mathbf{u}}{Dt} = \mathbf{f} + \nabla \cdot \boldsymbol{\sigma}. \tag{1.11}$$

2.4 The Cauchy Stress Tensor and the Navier–Stokes Equations

The wide array of mathematical models for vastly different types of materials and fluids reduce to a specification of the Cauchy stress tensor, $\boldsymbol{\sigma}$. In a classical elastic solid the stress tensor depends locally on the material deformation; in a classical viscous fluid the stress tensor depends locally on the *rate* of material deformation. Generally, however, and particularly for the complex fluids to be described in this book, $\boldsymbol{\sigma}$ may even evolve in a nonlinear and history-dependent way through a partial differential equation of its own.

The components of the stress tensor may be interpreted by considering a cubic volume as illustrated in Fig. 1.2. Since the traction (the force per unit area) on a

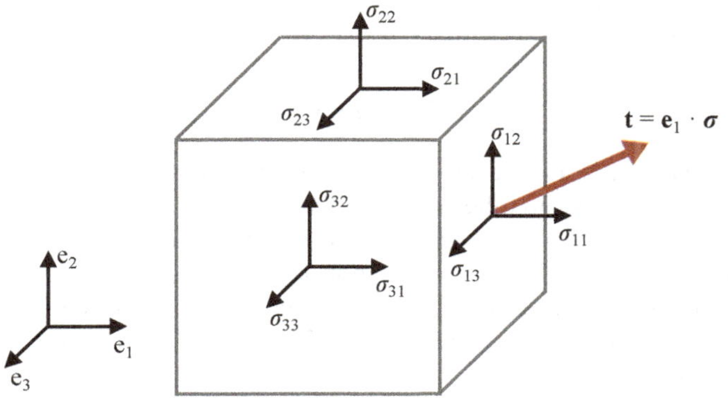

Fig. 1.2 The stress tensor $\boldsymbol{\sigma}$ contains all of the information about the surface tractions, save for the surface geometry. The traction (force per unit area) on the rightmost face of the cube, which is characterized by the outward pointing normal vector $\mathbf{e}_1$, is given by $\mathbf{t} = \mathbf{e}_1 \cdot \boldsymbol{\sigma} = \sigma_{11}\mathbf{e}_1 + \sigma_{12}\mathbf{e}_2 + \sigma_{13}\mathbf{e}_3$

surface is represented generally as $\mathbf{t} = \mathbf{n} \cdot \boldsymbol{\sigma}$, where $\mathbf{n}$ is the unit normal vector pointing out of the control volume, σ_{ij} represents the traction in the jth direction on a surface which is perpendicular to the ith direction. The traction on the rightmost boundary in Fig. 1.2, for instance, is given by $\mathbf{t} = \mathbf{e}_1 \cdot \boldsymbol{\sigma} = \sigma_{11}\mathbf{e}_1 + \sigma_{12}\mathbf{e}_2 + \sigma_{13}\mathbf{e}_3$.

The net force $\mathbf{F}$ and the torque $\mathbf{L}$ about a point $\mathbf{x}_0$ acting on an immersed body with boundary denoted by ∂S are given by integrating the traction over the surface:

$$\mathbf{F} = \int_{\partial S} \mathbf{n} \cdot \boldsymbol{\sigma}\, dS, \quad \mathbf{L} = \int_{\partial S} (\mathbf{x} - \mathbf{x}_0) \times (\mathbf{n} \cdot \boldsymbol{\sigma})\, dS. \tag{1.12}$$

In every case considered in this book, there are no body torques that may give rise to internal angular momentum. The consequence is a broad statement about the stress tensor; namely, for all materials studied in this book, the stress tensor is symmetric

$$\boldsymbol{\sigma} = \boldsymbol{\sigma}^T. \tag{1.13}$$

This result is recovered by evaluating the torque on a small control volume and imposing the conservation of angular momentum (see [3]). One important consequence of this fact is that the complete specification of the stress tensor in three dimensions requires the identification of only six components instead of nine.

The discussion thus far has made no assumptions about the specific fluid or material, and the relations above apply to any continuum model. Where the nature of the particular fluid of interest enters into the modeling is in the statement of a *constitutive law*, or a specification of the fluid response to deformation. This is achieved by establishing an equation for the evolution of the stress tensor $\boldsymbol{\sigma}$, which is commonly written as

$$\sigma = -p\mathbf{I} + \boldsymbol{\tau}, \tag{1.14}$$

where p is the pressure, $\mathbf{I}$ is the identity operator, and $\boldsymbol{\tau}$ is the *deviatoric stress tensor* which contains viscous and other stresses. The pressure may be of thermodynamic origin, but often is defined to represent the isotropic part of the stress tensor so as to render $\boldsymbol{\tau}$ traceless (though that is not done in much of this chapter). Many constitutive laws, including that which results in a *Newtonian* fluid model and the Navier–Stokes equations, relate the deviatoric stress to the local strain rate, which we now describe.

Consider the velocity field written in the spatial coordinates, $\mathbf{u}(\mathbf{x},t)$. Taking the convention that $(\nabla\mathbf{u})_{ij} = \partial u_j/\partial x_i$, the first terms in a Taylor expansion of the velocity field about a point $\mathbf{x}$ in space are given by

$$\mathbf{u}(\mathbf{x}+d\mathbf{x},t) = \mathbf{u}(\mathbf{x},t) + d\mathbf{x}\cdot\nabla\mathbf{u}(\mathbf{x},t) + O\left(|d\mathbf{x}|^2\right). \tag{1.15}$$

The gradient of the velocity field is usefully decomposed into its symmetric and antisymmetric parts, $\nabla\mathbf{u} = (\dot{\boldsymbol{\gamma}} + \boldsymbol{\omega})/2$, where

$$\dot{\boldsymbol{\gamma}} = \left(\nabla\mathbf{u} + (\nabla\mathbf{u})^T\right), \quad \boldsymbol{\omega} = \left(\nabla\mathbf{u} - (\nabla\mathbf{u})^T\right). \tag{1.16}$$

$\dot{\boldsymbol{\gamma}}$ is the (symmetric) rate-of-strain tensor, and $\boldsymbol{\omega}$ is the (antisymmetric) vorticity tensor. Consider a line element $d\mathbf{x}$ extending from a point $\mathbf{x}$ that evolves in a linear flow field. Then $d(d\mathbf{x})/dt = \mathbf{u}(\mathbf{x}+d\mathbf{x},t) - \mathbf{u}(\mathbf{x},t) = d\mathbf{x}\cdot(\dot{\boldsymbol{\gamma}} + \boldsymbol{\omega})/2$. Let us consider the response of the line element to the velocity gradient through the roles of the symmetric and antisymmetric parts separately. First, the action of $\dot{\boldsymbol{\gamma}}$ is best appreciated through its spectral decomposition. Since $\dot{\boldsymbol{\gamma}}$ is symmetric its eigenvectors are orthogonal, which we write as $\mathbf{d}_i$ for $i = 1,2,3$ (the *principle axes* of $\dot{\boldsymbol{\gamma}}$), and the eigenvalues $2\lambda_i$ associated with the principal axes are twice the *principal rates-of-strain*. Then we may represent the symmetric tensor $\dot{\boldsymbol{\gamma}}$ as $\Sigma_i(2\lambda_i)\mathbf{d}_i\mathbf{d}_i$, where $\mathbf{d}_i\mathbf{d}_i$ is a dyadic product, so if the flow gradient has only a symmetric part then

$$\frac{d}{dt}d\mathbf{x} = \frac{1}{2}d\mathbf{x}\cdot\dot{\boldsymbol{\gamma}} = \sum_i \lambda_i(d\mathbf{x}\cdot\mathbf{d}_i)\mathbf{d}_i. \tag{1.17}$$

A spherical control volume is thus instantaneously deformed by $\dot{\boldsymbol{\gamma}}$ to an ellipsoid along the principal axes of $\dot{\boldsymbol{\gamma}}$ with axis lengths indicated by the principal rates-of-strain. Meanwhile, the response of the line element $d\mathbf{x}$ to the antisymmetric part of the flow, $\boldsymbol{\omega}$, is a rigid body rotation,

$$\frac{d}{dt}d\mathbf{x} = \frac{1}{2}d\mathbf{x}\cdot\boldsymbol{\omega} = \frac{1}{2}(\nabla\times\mathbf{u})\times d\mathbf{x}, \tag{1.18}$$

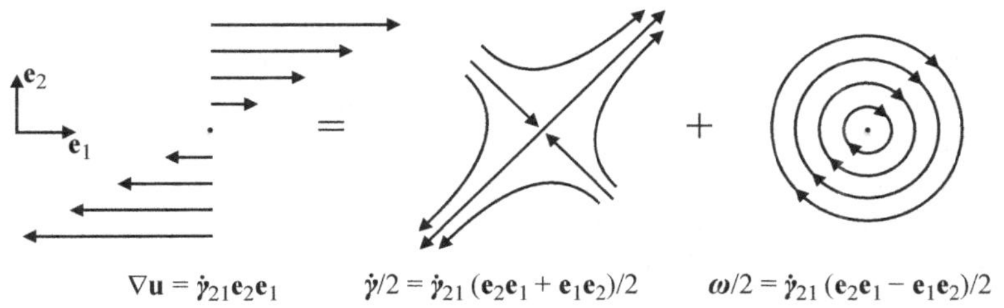

Fig. 1.3 A linear shear flow is decomposed into its extensional (symmetric) and rotational (antisymmetric) components, $\nabla\mathbf{u} = (\dot{\boldsymbol{\gamma}} + \boldsymbol{\omega})/2$

where $\nabla \times \mathbf{u}$ is the *vorticity*. An example of the decomposition of $\nabla\mathbf{u}$ into its symmetric and antisymmetric parts is shown in Fig. 1.3. A linear shear flow $\mathbf{u}(\mathbf{x},t) = \mathbf{x}\cdot(\dot{\gamma}_{21}\mathbf{e}_2\mathbf{e}_1) = (\dot{\gamma}_{21}y, 0, 0)$, where $\dot{\gamma}_{21}$ is a constant shear rate, contains both extensional (symmetric) and rotational (antisymmetric) features.

A classical (Newtonian) viscous fluid is defined to be that in which the deviatoric stress is linear in the rate of strain and the fluid is isotropic (there is no preferred direction in the fluid response to deformation). Given that the stress is symmetric, the most general form of the deviatoric stress tensor that satisfies these constraints reduces to a linear combination of $\dot{\boldsymbol{\gamma}}$ and $(\nabla\cdot\mathbf{u})\mathbf{I}$. Decomposing this general form into a traceless part and an isotropic part, the resulting constitutive relation is given by

$$\boldsymbol{\tau} = \mu\left(\dot{\boldsymbol{\gamma}} - \frac{2}{3}(\nabla\cdot\mathbf{u})\mathbf{I}\right) + \mu'(\nabla\cdot\mathbf{u})\mathbf{I}. \tag{1.19}$$

The coefficient of the traceless part of $\boldsymbol{\tau}$, or μ, is the fluid viscosity, while μ' is the dilational viscosity. In the event that the fluid is incompressible ($\nabla\cdot\mathbf{u} = 0$) then Eq. (1.19) reduces to $\boldsymbol{\tau} = \mu\dot{\boldsymbol{\gamma}}$ and the total stress tensor has the simple form

$$\boldsymbol{\sigma} = -p\mathbf{I} + \mu\dot{\boldsymbol{\gamma}}. \tag{1.20}$$

Let us now revisit the momentum balance equation. Inserting the stress above into Eq. (1.11), we obtain the equations

$$\rho\frac{D\mathbf{u}}{Dt} = -\nabla p + \mu\nabla^2\mathbf{u} + \mathbf{f}, \tag{1.21}$$

$$\nabla\cdot\mathbf{u} = 0. \tag{1.22}$$

Equations (1.21) and (1.22) are known as the incompressible Navier–Stokes equations.

2.5 *Dimensional Analysis and the Stokes Equations*

An important practice in the development of theory is to nondimensionalize the equations of motion, which reduces (often dramatically) the number of parameters that characterize the dynamics. Consider a flow with a characteristic time scale T, velocity scale U, and length scale L. For instance, U might be a swimming speed or a background flow speed, and L might be the approximate length of an immersed body or the gap width in a channel. Defining the dimensionless variables,

$$\mathbf{x}^* = \mathbf{x}/L, \quad \mathbf{u}^* = \mathbf{u}/U, \quad t^* = t/T, \quad p^* = Lp/(\mu U), \quad \mathbf{f}^* = L^2\mathbf{f}/(\mu U), \tag{1.23}$$

and inserting them into Eqs. (1.21) and (1.22), the dimensionless incompressible Navier–Stokes equations are obtained:

$$\mathrm{Re}\left(\mathrm{St}\frac{\partial \mathbf{u}^*}{\partial t^*} + \mathbf{u}^* \cdot \nabla \mathbf{u}^*\right) = -\nabla p^* + \nabla^2 \mathbf{u}^* + \mathbf{f}^*, \quad \nabla \cdot \mathbf{u}^* = 0. \tag{1.24}$$

Here we have introduced the dimensionless *Reynolds* and *Strouhal* numbers,

$$\mathrm{Re} = \frac{\rho U L}{\mu}, \quad \mathrm{St} = \frac{L}{UT}, \tag{1.25}$$

which characterize the flow. If the characteristic time scale is chosen to be the time for a velocity perturbation to be transported convectively by the flow, $T = L/U$, then $St = 1$. The Reynolds number indicates the relative importance of inertial effects to viscous dissipation in (1.24), and also gives the ratio between the time scale for a velocity perturbation to diffuse away due to viscosity, $\rho L^2/\mu$, and the convective time scale L/U.

The topics of interest in this book will focus on complex fluid flows at exceedingly small Reynolds numbers. For instance, in the fluid flow generated by the swimming of microorganisms, the relevant Reynolds number is on the order of 10^{-4}–10^{-2}. A common simplifying assumption is then to consider the idealized zero Reynolds number flow, resulting in the Stokes equations:

$$-\nabla p + \mu \nabla^2 \mathbf{u} + \mathbf{f} = \mathbf{0}, \quad \nabla \cdot \mathbf{u} = 0. \tag{1.26}$$

The linearity of the Stokes equations makes many methods of solution possible; in particular, Green's functions (fundamental singular solutions) may be derived and used to write representation formulae for the flow in terms of integrals over the fluid boundaries, and the Lorentz reciprocal theorem (see Chap. 8) may be used in many settings with tremendous effect. The reader is referred to [7] for a thorough discussion on the fundamental solutions and boundary-integral representations of Stokes flow and also to Chap. 11 where this approach is used in the study of blood flow.

In certain settings a more appropriate choice of characteristic time scale is related to the frequency of oscillation, ω, as may be relevant in rotational rheometers, or a frequency of undulation, as in flagellar locomotion. Taking $T = \omega^{-1}$, the Strouhal number is $St = L\omega/U$. Even for very small Reynolds numbers, $\mathrm{Re} \ll 1$, a sufficiently large frequency may result in $\mathrm{Re}\,\mathrm{St} = O(1)$, specifically, when $\rho L^2 \omega/\mu = O(1)$. This ratio is commonly referred to as the frequency Reynolds number. In this setting, the time scale for viscous diffusion of a velocity perturbation, $\rho L^2/\mu$, is commensurate with the time scale of oscillation. The resulting idealized equations are the *unsteady* Stokes equations, where the momentum balance equation in (1.26) is replaced by $\rho\,\mathbf{u}_t = -\nabla p + \mu \nabla^2 \mathbf{u} + \mathbf{f}$.

3 Generalized Newtonian Fluids

In the previous section we introduced the classical Newtonian constitutive model, Eq. (1.20), which is a linear relationship between the stress and velocity gradient in the fluid. This linear relationship can be viewed as the first term in a Taylor expansion of the true constitutive equation for the material in terms of small velocity gradients. The Newtonian approximation has been shown to work remarkably well for fluids consisting of small molecules, like water, liquid argon, etc., even at flow rates corresponding to fast and turbulent flows. Its success can be attributed to the separation of length and time scales in the flows of such fluids; realistic flows of Newtonian fluids do not alter the dynamics of individual constituents (atoms, molecules, etc.). In other words, typical intermolecular distances or velocity distributions of individual constituents even in very turbulent flows are the same as at rest, and, hence, the energy dissipation mechanism in the fluid, which is represented by viscosity in the Newtonian constitutive law, is not affected by the flow.

Only when the applied flows are capable of altering the local microstructure of the fluid might the classical Newtonian approximation fail to provide an adequate mathematical model of the dynamics. In Newtonian fluids this corresponds to velocity fields varying either across fluid parcels of order 10–100 particles or on time scales comparable to typical stress relaxation times. In simple fluids like liquid argon, the stress relaxation time scale is related to the typical time of molecular self-diffusion and is on the order of 10^{-13}–10^{-12} s [8]. Accessing either of these regimes requires very large velocity gradients that are very rarely achieved in natural or even experimental environments.

The situation is very different for solutions of colloidal particles, long flexible polymers, wormlike micelles, and similar *complex* fluids [9]. These particles are significantly larger than individual molecules of typical Newtonian fluids discussed above, and the time scales of stress relaxation in complex fluids are significantly longer than in their Newtonian counterparts and can easily be achieved in real-life situations. For example, in colloidal suspensions, while in dilute polymer solutions the stress relaxation time is proportional to the time required for a single polymer to regain its equilibrium configuration after being stretched, the Maxwell relaxation time, and is typically of order 10^{-3}–10^{0} s [9]. Although there are often multiple mechanisms of stress relaxation in complex fluids, one can use the longest relaxation time λ to form a dimensionless group $\mathrm{Wi} = \lambda\dot{\gamma}$—the Weissenberg number. Here, the shear rate $\dot{\gamma} = \sqrt{(\dot{\boldsymbol{\gamma}} : \dot{\boldsymbol{\gamma}})/2}$ is an invariant measure of the rate of strain in the fluid (see Sect. 3.1). For small velocity gradients, $\mathrm{Wi} \ll 1$, complex fluids obey the linear constitutive law, Eq. (1.20), and flow like Newtonian fluids at the same Reynolds number. When the Weissenberg number is comparable to or larger than unity, complex fluids exhibit non-Newtonian behavior and obey complicated constitutive models, often involving nonlinear dependence of the local stress on the velocity gradient and the deformation history of the fluid. In this section we focus on the simplest extension of Eq. (1.20) in which the flow only influences the instantaneous viscosity of the fluid, the so-called *generalized Newtonian model*. A general theory dealing with history-dependent properties of viscoelastic fluids will be discussed in the following sections.

3.1 Shear-Thinning and Shear-Thickening Fluids

A generalized Newtonian fluid is a phenomenological model that assumes that the applied flow only changes the dissipation rate in the fluid (i.e., its viscosity), but does not change the tensorial structure of the Newtonian constitutive model Eq. (1.20). The constitutive laws for this class of models can be written in the following general form:

$$\boldsymbol{\sigma} = -p\mathbf{I} + \eta(\dot{\boldsymbol{\gamma}})\dot{\boldsymbol{\gamma}}, \tag{1.27}$$

where $\eta(\dot{\boldsymbol{\gamma}})$ is the viscosity, made distinct from the Newtonian viscosity μ due to its possible dependence upon the fluid flow. First note that the local viscosity η can only depend on the invariants of the tensor $\dot{\boldsymbol{\gamma}}$, otherwise a similarity transform (a change of coordinate system) could change the value of the viscosity, which is unphysical. Also, it would be natural to require in the linear shear flow with a constant shear rate $\dot{\gamma}_{21}$ considered in the previous section, $\mathbf{u}(\mathbf{x},t) = (\dot{\gamma}_{21}y, 0, 0)$, that the viscosity should simply be a function of the scalar $\dot{\gamma}_{21}$. The second tensorial invariant of $\dot{\boldsymbol{\gamma}}$ is the lowest invariant that satisfies this condition, and we may write

$$\eta(\dot{\boldsymbol{\gamma}}) = \eta(\dot{\gamma}), \tag{1.28}$$

where

$$\dot{\gamma}^2 = \frac{1}{2}\dot{\boldsymbol{\gamma}} : \dot{\boldsymbol{\gamma}} = \frac{1}{2}(\nabla\mathbf{u} + \nabla\mathbf{u}^T)_{ij}(\nabla\mathbf{u} + \nabla\mathbf{u}^T)_{ji}, \tag{1.29}$$

summing over both repeated indices. The material properties of a generalized Newtonian fluid are determined entirely by the behavior of the function $\eta(\dot{\gamma})$. The simplest possibility, a monotonic function, results in:

- $\partial\eta/\partial\dot{\gamma} > 0$, a *shear-thickening* fluid, or
- $\partial\eta/\partial\dot{\gamma} < 0$, a *shear-thinning* fluid.

Shear-thickening fluids, as the name suggests, exhibit an increasing resistance to shear as the shear rate increases, while shear-thinning fluids exhibit the opposite behavior; "Oobleck" (cornstarch and water) and pastes are typical examples of the former type of fluids, while solution and melts of long flexible polymers and semidilute solutions of wormlike micelles are examples of the latter. Real materials can exhibit complicated combinations of the two, e.g., shear-thinning at low shear rates followed by shear-thickening at higher shear rates [10]. While these trends can easily be incorporated into a model for $\eta(\dot{\gamma})$, the generalized constitutive law, Eq. (1.27), is a strictly phenomenological model that mimics all the changes in the internal structure of the fluid due to the applied flow by a shear-dependent effective viscosity. The presence of shear-thinning and shear-thickening in the same material typically implies several competing mechanisms of stress creation and relaxation, and a naive model like Eq. (1.27) would most certainly fail in properly describing even simple flows of such fluids. Therefore, generalized Newtonian models should only be used in flows of complex fluids where there is a good reason to believe that the dynamics of principle concern are caused by the shear-induced changes in the viscosity of the fluid and only in the simplest of flows.

3.2 Carreau–Yasuda and Power-Law Fluids

One of the most popular models for shear-thinning fluids is the Carreau–Yasuda viscosity model:

$$\eta(\dot{\gamma}) = \eta_\infty + (\eta_0 - \eta_\infty)\left[1 + \left(\lambda\dot{\gamma}\right)^a\right]^{\frac{n-1}{a}} \qquad \textit{(Carreau–Yasuda)} \qquad (1.30)$$

that interpolates between the *zero-shear-rate viscosity* η_0 and the *infinite-shear rate viscosity* η_∞. A relaxation time λ sets the crossover shear rate: for $\dot{\gamma} < \lambda^{-1}$, the Carreau–Yasuda fluid exhibits, essentially, a Newtonian behavior with the viscosity η_0, while for higher shear rates its viscosity drops to $\eta_\infty < \eta_0$. The Carreau–Yasuda model contains two constants: the power-law index $n < 1$ that characterizes the degree of shear-thinning of the model and the constant a that sets the size and curvature of the crossover region between the Newtonian and shear-thinning behavior. A typical viscosity of the Carreau–Yasuda model is shown in Fig. 1.4.

For high shear rates, $\lambda\dot{\gamma} \gg 1$, the Carreau–Yasuda model can be simplified significantly to

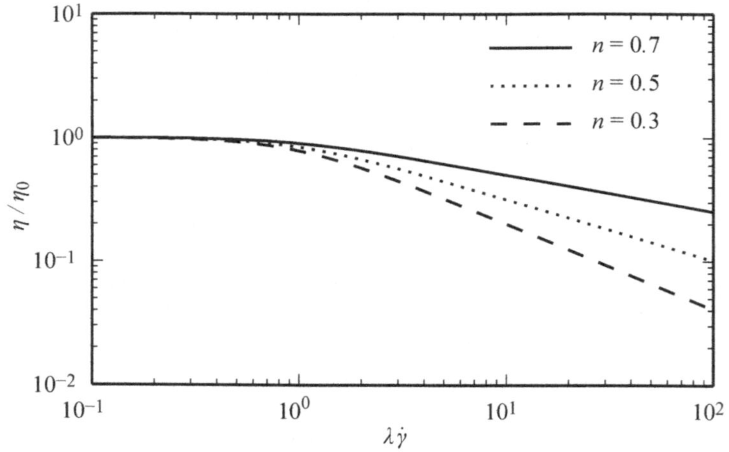

Fig. 1.4 Viscosity of the Carreau–Yasuda model for various values of the power index n with $\eta_\infty/\eta_0 = 10^{-3}$ and $a = 2$

$$\eta(\dot{\gamma}) \approx \eta_\infty + (\eta_0 - \eta_\infty)\left(\lambda\dot{\gamma}\right)^{n-1}. \tag{1.31}$$

The second term in this expression corresponds to the *power-law* model, which is closely related to the Carreau–Yasuda model, Eq. (1.30). To reduce the number of parameters, it is customary to write this term as $K\dot{\gamma}^{n-1}$, although the parameter K in this expression has the strange dimensions of Pa·s^n. Since the constitutive law corresponding to Eq. (1.31) is a sum of two contributions, a Newtonian term with the viscosity η_∞ and a power-law term, one can study the latter separately. Also, the relative magnitude of the Newtonian term in Eq. (1.31) is typically much smaller than the power-law contribution and can safely be neglected. This is the case, for example, in dilute polymer solutions where the Newtonian contribution in Eq. (1.31) corresponds to the viscosity of the solvent η_∞, while at moderate shear rates, the viscosity of the solution, $\eta_\infty + (\eta_0 - \eta_\infty)(\lambda\dot{\gamma})^{n-1}$, is typically several orders of magnitude larger.

In order to illustrate the typical features of flows of shear-thinning materials, we now consider flow in a pipe of a power-law fluid. We choose the cylindrical coordinate system with the z-direction along the axis of the pipe of radius R. The flow is assumed to be laminar, unidirectional, and axisymmetric, $\mathbf{u} = (0, 0, U(r))$, and the second tensorial invariant of $\dot{\boldsymbol{\gamma}}$ reduces to $\dot{\gamma} = |U'(r)|$, where the prime denotes the r-derivative. Combining the momentum balance equation (1.11), the generalized Newtonian constitutive law, Eq. (1.27), and the power-law model for the viscosity, $\eta(\dot{\gamma}) = K\dot{\gamma}^{n-1}$, we obtain the following equation of motion:

$$-\partial_z p + \frac{1}{r}\partial_r(r\,\sigma_{rz}) = 0, \tag{1.32}$$

where the shear stress σ_{rz} is given by

$$\sigma_{rz} = K U'(r) \left| U'(r) \right|^{n-1}. \tag{1.33}$$

The flow is driven by the applied constant pressure gradient $-\partial_z p = \Delta P/L = (P_{\text{inlet}} - P_{\text{outlet}})/L$, where L is the pipe length. Integrating Eq. (1.32) and requiring that the shear stress remains finite on the centerline $r = 0$, we obtain the following distribution of the shear stress in the cross section of the pipe

$$\sigma_{rz}(r) = -\sigma_w \frac{r}{R}, \tag{1.34}$$

where

$$\sigma_w = \frac{1}{2} \frac{\Delta P}{L} R \tag{1.35}$$

is the value of the shear stress at the wall. At the wall, we expect the fluid to satisfy the no-slip boundary condition, $U(R) = 0$, and therefore $U'(r)$ should be negative resulting in the following equation for the velocity:

$$K \left| U'(r) \right|^n = \sigma_w \frac{r}{R}. \tag{1.36}$$

Integrating this equation with the no-slip boundary condition, we obtain

$$U(r) = \left(\frac{\sigma_w}{K} \right)^{\frac{1}{n}} \frac{nR}{n+1} \left[1 - \left(\frac{r}{R} \right)^{\frac{n+1}{n}} \right]. \tag{1.37}$$

For the Newtonian case, $n = 1$, we have $K = \mu$, the Newtonian viscosity, and Eq. (1.37) reduces to the usual parabolic Hagen–Poiseuille profile:

$$U(r) = \frac{\Delta P R^2}{4 L \mu} \left[1 - \left(\frac{r}{R} \right)^2 \right]. \tag{1.38}$$

To demonstrate the effect of shear-thinning on the spatial profile in the pipe, we normalize Eq. (1.37) with the mean velocity in the pipe

$$\overline{U} = \frac{1}{\pi R^2} \int_0^{2\pi} d\theta \int_0^R U(r)\, r dr = \left(\frac{\sigma_w}{K} \right)^{\frac{1}{n}} \frac{nR}{3n+1}, \tag{1.39}$$

to obtain

$$\frac{U(r)}{\overline{U}} = \frac{3n+1}{n+1} \left[1 - \left(\frac{r}{R} \right)^{\frac{n+1}{n}} \right]. \tag{1.40}$$

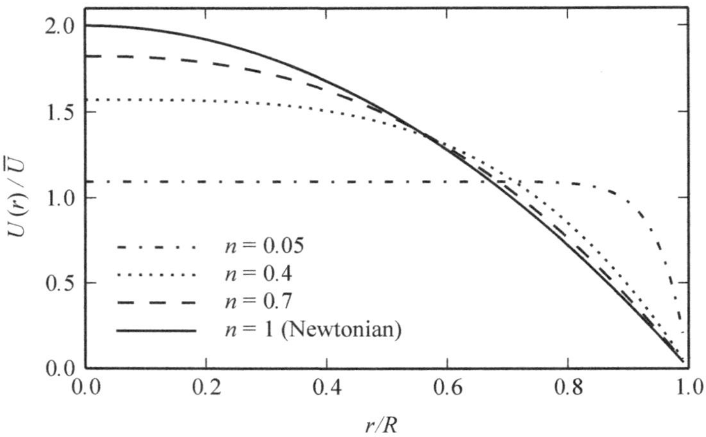

Fig. 1.5 The normalized velocity profile of a pressure-driven pipe flow, from Eq. (1.40), for various values of the power-law index n. As the fluid becomes more shear-thinning (decreasing n), the high-shear region of the flow moves progressively towards the wall and the region near the center of the pipe becomes more plug-like

The normalized velocity profile, Eq. (1.40), is shown in Fig. 1.5 for various values of the power-law index n. As the fluid becomes more shear-thinning (decreasing n), the high-shear region of the flow moves progressively towards the wall and the region near the center of the pipe becomes more plug-like. This is typical of all shear flows of shear-thinning fluids: they split into regions with high shear rates near boundaries where the local viscosity of the fluid is low and parts that move almost like solid bodies. As Eq. (1.40) suggests, in the pipe flow the tendency for shear to localize next to a boundary increases as n decreases until n reaches zero, at which point Eq. (1.40) becomes unphysical. This is a signal of a more general mechanical instability present in shear flows of extremely shear-thinning fluids, to which we now turn.

3.3 Mechanical Instability of Extremely Shear-Thinning Fluids

To demonstrate the origin of the mechanical instability mentioned above, we consider a plane Couette flow (linear shear flow) of a shear-thinning fluid. The fluid is confined between two parallel plates located at $y = 0$ and $y = h$ in a 2-dimensional Cartesian coordinate system $\mathbf{x} = (x, y)$. The base flow is given by $\mathbf{u}(\mathbf{x}) = (\dot{\gamma}_0 y, 0) = (u(y), 0)$, and the equations of motion and the constitutive equation are given by Eqs. (1.11) and (1.27). The upper wall moves with the velocity $\dot{\gamma}_0 h$ in its plane while the lower wall is kept stationary.

Consider an infinitesimal perturbation to the base profile of the following form:

$$\mathbf{u}(\mathbf{x},t) = (\dot{\gamma}_0 y, 0) + (\delta u(y,t), 0). \tag{1.41}$$

This perturbation deforms the original profile but does not change the unidirectional nature of the flow. An equation of motion for the perturbation reduces to

$$\rho \frac{\partial}{\partial t} \delta u(y,t) = \frac{\partial}{\partial y} \left\{ \left(\dot{\gamma}_0 + \frac{\partial}{\partial y} \delta u(y,t) \right) \eta \left(\dot{\gamma}_0 + \frac{\partial}{\partial y} \delta u(y,t) \right) \right\}. \tag{1.42}$$

Assuming the perturbation to be small relative to the background shear flow, a linearization of this equation returns:

$$\rho \frac{\partial}{\partial t} \delta u(y,t) = \left(\eta(\dot{\gamma}_0) + \dot{\gamma}_0 \frac{\partial \eta}{\partial \dot{\gamma}}(\dot{\gamma}_0) \right) \frac{\partial^2}{\partial y^2} \delta u(y,t). \tag{1.43}$$

The no-slip boundary conditions are already satisfied by the base profile, so that the perturbation must have $\delta u(0,t) = \delta u(h,t) = 0$. Therefore, without loss of generality, the perturbation can be written as

$$\delta u(y,t) = \sum_{m=1}^{\infty} \delta u_m e^{\alpha_m t} \sin \frac{m\pi y}{h}. \tag{1.44}$$

Here the δu_m are unknown coefficients of the expansion and α_m is an eigenvalue associated with the Fourier mode m. If the real part of α_m is positive, the corresponding Fourier mode will grow exponentially in time, indicating a loss of stability of the base flow. Substituting this expansion into the equation of motion we obtain

$$\alpha_m = -\left(\frac{m\pi}{h} \right)^2 \left(\eta(\dot{\gamma}_0) + \dot{\gamma}_0 \frac{\partial \eta}{\partial \dot{\gamma}}(\dot{\gamma}_0) \right). \tag{1.45}$$

For a power-law fluid, $\eta(\dot{\gamma}) = K\dot{\gamma}^{n-1}$, the term in parentheses reduces to $nK\dot{\gamma}_0^{n-1}$, and hence α_m is positive for $n < 0$. In other words, steady shear flows of shear-thinning fluids with a power-law steeper than -1 are unstable and cannot be realized. More generally, Eq. (1.45) implies that any shear flow is unstable if its shear stress σ_{12} decreases with $\dot{\gamma}$, i.e.,

$$\frac{\partial \sigma_{12}}{\partial \dot{\gamma}} < 0. \tag{1.46}$$

Generally, in dilute polymer solutions this condition is never satisfied and steady shear flows are possible for these fluids. However, linear shear flows of semidilute wormlike micellar solutions have been demonstrated to split into piecewise linear shear flow with regions of different shear rates. This phenomenon of *shear-banding* is well studied and is usually attributed to the region of the flow where Eq. (1.46) is satisfied. Comprehensive reviews of shear-banding can be found elsewhere [11, 12].

4 Differential Constitutive Equations for Viscoelastic Fluids

In the previous section we introduced the generalized Newtonian fluid model, a simple extension of the linear relation between the stress and velocity gradient in the classical Newtonian constitutive law. While exhibiting shear-dependent viscosity these fluids are essentially Newtonian in the following aspects: the structure of the stress tensor of generalized Newtonian fluids in a particular flow is the same as in their Newtonian counterparts, and their velocity fields adjust instantaneously to changes in stresses. Many complex fluids behave quite differently. One of the key features of viscoelastic fluids is the presence of memory; stresses in such fluids depend on the flow history. Another is stress anisotropy. Generally, a viscoelastic fluid generates stresses that are absent in a Newtonian fluid subjected to the same deformation history.

The consequences of these features are dramatic: viscoelastic fluids do not flow like their Newtonian counterparts. In this section, we develop a mathematical framework that will allow us to incorporate memory and stress anisotropy into constitutive equations for viscoelastic fluids. Unfortunately, there is no single model that describes all viscoelastic fluids similar to the Navier–Stokes equations for Newtonian flows. Instead, one usually chooses a model that is known to describe a particular type of fluid microstructure in a particular type of flow. In this section we introduce several popular models used for polymer solutions and discuss their physical interpretations and the domains of their applicability.

4.1 Linear Maxwell Fluids and Kelvin–Voigt Solids

Viscoelastic fluids are materials that exhibit both viscous and elastic responses to forces. The distinction between viscous and elastic materials is best illustrated by their responses to a sudden deformation: stresses created in an elastic material stay constant in time for as long as the deformation is present, while stresses in a viscous fluid dissipate on a time scale governed by its viscosity. For example, a bow is stressed as long as it is strung by a bowstring, while in spilled water all stresses disappear once the fluid comes to rest. Essentially, whether a material is fluid-like or solid-like is determined by its longtime response to a deformation.

To explore the distinction between the two types of responses, consider a simple shear deformation where adjacent layers of a material are shifted impulsively in the same direction along their planes relative to each other. The strength of this deformation can be characterized by its gradient, denoted by γ, which for small

Fig. 1.6 Graphical depiction of two types of material response: (*left*) a dashpot (viscous fluid) and (*right*) a spring (elastic solid)

Fig. 1.7 Illustrations of the one-dimensional (*left*) Maxwell and (*right*) Kelvin–Voigt linear viscoelastic models

displacements is approximated as the ratio of the total relative shift between two layers to the distance between them. The shear stress σ created in an elastic solid by such a deformation obeys Hooke's law and can be written as

$$\sigma = G\gamma, \tag{1.47}$$

where G is the elastic constant of the material, or *shear modulus*. Meanwhile, the constitutive equation for a Newtonian viscous fluid is a linear relationship between the stress and velocity gradient, Eq. (1.20), and, adopted to the case of present interest, reads

$$\sigma = \eta\,\dot{\gamma}. \tag{1.48}$$

Here, as before, η is the viscosity of the fluid, and the dot denotes a time derivative. Note the distinction between $\dot{\gamma}$, the velocity gradient, and γ, the displacement gradient. Due to obvious similarities, linear solids and liquids are often denoted graphically by springs and dashpots (shock-absorbing devices based on viscous fluids used, for example, to prevent doors from slamming); see Fig. 1.6. The simplest viscoelastic material is a linear combination of the two types of material responses discussed above. In direct analogy with electric circuits, one can think of either *serial* or *parallel* connection between the basic elements from Fig. 1.6, and the two possible combinations are shown in Fig. 1.7. Each combination should be thought of as a fluid with both elastic and viscous properties.

The serial connection of a spring and a dashpot is a viscoelastic fluid, while the parallel connection is a viscoelastic solid. To demonstrate this we consider both types of viscoelastic elements subjected to a fixed displacements of their ends. In the serial connection, both the spring and the dashpot are stretched initially. However, the displacement of the spring can be redistributed to the dashpot, keeping the total displacement constant, and resulting in the absence of stress in this material at long times, since $\sigma \sim \dot{\gamma}$ for a dashpot. On the contrary, the parallel connection remains under stress for as long as $\gamma \neq 0$, as evident from Fig. 1.7. Historically, these models have been called the *Maxwell fluid* and *Kelvin–Voigt solid* models, and they are the simplest models of viscoelastic materials. Alternatively, the same models are sometimes referred to as *solid-like liquids* and *liquid-like solids*, where the last word of the name identifies the model's behavior at long times.

To derive constitutive equations for the Maxwell and Kelvin–Voigt materials, we introduce the total deformation γ and the total stress σ for each model. The corresponding deformations and stresses of the spring and dashpot are denoted by γ_s and γ_d, and σ_s and σ_d, correspondingly. We note here that only the total deformation and stress are measurable quantities, while the deformations and stresses of the springs and dashpots are auxiliary variables that are used to describe internal mechanisms of stress creation and dissipation within each material. The present goal is to find a constitutive relation between σ and γ for both models. Continuing the analogy with electric circuits, we observe that $\gamma = \gamma_s + \gamma_d$ and $\sigma = \sigma_s = \sigma_d$ for the Maxwell fluid, while $\gamma = \gamma_s = \gamma_d$ and $\sigma = \sigma_s + \sigma_d$ for the Kelvin–Voigt solid. Using Eqs. (1.47) and (1.48), we obtain

$$\sigma + \frac{\eta}{G}\dot{\sigma} = \eta\,\dot{\gamma} \qquad \textit{(Maxwell fluid)}, \tag{1.49}$$

$$\sigma = G\gamma + \eta\,\dot{\gamma} \qquad \textit{(Kelvin–Voigt solid)}. \tag{1.50}$$

While neither the Kelvin–Voigt or Maxwell linear models are generally adequate for describing real materials, they are prototype models for systems like polymer brushes grafted on a surface and dilute polymer solutions, respectively. Since the scope of this chapter is complex fluids, we focus our attention on the Maxwell model.

The Maxwell model, Eq. (1.49), can be formally solved to yield

$$\sigma(t) = \frac{1}{\lambda}\int_{-\infty}^{t} e^{-\frac{t-t'}{\lambda}}\,\eta\,\dot{\gamma}(t')\,dt', \tag{1.51}$$

where we have introduced the Maxwell relaxation time $\lambda = \eta/G$. As can be seen from the solution, the stress created by a steplike deformation relaxes exponentially on the time scale λ indicating viscous-fluid-like properties, while at short times, $\sigma(t) \sim \eta\gamma(t)/\lambda$ and the Maxwell material is solid-like.

Since Eq. (1.49) is linear, its behavior is easily analyzed in terms of its response to a periodic deformation with a frequency ω. Time evolution of the stress for arbitrary time-dependent deformations may then be reconstructed through the evolution of decoupled Fourier coefficients. Integrating Eq. (1.51) with only one frequency of deformation, $\gamma(t) = \gamma_0 \sin\omega t$, we obtain

$$\sigma(t) = \gamma_0\eta\omega\frac{\cos\omega t + \lambda\omega\sin\omega t}{1+(\lambda\omega)^2} = \frac{\eta}{1+(\lambda\omega)^2}\dot{\gamma}(t) + G\frac{(\lambda\omega)^2}{1+(\lambda\omega)^2}\gamma(t). \tag{1.52}$$

Equation (1.52) demonstrates that the Maxwell model exhibits stress response both in and out of phase with the applied deformation. Comparing this expression to Eqs. (1.47) and (1.48), we conclude that the stress response can be interpreted in terms of a frequency-dependent viscosity, $\tilde{\eta}(\omega)$, and shear modulus, $\tilde{G}(\omega)$, where

$$\tilde{\eta}(\omega) = \frac{\eta}{1+(\lambda\omega)^2}, \qquad \tilde{G}(\omega) = G\frac{(\lambda\omega)^2}{1+(\lambda\omega)^2}. \tag{1.53}$$

Once again, at short times ($\lambda\omega \gg 1$), the Maxwell model behaves like a solid with the shear modulus $\tilde{G}(\omega) \approx G$, while at long times ($\lambda\omega \ll 1$) it behaves as a viscous fluid with the viscosity $\tilde{\eta}(\omega) \approx \eta$. The crossover between the two regimes occurs when the time scale of deformation is similar to the time scale of relaxation, $\omega^{-1} \sim \lambda$.

Equations (1.52) and (1.53) form the theoretical basis of linear rheology. For very small deformation amplitudes γ_0, even very nonlinear viscoelastic materials are expected to obey Eq. (1.49), and measuring the in- and out-of-phase response of the shear stress $\sigma(t)$ allows one to determine the viscosity, elastic modulus, and Maxwell relaxation time of the fluid. Linear rheological measurements are usually interpreted in terms of the complex modulus, $G^*(\omega) = G'(\omega) + iG''(\omega)$, defined for the case considered here by $\sigma(t) = \Im\left(G^*(\omega)\gamma_0 e^{i\omega t}\right)$, where $\Im(\cdot)$ denotes the imaginary part of its complex argument. Commercial rheometers readily provide the storage and loss moduli $G'(\omega)$ and $G''(\omega)$ as functions of ω (the so-called frequency sweep), and for the Maxwell model the relaxation time can be determined as $\lambda = \omega_0^{-1}$, with ω_0 as the frequency where $G'(\omega_0) = G''(\omega_0)$. The other parameters are then determined by fitting the low-frequency behaviors of G' and G''. In reality, however, the Maxwell model is often insufficient to describe even the linear rheology of polymer solutions and G' and G'' do not cross due to additional dissipation mechanisms that will be discussed later in this section. Nevertheless, it is a very useful minimal model that sets the stage for more complete theories.

4.2 *Objectivity and Convected Derivatives*

In Sect. 4.1 we introduced the linear Maxwell model that combines viscous and elastic responses to deformations. For the simplest case of linear shear that model was written as $\sigma + \lambda\dot{\sigma} = \eta\,\dot{\gamma}$, where σ is the shear stress and $\dot{\gamma}$ is the shear rate. To generalize this model for arbitrary flows, it would seem that one would only need to rewrite this equation in terms of the stress $\boldsymbol{\sigma}$ and velocity gradient $\nabla\mathbf{u}$ tensors as

$$\boldsymbol{\sigma} + \lambda\frac{\partial\boldsymbol{\sigma}}{\partial t} = \eta\dot{\boldsymbol{\gamma}}. \tag{1.54}$$

However, this equation suffers from a serious physical problem: it is not frame-invariant. To demonstrate this, assume that we perform the same experiment twice:

once in a stationary lab frame, the other on a (very smoothly operated) train moving with a constant velocity $\mathbf{u}_0$ with respect to the lab frame. The stress components of the moving-frame experiment can be written in terms of the lab frame coordinates as $\sigma_{ij}(\mathbf{x}+\mathbf{u}_0 t, t)$ and their time derivatives become

$$\frac{\partial}{\partial t}\sigma_{ij}(\mathbf{x}+\mathbf{u}_0 t, t) = \frac{\partial \sigma_{ij}}{\partial t} + \mathbf{u}_0 \cdot \nabla \sigma_{ij}. \tag{1.55}$$

Obviously, both experiments should be described by the same equations since a constant velocity added to each point in the fluid does not result in any velocity gradients; hence no additional stresses should be created in the fluid. However, the time derivatives in Eqs. (1.54) and (1.55) differ by a term proportional to $\mathbf{u}_0$. Clearly, this indicates that Eq. (1.54) is not frame-invariant and *it should not be used*.

The problem, as we can see from this example, is that the time derivatives of individual components of the stress tensor do not form a tensor themselves, i.e., the generalization $\partial\sigma/\partial t \rightarrow \partial\boldsymbol{\sigma}/\partial t$ is unphysical since it does not lead to a frame-invariant equation. This problem is reminiscent of the argument that led to the introduction of the material derivative in Eq. (1.2). There we showed that the frame-invariant time derivative of a vector field embedded in a moving fluid is given by D/Dt. Our goal now is to derive a similar expression for a second-rank tensor. A detailed treatment of this derivation can be found in [13–15].

Recall the reference and current configurations described in Sect. 2.1, illustrated again in Fig. 1.8. We will make use of the curvilinear coordinate system defined by the material (Lagrangian) coordinates, $Oa^1a^2a^3$, which moves and deforms with the fluid (the so-called *convected* frame). The new coordinates are related to the fixed Cartesian system $Ox_1x_2x_3$ by $\mathbf{x} = \mathbf{x}(\mathbf{a},t)$, and the relation is assumed to be invertible. Recall the choice (without loss of generality) to set $\mathbf{x}(\mathbf{a},0) = \mathbf{a}$, so that the convected and Cartesian frames are in alignment at $t = 0$; while this is a useful illustration for how the convected frame moves and deforms, we will only make use of the convected and Cartesian frames in the current configuration.

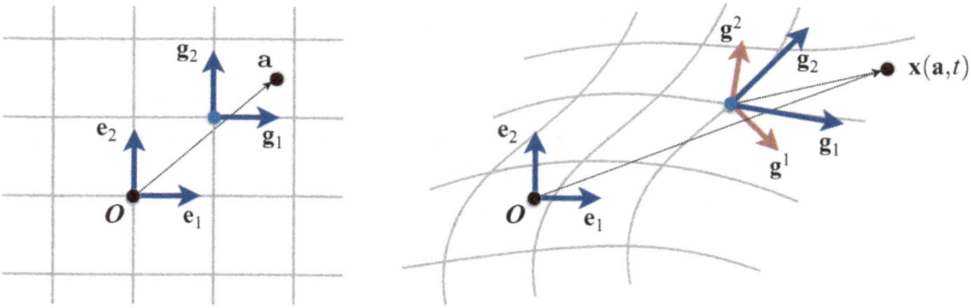

Fig. 1.8 The curvilinear, *convected* frame, has base vectors $\{\mathbf{g}_i\}$ which move and deform with the material deformation. The base vectors are aligned with the Cartesian basis $\{\mathbf{e}_i\}$ in the reference configuration (*left*). The reciprocal vectors $\{\mathbf{g}^i\}$ form an orthonormal basis and satisfy $\mathbf{g}^i \cdot \mathbf{g}_j = \delta_{ij}$. While a useful illustration, the curvilinear and Cartesian frames are only used in the current configuration (*right*) to derive objective time derivatives

Consider an arbitrary point in space, $\mathbf{x}$, which may be represented in terms of base vectors in a given, fixed curvilinear coordinate system, $\{\mathbf{g}_i\}$, as

$$\mathbf{x} = \sum_{i=1}^{3} a^i \mathbf{g}_i. \tag{1.56}$$

The base vectors may then be defined as

$$\mathbf{g}_i = \frac{\partial \mathbf{x}}{\partial a^i}, \tag{1.57}$$

which change both in length and orientation along with the material and are tangential to the lines of constant material coordinates as can be seen from their definition. In a curvilinear system we may also introduce reciprocal vectors, $\{\mathbf{g}^i\}$, that are orthogonal to the base vectors, $\{\mathbf{g}_i\}$, in the following sense: $\mathbf{g}^i \cdot \mathbf{g}_j = \delta_{ij}$, where δ_{ij} is the Kronecker delta. This implies that the reciprocal vectors are orthogonal to planes spanned by two base vectors. They can be constructed, for example, by the usual orthogonalization procedure: $\mathbf{g}^i = \mathbf{g}_j \times \mathbf{g}_k / (\mathbf{g}_i \cdot (\mathbf{g}_j \times \mathbf{g}_k))$ for all cyclic permutations of $(i,j,k) = (1,2,3)$ compatible with the right-handed coordinate system, and $\mathbf{g}_i \cdot (\mathbf{g}_j \times \mathbf{g}_k) = 1$ in an incompressible flow. Here we choose the reciprocal vectors

$$\mathbf{g}^i = \frac{\partial a^i}{\partial \mathbf{x}}, \tag{1.58}$$

which satisfy all the requirements listed above.

Observe that any point $\mathbf{x}$ may be expressed in terms of the Cartesian system, $\mathbf{x} = \sum_{j=1}^{3} x_j \mathbf{e}_j$ (see Fig. 1.8), so that

$$\mathbf{g}_i = \frac{\partial \mathbf{x}}{\partial a^i} = \frac{\partial}{\partial a^i} \sum_{j=1}^{3} x_j \mathbf{e}_j = \sum_{j=1}^{3} F_{ji} \mathbf{e}_j, \tag{1.59}$$

where

$$\mathbf{F} = \frac{\partial \mathbf{x}}{\partial \mathbf{a}} = \sum_{i,j=1}^{3} \frac{\partial x_i}{\partial a^j} \mathbf{e}_i \mathbf{e}_j \tag{1.60}$$

is the deformation gradient tensor introduced in Sect. 2.1. In a similar fashion, we may write

$$\mathbf{g}^i = \frac{\partial a^i}{\partial \mathbf{x}} = \sum_{j=1}^{3} \left(\mathbf{e}_j \frac{\partial}{\partial x_j} \right) a^i = \sum_{j=1}^{3} (F^{-1})_{ij} \mathbf{e}_j, \tag{1.61}$$

where

$$\mathbf{F}^{-1} = \frac{\partial \mathbf{a}}{\partial \mathbf{x}} = \sum_{i,j=1}^{3} \frac{\partial a^i}{\partial x_j} \mathbf{e}_i \mathbf{e}_j. \tag{1.62}$$

The deformation gradient tensors, $\mathbf{F}$ and $\mathbf{F}^{-1}$, are both expressed in terms of the fixed Cartesian basis, and naturally $\mathbf{F}^{-1} \cdot \mathbf{F} = \mathbf{F} \cdot \mathbf{F}^{-1} = \mathbf{I}$.

Next we show how $\mathbf{F}$ and $\mathbf{F}^{-1}$ change in time. Consider the time derivative of $\mathbf{F}$,

$$\frac{d}{dt} F_{ij} = \frac{d}{dt} \frac{\partial x_i}{\partial a^j} = \frac{\partial u_i}{\partial a^j} = \sum_{k=1}^{3} \frac{\partial u_i}{\partial x_k} \frac{\partial x_k}{\partial a^j} = \left[(\nabla \mathbf{u})^T \cdot \mathbf{F} \right]_{ij}, \tag{1.63}$$

or $d\mathbf{F}/dt = (\nabla \mathbf{u})^T \cdot \mathbf{F}$. The time evolution of $\mathbf{F}^{-1}$ can be obtained by taking the time derivative of the orthogonality condition, $\mathbf{F} \cdot \mathbf{F}^{-1} = \mathbf{I}$, which gives

$$\frac{d\mathbf{F}}{dt} \cdot \mathbf{F}^{-1} + \mathbf{F} \cdot \frac{d\mathbf{F}^{-1}}{dt} = 0. \tag{1.64}$$

Using $d\mathbf{F}/dt = (\nabla \mathbf{u})^T \cdot \mathbf{F}$, we obtain

$$\frac{d\mathbf{F}^{-1}}{dt} = -\mathbf{F}^{-1} \cdot (\nabla \mathbf{u})^T. \tag{1.65}$$

We are now set to derive the frame-invariant time derivative of the stress tensor or any other second-rank tensor embedded in a moving fluid. The total stress tensor $\boldsymbol{\sigma}$ may be represented alternately in terms of its Cartesian, covariant, or contravariant components:

$$\boldsymbol{\sigma} = \sum_{m,n=1}^{3} \sigma_{mn} \mathbf{e}_m \mathbf{e}_n = \sum_{i,j=1}^{3} \hat{\sigma}^{ij} \mathbf{g}_i \mathbf{g}_j = \sum_{i,j=1}^{3} \hat{\sigma}_{ij} \mathbf{g}^i \mathbf{g}^j. \tag{1.66}$$

Since the base vectors $\mathbf{g}_i$ and the reciprocal vectors $\mathbf{g}^i$ can be expressed in terms of Cartesian unit vectors, Eqs. (1.59) and (1.61), the co- and contravariant components of $\boldsymbol{\sigma}$ can be expressed through its Cartesian components as

$$\sigma_{mn} = \sum_{i,j=1}^{3} \hat{\sigma}^{ij} F_{mi} F_{nj} = \sum_{i,j=1}^{3} \hat{\sigma}_{ij} F_{im}^{-1} F_{jn}^{-1}. \tag{1.67}$$

These relations can be inverted to read

$$\hat{\sigma}^{ij} = \sum_{m,n=1}^{3} F_{im}^{-1} F_{jn}^{-1} \sigma_{mn} = [\mathbf{F}^{-1} \cdot \boldsymbol{\sigma} \cdot \mathbf{F}^{-T}]_{ij}, \tag{1.68}$$

$$\hat{\sigma}_{ij} = \sum_{m,n=1}^{3} F_{mi} F_{nj} \sigma_{mn} = [\mathbf{F}^T \cdot \boldsymbol{\sigma} \cdot \mathbf{F}]_{ij}, \tag{1.69}$$

where $\mathbf{F}^{-T} = (\mathbf{F}^{-1})^T$, and the subscripts on the right-hand side indicate components in the Cartesian coordinate system. We notice that these equations are, in fact, definitions of co- and contravariant components of a second-rank tensor, as they state the transformation laws for the components upon a change of the coordinate system. Any set of quantities that obey these transformation laws form a second-rank tensor and are thus frame-invariant.

Let us now calculate the time derivative of the co- and contravariant components of the stress tensor $\boldsymbol{\sigma}$. From Eq. (1.68),

$$\frac{d\hat{\sigma}^{ij}}{dt} = \left[\frac{d\mathbf{F}^{-1}}{dt} \cdot \boldsymbol{\sigma} \cdot \mathbf{F}^{-T} + \mathbf{F}^{-1} \cdot \frac{d\boldsymbol{\sigma}}{dt} \cdot \mathbf{F}^{-T} + \mathbf{F}^{-1} \cdot \boldsymbol{\sigma} \cdot \frac{d\mathbf{F}^{-T}}{dt} \right]_{ij}, \tag{1.70}$$

and, after using Eq. (1.65) and rearranging, we obtain

$$\frac{d\hat{\sigma}^{ij}}{dt} = \left[\mathbf{F}^{-1} \cdot \left(\frac{\partial \boldsymbol{\sigma}}{\partial t} + \mathbf{u} \cdot \nabla \boldsymbol{\sigma} - (\nabla \mathbf{u})^T \cdot \boldsymbol{\sigma} - \boldsymbol{\sigma} \cdot \nabla \mathbf{u} \right) \cdot \mathbf{F}^{-T} \right]_{ij}, \tag{1.71}$$

where we have used Eq. (1.2) for the total time derivative of $\sigma_{mn}(\mathbf{x}(\mathbf{a},t),t)$. We observe here that Eq. (1.71) has the same structure as the transformation laws, Eq. (1.68), and therefore $d\hat{\sigma}^{ij}/dt$ and the terms in the brackets on the right hand side are components of a second-rank tensor in the corresponding coordinate systems. Therefore,

$$\overset{\triangledown}{\boldsymbol{\sigma}} \equiv \frac{\partial \boldsymbol{\sigma}}{\partial t} + \mathbf{u} \cdot \nabla \boldsymbol{\sigma} - (\nabla \mathbf{u})^T \cdot \boldsymbol{\sigma} - \boldsymbol{\sigma} \cdot \nabla \mathbf{u} \tag{1.72}$$

is a second-rank tensor, and it is a time derivative of the second-rank tensor $\boldsymbol{\sigma}$ embedded in a fluid with the velocity field $\mathbf{u}$. In a similar fashion, considering $d\hat{\sigma}_{ij}/dt$, we arrive at another tensorial formulation of the full-time derivative:

$$\overset{\vartriangle}{\boldsymbol{\sigma}} \equiv \frac{\partial \boldsymbol{\sigma}}{\partial t} + \mathbf{u} \cdot \nabla \boldsymbol{\sigma} + \boldsymbol{\sigma} \cdot (\nabla \mathbf{u})^T + \nabla \mathbf{u} \cdot \boldsymbol{\sigma}. \tag{1.73}$$

Equations (1.72) and (1.73) define the so-called *upper-convected* and *lower-convected* derivatives of a second-rank tensor. Time derivatives in constitutive relations are generally in one of these two forms in order to ensure that the relation is frame-invariant; other choices of frame-invariant time derivatives are possible (e.g., the so-called *Jaumann* or corotational derivative), but the most popular polymeric constitutive equations are formulated in terms of the upper- and lower-convected derivatives only.

4.3 Canonical Nonlinear Differential Constitutive Equations

4.3.1 A Cooking Recipe

As we have seen in the previous sections, not all combinations of stress and velocity gradient tensors result in physically meaningful equations. It was demonstrated, for example, that a term $d\sigma_{mn}/dt$, where σ_{mn} are the Cartesian coordinates of the stress tensor, can only enter a constitutive equation in a combination with other terms given by the upper- or lower-convected derivatives, Eqs. (1.72) and (1.73). Here we extend this argument and present general principles for formulating a physically admissible constitutive equation. For polymer flows these principles were first formulated by J. Oldroyd and are extensively discussed by Bird et al. [13]. They consist of three main requirements:

- An admissible equation should be frame-invariant.
- The stress tensor $\boldsymbol{\sigma}(t)$ can depend only on the past deformations, $t' < t$, and not on the future $t' > t$.
- Equations should be local in space, i.e., stresses should not depend on the stresses and velocities in the neighboring fluid elements, save through their continuity at the interfaces.

These conditions severely restrict the form of an admissible constitutive equation. Essentially, they imply that such an equation can only be written in terms of functions of frame-invariant combinations of stress and velocity gradient tensors (or, more generally, deformation tensors), as well as their spatial gradients and convected time derivatives. Unfortunately, while eliminating a large number of possible equations, these conditions do not sufficiently restrict the form of the constitutive relation and there is no unique equation that describes viscoelastic polymer solutions similar to the Navier–Stokes equation for Newtonian fluids.

When modeling polymer solutions there are two classes of modeling strategies that one can adopt. The first approach is based on a combination of field-theoretical/symmetry arguments and experimental input. As a first step, one selects a particular order of approximation; for instance, only terms that are at most quadratic in the stress and velocity gradient tensors are considered. Next, the constitutive relation is expressed as a linear combination of these terms with unknown coefficients. Finally, one uses experimentally determined rheological properties in various types of flows to determine whether the equation is sufficient to describe the behavior observed and to fix the values of the unknown coefficients. There is a degree of art involved in this procedure since it is a priori unclear which allowed terms should be included in the model constitutive equation. However, at moderately weak and slow deformations, one would expect only moderate stresses and, hence, the approach outlined above can be seen as using a Taylor expansion to construct successive approximations to the true constitutive law.

The second modeling strategy is to build upon a kinetic theory. Assuming a particular model for polymer molecules, their self-interactions and interactions with other molecules, and their behaviors under flow, one can derive a hydrodynamic equation relating the macroscopic stress and velocity gradient tensors (albeit often only by using rather uncontrolled approximations). Obviously, this approach cannot produce an equation that is not generated through the first modeling strategy. Indeed, this would imply that the "new" terms somehow do not satisfy the admissibility conditions outlined above and, hence, the resulting equation is unphysical. The strength of this approach is that it provides a microscopic basis for the arbitrary coefficients introduced by the first approach. It also allows one to build up intuition as to how particular molecular models project onto macroscopic constitutive equations.

In the next section we will discuss several popular models constructed with the first strategy, while in Sect. 4.4 we will use the second strategy and develop a macroscopic model for a dilute solution of noninteracting dumbbells using a kinetic theory.

4.3.2 Constitutive Equations from Field-Theoretical and Symmetry Arguments

The simplest class of equations for viscoelastic solutions involves the expression of the stress tensor as a sum of all admissible combinations of the velocity gradient tensor. Depending on the highest algebraic power of the velocity gradient tensor involved, they are called the *second-order fluid*, *third-order fluid*, etc. For example, the deviatoric stress of the second-order fluid is given by

$$\boldsymbol{\tau} = \eta\dot{\boldsymbol{\gamma}} + b_2\overset{\nabla}{\dot{\boldsymbol{\gamma}}} + b_{11}\dot{\boldsymbol{\gamma}}\cdot\dot{\boldsymbol{\gamma}} \qquad \textit{(second-order fluid)}, \tag{1.74}$$

where η is the total viscosity of the solution, b_2 and b_{11} are material constants, and the triangle denotes the upper-convected derivative, Eq. (1.72). As is clear by observation of Eq. (1.74), the highest-order nonlinearity is quadratic in the velocity gradient, accounting for the name of the model. We also see that model materials of this class exhibit nonlinear responses to applied deformations but have no memory of past stresses and therefore should only be used in situations where the flow changes on time scales significantly longer than the polymer relaxation time. In fact, as we will show later on in Sect. 6, the second-order model is unphysical in any time-dependent flow and should only be used in weak stationary flows.

The simplest equations that take the relaxation of the stress into account are produced by writing a frame-invariant analogue of the linear Maxwell model, already discussed in Sects. 4.1 and 4.2. By choosing either the upper- or lower-convected derivative for the full-time derivative in Eq. (1.54) we arrive at

$$\boldsymbol{\sigma} = -p\mathbf{I} + \boldsymbol{\tau}_p, \tag{1.75}$$

where the polymeric contribution to the stress $\boldsymbol{\tau}_p$ obeys

$$\boldsymbol{\tau}_p + \lambda \overset{\nabla}{\boldsymbol{\tau}}_p = \eta_p \dot{\boldsymbol{\gamma}} \qquad \textit{(upper-convected Maxwell)}, \tag{1.76}$$

$$\boldsymbol{\tau}_p + \lambda \overset{\Delta}{\boldsymbol{\tau}}_p = \eta_p \dot{\boldsymbol{\gamma}} \qquad \textit{(lower-convected Maxwell)}. \tag{1.77}$$

Here λ is the polymer relaxation time introduced in Sect. 4.1 and η_p is the polymer contribution to the viscosity. Alternatively, one can use a linear combination of the upper- and lower-convected derivatives in the Maxwell model to obtain the so-called *Johnson–Segalman* equation

$$\boldsymbol{\tau}_p + \lambda \left(\frac{1+a}{2} \overset{\nabla}{\boldsymbol{\tau}}_p + \frac{1-a}{2} \overset{\Delta}{\boldsymbol{\tau}}_p \right) = \eta_p \dot{\boldsymbol{\gamma}} \qquad \textit{(Johnson–Segalman)}. \tag{1.78}$$

The *slip parameter* a ($a \in [-1, 1]$) sets the relative importance of the two objective time derivatives derived in Sect. 4.2. Despite their apparent similarities, Eqs. (1.76)–(1.78) produce very different rheological predictions. While the UCM model can capture the properties of many dilute polymer solutions to a relatively good first approximation, the rheological predictions of the lower-convected Maxwell (LCM) model are in strong qualitative disagreement with experimental observations and this model is not generally used. The Johnson–Segalman model predicts non-monotonic behavior of the shear stress with the shear rate in simple shear flows for a wide range of the model parameters and is also not usually used to describe polymeric systems. Instead, it is often employed as a model for shear-banding in wormlike micellar solutions together with the Giesekus model to be introduced shortly [11, 12].

Often a Newtonian stress with a viscosity η_s is added to the total stress in Eq. (1.75):

$$\boldsymbol{\sigma} = -p\mathbf{I} + \eta_s \dot{\boldsymbol{\gamma}} + \boldsymbol{\tau}_p. \tag{1.79}$$

If $\boldsymbol{\tau}_p$ obeys the UCM model, the resulting set of equations is called the *Oldroyd-B* model. The Oldroyd-B model is often formulated in terms of the total deviatoric stress, $\boldsymbol{\tau} = \eta_s \dot{\boldsymbol{\gamma}} + \boldsymbol{\tau}_p$, which satisfies

$$\boldsymbol{\tau} + \lambda \overset{\nabla}{\boldsymbol{\tau}} = \eta \left(\dot{\boldsymbol{\gamma}} + \lambda_r \overset{\nabla}{\dot{\boldsymbol{\gamma}}} \right) \qquad \textit{(Oldroyd-B)}, \tag{1.80}$$

where $\eta = \eta_s + \eta_p$ is the total viscosity. The so-called *retardation* time λ_r is not an independent time scale, but is in fact a combination of the Maxwell relaxation time λ and the solvent and polymeric viscosities, $\lambda_r = \lambda(\eta_s/\eta)$. One drawback of the viscoelastic models above is that tensile stresses can grow without bound in extensional flows. As will be made clear in Sect. 4.4, this can be understood as a continuous stretching of polymers in the flow and an unbounded Hookean stress response.

In the spirit of including all possible tensorial invariants up to a particular order, one can generalize this equation to the so-called *Oldroyd 8-constant* model given by the following constitutive equation:

$$\boldsymbol{\tau} + \lambda_1 \overset{\triangledown}{\boldsymbol{\tau}} + \lambda_2 (\dot{\boldsymbol{\gamma}} \cdot \boldsymbol{\tau} + \boldsymbol{\tau} \cdot \dot{\boldsymbol{\gamma}}) + \lambda_3 \mathrm{Tr}(\boldsymbol{\tau})\, \dot{\boldsymbol{\gamma}} + \lambda_4 (\boldsymbol{\tau} : \dot{\boldsymbol{\gamma}})\mathbf{I}$$
$$= \eta \left(\dot{\boldsymbol{\gamma}} + \lambda_5 \overset{\triangledown}{\dot{\boldsymbol{\gamma}}} + \lambda_6 \dot{\boldsymbol{\gamma}} \cdot \dot{\boldsymbol{\gamma}} + \lambda_7 \left(\dot{\boldsymbol{\gamma}} : \dot{\boldsymbol{\gamma}} \right) \mathbf{I} \right) \quad \textit{(Oldroyd 8-constant)}, \tag{1.81}$$

where $\mathrm{Tr}(\mathbf{A})$ is the trace of $\mathbf{A}$ and $\mathbf{A} : \mathbf{B} = \sum_{i,j} A_{ij} B_{ji}$ as before. The time scales $\lambda_1 \dots \lambda_7$ and the total viscosity η are the model parameters that are usually determined from experiments. This model covers a wide range of possible rheological predictions and, in principle, can be used to describe a variety of viscoelastic systems. In practice, this is prevented by the large number of model parameters that usually cannot all be fixed by standard rheological measurements. Even in theoretical studies, determining the predictions of the Oldroyd 8-constant model requires a scan of a very large space of possible parameter values and is also not practical. The Oldroyd-B model is a special case of the Oldroyd 8-constant model, partially capturing numerous important viscoelastic phenomena but with many fewer parameters, and is a popular model among experimentalists and theoreticians.

Another class of models is formed by adding various terms nonlinear in $\boldsymbol{\tau}_p$ to the UCM model, Eq. (1.76). One example of such models is the Giesekus equation,

$$\boldsymbol{\tau}_p + \lambda \overset{\triangledown}{\boldsymbol{\tau}}_p + \alpha \frac{\lambda}{\eta_p} \boldsymbol{\tau}_p \cdot \boldsymbol{\tau}_p = \eta_p \dot{\boldsymbol{\gamma}} \qquad \textit{(Giesekus)}. \tag{1.82}$$

Here α is a dimensionless model parameter that should be kept smaller than $1/2$ to avoid a non-monotonic dependence of the shear stress on the shear rate in simple shear flows. Another example is given by the *Phan–Thien–Tanner* (PTT) model,

$$f(\boldsymbol{\tau}_p)\, \boldsymbol{\tau}_p + \lambda \overset{\triangledown}{\boldsymbol{\tau}}_p = \eta_p \dot{\boldsymbol{\gamma}} \qquad \textit{(PTT)}, \tag{1.83}$$

where $f(\boldsymbol{\tau}_p)$ is a nonlinear function that can be chosen either in its exponential or, more commonly, in its linear form:

$$f(\boldsymbol{\tau}_p) = \begin{cases} \exp\left\{ \dfrac{\lambda \varepsilon}{\eta_p} \mathrm{Tr}(\boldsymbol{\tau}_p) \right\} & \text{(exponential)}, \\[2ex] 1 + \dfrac{\lambda \varepsilon}{\eta_p} \mathrm{Tr}(\boldsymbol{\tau}_p) & \text{(linear)}. \end{cases} \tag{1.84}$$

The dimensionless parameter ε controls how fast the effective polymeric viscosity and the relaxation time decrease with the stress.

Other constitutive relations were developed to correct the unphysical behavior in the Oldroyd-B and similar models, the so-called finite-extensibility-nonlinear-elastic (FENE) models. These models are comprised of various approximations to the kinetic theory of the FENE model to be discussed in the following section. The two most commonly used models of this type are the FENE-CR model (suggested by Chilcott and Rallison [16]) and the FENE-P model (a Gaussian closure of the kinetic theory model suggested by Peterlin [17]):

$$\boldsymbol{\tau}_p + \lambda \overset{\nabla}{\left(\frac{\boldsymbol{\tau}_p}{f(\boldsymbol{\tau}_p)} \right)} = \eta_p \dot{\boldsymbol{\gamma}} \qquad (\textit{FENE-CR}) \tag{1.85}$$

and

$$\boldsymbol{\tau}_p + \lambda \overset{\nabla}{\left(\frac{\boldsymbol{\tau}_p}{f(\boldsymbol{\tau}_p)} \right)} = \frac{\eta_p}{f(\boldsymbol{\tau}_p)} \dot{\boldsymbol{\gamma}} - \eta_p \frac{D}{Dt} \left(\frac{1}{f(\boldsymbol{\tau}_p)} \right) \mathbf{I} \quad (\textit{FENE-P}). \tag{1.86}$$

In the equations above the upper-convected derivatives act on the entire parenthetical expressions, and the function f is given by

$$f(\boldsymbol{\tau}_p) = 1 + \frac{\lambda}{\eta_p L^2} \mathrm{Tr}(\boldsymbol{\tau}_p), \tag{1.87}$$

where L is a dimensionless parameter related to the maximum possible extension of polymer chains. There are several versions of these models in the literature, but in the limit of large L they all reduce to Eqs. (1.85) and (1.86) above.

Finally we introduce the *Rolie-Poly* model, which in its simplest form is given by

$$\boldsymbol{\tau}_p + \lambda \overset{\nabla}{\boldsymbol{\tau}}_p = \eta_p \dot{\boldsymbol{\gamma}} - \frac{2}{3} \lambda (\boldsymbol{\tau}_p : \nabla \mathbf{u}) \left(\mathbf{I} + (1+\varepsilon) \frac{\lambda}{\eta_p} \boldsymbol{\tau}_p \right) \quad (\textit{Rolie-Poly}). \tag{1.88}$$

Here again, ε is a dimensionless parameter. The Rolie-Poly model is a relatively new constitutive relation and, as such, has not been studied as much as the other models described in this section. However, it is based on our most detailed molecular picture of polymer solutions and is believed to be one of the best models for concentrated polymeric systems [18].

In this section we have only listed a few popular constitutive models without discussing their physical implications. Some basic predictions of the Oldroyd-B model and how they differ from their Newtonian counterparts will be discussed in Sect. 5. But first, let us turn to the second strategy for developing constitutive laws: using kinetic theories of the polymeric structure and dynamics to build from the ground up.

4.4 A Kinetic Theory: The Linear Elastic Dumbbell Model

In this section we develop a basic kinetic theory for very dilute polymer solutions. Our goal here is to give the reader a taste of how microscopic dynamics of model polymers project onto a particular constitutive equation. In this way we will re-derive the UCM model discussed in Sect. 4.3 and provide relationships between the polymeric viscosity and the relaxation time of the UCM equation and the microscopic properties of polymers.

We begin by considering one of the simplest models of polymer molecules: dumbbells consisting of two beads connected by an elastic spring. Much more intricate models are considered at great length in other texts (see [19–21]), but this simple example remains instructive. The solution of polymers in solvent is assumed to be so dilute that the dumbbells do not affect each other through either hydrodynamic interactions or intermolecular forces. In this setting it is sufficient to consider a single polymer molecule in a background fluid flow, and the total contribution of polymer molecules to the stress of the fluid will simply be a sum of individual contributions. We assume that a spring cannot be bent, i.e., it is always oriented along the line connecting the beads of a dumbbell.

The calculation to follow hinges on a critical separation of scales. Even in very turbulent flows, spatial variations of the velocity field occur over much longer scales than the very small length of a single polymer, and the velocity field "seen" by a single polymer is safely assumed to be linear. Moreover, and importantly, the spatial variation of the velocity field occurs over much longer scales than *very many* polymers. Therefore, in deducing an averaged effect of polymers on the fluid rheology, we need only consider the dynamics of a suspension of polymers in a single linear background flow,

$$\mathbf{u}(\mathbf{x},t) = \mathbf{u}_0 + \mathbf{x}\cdot\mathbf{A}, \tag{1.89}$$

where $\mathbf{u}_0$ is a constant velocity vector and $\mathbf{A} = \nabla\mathbf{u}$ is the (constant) velocity gradient tensor. Finally, as discussed in Sect. 3, there is a significant separation of time and length scales between the polymer and the solvent molecule dynamics, and therefore the polymers are assumed not to disturb the equilibrium properties of the solvent molecules. These are the same assumptions made when considering the Brownian motion of a large particle in a solvent (see Chap. 3). In what follows, the solvent is treated as a heat bath with a large number of degrees of freedom kept at a constant temperature T.

4.4.1 Dumbbell Dynamics and the Smoluchowski Equation

The dumbbell dynamics are found by balancing the forces acting on the two beads. For illustration the beads are assumed to be spherical, and we will not consider the balance of torques. Since the Reynolds number associated with a bead's motion is

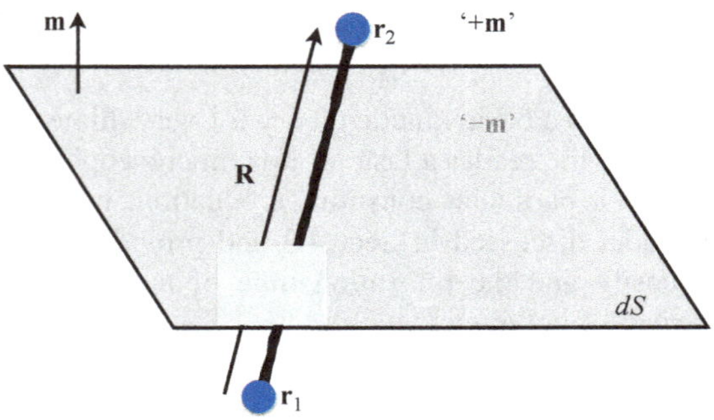

Fig. 1.9 A schematic of the model dumbbell studied in Sect. 4.4 as it passes through an imaginary planar surface with normal vector $\mathbf{m}$. The vector from the first bead (lying in the "$-\mathbf{m}$" half-space) to the second bead (lying in the "$+\mathbf{m}$" half-space) is denoted by $\mathbf{R} = \mathbf{r}_2 - \mathbf{r}_1$. The force exerted by the elastic spring on the first bead is denoted by $\mathbf{F}_s$ and on the second by $-\mathbf{F}_s$

exceptionally small in accordance with its small size, inertia may be safely neglected (see Sect. 2.5). The beads, labeled as "1" and "2", have respective positions $\mathbf{r}_1$, $\mathbf{r}_2$ and velocities $\dot{\mathbf{r}}_1$, $\dot{\mathbf{r}}_2$ (see Fig. 1.9), and the equations of motion are given by

$$\zeta\left(\mathbf{u}\left(\mathbf{r}_1\right) - \dot{\mathbf{r}}_1\right) + \mathbf{F}_s - k_B T\, \nabla_{\mathbf{r}_1} \ln \Psi = 0, \tag{1.90}$$

$$\zeta\left(\mathbf{u}\left(\mathbf{r}_2\right) - \dot{\mathbf{r}}_2\right) - \mathbf{F}_s - k_B T\, \nabla_{\mathbf{r}_2} \ln \Psi = 0, \tag{1.91}$$

where the subscript on the gradient operator indicates the variables over which the derivatives are taken, and $\mathbf{u}(\mathbf{x},t) = \mathbf{u}(\mathbf{x})$ from Eq. (1.89). The first term in each equation is the viscous drag on each bead as it moves relative to the flow; the drag coefficient ζ is given by the Stokes drag law, $\zeta = 6\pi\eta_s a$, where η_s is the solvent viscosity and a is the radius of each bead—we have neglected the hydrodynamic interactions between the two beads. The force exerted by the elastic spring on the first bead is denoted by $\mathbf{F}_s$ and on the second by $-\mathbf{F}_s$. The last terms in Eqs. (1.90) and (1.91) are thermodynamic forces exerted by the fluid on the beads, where k_B is the Boltzmann constant. This force has entropic origins and is written in terms of the distribution function $\Psi\left(\mathbf{r}_1, \mathbf{r}_2, t\right)$ that gives the probability of finding the first bead at $\mathbf{r}_1$ and the second bead at $\mathbf{r}_2$ at time t. Below we will provide an intuitive explanation for this particular form of the thermodynamic force.

It is convenient to introduce the position of the center of mass of each dumbbell, $\mathbf{X} = \left(\mathbf{r}_1 + \mathbf{r}_2\right)/2$, and the end-to-end vector, $\mathbf{R} = \mathbf{r}_2 - \mathbf{r}_1$. Since we have assumed a constant velocity gradient, space is homogeneous in our problem (i.e., the polymer "sees" the same velocity gradient $\mathbf{A}$ at each point in space), so that the probability distribution must be independent of the location of the dumbbell. The probability density function may then be written instead as $\Psi = \Psi(\mathbf{R},t)$. Since Ψ is a probability density we require $\int \Psi\left(\mathbf{R},t\right) d^3\mathbf{R} = 1$, and it is assumed to decay sufficiently fast for large $|\mathbf{R}|$.

Adding and subtracting Eqs. (1.90) and (1.91) and using Eq. (1.89), we find

$$\dot{\mathbf{X}} = \mathbf{u}(\mathbf{X}), \tag{1.92}$$

$$\dot{\mathbf{R}} = \mathbf{R} \cdot \mathbf{A} - \frac{2}{\zeta}\mathbf{F}_s(\mathbf{R}) - \frac{2k_BT}{\zeta}\nabla_{\mathbf{R}} \ln \Psi. \tag{1.93}$$

The center of mass is advected by the background flow, while the end-to-end vector evolves in time due to the gradient of the background flow, the spring force, and the thermal fluctuations.

Our next step is to derive an evolution equation for the distribution function Ψ. While such an equation may be derived using a nearly identical approach to that described in Sect. 2.2 (see Chap. 9 for more details), it is more instructive to provide an analogy with the diffusion equation that will allow us to better understand the origin of the terms in Eqs. (1.90) and (1.91). Consider a concentration field $c(\mathbf{x},t)$ of particles suspended in a fluid. The behavior of the concentration is governed by a diffusion equation,

$$\frac{\partial c}{\partial t} + \nabla \cdot \mathbf{J} = 0, \tag{1.94}$$

where the flux $\mathbf{J}$ is given by the familiar Fick's law, $\mathbf{J} = -D\nabla c$, with D a diffusion constant. If an additional force $\mathbf{F}$ is acting on the particles it creates an extra flux $c\mathbf{v}$, where the velocity $\mathbf{v}$ is given by the balance of the force $\mathbf{F}$ and the viscous drag $-\zeta\mathbf{v}$ acting on each particle, and then the total flux can be written as

$$\mathbf{J} = -D\nabla c + \frac{c}{\zeta}\mathbf{F} = \frac{c}{\zeta}\Big(- \zeta D \nabla \ln c + \mathbf{F}\Big). \tag{1.95}$$

Invoking the fluctuation-dissipation theorem one obtains the Stokes-Einstein relation, $\zeta D = k_BT$ (see Chap. 3), and the term in parentheses may be identified as the total force acting on the particle. The first term is an entropic force that acts to remove any concentration gradients in the solution and has the same form as we have used in Eqs. (1.90) and (1.91) if we identify the concentration field c with the probability distribution Ψ.

Analogously, the equation of probability conservation takes the form

$$\frac{\partial \Psi}{\partial t} + \nabla_{\mathbf{R}} \cdot \big(\dot{\mathbf{R}}\Psi\big) = 0, \tag{1.96}$$

which is known as the Smoluchowski equation. Upon insertion of Eq. (1.93) into Eq. (1.96) we obtain

$$\frac{\partial \Psi}{\partial t} + \nabla_{\mathbf{R}} \cdot \left\{ \left(\mathbf{R} \cdot \mathbf{A} - \frac{2}{\zeta} \mathbf{F}_s(\mathbf{R}) - \frac{2k_B T}{\zeta} \nabla_{\mathbf{R}} \ln \Psi \right) \Psi \right\} = 0. \tag{1.97}$$

The Smoluchowski equation is nonlinear and in general can only be solved numerically.

4.4.2 The Special Case of the Hookean Dumbbell

Although Eq. (1.97) is analytically intractable in general, for a simple Hookean spring force $\mathbf{F}_s = K\mathbf{R}$, with K a spring constant, a constitutive equation may be derived without knowing the exact form of Ψ. To accomplish this we will require an equation of motion for the average dyadic product of two end-to-end vectors:

$$\langle \mathbf{RR} \rangle \equiv \int \mathbf{RR}\, \Psi(\mathbf{R}, t)\, d^3\mathbf{R}. \tag{1.98}$$

Here the angle brackets denote an ensemble average with the distribution function Ψ. The desired equation of motion is readily obtained by multiplying Eq. (1.97) by $\mathbf{RR}$, taking the ensemble average, and using the following identities (see [21]):

$$\int \mathbf{RR} \frac{\partial \Psi}{\partial t} d^3\mathbf{R} = \frac{\partial}{\partial t} \langle \mathbf{RR} \rangle, \tag{1.99}$$

$$\int \mathbf{RR}\, \nabla_{\mathbf{R}} \cdot (\mathbf{R} \cdot \mathbf{A} \Psi)\, d^3\mathbf{R} = -\mathbf{A}^T \cdot \langle \mathbf{RR} \rangle - \langle \mathbf{RR} \rangle \cdot \mathbf{A}, \tag{1.100}$$

$$\int \mathbf{RR}\, \nabla_{\mathbf{R}} \cdot (\mathbf{R} \Psi)\, d^3\mathbf{R} = -2 \langle \mathbf{RR} \rangle, \tag{1.101}$$

$$\int \mathbf{RR}\, \nabla_{\mathbf{R}} \cdot (\nabla_{\mathbf{R}} (\ln \Psi) \Psi)\, d^3\mathbf{R} = 2\mathbf{I}, \tag{1.102}$$

and assuming that the spring force acts only along the axis of the dumbbell, $\mathbf{F}_s = F_s(\mathbf{R})\mathbf{R}$, the resulting equation for the evolution of $\langle \mathbf{RR} \rangle$ is

$$\overset{\nabla}{\langle \mathbf{RR} \rangle} = \frac{4k_B T}{\zeta} \mathbf{I} - \frac{4}{\zeta} \langle \mathbf{RF}_s \rangle. \tag{1.103}$$

In deriving this equation we have used the assumption that Ψ decays sufficiently fast for large $|\mathbf{R}|$ to neglect the boundary terms during the integration by parts. Equation (1.103) implies that in equilibrium, in the absence of flow, we have

$$\langle \mathbf{RF}_s \rangle_{\text{equil}} = k_B T \mathbf{I}. \tag{1.104}$$

4.4.3 Completing the Picture: The Upper-Convected Maxwell Model

We now have the ingredients needed to calculate the polymer contribution to the stress tensor, which will involve measuring the number of dumbbells crossing a given surface and identifying the forces from the stretched dumbbells with the polymeric fluid stress.

To begin, consider an imaginary planar surface in the fluid with area dS and normal vector $\mathbf{m}$ as illustrated in Fig. 1.9. According to the definition given in Sect. 2.4, the traction from the fluid acting on the surface facing in the direction of $\mathbf{m}$ is given by $\mathbf{m}\cdot\boldsymbol{\tau}_p$, which is positive when the resultant force points into the same half-space as $\mathbf{m}$. Consider now a dumbbell with the end-to-end distance $\mathbf{R}$ with its first bead in the "$-\mathbf{m}$" half-space and its second bead in the "$+\mathbf{m}$" half-space, as shown in Fig. 1.9. A dumbbell with the end-to-end distance $\mathbf{R}$ can span both sides of the plane only if its second bead is within a distance $\mathbf{m}\cdot\mathbf{R}$ of the plane and, then depending on that distance, only for a certain range of $\mathbf{R}$. Let n be the number density of dumbbells in the solution. The total number of dumbbells straddling the imaginary surface is then given by $n\,dS\,(\mathbf{m}\cdot\mathbf{R})$.

In the convention we have introduced for the elastic spring, its force $\mathbf{F}_s$ acts on the first bead ($-\mathbf{F}_s$ acts on the second bead). The local force balance on the "$+\mathbf{m}$"-side of the surface implies that the traction on the surface at the point where the end-to-end vector crosses it is equal to $\mathbf{F}_s$. Since the probability of finding a dumbbell in such a configuration is given by $\Psi(\mathbf{R},t)\,d^3\mathbf{R}$, the traction due to this configuration is given by

$$\mathbf{t}^+ = \int_{\mathbf{R}\cdot\mathbf{m}>0} n\,dS\,(\mathbf{m}\cdot\mathbf{R})\,\mathbf{F}_s\,\Psi(\mathbf{R},t)\,d^3\mathbf{R} = n\,dS\,\mathbf{m}\cdot\int_{\mathbf{R}\cdot\mathbf{m}>0}\mathbf{R}\mathbf{F}_s\,\Psi d^3\mathbf{R}, \qquad (1.105)$$

where the restriction of the integration domain to $\mathbf{R}\cdot\mathbf{m}>0$ ensures that the beads are in the configuration depicted in Fig. 1.9. On the opposite face of the surface, a similar argument yields the traction there:

$$\mathbf{t}^- = n\,dS\,\mathbf{m}\cdot\int_{\mathbf{R}\cdot\mathbf{m}<0}\mathbf{R}\mathbf{F}_s\,\Psi d^3\mathbf{R}. \qquad (1.106)$$

The total traction acting on the imaginary surface is given by the sum of the two tractions above, $\mathbf{t}=\mathbf{t}^+ + \mathbf{t}^- = n\,\mathbf{m}\cdot\langle\mathbf{R}\mathbf{F}_\mathbf{s}\rangle\,dS$. Since the same traction can be written as $\mathbf{m}\cdot\boldsymbol{\tau}_p\,dS$, the polymeric contribution to the stress tensor must be equal to

$$\boldsymbol{\tau}_p = n\langle\mathbf{R}\mathbf{F}_s\rangle - 2nk_BT\mathbf{I}. \qquad (1.107)$$

The last term is the isotropic pressure of the ideal gas of the beads, where $2n$ is the number density of the beads, and in the absence of flow this stress tensor does not vanish. Indeed, when $\mathbf{u}=0$,

$$(\boldsymbol{\tau}_p)_{\text{equil}} = n\langle\mathbf{R}\mathbf{F}_s\rangle_{\text{equil}} - 2nk_BT\mathbf{I} = -nk_BT\mathbf{I}, \qquad (1.108)$$

using Eq. (1.104). The flow-induced polymeric contribution to the stress tensor is then

$$\boldsymbol{\tau}_p = n\langle \mathbf{R}\mathbf{F}_s\rangle - 2nk_BT\mathbf{I} - (\boldsymbol{\tau}_p)_{\text{equil}} = n\langle \mathbf{R}\mathbf{F}_s\rangle - nk_BT\mathbf{I}. \tag{1.109}$$

Equations (1.97) and (1.109) are the key results of the kinetic theory for dilute solutions of polymers. For a particular choice of the spring law, $\mathbf{F}_s = F_s(\mathbf{R})\,\mathbf{R}$, one would need to solve Eq. (1.97) for the end-to-end distribution function, find the average $\langle \mathbf{R}\mathbf{F}_s\rangle$, and use that result in Eq. (1.109) to find the stress. For the particular case of the Hookean spring law, $\mathbf{F}_s = K\mathbf{R}$, we have $\langle \mathbf{R}\mathbf{F}_s\rangle = K\langle \mathbf{R}\mathbf{R}\rangle$, and we can use Eqs. (1.109) and (1.98) to eliminate $\langle \mathbf{R}\mathbf{R}\rangle$. At last we have reached the final result. Identifying a relaxation time $\lambda = \zeta/(4K)$ and polymer viscosity $\eta_p = \lambda nk_BT$, the polymeric stress above satisfies the following equation:

$$\boldsymbol{\tau}_p + \lambda \overset{\nabla}{\boldsymbol{\tau}}_p = \eta_p \dot{\boldsymbol{\gamma}}. \tag{1.110}$$

As discussed in Sect. 4.3, kinetic theories cannot produce new types of constitutive equations. Instead, they provide connections between a particular type of molecular theory and a constitutive law and give expressions for the parameters in terms of molecular properties. In the particular case considered here, we have shown that a dilute solution of Hookean dumbbells is described by the UCM model.

The reader should take care to note that the opposite point of view is incorrect: the fact that a particular solution is well described by the UCM model does not necessarily imply that one is dealing with a very dilute solution that consists of approximately Hookean springs. The reason why this statement is incorrect is that the UCM model is one of the simplest frame-invariant models, and many types of kinetic theories project (at least for weak flows) onto that model. In other words, for small deformations, a constitutive equation for polymer solutions almost cannot be anything else save for a few special cases. One possible way of thinking about this is based on one of the central concepts from solid state physics. There it is shown that excitations of a complicated lattice of point-like masses connected by elastic springs can be described as noninteracting degrees of freedom (phonons) that all perform independent harmonic motion with various frequencies. Although these degrees of freedom involve many particles moving in a complicated fashion, these effective degrees of freedom are decoupled from each other. In a similar fashion, for small deformations, one can think of an entangled polymer solution as a dilute solution of *effectively noninteracting* elastic degrees of freedom, each of which involves a significant number of polymers. In turn, this dilute "solution" of the effective degrees of freedom corresponds to the UCM model where the role of dumbbells is played by the collective excitations (normal modes).

5 Material Properties of Viscoelastic Fluids

5.1 Normal Stress Differences

In this section we will discuss the rheological predictions of some of the constitutive models introduced in this chapter. We will show that in a viscoelastic fluid, unlike in a Newtonian fluid, even a two-dimensional shear flow commonly gives rise not only to the off-diagonal elements of the stress tensor (shear stresses), but also to the diagonal components, the normal stresses. Many of the surprising phenomena seen in the flow of complex fluids, and in viscoelastic fluids in particular, can be understood by an examination of these normal stresses and the normal stress differences. We will also show that in shear flows with curved streamlines, normal stresses generate extra forces that are directed towards the center of curvature, pushing fluid elements from their streamlines.

The general mechanism of normal stress development, if not the details which are fluid dependent, is simple to understand and is illustrated in Fig. 1.10a. Polymers are stretched and rotated under the action of the local shear and tend on average to align with the streamlines, while the entropic forces acting to return the molecule to its undisturbed conformation lead to an extra tension in the direction of the flow. Many physical effects attributed to fluid elasticity can be qualitatively understood immediately through this simple concept alone. One of the most famous examples is the Weissenberg effect (viscoelastic rod climbing), illustrated in Fig. 1.10b. A rotating rod in a fluid produces circular streamlines in the flow. In a viscoelastic fluid, polymers align with and stretch along streamlines and respond with a so-called hoop stress (a "strangulation" of the immersed rod, illustrated in Fig. 1.10c, to be discussed). Absent an upper boundary, this response drives the fluid up the rod and out of the bulk. Similar reasoning also accounts for die swell in fluid extrusion and a great number of other fluid phenomena (see a wonderful gallery in [22]). In addition to large-scale collective effects, the presence of normal stress differences in flow can be important on smaller scales as well: cells and other soft biological matter may experience extra polymeric stresses that lead to deformation or possibly rupture.

Each of the many constitutive laws introduced in this chapter comes with its own predictions of normal stress differences, which we shall now investigate using the linear shear flow $\mathbf{u}(\mathbf{x},t) = \mathbf{x} \cdot (\dot{\gamma}\,\mathbf{e}_2\mathbf{e}_1) = (\dot{\gamma}y, 0, 0)$. The first and second normal stress differences, N_1 and N_2, and their coefficients, Ψ_1 and Ψ_2, are defined as

$$N_1 = \tau_{11} - \tau_{22} = \Psi_1 \dot{\gamma}^2, \quad N_2 = \tau_{22} - \tau_{33} = \Psi_2 \dot{\gamma}^2. \tag{1.111}$$

The first normal stress difference monitors the variation in normal stress between the direction of flow (where tension along streamlines is expected as illustrated in Fig. 1.10) and the direction of shear. The first normal stress difference is usually positive in viscoelastic flows, $\Psi_1 > 0$. The second normal stress difference monitors the stress difference in the two directions normal to the flow direction and is,

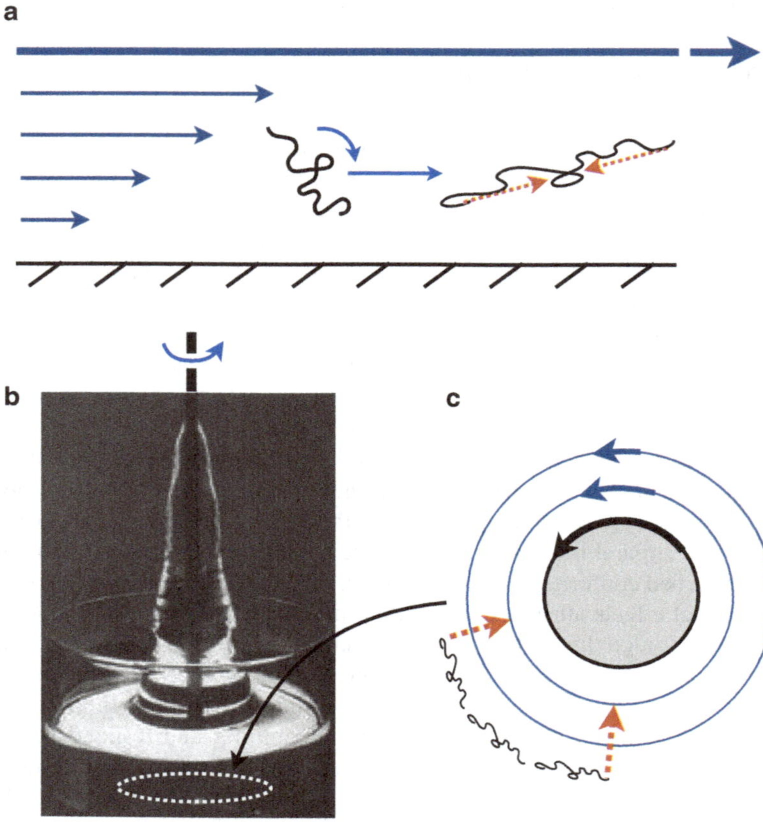

Fig. 1.10 (**a**) An illustration of the mechanism of normal stress differences in viscoelastic fluids. A shear flow rotates and stretches polymers along streamlines creating anisotropic elastic stresses. (**b**) The Weissenberg effect (rod climbing): a rotating rod inside of a viscoelastic fluid excites an upward climb of fluid, unlike in a Newtonian flow (adapted from [22] with permission). (**c**) The Weissenberg effect is explained by normal stress differences, here by the development of hoop stresses along curved streamlines, leading to "strangulation" and an upward ascent

generally, negative and very small compared to the first normal stress difference. Normal stress differences in a general flow may be pronounced near boundaries, including the boundaries of immersed bodies, as the no-slip velocity boundary condition and/or stagnation points of the flow generally introduce a shear flow local to the boundary surface.

In a Newtonian shear flow the pressure is constant, $p = p_0$ [found by inserting the velocity field into Eqs. (1.21) and (1.22)], and the deviatoric stress is

$$\boldsymbol{\tau} = \mu\dot{\boldsymbol{\gamma}} = \mu\dot{\gamma}(\mathbf{e}_2\mathbf{e}_1 + \mathbf{e}_1\mathbf{e}_2), \tag{1.112}$$

so that clearly $\Psi_1 = \Psi_2 = 0$. In fact there are no normal stresses whatsoever outside of the constant isotropic pressure in this case, let alone normal stress differences. It is a short exercise to find that there are no normal stress differences associated with the generalized Newtonian fluids introduced in Sect. 3, but this is in keeping with our physical explanation of the source of normal stress differences described above. The linearly viscoelastic fluids introduced in Sect. 4 also do not predict normal stress differences in a shear flow. In order to capture or predict normal stress differences, we must look to the nonlinear constitutive models of viscoelastic fluids. Interestingly, as we shall see below, the normal stresses are generated by the nonlinear terms in the convected derivatives introduced to ensure frame-invariance, revealing a deep connection between geometry and mechanical properties.

Consider the second-order fluid model described in Sect. 4.3. The coefficients in the constitutive law, Eq. (1.74), are in fact directly related to the normal stress differences. Inserting the steady shear flow above into the constitutive relation we find that

$$\boldsymbol{\tau} = \eta\dot{\gamma}\left(\mathbf{e}_2\mathbf{e}_1 + \mathbf{e}_1\mathbf{e}_2\right) - 2b_2\dot{\gamma}^2\mathbf{e}_1\mathbf{e}_1 + b_{11}\dot{\gamma}^2\left(\mathbf{e}_1\mathbf{e}_1 + \mathbf{e}_2\mathbf{e}_2\right). \tag{1.113}$$

Unlike in a Newtonian fluid, the nonlinear terms in the second-order model allow for nonzero normal stress differences in the fluid, $\Psi_1 = -2b_2$ and $\Psi_2 = b_{11}$. Hence, if the viscosity and normal stress differences (the *viscometric functions*) have been measured for a particular fluid, they can be used to specify this particular constitutive model directly as

$$\boldsymbol{\tau} = \eta\dot{\boldsymbol{\gamma}} - \frac{\Psi_1}{2}\overset{\triangledown}{\dot{\boldsymbol{\gamma}}} + \Psi_2\dot{\boldsymbol{\gamma}}\cdot\dot{\boldsymbol{\gamma}}. \tag{1.114}$$

We also see from Eq. (1.113) that the second-order fluid model has no shear-dependent viscosity $\tau_{12}/\dot{\gamma} = \eta$ (constant). Shear-dependent viscosity can be captured at the next order in the retarded-motion expansion, (the third-order fluid model), which has the same normal stress differences as in the second-order model. It is common in the literature to see the approximation $-\Psi_2/\Psi_1 = 1/2$, which overestimates the ratio's value as observed in experiments with most fluids, but is of great use in improving the mathematical tractability of the model. For in this case, the effect of viscoelasticity at first order variation away from the Newtonian flow is simply to modify the pressure and not the fluid velocity field (see [13]).

Next, consider the Oldroyd-B model fluid, Eq. (1.80):

$$\boldsymbol{\tau} + \lambda\overset{\triangledown}{\boldsymbol{\tau}} = \eta\dot{\boldsymbol{\gamma}} + \eta_s\lambda\overset{\triangledown}{\dot{\boldsymbol{\gamma}}}. \tag{1.115}$$

Assuming the same steady shear flow, the individual components of the stress are found to be

$$\boldsymbol{\tau} = 2\lambda\eta_p\dot{\gamma}^2\mathbf{e}_1\mathbf{e}_1 + \eta\dot{\gamma}(\mathbf{e}_1\mathbf{e}_2 + \mathbf{e}_2\mathbf{e}_1) \tag{1.116}$$

(recall that $\eta = \eta_s + \eta_p$). Hence, the first normal stress difference coefficient is $\Psi_1 = 2\lambda\eta_p (\geq 0)$. The first normal stress difference is linear in the polymer relaxation time, λ, and vanishes in the limit $\eta_p \to 0$. In the Oldroyd-B model there are no transverse normal stresses in a shear flow, $\tau_{22} = \tau_{33} = 0$, and then trivially $\Psi_2 = 0$. In this model as well we see that the viscosity is always equal to the zero-shear-rate viscosity, $\tau_{12}/\dot{\gamma} = \eta$ (constant). This prediction is inconsistent with experimental observations, in that the viscosity of real polymer solutions often exhibits shear-thinning. Shear-thinning is, however, successfully captured by other nonlinear models, including the FENE-P, Giesekus, and PTT models discussed in Sect. 4.3.

As we have demonstrated above, the only nontrivial component of the stress tensor that appears in the second-order and Oldroyd-B fluids is τ_{11}. Within the kinetic theory approach, Eq. (1.109) allows us to conclude that this stress component is generated by a nonzero component of the end-to-end tensor $\langle R_1 R_1 \rangle$. In turn, this implies that the polymers are stretched and oriented in the flow direction as described at the beginning of this section. From symmetry arguments, changing the direction of the shear, $\dot{\gamma} \to -\dot{\gamma}$, does not change the polymer stretch and orientation. Hence, τ_{11} and N_1 in general should depend on an even power of $\dot{\gamma}$. As we see in the Oldroyd-B model from Eq. (1.116), $N_1 \sim \dot{\gamma}^2$.

Note that the picture of polymers being stretched and aligned in the direction of flow is only accurate *on average*. Recent simulations [23, 24] of individual dumbbells in shear flows suggest that a dumbbell performs the motion that is similar to the Jeffery orbit of a rod in shear flows [9]: most of the time the dumbbell is oriented at an angle with respect to the flow direction, but it periodically tumbles out of this configuration. The relative time spent tumbling and the angle with respect to the flow direction decrease with the shear rate, while its stretch increases with $\dot{\gamma}$, giving support to the coarse-grained molecular picture discussed above.

5.2 Normal-Stress Measurements

There are a number of ways in which the normal stress differences may be measured in the lab. Oftentimes, the only significant component of the normal stresses in a two-dimensional linear shear flow is τ_{11}. In order to measure it, one has to somehow access the forces exerted by the fluid in the flow direction without disturbing the velocity profile. This is very difficult technically and instead one usually takes advantage of the coupling between the normal stresses and the forces on the boundaries that exist in some curvilinear geometries. One of the most common devices for measuring the material properties of a fluid, or *rheometers*, is the *cone-and-plate* apparatus illustrated in Fig. 1.11. In this device a cone is rotated above a horizontal plate with the fluid under investigation filling the gap between them. The fluid meniscus is exposed to the atmosphere.

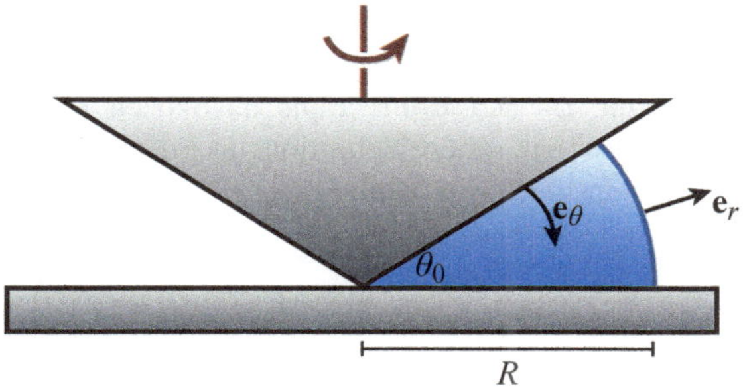

Fig. 1.11 Illustration of a common cone-and-plate rheometer for measuring material properties. The meniscus of the fluid is exposed to the surrounding atmosphere (*the curved blue line*). The geometry is special because the shear rate is uniform throughout the sample

In order to most easily connect the measurements of the cone-and-plate rheometer to the constitutive equations we will assume that the cone touches the plate at a point. The gap between the cone and the plate has the shape of a spherical segment with a small opening angle θ_0, which typically measures only a few degrees. The most convenient way to describe this geometry is to introduce a spherical coordinate system (r, θ, ϕ), with r measuring the distance from the point of contact between the cone and plate. The angle θ is measured from the axis of rotation and the angle ϕ—around that axis, see Fig. 1.11. The fluid fills the gap up to $r = R$. Given the rotational symmetry, the fluid velocity at each point has only one component, u_ϕ, that depends only on the local distance between that point, the cone, and the plate; that position in the gap is set by θ.

The advantages of this geometry are that it is curvilinear (we shall see below why this is important) and that it has a constant shear rate. Indeed, since the angle θ_0 is small, the azimuthal velocity of a point on the cone is given by ωr, where ω is the angular velocity of the cone and r is the radial position of the point. The distance between the cone and the plate at that point is given by $r\theta_0$, which results in the shear rate at that radial distance, ω/θ_0, independent of r! Since the shear rate is the same everywhere, the polymeric stresses may be assumed position-independent in the sample, similar to the case of linear shear considered above.

In what follows, we do not employ any constitutive relation between the polymeric stress and the velocity gradient and only consider the momentum balance equation. Since, typically, polymeric fluids are rather viscous and the measurements are performed at relatively low flow velocities (though the velocity gradients are not small), fluid inertia is generally neglected. The radial component of the momentum balance equation in spherical coordinates then reads

$$-\frac{\partial p}{\partial r} + \frac{1}{r^2}\frac{\partial}{\partial r}\left(r^2 \tau_{rr}\right) - \frac{\tau_{\theta\theta} + \tau_{\phi\phi}}{r} = 0, \tag{1.117}$$

where $\boldsymbol{\tau}$ is the deviatoric stress tensor that has both polymeric and Newtonian solvent contributions and p is the pressure. Using the definition of the first and second normal stress differences, N_1 and N_2, we can write

$$N_1 = \tau_{\phi\phi} - \tau_{\theta\theta}, \qquad N_2 = \tau_{\theta\theta} - \tau_{rr}, \tag{1.118}$$

where the ϕ- and θ-directions have been identified as the flow and gradient directions. Using these definitions in Eq. (1.117) and employing the fact that $\boldsymbol{\tau}$ is constant in space, we obtain

$$-\frac{\partial p}{\partial r} = \frac{2N_2 + N_1}{r}. \tag{1.119}$$

This result is the mathematical foundation of the hoop stresses discussed in the last section: in curved geometries, the tension along the flow lines N_1 is balanced by an inward-pointing pressure gradient that grows in the direction of the origin. In an open geometry, this pressure gradient can result in the rod-climbing effect shown in Fig. 1.10b. Integrating Eq. (1.119), we have

$$-p(r) = -p(R) + (2N_2 + N_1)\ln\frac{r}{R}, \tag{1.120}$$

where the value of the pressure at the meniscus of the cone and plate rheometer, $p(R)$, is an unknown integration constant to be determined. Since the meniscus is in equilibrium, we require that the force per unit area acting on it from the fluid perpendicular to its surface is balanced by the atmospheric pressure p_{atm} from the outside

$$-p(R) + \tau_{rr}(R) + \frac{2\alpha}{R} = p_{\text{atm}}. \tag{1.121}$$

The third term on the l.h.s. is the Laplace pressure under a curved surface and α is the surface tension. Note that the meniscus is the surface of a spherical segment and its principle radii of curvature R_1 and R_2 are the same, $R_1 = R_2 = R$. Force balance, Eq. (1.121), allows us to determine the unknown constant $p(R)$.

The normal force exerted on the plate by the fluid is given by

$$F = 2\pi \int_0^R r\left[-p(r) + \tau_{\theta\theta}\right] dr + 2\pi\alpha R. \tag{1.122}$$

The first term is the local fluid stress normal to the plate, $-p(r) + \tau_{\theta\theta}$, integrated over its surface, and the last term is the contribution of the line tension of the meniscus on the plate. The total stress perpendicular to the plate can be rearranged as

$$-p(r) + \tau_{\theta\theta} = (2N_2 + N_1)\ln\frac{r}{R} + p_{\text{atm}} + N_2 - \frac{2\alpha}{R}, \tag{1.123}$$

where we have made use of Eqs. (1.120) and (1.121). After integration, we obtain

$$F = \pi R^2 p_{\text{atm}} - \frac{\pi R^2}{2} N_1 . \tag{1.124}$$

The force acting on the plate from the atmosphere is simply $\pi R^2 p_{\text{atm}}$, and the excess force ΔF is, finally,

$$\Delta F = -\frac{\pi R^2}{2} N_1 . \tag{1.125}$$

This expression provides a simple way of measuring the first normal stress difference by measuring the total force exerted on the plate when the cone is rotating. We note here that this is only possible because of the curvilinear geometry of this setup. As can be seen from Eq. (1.117) or Eq. (1.119), in the limit of a linear shear that is attained by $r \to \infty$, the normal stresses decouple from the pressure gradient and one cannot access N_1 by measuring the force on the plate. A similar argument shows that the normal stresses cannot be measured in the Taylor–Couette geometry (flow in the gap between two rotating coaxial cylinders), but are accessible in another curvilinear geometry, the plate-and-plate setup, that is essentially the cone-and-plate rheometer with the cone replaced by another plate. In that setup, however, the shear rate is not constant everywhere in the sample and it is often difficult to interpret measured quantities.

In practice there are many sources of experimental error, and great care must be taken in interpreting the measured data correctly. Fluid and instrument inertia, secondary flows, elastic instabilities, slip, and many other real issues can lead to Newtonian samples appearing to be complex and vice versa. The derivation of the force exerted by the fluid on the plate, the quantity that is measured by commercial rheometers, helps us to identify potential sources of errors in such experiments. One contribution to the experimental error is due to the instrumental resolution of the pressure transducer used to measure the force, and it is typically rather small. More importantly, the effects of the surface tension in the stress continuity at the meniscus, Eq. (1.121), and the force due to the line tension in Eq. (1.122) strongly depend on the contact angle between the fluid and the walls of the rheometer. In fact, the contact angle between the fluid and the walls of the rheometer can vary from experiment to experiment giving rise to an extra force on the plate that would depend on how the gap was filled with the fluid. Obviously, this extra force has nothing to do with the normal stresses and is a major source of nonsystematic experimental errors. These challenges and others, along with the techniques for reducing errors, are the topic of Chap. 6.

5.3 Other Flows

Besides the simple shear flows discussed above, there are many ways of assessing the material properties of viscoelastic fluids. A natural generalization of the steady two-dimensional shear flow introduces time dependence, $\mathbf{u} = \mathbf{x} \cdot (\dot{\gamma}(t)\mathbf{e}_2\mathbf{e}_1) = (\dot{\gamma}(t)y, 0, 0)$. Using the Oldroyd-B constitutive law as an illustrative example, it is a simple exercise to show that the stress has the form $\boldsymbol{\tau}(t) = \tau_{11}(t)\mathbf{e}_1\mathbf{e}_1 + \tau_{12}(t)(\mathbf{e}_1\mathbf{e}_2 + \mathbf{e}_2\mathbf{e}_1)$ and that the stress components satisfy the following coupled differential equations:

$$\left(1 + \lambda \frac{d}{dt}\right) \tau_{11} - 2\lambda \dot{\gamma}(t) \tau_{12} = -2\eta_s \lambda \dot{\gamma}^2(t), \tag{1.126}$$

$$\left(1 + \lambda \frac{d}{dt}\right) \tau_{12} = \left(\eta + \lambda \eta_s \frac{d}{dt}\right) \dot{\gamma}(t). \tag{1.127}$$

From these expressions the normal stress differences may be predicted for arbitrary time-dependent shear flows. Many experimental apparatuses, including the popular cone and plate, plate and plate, and capillary rheometers, are designed to impose the time dependence of the shear rate and infer material parameters by the forces and torques found in response. Such tests include steady shear, $\dot{\gamma}(t) = \dot{\gamma}_0$; stress growth and relaxation, $\dot{\gamma}(t) = \dot{\gamma}_0 H(t)$ and $\dot{\gamma}(t) = \dot{\gamma}_0 H(-t)$ with $H(t)$ the Heaviside function; step strain, $\dot{\gamma}(t) = \dot{\gamma}_0 (H(t) - H(t - t_0))$ with $t_0 > 0$; and small-amplitude oscillatory shear, $\dot{\gamma} = \dot{\gamma}_0 \sin(\omega t)$ (as discussed in Sect. 4.1). Another common test uses a creep flow, in which an impulsive and constant shear stress is applied to the material and the time dependence of the fluid response is observed. Another method of rheology growing in popularity makes use of large-amplitude oscillatory shear [25–27]. An entirely different approach to measuring material properties makes use of the fluctuations of small probes at the microscale and is termed microrheology. Theoretical microrheology is discussed extensively in Chap. 3 and used as a basis to study the material properties of membranes in Chap. 4.

Other rheometers have been designed to measure the stress-strain responses in another important rheological flow, an extensional flow. A pure extensional flow is written as $\mathbf{u}(\mathbf{x}, t) = \mathbf{x} \cdot (\dot{\varepsilon}\mathbf{e}_1 - \dot{\varepsilon}(\mathbf{e}_2 + \mathbf{e}_3)/2) = \dot{\varepsilon}(x, -y/2, -z/2)$, where $\dot{\varepsilon}$ is the rate of extension, so that $\dot{\boldsymbol{\gamma}} = \dot{\varepsilon}\,(2\mathbf{e}_1\mathbf{e}_1 - \mathbf{e}_2\mathbf{e}_2 - \mathbf{e}_3\mathbf{e}_3)$. This flow is called extensional since two material points originally close to each other will be separated exponentially in time by this flow as can be seen from the kinematic equations

$$\dot{x}(t) = \dot{\varepsilon} x, \qquad \dot{y}(t) = -\frac{\dot{\varepsilon}}{2} y, \qquad \dot{z}(t) = -\frac{\dot{\varepsilon}}{2} z, \tag{1.128}$$

where a dot denotes a time derivative. A commonly measured material response in this setting is called the "extensional viscosity," defined as

$$\bar{\eta} = \frac{\tau_{11} - \tau_{22}}{\dot{\varepsilon}}, \tag{1.129}$$

though the use of the term in time-dependent flows is "fraught with danger" [28]. In a Newtonian fluid, with $\boldsymbol{\tau} = \mu\dot{\boldsymbol{\gamma}}$, we have simply that $\bar{\eta} = 3\mu$. In a second-order fluid we have

$$\bar{\eta} = 3\left(\eta + (b_{11} - b_2)\,\dot{\varepsilon}\right), \tag{1.130}$$

and in an Oldroyd-B fluid,

$$\bar{\eta} = 3\frac{\eta - \eta_s\lambda\dot{\varepsilon} - 2\eta_s\lambda^2\dot{\varepsilon}^2}{(1 - 2\lambda\dot{\varepsilon})\,(1 + \lambda\dot{\varepsilon})}. \tag{1.131}$$

This expression indicates that the stresses grow very rapidly with $\dot{\varepsilon}$ since polymers oppose exponential separation of their ends and, in the case of the Oldroyd-B model, Eq. (1.131), even diverge for $\lambda\dot{\varepsilon} = 1/2$. This phenomenon, sometimes miscalled the *coil-stretch transition*, is the consequence of the unrealistic behavior of the underlying Hookean force law for the dumbbells—their ends can be separated without limits producing very large stresses. Finite extensibility of polymer chains, or other nonlinear mechanisms presented in models like FENE-P, Giesekus, PTT, and others, cures this problem while still exhibiting a rapid growth of $\bar{\eta}$ with $\dot{\varepsilon}$. However, in all of the models above, large extensional stresses and its gradients have proven to be very problematic in the computation of highly elastic flows. Together with the loss of positive definiteness discussed in Sect. 6, it forms the basis of the High-Weissenberg-Number Problem in computational complex fluids in both two and three dimensions. This topic is addressed in detail in Chap. 10.

6 Final Words of Caution: A Health Warning

Modeling complex fluids can be a tricky business. The primary challenges when dealing with the constitutive models described in this chapter generally arise due to their strongly nonlinear nature. Unless these models are used to study simple, steady flows, it is generally impossible to derive analytical solutions. Instead, one is faced with making perhaps severe analytical approximations or performing time-dependent direct numerical simulations. While both strategies have proven to be fruitful in understanding complicated flows of viscoelastic fluids, they both open the door to serious potential pitfalls. To conclude this chapter we will describe several typical problems that can arise in *approximating* solutions to viscoelastic equations of motion, whether the approximation be analytical or numerical.

The Linear Maxwell Model Is Not Objective One of the more common mistakes made is the inappropriate use of the linear Maxwell model, Eq. (1.54), in actual calculations. It is often argued that studying this model allows one to understand how fluid memory affects the flow, as opposed to normal stress differences, shear-thinning, and other effects that arise from various nonlinear terms in the constitutive models. The trouble with this approach is that the linear Maxwell model, Eq. (1.54), is not frame-invariant and none of the conclusions drawn from studying this model are guaranteed to be physical. The only way to check whether its predictions are physical is to perform an analysis of the full original constitutive model and compare the two results, at which point the analysis of Eq. (1.54) will have become unnecessary. Unless the original constitutive model reduces exactly to the linear Maxwell model (as is the case with the shear-stress equation for the small-amplitude oscillatory shear flow[1]), the use of the linear Maxwell model should be forbidden!

Time-Dependent Flows in Weakly-Nonlinear Viscoelastic Fluids Are Unstable A second common problem arises in studying weakly nonlinear flows of viscoelastic systems. Consider, for instance, the complete Oldroyd-B model equations (see Sect. 4.3):

$$\rho\left(\frac{\partial \mathbf{u}}{\partial t}+\mathbf{u}\cdot\nabla\mathbf{u}\right)=-\nabla p+\nabla\cdot\boldsymbol{\tau}, \tag{1.132}$$

$$\boldsymbol{\tau}+\lambda\overset{\nabla}{\boldsymbol{\tau}}=\eta\left(\dot{\boldsymbol{\gamma}}+\lambda(\eta_s/\eta)\overset{\nabla}{\dot{\boldsymbol{\gamma}}}\right), \tag{1.133}$$

$$\nabla\cdot\mathbf{u}=0. \tag{1.134}$$

A weakly nonlinear flow is a situation wherein the flow changes on time scales much longer than the relaxation time, λ, and therefore λ is in some sense small and can be used as an expansion parameter. It is generally a bad practice to perform a Taylor expansion in a dimensional variable; a better expansion parameter might be λ/τ_0, where τ_0 is the typical time scale set by the flow, which becomes apparent when Eqs. (1.132)–(1.134) are written in dimensionless form. Nevertheless, formally we might write

$$\boldsymbol{\tau}=\boldsymbol{\tau}^{(0)}+\lambda\boldsymbol{\tau}^{(1)}+\mathbf{O}(\lambda^2). \tag{1.135}$$

Substituting this expression into Eq. (1.133), at leading order we recover a Newtonian constitutive relation:

$$\boldsymbol{\tau}^{(0)}=\eta\dot{\boldsymbol{\gamma}}, \tag{1.136}$$

(recall that $\eta=\eta_s+\eta_p$), and

$$\boldsymbol{\tau}^{(1)}=-\overset{\nabla}{\boldsymbol{\tau}}^{(0)}+\eta_s\overset{\nabla}{\dot{\boldsymbol{\gamma}}}=-\eta_p\overset{\nabla}{\dot{\boldsymbol{\gamma}}}. \tag{1.137}$$

[1]Note, however, that this coincidence is only partial: for example, equations for the normal components of the stress tensor *do not* reduce to the linear Maxwell equations in the same geometry.

Truncating the $O\left(\lambda^2\right)$ terms in Eq. (1.137), we are thus left with a particular case of the second-order model, Eq. (1.74), where $b_1 = -\lambda\eta_p$ and $b_{11} = 0$. With this in mind, let us consider the more general constitutive relation given by Eq. (1.74).

Consider a two-dimensional shear flow given by $\mathbf{u} = (u(y,t), 0)$. Upon insertion into the momentum balance equation Eq. (1.132), and using the second-order fluid model, Eq. (1.74), an equation for the evolution of $u(y,t)$ is found:

$$\rho\frac{\partial u}{\partial t} = \frac{\partial}{\partial y}\tau_{xy}, \tag{1.138}$$

where the shear stress is given by

$$\tau_{xy} = \eta\frac{\partial u}{\partial y} + b_2\frac{\partial}{\partial t}\frac{\partial u}{\partial y}. \tag{1.139}$$

Combining these two equations we find that

$$\rho\frac{\partial u}{\partial t} = \eta\frac{\partial^2 u}{\partial y^2} + b_2\frac{\partial}{\partial t}\frac{\partial^2 u}{\partial y^2}. \tag{1.140}$$

Just as in Sect. 3.3, we take no-slip boundary conditions at the walls of a channel located at $y = 0$ and $y = h$ and write the flow velocity as a Fourier series:

$$u(y,t) = \sum_{m=1}^{\infty} u_m e^{\alpha_m t} \sin\frac{m\pi y}{h}. \tag{1.141}$$

The growth rates α_m associated with each mode are obtained by substituting this expression into Eq. (1.140), revealing

$$\alpha_m = \frac{-\eta}{b_2 + \rho h^2/(m\pi)^2}. \tag{1.142}$$

For polymer solutions b_2 is typically negative (recall its value based on the Oldroyd-B model, $b_2 = -\lambda\eta_p$), so that α_m is positive for sufficiently large m. This implies that a steady shear flow of a second-order fluid is unstable to short-wavelength perturbations and cannot be realized. In the case of negligible inertia, achieved in the above by setting $\rho = 0$, Eq. (1.142) predicts that *all* Fourier modes are unstable. The implications of this result are profound: it shows that an approximation of slow flows, or, in other words, Taylor expansions of the stress in terms of the relaxation time cannot be used in time-dependent flows where any shear component would result locally in a linear instability and exponential growth of the stress. Hence, the second-order fluid and similar approximations should generally not be used to study time-dependent flows!

The Conformation Tensor Must Remain Positive-Semidefinite Finally, we comment on a problem that is often encountered in the numerical solution of viscoelastic flows. We will base our discussion on the UCM model for simplicity, but the conclusions we will reach are much more general. In the kinetic theory outlined in Sect. 4.4, we concluded that the polymeric contribution $\boldsymbol{\tau}_p$ to the stress tensor is related to the dyadic tensor $\langle \mathbf{RR} \rangle$ by

$$\boldsymbol{\tau}_p = nK\langle \mathbf{RR} \rangle - \frac{\eta_p}{\lambda}\mathbf{I}, \qquad (1.143)$$

where $\mathbf{R}$ is the end-to-end distance of a polymer molecule that was represented by a dumbbell in Sect. 4.4. To obtain this expression we have used Eq. (1.109) and the molecular expressions for η_p and λ obtained in Sect. 4.4. The tensor $\langle \mathbf{RR} \rangle$ in its dimensionless form is often referred to as the *conformation tensor*. At a given point in an arbitrary flow, the $\langle \mathbf{RR} \rangle$-tensor can be diagonalized, $\langle \mathbf{RR} \rangle = \mathrm{diag}\left(R_1^2, R_2^2, R_3^2\right)$, where R_1, R_2, and R_3 are the projections of the end-to-end vector on the corresponding coordinate axes. Since the diagonal entries are the *squares* of these projections, they cannot be negative in any flow if the $\langle \mathbf{RR} \rangle$-tensor is to remain physical. Since, by construction, R_1^2, R_2^2, and R_3^2 are the eigenvalues of the $\langle \mathbf{RR} \rangle$-tensor at the considered point in space and time, this statement translates into the requirement that the eigenvalues of the $\langle \mathbf{RR} \rangle$-tensor always remain nonnegative in the whole domain considered.

Although we have introduced this requirement based on the kinetic theory, it is more general. Even if nothing is stated about the physical meaning of the $\langle \mathbf{RR} \rangle$-tensor, it can be formally introduced through, say, Eq. (1.143) (if one deals with the Oldroyd-B model) and it is then possible to prove that if at time t_0 the eigenvalues of $\langle \mathbf{RR} \rangle$ are nonnegative everywhere in the domain, they remain nonnegative for all later times. This property is often referred to as the *evolutionary* nature of the corresponding constitutive equation and has been proven for the Oldroyd-B (UCM), Giesekus, and other models [29]. Since the rest state with no stress is clearly positive-semidefinite (its eigenvalues are not negative), any time evolution starting from this state should remain positive-semidefinite. Unfortunately, this is not the case in both simulations and analytic calculations involving approximations.

Often, due to either accumulation of numerical errors or a severe approximation the conformation tensor may develop negative eigenvalues and become unphysical. Unfortunately, the constitutive equations that have been used in this chapter do not provide a clear indicator of when this will happen (i.e., the stress values do not suddenly diverge at this point or similar). In order to ensure that the results are physical it is therefore advisable to check that the conformation tensor is positive-semidefinite in the whole domain at each time-step in simulations or at the end of analytical calculations. If the conformation tensor is ever found not to be positive-semidefinite, the resulting calculation should not be trusted!

7 Conclusion

In this chapter we presented the fundamental concepts in continuum mechanics and laid the foundations for the mathematical modeling of complex fluids. Numerous constitutive models were introduced, each used to describe complex fluid phenomena such as shear-dependent viscosity and viscoelasticity at varying levels of sophistication. The importance of frame-invariance was stressed in the path to developing mathematically and physically sound nonlinear models including the upper-convected Maxwell (UCM) and Oldroyd-B models and others. Kinetic theory was used as an alternate means of deriving a constitutive law, namely the UCM model, from the ground up. Finally, normal stress differences were discussed, and warnings were given about common dangers encountered in the mathematical modeling of complex fluids.

There are great challenges that remain in the study of complex fluid flows in biological systems. The mathematical modeling of real biological materials by a careful selection of constitutive relation remains problem dependent and is a delicate art. Some of the most popular constitutive laws, such as the Oldroyd-B model of viscoelastic fluids, still present challenges to mathematical analysis and even numerical simulation of highly elastic fluid flows. We may have made great strides in understanding how complex fluid flows change the behavior of immersed soft biological structures, from individual cells to motile microorganisms, but we have barely scratched the surface when it comes to understanding the evolution of biological materials and organisms in the context of non-Newtonian fluid environments. There is much yet to learn in this very exciting convergence of fields.

References

1. G.K. Batchelor, *An Introduction to Fluid Dynamics* (Cambridge University Press, Cambridge, 2000)
2. L.D. Landau, E.M. Lifshitz, *Fluid Mechanics* (translated from Russian by J.B. Sykes and W.H. Reid), vol. 6 (Butterworth-Heinemann, Oxford, 1987)
3. L.G. Leal, *Advanced Transport Phenomena: Fluid Mechanics and Convective Transport Processes*, vol. 7 (Cambridge University Press, Cambridge, 2007)
4. C. Pozrikidis, *Introduction to Theoretical and Computational Fluid Dynamics* (Oxford University Press, Oxford, 2011)
5. D.J. Acheson, *Elementary Fluid Dynamics* (Oxford University Press, Oxford, 1990)
6. S. Childress, *An Introduction to Theoretical Fluid Mechanics* (Courant Lecture Notes) (American Mathematical Society, Courant Institute of Mathematical Sciences at New York University, 2009)
7. C. Pozrikidis, *Boundary Integral and Singularity Methods for Linearized Viscous Flow* (Cambridge University Press, Cambridge, 1992)
8. I.Z. Fisher, *Statistical Theory of Liquids* (University of Chicago Press, Chicago, 1961)
9. R.G. Larson, *The Structure and Rheology of Complex Fluids* (Oxford University Press, Oxford, 1999)

10. D.R. Foss, J.F. Brady, J. Rheol. **44**, 629 (2000)
11. M.E. Cates, S.M. Fielding, Adv. Phys. **55**, 799 (2006)
12. S. Lerouge, J.F. Berret, Adv. Polym. Sci. **230**, 1 (2010)
13. R.B. Bird, C.F. Curtiss, R.C. Armstrong, O. Hassager, *Dynamics of polymeric liquids*, vol. 1, *Fluid Mechanics*, 2nd edn. (Wiley, New York, 1987)
14. A. Lodge, *Elastic Liquids: An Introductory Vector Treatment of Finite-Strain Polymer Rheology* (Academic, New York, 1964)
15. L. Sedov, *Introduction to the Mechanics of a Continuous Medium*. ADIWES International Series in the Engineering Sciences (Addison-Wesley, Reading, 1965)
16. M. Chilcott, J. Rallison, J. Non-Newtonian Fluid Mech. **29**, 381 (1988)
17. A. Peterlin, J. Polym. Sci. Part B Polym. Phys. **4**, 287 (1966)
18. A.E. Likhtman, R.S. Graham, J. Non-Newtonian Fluid Mech. **114**, 1 (2003)
19. R.B. Bird, C.F. Curtiss, R.C. Armstrong, O. Hassager, *Dynamics of polymeric liquids*, vol. 2, *Kinetic Theory*, 2nd edn. (Wiley, New York, 1987)
20. M. Doi, S. Edwards, *The Theory of Polymer Dynamics* (Oxford University Press, Oxford, 1986)
21. R.G. Larson, *Constitutive Equations for Polymer Melts and Solutions* (Butterworths, Boston, 1988)
22. D.V. Boger, K. Walters, *Rheological Phenomena in Focus* (Elsevier, Amsterdam, 1993)
23. A. Celani, A. Puliafito, K. Turitsyn, Europhys. Lett. **70**, 464 (2005)
24. E. Sultan, J.W. van de Meent, E. Somfai, A.N. Morozov, W. van Saarloos, Europhys. Lett. **90**, 99–121 (2010)
25. A.J. Giacomin, J.M. Dealy, *Techniques in Rheological Measurement* (Springer, New York, 1993)
26. R.H. Ewoldt, A. Hosoi, G.H. McKinley, J. Rheol. **52**, 1427 (2008)
27. K. Hyun, M. Wilhelm, C.O. Klein, K.S. Cho, J.G. Nam, K.H. Ahn, S.J. Lee, R.H. Ewoldt, G.H. McKinley, Prog. Polym. Sci. **36**, 1697 (2011)
28. C.J. Petrie, J. Non-Newtonian Fluid Mech. **137**, 15 (2006)
29. M.A. Hulsen, J. Non-Newtonian Fluid Mech. **38**, 93 (1990)

Chapter 2
Complex Fluids and Soft Structures in the Human Body

Paula A. Vasquez and M. Gregory Forest

Abstract The human body is a composite of diverse materials able to perform specific functions. Few of these materials are simple liquids or solids; rather they share both liquid-like and solid-like properties. The material world between liquids and solids is unlimited and exploited by Nature to form complex fluids and soft structures with properties that are tuned to perform highly specialized functions. This chapter will briefly summarize the diversity of materials in the human body, and then drill deeper into one complex fluid (mucus, which coats every organ in the body) and one soft structure (an individual cell), and their remarkable properties. Some progress in characterizing these materials and modeling their functional properties, by others and our research group, will be presented with the take home message that we are in the early stages of interpreting experimental data and building predictive models and simulations of biological materials.

1 Introduction

This review chapter takes a *multidisciplinary* point of view toward complex fluids, or soft matter, in the human body. This multidisciplinary approach is required since the remarkable properties and functions of biological complex fluids are hard to resolve from a singular disciplinary approach. Examples of disciplines encountered in the study of biological fluids are systems biology, molecular biology, applied mathematics, medical biology, chemistry, physics, computer science, and several aspects of engineering. In this chapter, we highlight two biological fluids, lung mucus and single cells, and use them to illustrate the challenges faced in faithfully modeling their function and behavior.

P.A. Vasquez (✉)
University of South Carolina, Columbia, SC 29208, USA
e-mail: paula@math.sc.edu

M.G. Forest
University of North Carolina at Chapel Hill, Chapel Hill, NC 27599, USA
e-mail: forest@unc.edu

S.E. Spagnolie (ed.), *Complex Fluids in Biological Systems*, Biological and Medical Physics, Biomedical Engineering,
DOI 10.1007/978-1-4939-2065-5_2

A fundamental goal of systems biology is to understand the mechanisms underlying material performance, e.g., clearance of mucus from lung airways, intracellular organization during phases of a cell cycle, and motility of cells in one environment versus another. What these mechanisms have in common is that they are emergent processes from the collective behavior of many (perhaps thousands) molecular components. The challenge in systems biology and molecular biology is to describe the dynamic network that results from the complex interactions among molecular constituents. Such interactions dictate material properties and functions, such as the distribution of mucins, other proteins and ions that determine the flow and diffusive transport properties of mucus, how dynein motors control spatial extent and polarity of the mitotic spindle, and how intracellular structures, activating and deactivating proteins, and remodeling processes conspire to achieve cell motility.

The applied mathematical goal is to capture the molecular constituents and their interactions in a modeling and simulation toolkit that reproduces collective, organized behavior and thereby reveals biological mechanisms. An understanding of how mechanisms work is prerequisite to understanding failure and strategies to recover from compromised functionality. Systems biology has traditionally averaged over molecular details, positing continuum-scale balance laws for observable macroscopic properties. From a bottom-up approach, coarse-graining methods, such as projection onto moments of distributions for molecular constituents, often lead to continuum equations that resemble and validate continuum-scale models and give molecular meaning to their coefficients. These connections between molecular kinetic and continuum models have been studied for many model complex fluid systems within the statistical physics literature. For polymers, the books by Bird, Curtiss, Armstrong, and Hassager [1], Beris and Edwards [2], Larson [3], and Rubinstein and Colby [4] give excellent treatments.

A medical biology challenge is to detect and quantify the sources of disruptions in normal material functionality, e.g., a genetic defect that disrupts an ion channel in cystic fibrosis or DNA damage that leads to a runaway cascade in a cell cycle. Once understood, these insights explain symptoms (e.g., dehydrated lung mucus, proliferation of cell division), point to diagnostics for disease progression, and focus medical treatment on the sources of compromised function.

The engineering and clinical challenge is to design health solutions to restore function, aided by validated models and predictive simulations to test outcomes of alternative therapies. The solution could likewise be molecular (e.g., drugs specific to cellular pathways, gene therapy, DNA repair) or systems level (e.g., a percussive therapy), or both.

Each challenge above provides research opportunities in complex biological fluids for experimental and theoretical scientists. The aim of this chapter is to give young researchers an insight into fascinating aspects of biological complex fluids through the lens of our experiences with colleagues spanning all of the above disciplines. Subsequent chapters in this book will focus and go into significant detail on:

- Diverse biological fluids
- Methods to experimentally probe their behavior
- Observations and data afforded by advanced instrumentation
- Progress in theory, modeling, and direct numerical simulation

There are open problems as far as one can foresee in the quest to understand, measure, characterize, model, or predict biological soft matter behavior, a point that will become clear throughout this and other chapters. Two materials are discussed in this chapter as already mentioned: mucus and a single cell. *Mucus* is a barrier complex fluid, coating every human organ, whose components and properties are considered to be stationary on short time scales (perhaps hours depending on the organ). A *single cell* is an assembly of diverse structural components that are locally in time (perhaps minutes) spatially organized yet undergoing continuous activation by molecular events that maintain a living cell as a nonstationary complex fluid mixture.

We will highlight lung mucus in particular, a remarkable functional material that is continuously forced toward the larynx by coordinated cilia and asymmetric air drag from breathing and cough, and other mechanisms like chest cavity pumping in tapered, deformable airways and surfactant gradients in the deep lung airways. Lung mucus has a dual role: to trap and to clear airborne pathogens, and it is biochemically tuned to perform both tasks. While lung mucus constituents are locally stationary, airway mucus in healthy humans is in a state of continuous forcing at diverse length scales, frequencies, and force scales by the clearance mechanisms noted just above. In this way, transport of mucus is dictated by its nonequilibrium properties, and we have yet to determine thresholds of nonlinear behavior from all physiological forcing conditions and airway geometries. In this chapter we will briefly discuss nonlinear viscoelastic behavior in the context of lung mucus and we refer the reader to Chap. 6 for a more detailed discussion of nonlinear viscoelastic behavior. While Chap. 6 addresses metrics of nonlinearity in macrorheology, a microrheological alternative that can be applied to microliter volumes to characterize mucus and substructures inside of cells is still needed.

Chapter 3 addresses active microrheology, which we also touch upon below to explore whether a single cilium is capable of forcing airway mucus into a nonlinear response regime. Chapters 7 and 8 are related to cilia-mucus interactions in that cilia can be viewed as "swimmers" that penetrate and pass through mucus during the power stroke. A fundamental challenge remains to understand the mechanism for the transfer of energy spent in a cilium stroke cycle into transport of the mucus layer: is it simply a momentum transfer, or is there a deformation of the microstructure that tugs on the entangled polymer network and pulls on the mucus like a carpet layer? For medical biologists and clinicians, it is critical to understand that mucus changes, biochemically and functionally, over time scales of disease progression and what those changes are, to guide potential remedies and therapies. We will address this issue in more detail below.

Living cells are inherently active materials, to be distinguished from other biological materials such as mucus, which are passive materials undergoing active forcing at free interfaces. The molecular machinery inside a living cell is constantly working (e.g., molecular motors, polymerization and depolymerization, biochemical reactions), generating forces that easily compete with thermal fluctuations that would still be present even if the molecular machinery was switched off. This means that single living cells are maintained out of equilibrium by chemical and mechanical processes, that these processes change during the cell cycle, that probes inserted into the cell may interrupt passive and active cell behavior, and that the influence of active forcing is spatially heterogeneous. For example, if an activating protein species binds at the bilipid membrane or in the cellular cortex to cause a local contraction, that deformation propagates. Similar mechanochemical processes are taking place throughout the cell. Chapters 5–8 highlight the challenges in characterizing material properties of cellular components in the presence of often unknown or uncharacterized active forces. Chapters 9 and 10 give further illustrations of other active materials ranging from bacterial suspensions to motor proteins.

With respect to biological materials, the first aim and challenge is to prescribe a series of tests and experiments to characterize the material of interest *in physiologically relevant conditions* (e.g., lung mucus at different disease states) or to observe the material of interest in biologically relevant conditions (e.g., yeast cells at specific phases of the cell cycle). Of course one has to have sufficient experience and data to compare these outcomes with other complex fluid species and with diverse samples of the fluid of interest from "normal" and "dysfunctional" sources. This phase could be described as comparative rheology or biology, where the measurements or observational data of a particular specimen are compared to known or possibly benchmark specimens of similar origin. If there is sufficient data, inferences can be made by performing various statistical tests to decide if a given specimen lies within certain percentiles or standard deviations from a population mean.

The second challenge is to build predictive model capabilities. The aim is to develop quantitatively accurate mathematical models that extrapolate beyond experimental controls and that can be coupled to physiological forces and in vivo geometry to predict complex fluid behavior and function. For instance, design and evaluation of drug or physical therapies for lung disease would be far more efficient with accurate, predictive models for airway mucus flow and particle diffusion within mucus layers. Cellular abnormalities likewise would be better understood and designer molecules for cell repair would be more efficient with accurate predictive models.

The first step, characterization, is the purview of rheology: the study of how materials deform or flow due to applied loads. This subject is reviewed in Chap. 1; subsequent chapters give detailed experimental, theoretical, and computational insights into biological fluid behavior when forces and strains are applied at macroscopic scales (macrorheology) and at microscopic scales (microrheology). The second step, modeling, relies upon fundamental conservation laws, physical, chemical, and mathematical principles to posit a predictive model (governing equations, boundary conditions, and initial data) that when properly parameterized accurately reproduces experimental observations. An inverse problem can therefore be solved to infer all model parameters from wisely chosen experiments and data. Given the model and parameter fittings, accurate simulations of the model should be feasible and able to predict behavior under conditions more general than the characterization experiments. These goals are far from complete for almost all complex fluids in the human body, although progress has been made. For example, remarkable advances have been made in the modeling of tear films [5], the mathematical description of soft tissues like articular cartilage [6], and modeling blood flow discussed in Chap. 11.

To be honest at the outset, the novice to complex fluids is forewarned that the world of materials that are neither simple viscous fluids nor simple solids encompasses such diversity that there is no universal class of models to begin with. In fact, there are few complex fluids for which predictive models exist, have been validated, and used for biological or biomedical applications. We will mention some examples, and the other contributors to this volume provide more specific details. Significant research opportunities in this area lie in: the derivation of accurate, computationally feasible models for physiologically relevant complex fluids; characterization in terms of linear and nonlinear constitutive laws and parameter inference from experimental data; and predictive simulations, especially those that might support health assessments and therapeutic applications.

The term "fluid" in complex fluid is potentially misleading, since it conjures images of hydrodynamics to many physical scientists. However, the complexity of complex fluids lies in the intimate coupling between flow or deformation and the microstructure of the material. In complex fluids, the dynamics of the flow or deformation field is on equal footing with the dynamics of the microstructure. Indeed in rheological experiments, instruments are designed either (1) to control flow or deformation to learn how the microstructural stresses respond in approximately linear flows or deformations in the relevant geometry of the instrument; (2) to control stress and study the flow or deformational response. The classical text by John Ferry [7], or the more recent text by Chris Macosko [8], are excellent sources for rheological instruments and how to characterize materials based on experimental data. For predictive purposes, outside of such controlled experiments, one must have constitutive laws with material parameters that are inferred from rheological experiments and then simulated or analyzed to predict more general behavior. In this respect we refer the reader to Chaps. 1 and 6.

The term "soft matter" is more appropriate and indeed is gaining traction as a descriptive term that encompasses biological as well as synthetic materials that are neither simple viscous fluids nor simple solids. More traditional descriptors are viscoelastic or non-Newtonian. The complexity of soft matter is revealed by the memory exhibited in equilibrium and nonequilibrium responses of soft matter. This memory is evident macroscopically (press on your flesh, and it recovers on a time scale that doctors use to assess excess water retention), and with appropriate instrumentation, memory is evident down to the scales of the microstructure. A generic challenge in soft matter is to determine the time scales of memory and their influence on their transport properties: how a sample deforms or flows under forcing, how the microstructure fluctuates and diffuses in thermal equilibrium, and how foreign particles of diverse size, shape, and surface chemistry diffuse within the material. Indeed, the field of passive microrheology (see reviews by Waigh [9], Squires and Mason [10], Chen et al. [11] and Crocker and Hoffman [12], and Chap. 3) aspires to measure the fluctuations of probe microscopic particles and exploit a generalized fluctuation-dissipation theorem (FDT) to infer the complex dissipative properties. The remarkable FDT states that the memory spectrum of the microstructure in the linear response regime is revealed from the colored noise of the fluctuations of probe particles. For biological materials, including the internal structure of single cells or lung mucus, there are unknown length scales and heterogeneity of the microstructure, so it is not sufficient to probe with one particle or at one location [13, 14]. This is why the techniques of microrheology have been developed and honed on benchmark complex fluids such as colloids or pure homogeneous solutions of naked DNA as explained in Chap. 5. The challenges in biological fluids are addressed in many other articles within this volume, as well as in the specific sections below on cells and mucus.

The integration of biology and medicine with mathematical modeling and computational simulations is moving the life sciences toward new frontiers where physiological and pathological information from living organisms can be quantitatively described in silico [15]. In this chapter, we review the properties of fluids and soft matter encountered in the human body. As new methods are applied to increasingly complex biological process, our understanding of the mechanistic role of these biological materials has grown. However, there is still much to be learned in order to apply and advance the available tools toward predictive medicine. Here, and in the subsequent chapters, some of the challenges faced in the mathematical modeling of biological fluids and soft matter are reviewed.

1.1 Biological Materials in the Human Body

Sciences like rheology [1, 3] and biomechanics [16] study biological materials to find relationships between forces and deformations or flows. In the human body, atoms and molecules are organized into cells, tissues, organs, and individual organisms. As a result, forces, deformations, and flows can originate within the individual organisms or around them and include a wide range of time and length scales.

Human biological materials include tissues, organs, blood, plasma, skin, DNA, RNA, proteins, cells, mucus, saliva, and other body fluids. Some of the common characteristics of these materials are [17]:

- They are composites, containing both inorganic and organic components.
- They are able to self-assemble.

- They are multifunctional, with the ability to change their characteristics tailored to a specific function.
- They are hierarchically organized at the atomistic, molecular, and larger scales.
- Many properties are length and time dependent and can vary significantly across various scales.

As an example, consider the flow of blood in the microvascular network. The rheological properties of blood are dependent on shear rate, shear history, and the dimensions and geometry of the system in which it is contained. The apparent viscosity of blood measured in tubes with diameters ~200 μm shows a precipitous decrease with decreasing diameter, reaching a minimum at diameters of ~5–7 μm, corresponding to the diameter of capillary blood vessels [18]. That is, blood is a shear-thinning material, whose nonlinear response depends on the particulars of the flow properties in vivo and is highly specialized to meet the needs of the specific organ or tissue. Current advances on modeling blood flow as well as a comprehensive discussion of the constituents of blood are addressed in Chap. 11. There is even evidence that viscoelasticity of blood is relevant to the fluid-structure interactions in the highly dynamic conditions in heart chambers, coronary arteries, and valves [private discussions with Boyce Griffith, NYU and UNC].

1.1.1 Mathematical Modeling of Biological Materials

Above the atomic and molecular level, cells, tissues, organs, and organisms can be considered as continua and have traditionally been described by classical mechanics. In this way, descriptions of these systems are derived from fundamental physical laws like conservation of mass, moment, and energy, together with the respective constitutive equations for the material.

In their simplest form, constitutive equations relate forces and deformations, or more precisely, stresses and either strains or velocity fields. Depending on the experimental of physiological conditions the input can either be the stress or the strain (or velocity). For example, in active microrheology, the input is the strain rate if the particle is moved with a constant velocity, such as in optical tweezer experiments. Alternatively, the particle can be driven with a constant force, such as in magnetic bead rheology. Determining the stress-strain relationship is then crucial in the understanding of these materials through the formulation of constitutive equations.

The basis to solve these problems is the coupling of the conservation of mass and momentum equations for fluid density ρ and velocity field $\mathbf{u}$, with the appropriate constitutive equation for the stress, τ,

$$\frac{\partial \rho}{\partial t} + \nabla \cdot (\rho \mathbf{u}) = 0, \tag{2.1}$$

$$\rho \left(\frac{\partial \mathbf{u}}{\partial t} + (\mathbf{u} \cdot \nabla) \mathbf{u} \right) = -\nabla p + \nabla \cdot \tau + \mathbf{F}. \tag{2.2}$$

Under isothermal conditions, simple liquids like water or honey behave as Newtonian viscous fluids. In this case, the stress is directly proportional to the rate of strain ($\dot{\gamma} = \nabla\mathbf{u} + (\nabla\mathbf{u})^T$) and the constant of proportionality is the viscosity (η), which measures resistance to flow:

$$\tau = \eta\dot{\gamma}. \tag{2.3}$$

For elastic solids, the simplest relation is that of an isotropic Hookean solid where the stress is directly proportional to the strain (γ, where $\partial\gamma/\partial t = \dot{\gamma}$) and the constant of proportionality is the modulus (G), which measures the stiffness of the material. The simplest elastic constitutive law is

$$\tau = G\gamma. \tag{2.4}$$

Most biological materials exhibit characteristics of both viscous fluids and elastic solids. The nature of the response (more viscous-like or more solid-like) depends on the magnitude of the imposed deformation or forces or on the time scale at which the input is being imposed. These types of fluids are known as viscoelastic. Chapter 1 provides a more in-depth introduction to modeling and constitutive laws.

The simplest viscoelastic constitutive laws are the linear viscoelastic models, where the relation between the stress and the strain is linear, with a combination of both viscous and elastic terms. For example, when a spring and a dashpot are combined in series, the force on both units is the same, while the total deformation is the sum of the individual deformations. With these conditions, one can easily find the constitutive equation for this toy mechanical model as

$$\tau + \frac{\eta}{G}\frac{d\tau}{dt} = \eta\dot{\gamma}. \tag{2.5}$$

Other linear mechanical models can be formulated by different configurations of dashpots and spring units. For a detailed review of these models, we refer the reader to the book by Tschoegl [19].

Besides differential models, as the one given by Eq. (2.5), constitutive equations can also be represented by integral models. For example, continuing with our spring-dashpot model, integration of Eq. (2.5) with respect to time gives

$$\tau(t) = \int_{-\infty}^{t} G\exp(-(t-t')/\lambda)\dot{\gamma}(t')dt', \tag{2.6}$$

where the stress is related to the strain history by a kernel, which in the case of Eq. (2.5) is $G(t) = G\exp(-t\lambda)$, and $\lambda = \eta/G$ is the relaxation time.

As mentioned above, most biological materials are viscoelastic, but even more than that the relation between the stress and the strain is not linear for all conditions. Some simplifications can be applied to use linear viscoelastic equations to describe a material. For example in passive microrheology, the motion of imbedded

"Brownian" probes is followed to gain insight into the viscoelastic properties of the material. The fluctuations of the particle arise from fluctuations of the microstructure and any solvent, and a generalization of the FDT to generalized Langevin equations (GLEs) is exploited to infer both viscous and elastic moduli of the material. For a detailed description see Chap. 3.

Another approach is to apply small stresses or deformations to the sample, guaranteeing that the strain-stress relation is linear. One such experimental approach is the small amplitude oscillatory shear (SAOS) where a sinusoidal stress with small amplitude is applied to the sample, for example, using commercial rheometers [8]. In this setup, given a strain of the form

$$\gamma = \gamma_0 \sin(\omega t), \tag{2.7}$$

Equation (2.5) can be solved to show that the stress is given by

$$\tau(t) = G\gamma_0 \left[-\frac{\lambda\omega}{1+(\lambda\omega)^2} e^{-t/\lambda} + \frac{(\lambda\omega)^2}{1+(\lambda\omega)^2} \sin(\omega t) + \frac{\lambda\omega}{1+(\lambda\omega)^2} \cos(\omega t) \right]. \tag{2.8}$$

At the steady state, $t \to \infty$, the stress becomes the sum of two functions: one proportional to $\sin(\omega t)$, i.e., the strain, and one proportional to $\cos(\omega t)$, i.e., the strain rate. Because of their relevance to elastic and viscous behavior, these functions are known as the elastic (or storage) modulus $G'(\omega)$ and the viscous (or loss) modulus $G''(\omega)$. For the so-called upper convected Maxwell (UCM) linear viscoelastic constitutive law, these functions are given by

$$G'(\omega) = G\frac{(\lambda\omega)^2}{1+(\lambda\omega)^2}, \qquad G''(\omega) = G\frac{\lambda\omega}{1+(\lambda\omega)^2}. \tag{2.9}$$

In this way, if a material obeys $G' > G''$, it is said to be more elastic than viscous and if $G'' > G'$, the material is more viscous than elastic. For a detailed review of these types of responses and their extensions to nonlinear regimes, see Chap. 6.

Yet another set of experiments are known as creep measurements, where a constant stress is imposed and the resulting strain is measured. For a Hookean solid, the strain is directly proportional to the stress so that the resulting strain history is like the one shown by the purple circles in Fig. 2.1. For a viscous fluid, the stress is proportional to the time derivative of the strain, with the resulting strain described by a straight line as shown by the blue squares in Fig. 2.1. Viscoelastic materials exhibit both viscous and elastic behavior. For example, some experiments in cells [20] have shown that their behavior is that of a viscoelastic fluid shown in red triangles in Fig. 2.1. For a further discussion on rheological tests, we refer the reader to Chaps. 1 and 6.

The analysis of the material properties based on linear viscoelastic measurements have provided great insight into the behavior and function of these materials. However, as mentioned above under physiological conditions these materials are

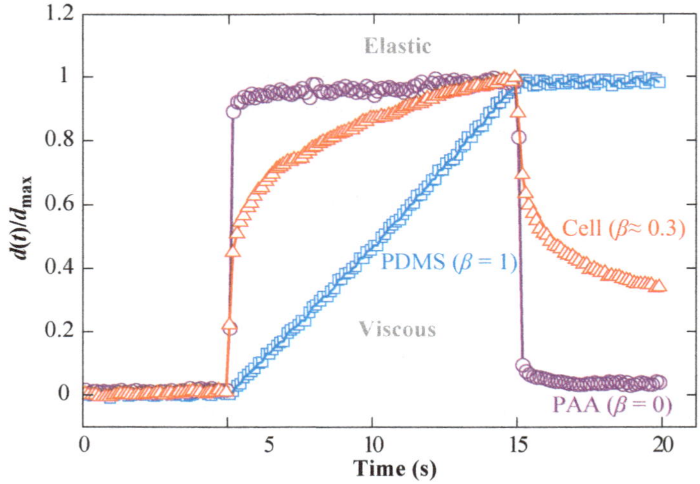

Fig. 2.1 Experimental creep response: the material is loaded at time $t = 5$ s with a constant force F and the displacement, $d(t)$, is measured. The figure shows typical responses of three types of materials: elastic material (*purple curve*) polyacrylamide-bis-acrylamide (*PAA*) hydrogel; viscous material (*blue curve*) polydimethylsiloxane (*PDMS*) silicone oil; and cellular material (*red curve*) F9 embryonic carcinoma cell. The creep function is defined as $J(t) = d(t)/F = j_0(t/t')^{\beta}$, where t' is a characteristic time scale of the experiment, the prefactor j_0 characterizes the softness or compliance of the material, and the power-law exponent β represents the type of material; $\beta = 0$ for elastic solids, $\beta = 1$ for viscous fluids, and $0 < \beta < 1$ for viscoelastic materials. Figure from [20]

often under nonlinear conditions. Consequently, experimental tests and models such as the ones we have discussed thus far fail to capture features relevant to the physiological functions of many biological materials.

1.1.2 Nonlinear Viscoelasticity

Chapter 1 provides a detailed description of modeling hydrodynamics in viscoelastic materials. Here we highlight the main challenges involved in the formulation and modeling of nonlinear viscoelastic behavior; Chaps. 6 and 10 should also be consulted.

The main characteristic of viscoelastic materials is that the relationship between stress and strain depends upon their deformation history. In particular, their rheological properties are dictated by the evolution of the conformation of microstructures. One of the main challenges is how to develop methods to connect the configuration at the microscopic level to the dynamics involved at macroscopic length scales. However, this need to bridge disparate length and time scales presents numerous challenges. As an illustration, consider the range of length and time scales encountered in a typical polymeric system. Characteristic times can span from $O(10^{-13}\,\mathrm{s})$ for bond vibrations to seconds or minutes for the relaxation of chain orientation to hours for glassy states and phase separation. Similarly, relevant length scales vary from angstroms for bond lengths to nanometers for average chain length to micrometers or larger for experimental and industrial processes.

Depending on the level of investigation, one can model these systems by using stochastic differential equations, Fokker–Planck-type equations or macroscopic constitutive relations that can be differential equations, integral equations, or a combination of these two. The choice of level of description depends on the objective, the questions one expects to answer, and the computational capabilities one has when it comes to solving the resulting coupled system of equations. For different macroscopic constitutive equations of viscoelastic materials, the reader is referred to Larson's book [21]. For a stochastic modeling of the materials, we refer to Ottinger's book [22] and for a review on multiscale methods to the article by Keunings [23]. In addition, Chap. 10 and the book by Owens and Phillips [24] discuss different numerical approaches used in the simulation of viscoelastic materials.

The kind of nonlinear behavior and how to tease out such behavior depends greatly on the type of system under study and the particular information one expects to gain with modeling. For the scope of this chapter, we will focus on two exemplary materials: mucus and cells. For other behaviors and fluids the reader is pointed to other chapters in this book.

Under in vivo conditions, these systems are often subjected to large deformations and/or stresses rendering the stress-strain relationship nonlinear and time dependent. The adequate formulation of constitutive equations for these materials is challenging. In addition, probing the validity of such equations requires testing of the materials in controlled conditions, e.g., using the protocols and instruments of rheology. Experiments involving biological fluids and soft matter can be difficult either because the sample cannot be isolated for testing, or the available volumes are too small or too "soft" for many devices, or because it is difficult to keep the specimen in normal living conditions (e.g., controlled temperature and humidity). Additional perspectives are given in Chaps. 3–6.

Other challenges faced in the modeling of biological fluids include:

- The role of geometry and confinement. The material may reside in a complex, nonstationary, 3D geometry, e.g., mucus in deformable lung airways or a cell within a living tissue or exposed to a foreign substrate. Numerical modeling of in vivo material behavior requires solvers that adapt to these dynamic, complex geometries.
- Since many biological processes are nonlinear, the constitutive relations that describe material behavior can include mathematical terms with complex functional forms, e.g., memory over a broad frequency spectrum.
- Material response depends on a large number of variables. For example, blood flow can be dramatically different depending on capillary diameter, compliance of surrounding tissue or capillary walls, temperature, pressure, stresses, and heart rate. For a detailed discussion see Chap. 11.

- Most biological materials are heterogeneous, consisting of substructures with markedly different properties. Modeling therefore must either explicitly resolve the heterogeneity (e.g., phase-field modeling of single cells or dynamics of molecular constituents of mucus) or posit homogenized models with effective material parameters that average over the microstructural heterogeneity.
- Biological materials are often anisotropic so that their behavior is directionally dependent. For example, the filamentous cortex in cells is aligned as opposed to randomly oriented.
- Some biological systems change their properties during a process in response to imposed stresses. In this way, there can be dramatic changes in the properties over both time and position. Modeling efforts must either model the substructures and their dynamics or capture these effects through coarse-grained parameters and their space-time variations.
- The development of accurate, meaningful boundary conditions is a major challenge, including fluid-structure conditions, adherence versus slip at interfaces, stress versus strain versus velocity boundary conditions, and their compatibility with models. All of these issues weigh heavily on the choice of numerical algorithms, and often potential boundary conditions need to be tested against experimental data to determine their validity.

2 Mucus in the Human Body

In the human body, mucus covers the luminal surface of the gastrointestinal (GI) [25], respiratory [26], and reproductive [27] tracts. Mucus also coats eyes, the epithelium of the nose, mouth, and salivary glands, as well as the peritoneal surface of intra-abdominal organs [28]. All mucus is not made equal; it is biochemically tuned for diverse barrier and flow transport properties depending on which organ it coats. For example, mucus acts as a lubricant, as a moisture barrier to prevent dehydration of underlying tissues, as a chemical barrier to prevent gastric acids from destroying tissues, and as a diffusional barrier to pathogens and airborne particulates. In many organs, mucus flows to clear its trapped contents, with the flow generation mechanisms as diverse as mucus itself, e.g., blinking eyelids, gravitational drainage in the reproductive tract, airway cilia, airdrag from breathing, coughing, or sneezing. Typical human mucus co-regulates diffusion of trapped pathogens and clearance of the trapped load. Since mucus is being swept away continuously, the particular organs or epithelial tissues are likewise continuously manufacturing a distribution of high-molecular-weight mucin molecules and controlling water content via ion-nucleotide feedback mechanisms [29, 30]. These molecules collectively endow mucus with its ability to recognize (in the sense of control over their diffusion) and discard (by clearance of the mucus layer) particles ranging from O(10 nm) antibodies to O(100 nm) viruses to $O(1) - O(10)$ micron bacteria and any number of environmental particulates.

Disease states are often associated with a breakdown in "normal" properties of the mucus barrier [31], from stomach ulcers to chronic obstructive pulmonary disease (COPD). While mucus is ubiquitous in the human body, and therefore fundamental to human health, the links between the molecular constituents of mucus, the mechanisms by which they are continuously replenished to maintain their critical functions, and indeed the key microstructural features and variations of mucus throughout the body that endow such diverse functional properties remain active areas of research. It is fair to say that mucus is now "hot," with a resurgence of attention since the landmark paper on airway mucus by Knowles and Boucher [32].

Mucus functions as a viscoelastic, dynamic, semipermeable barrier that protects organs and epithelial tissue by selectively trapping and discarding pathogens, toxins, and other particulates [32]. At the same time, mucus layers allow the flux of water, gases, and nutrients that are transported through epithelial cells and distributed inside the body [28, 33]. In performing these functions mucus is constantly secreted, shed, and digested, recycled, or discarded. The mechanical and chemical properties of mucus are critical to its functional specificity, as it coats every organ and surface not covered by skin or nails. These diffusive, flow, and lubricating properties vary not only across organs but among individuals, with age, and physiological and pathological conditions [28, 32, 33].

Given the fundamental role of mucus in human health, it is perhaps surprising how much is known yet how little is understood about its biophysical and rheological properties. However, the diversity of mucus across the body and across populations, coupled with the difficulty in procuring samples (extremely low-volume samples) and the sensitivity of mucus to handling, presents major challenges in the experimental and theoretical characterization of human mucus. It is only in the recent past that the heightened awareness of the role of mucus in human health has converged with new instrumentation and new theoretical advances, in particular the emergence of the field of microrheology, together with the historical shift of the physical and mathematical sciences toward the biological and biomedical sciences.

Next, we review the main composition of human mucus and how it affects its flow and diffusive rheology. We discuss one of the main mechanisms for mucus clearance, coordinated ciliary beating, and review modeling approaches. Finally, we discuss diffusion of particulates in mucus.

2.1 Mucus Composition

A typical "healthy" mucus sample consists of 90–95 % (by mass) of water, 2–5 % high-molecular-weight glycoproteins (mucins), 1–2 % lipids, 1 % salts, and 0.02 % of DNA and other molecules [34]. Despite its dominant water content, mucus readily exhibits both viscous and elastic behavior that may vary dramatically with frequency (or shear rate), with amplitude of forcing (imposed strain or stress), and with the length scale of forcing. These viscoelastic properties are biochemically regulated by the relative concentration of, and interplay between, the components listed above.

Some key studies point to a classification of mucus as a physical gel, distinct from an entangled polymeric material [35–37]. This gel quality derives primarily from mucins and other low-molecular-weight proteins that form a three-dimensional network or gel matrix [35]. Mucins are negatively charged, glycosylated proteins that are continuously synthesized and secreted to replenish the mucus layer. Mucins can be divided into secreted (gel-forming and non-gel-forming) and membrane-anchored. Gel-forming mucins are a complex group of high-molecular-weight, polymeric glycoproteins. The main gel-forming mucins present in human mucus are MUC2, MUC5AC, MUC5B, and MUC6 [38–42]. Mucins also contain cysteine-rich domains where no glycosylation is present. These "naked" domains have hydrophobic properties [38, 43, 44]. To avoid contact with water, the hydrophobic portions of the molecules form dynamic, physical mucin-cross-links. This "intertwining" of mucins with other musins and biomolecules presents in the mucus constitute the gel matrix. Among other things, the density of cross-links controls the characteristics of a gel network. In a 2–5 wt% mucus gel, each mucin molecule overlaps 10–100 other mucin [28, 45]. In addition, other types of intermolecular interactions of different characteristic time and length scales contribute to the formation and strength of the mucus gel network. In particular, electrostatic interactions [39], hydrophobic interactions [46], and calcium-mediated interactions [47] have a well-documented role. Figure 2.2 depicts a mucin network including hydrophobic interactions and strong disulfide bonds (s-s). For extended reviews on mucus gelation see [36, 45, 48–52].

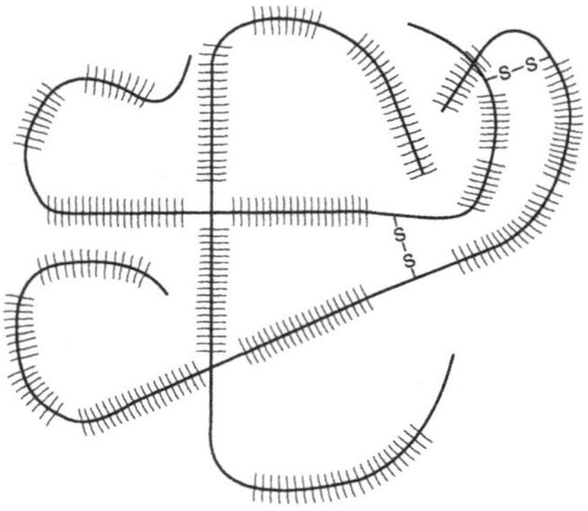

Fig. 2.2 Mucin molecules are made up of a peptide backbone (*solid curves*) with glycosylated regions (*perpendicular cross hairs*) and naked regions. These naked regions allow mucins to interact with other mucins and proteins through weak non-covalent bonds. These bonds together with strong disulfide bonds (s-s in the figure) result in the formation of a cross-linked network, and, at typical mucin concentrations, a high density of entanglements. Figure from [43]

2.2 *Mucus Viscoelasticity*

In general, all mucus secretions are viscoelastic; however, the absolute values of their viscous and elastic moduli and the stresses they are exposed to in vivo vary significantly depending on what organ, epithelial tissue, or cell culture the mucus is harvested from. Furthermore, the viscoelastic moduli from a single organ of a particular person evolve with environmental exposure (e.g., altitude), disease, and disease progression. To illustrate this, consider mucus in the stomach. Taylor and coworkers [53] showed that, in the stomach, two physically distinct mucus secretions are produced. One is a "shear-resistant" mucus gel that forms a protective mechanical barrier and the other is a "shear-compliant" secretion, which transforms into a viscous liquid when subjected to even low mechanical shear stress and acts as a lubricant facilitating the movement of solid matter through the gut during the digestive processes.

John Sheehan [54] led the effort to identify which mucin molecules were primarily distributed in sputum and airway mucus compared to the mucin distribution and concentrations in the periciliary liquid layer (PCL) of lung airways. It is remarkable to recognize that goblet and other cells in the epithelium manufacture all these molecules, and thus all are shed from the epithelial surface. Thus, they choose by physical and chemical affinities to reside in the PCL or mucus layers. Sheehan's goal was to biochemically and biophysically characterize the functionalities of mucus and PCL layers in airways and to understand their evolution during aging, environmental conditions, and disease progression [34, 42, 44, 47, 48, 51, 54–56].

As is the case with other viscoelastic materials, the interplay between viscous and elastic properties directly affects the transport capabilities of mucus. However, other physical properties play an important role in the function of mucus. For instance, adhesiveness and wettability govern the properties of the interface between mucus and the epithelial surface [57]. Optimal conditions for the clearance and lubricant properties of mucus require that both wettability and adhesiveness are high enough to prevent flow of mucus under body forces (gravity) but low enough to mobilize mucus by ciliary beating and other air–liquid pumping mechanisms.

Although the mucin network primarily governs the rheological characteristics of mucus, other biochemical constituents such as proteins, proteoglycans, lipids, DNA and cellular debris affect the properties of mucus by modifying gel formation and strength. In the same manner, other macromolecular components of exogenous or pharmaceutical origin can influence the viscoelastic properties of mucus and as such alter its functional properties. Below we give several examples where changes in rheological properties of mucus are induced by factors other than the self-dynamics of the mucin network.

- Rheological measurements of human tears reveal shear-thinning viscoelastic properties. However, contrary to other types of mucus, the specific mucins and their concentration in tears are insufficient to produce the observed degree of non-Newtonian behavior. Gouveia and Tiffany [58] showed that if the lipids in tears are removed, the viscous response becomes Newtonian (shear-independent).

- Airway mucus is normally degraded by proteases [32]. Innes and coworkers [59] reported that an excess of plasma proteins present in acute asthma patients inhibits the degradation of mucins in a protease-dependent manner. This results in changes in the viscoelastic properties of mucus that reduces clearance, resulting in mucus plugs occluding the airway. Why clearance is reduced is an open question, relating back to the need for predictive models that evaluate clearance efficiency versus viscoelastic characterization and comprehensive experiments and theory to provide viscoelastic characterization.
- Dynamic light scattering and bulk rheology measurements reveal that gastric mucin solutions undergo a pH-dependent sol-gel transition from a viscoelastic solution at neutral pH ($\sim$7) to a soft viscoelastic gel in acidic conditions (pH$<$7), with the transition occurring near a pH of 4 [45, 49].
- Several studies showed that the presence of salts in mucin solutions greatly affects their rheological properties [46, 49, 60]. In particular, the concentration of salts is correlated with decrease in mucin gel strength. This is consistent with other studies showing that inhalation of hypertonic saline increases mucociliary clearance (MCC) [61] and aids in mucus clearance for cystic fibrosis patients [62, 63].

We are still far from a synthesis of the biochemical basis that conveys mucus with its functional properties, still far from a viscoelastic classification of mucus specific to human organs or epithelial surfaces, still far from quantifying how mucus viscoelasticity varies during the life of a healthy human, and still far from quantifying how mucus viscoelasticity evolves with disease. However, the importance of the viscoelastic properties of mucus for many physiological functions is undeniable [38, 57–59, 64–68], nor can one deny the biomedical potential of tools that selectively modify mucus viscoelasticity [69–72]. Note that the awareness that mucus viscoelasticity is the controlling factor for barrier and clearance properties, and that efficiency of the diffusive barrier to specific foreign particles and of transport from specific clearance mechanisms, is relatively new in biology and medicine. This recognition can only be exploited effectively if the science and engineering tools are developed and implemented in physiologically relevant and clinically relevant conditions.

2.2.1 Rheological Characterization

As discussed above, the rheology of mucus is determined by its composition and structure. However, rheological properties are dynamic rather than static measurements. In this sense, it is important to recognize that the rheology of mucus also depends on its interaction with dynamic forces like ciliary beating and airflow, eye blinking, gut contractions, etc. Thus, as with any other viscoelastic material, the rheological testing of mucus gels yields different results with different applied stresses or deformation frequencies.

- *Dynamic Moduli.* One way of classifying the rheological response of viscoelastic materials is through the frequency-dependent storage ($G'(\omega)$) and loss ($G''(\omega)$) moduli. In Fig. 2.3, the dynamic moduli of mucus from different sources have been plotted. The wide range of values observed in the figure arises from differences in measurement methods, functions and composition of the different types of mucus secretions, and samples obtained from normal and pathological individuals. Because of the dependence of mucus viscoelastic properties on the driving frequency, in the studies of the rheology of airway mucus, frequencies need to be chosen to mimic different transport mechanisms. Examples are the low-stress, high-frequency forces applied by beating cilia (~10 Hz or 62 Rad/s); the low-stress, low-frequency extrusion of secretions from glands (~1,z or 6 Rad/s); or the high-velocity, high-stress forces imposed by cough or high-frequency ventilation or percussive therapies.

  It is important to note that the viscosity of human respiratory mucus has often been given as 12–15 Pa-s with a relaxation time of about 40 s and elastic modulus of 1 Pa. These values are said to represent an optimal rheological profile for MCC [33]. However, it becomes apparent in Fig. 2.3 that such a classification of mucus viscoelastic properties in terms of a single value of the moduli fails to capture the complex spatial and temporal interactions of mucus with physiological forcing.
- *Spinnability.* Another rheological measurement used to determine mucus properties like adhesion and elasticity is spinnability. Spinnability is characterized by the mucus ability to be drawn into long threads under the effect of traction, which measures the cohesive forces that hold the mucus together. A typical measurement is performed with a given mucus volume (typically 30-μL) stretched at some velocity (e.g., 10 mm/s). An electric signal conducted through the sample is interrupted at the point where the stretched thread is broken. This measured distance is reported as the spinnability of the sample. The spinnability of normal respiratory mucus ranges from 40 to 100 mm and becomes less for sputum [77, 78]. In a model study, a high spinnability was found to correlate inversely with cough clearance [79]. Table 2.1 shows several values of the spinnability of respiratory mucus.

2.2.2 Modeling Mucus Rheology

It is clear that, to faithfully characterize mucus in terms of predictive models, one must consider the dynamics of the gel network together with a wide range of interactions and conditions that affect the properties of mucus. One of the main challenges in the study of the rheological properties of mucus resides in the vast set of factors affecting such properties in vivo. Mucus harvested from human bronchial epithelial (HBE) cultures provides a model respiratory mucus free of inhaled infectious and inflammatory materials [56, 83, 84]. Here we show passive microrheology results for the linear dynamic moduli of HBE mucus and then use a canonical viscoelastic model, the UCM model, as a mode basis to fit the data. For a detailed review of microrheology methods see Chap. 3.

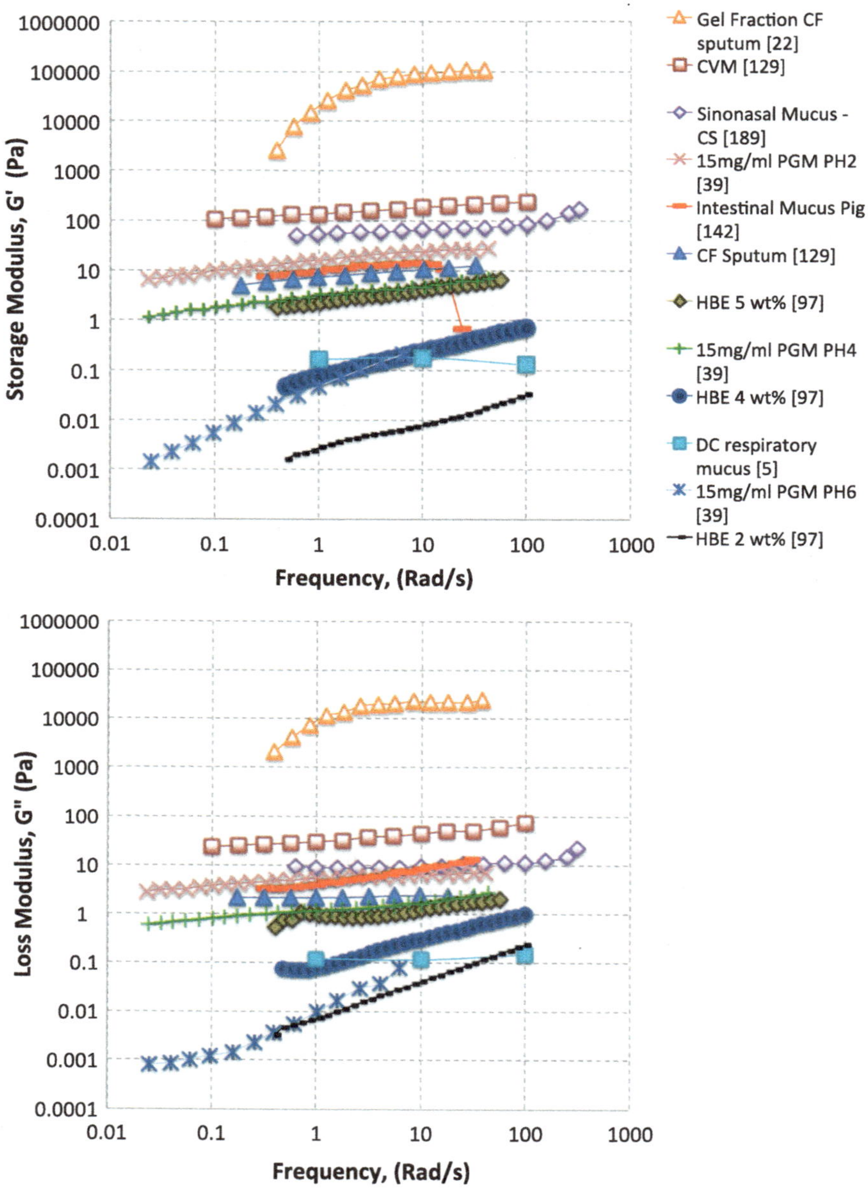

Fig. 2.3 Comparison of the viscoelastic properties of different types of mucus. Data points are a compilation of viscoelastic properties of mucus reported in the literature including: HBE mucus, [73], cystic fibrosis (*CF*) sputum [33], gel fraction of CF sputum [74], cervicovaginal mucus (*CVM*) [33], sinonasal mucus in patients with chronic sinusitis (*CS*) [68], pig gastrointestinal mucus (*PGM*) [49, 75], and sputum obtained by direct collection (*DC*) [76]

Table 2.1 Values of spinnability for three different types of respiratory mucus

Source of mucus	Spinnability (mm)	Reference
Endotracheal tube technique	35.7 ± 17.5	[80]
CF sputum	12.6 ± 2.99	[81]
Patients with bronchiectasis	11.6 ± 0.4	[82]

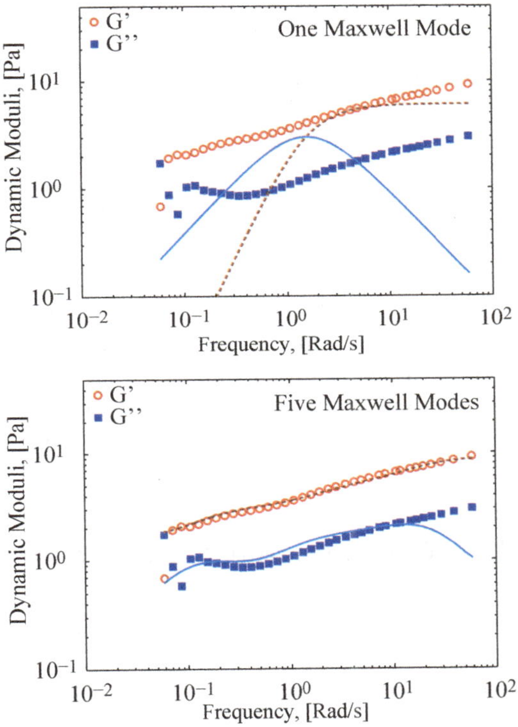

Fig. 2.4 Fittings of 5 wt% HBE mucus to the UCM model with one relaxation time (*top*) and five relaxation times (*bottom*). Data courtesy of David Hill, Cystic fibrosis Pulmonary Research and Treatment Center, The University of North Carolina at Chapel Hill

The moduli of a multimode UCM model, consisting of a linear superposition of UCM modes, are the summation of the moduli, given in Eq. (2.9):

$$G'(\omega) = \sum_i G_{0,i} \frac{(\lambda_i \omega)^2}{1 + (\lambda_i \omega)^2}, \qquad G''(\omega) = \sum_i G_{0,i} \frac{\lambda_i \omega}{1 + (\lambda_i \omega)^2}. \tag{2.10}$$

Figure 2.4 shows fittings to a 5 wt% HBE culture sample. The presence of multiple relaxation modes describing the linear dynamic response of mucus is evident.

A viscoelastic material is said to behave like a liquid at a given frequency if it dissipates more energy than it stores, i.e., the loss modulus is greater than the storage modulus ($G'' > G'$) at that frequency. Similarly, the material behaves elastically (or gel-like) at a given frequency if $G'' < G'$. In the UCM model, the relaxation time, λ, marks a transition from liquid-like to elastic-like, as seen in the right top in Fig. 2.4 where the blue solid curve (G'') intersects the dashed red curve at $\lambda = 10\,\text{rad/s}$.

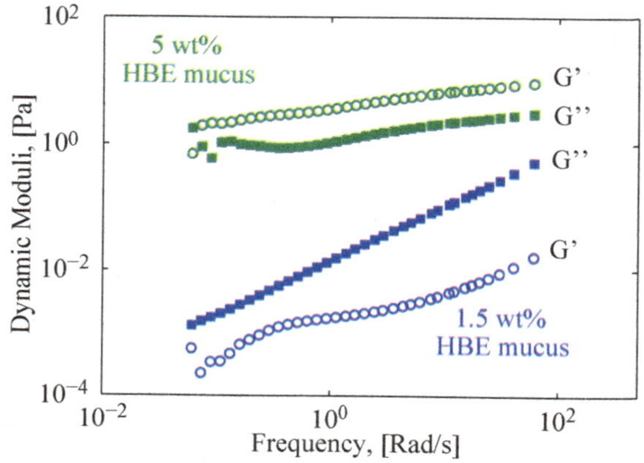

Fig. 2.5 Storage (G') and loss (G'') moduli for two concentrations of HBE mucus: 1.5 wt% (*blue*) and 5 wt% (*green*). When $G' < G''$ mucus dissipates more energy than it stores, behaving like a liquid. When $G' > G''$ the storage of energy is greater than the dissipation and mucus behaves like an elastic gel. Data adapted from [73]

Across physiological frequencies, e.g., from 0.1–100 Hz, HBE mucus has been shown to behave like a viscoelastic fluid ($G'' > G'$) at low mucin concentrations and like a viscoelastic gel ($G'' < G'$) at larger concentrations as shown in Fig. 2.5. The concentration at which the transition from a liquid to a gel occurs is called the gel point (GP). Recent studies show that the GP for HBE mucus is around a concentration of 4 wt% [73].

2.3 *Respiratory Mucus Clearance*

The airway epithelium is covered with a layer of fluid called the airway surface layer (ASL) composed of PCL and a mucus layer; see Fig. 2.6. In the human lungs, the mucus layer is believed to be ∼2–70 μm thick [85]. This thickness is determined by the balance between the rate of secretion and rate of degradation and shedding [28]. Mucus is produced at a resting rate of 0.5–1 ml of mucus per square centimeter of epithelial tissue surface over a 24-h period [43]. Studies of mucus transport velocity in vivo show that typical mucus transport rates in the trachea are 7–14 mm/min [86]. According to several in vivo studies, airway secretions loaded with organic and inorganic matter are typically cleared within 6 h [87]. This means that clearance of the entire tracheobronchial tree is mainly completed within 24 h [88]. In healthy individuals, the rate of mucus secretion is carefully balanced by mucus clearance.

2.3.1 Mucociliary Clearance

The human airway surface coating consists of an overlaying gel-like mucus layer and a lower PCL, which protects the epithelial surface from inhaled pathogens and particulates contained in mucus, as shown in Fig. 2.6. To prevent infection or

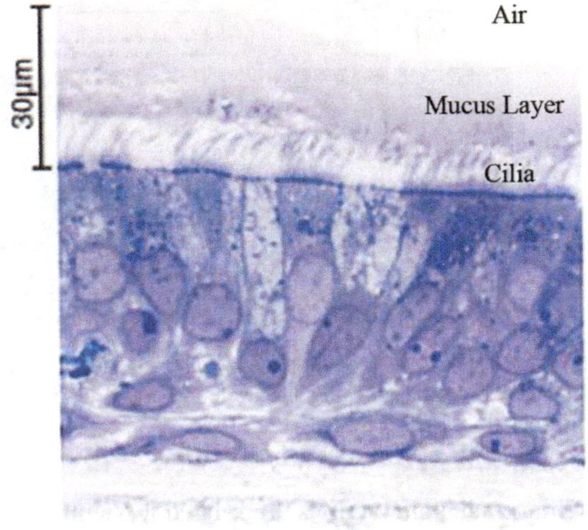

Fig. 2.6 Visualization of ASL surface setup. The cilia is in the lower PCL and it is in constant contact with the mucus gel layer. Figure adapted from [89]

inflammation, airway mucus must be cleared by a combination of coordinated cilia and airdrag; clearance requires a balance between the rheological properties of the mucus, PCL properties and volume, ciliary beat frequency and cilium length [32], and air–liquid transport [90, 91]. The PCL in healthy conditions is $\sim 7\,\mu$m thick [92, 93]. The thickness of the luminal mucus layer varies throughout the respiratory tract, increasing from distal to proximal airways [93, 94].

The continuous ciliary beating propels mucus, in a proximal direction, up and out of the lung [28, 32, 43]. The coordinated beating of cilia can propel the mucus layer at reported speed of 1–10 mm/min [28, 95, 96]. During the effective stroke, the ciliary tips penetrate into the mucus layer, and, during the recovery stroke, they withdraw from this layer, as seen in Fig. 2.7 [97, 98]. Thus, most of the ciliary motions occur within the PCL, and only the tips of the cilia sweep against the mucus gel, thereby optimizing the propulsive force of the ciliary beating in MCC. When cilia penetrate mucus, they tend to bend backwards and if this bend is too great because of excessive mucus viscosity the system does not work effectively [99]. One can perform a back-of-the-envelope calculation to determine the dissipative losses per unit volume of the cilia beating in the PCL. These loses are proportional to $\eta_{PCL}(\omega L \delta_{PCL})^2$, where η_{PCL} is the viscosity and δ_{PCL} the thickness of the PCL, L is the length of the cilium, and ω the frequency of the beat. As an illustration, it has been shown that defects like cilium damage, reduced cilia beat frequency, and reduction of cilia length are factors that hinder MCC in smokers [100].

Phase shifts between beating cilia result in patches of adjacent cilia that coordinate in the propulsion of mucus [98, 102]. The mechanism of mucus transport by cilia is a biological illustration of active microrheology discussed in Chap. 3.

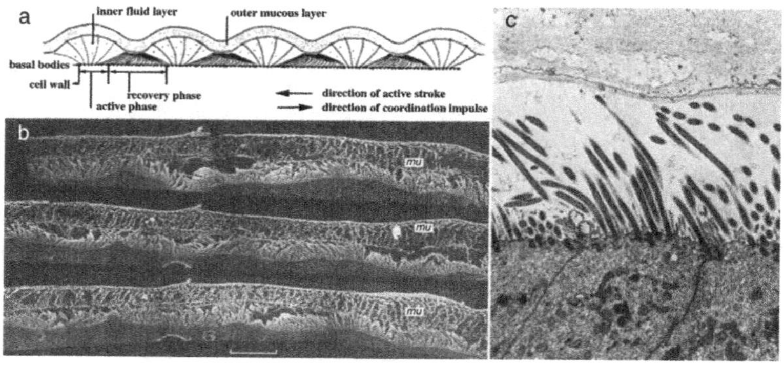

Fig. 2.7 Two layer hypothesis of airway secretions, showing the profile appearance of the cilia at successive stages of the beat cycle and their relationship to the overlying mucus (mu). Figure from [101]

Cilia-mucus interactions are related to swimmer-like propulsion in complex fluids, discussed in Chaps. 7 and 8, with a role reversal: the cilia aim to move the fluid past them rather than propel themselves. In order to produce a synchronous, wave-like movement of cilia (a so-called metachronal wave), it is widely believed that the PCL should have a low viscosity, which we argue against below. The characteristics of the synchronous metachronal wave are determined by the size and spacing of the cilia patches as well as the viscoelastic properties of the PCL. In addition, the thickness of the periciliary layer is critical for effective propulsion of mucus [98, 103]. The view that the PCL is a low-viscosity layer does not recognize that this layer is rich in many of the same mucin molecules that comprise mucus. A more recent view of the PCL has been proposed by our colleagues Button et al. [55]. The authors proposed a gel-on-brush model of the airway surface where the PCL is stabilized by osmotic effects and intermolecular repulsions. In addition, the macromolecules tethered to the cilia form a mesh that prevents large molecules and inhaled particles from penetrating the PCL. In general, if the periciliary layer is too shallow the cilia will not be able to perform a recovery stroke, and if it is too deep the cilia will not reach the mucus layer and the mucociliary action will be uncoupled [55]. Volume and composition of the PCL is mainly controlled by two mechanisms [104–106]: active ion transport and the continuous replenishment of loss of water by evaporation. For a more detailed discussion of the composition of the PCL we refer the reader to [55] and references therein. Finally, some reported characteristics of cilia are given in Table 2.2.

MCC relies not only on coordinated ciliary activity but also on specific rheological properties of mucus. It is widely believed that mucus viscoelasticity is optimized for clearance by coordinated cilia, yet as in any biological function, secondary mechanisms must exist such as tidal breathing and coughing. Indeed, in conditions such as the rare genetic disorder called primary ciliary dyskinesia or advanced lung disease, cilia are either asynchronous or unable to penetrate mucus, and air–liquid pumping is the primary mucus clearance mechanism [109]. Deep

Table 2.2 Reported properties of cilia

Quantity	Value
Cilium height[a]	5–8 μm
Cilium diameter[a]	0.15–0.3 μm
Frequency of beat[a]	10–20 Hz (60–120 Rad/s)
Wave length of metachronal wave[a]	20–40 μm
Cilia spacing[a]	0.3–0.4 μm
Density of cilia[a]	6–10 per μm^2
Number of cilia per cell[a]	200–400
Duration of effective stroke[b]	10 ms
Percentage effective stroke in beat cycle[b]	20 %
Duration of rest[b]	13 ms
Percentage rest in beat cycle[b]	26 %
Duration of recovery stroke[b]	27 ms
Percentage recovery stroke in beat cycle[b]	54 %
Speed of a tip cilium during effective stroke[b]	1 mm/s

References [a][107], [b][108]

in the lung, surfactants are prevalent whereas cilia density is very low and air is essentially stagnant; surfactant gradients play a significant role in clearing deep lung particles, a very slow mechanism relative to other mucus clearance modes. Under normal conditions, when the tip of a cilium engages the surface of the mucus layer, it sweeps with a shearing motion in the power stroke that is fast (recall cilia beat cycles are 10–15 Hz) and acts over micron scales. Silberberg argues that the elastic characteristic of the mucus gel dominates the efficiency of transport [26]. Hence, it is argued that cilia can only transport mucus if it has the proper viscoelasticity.

It turns out that frog palates are a good model system to evaluate MCC. Although the palate stops secreting mucus some time after excising the palate, the cilia continue to beat, allowing the placement of mucus samples or mucus simulants to be transported by cilia [110, 111]. With this setup, it has been shown that ciliary beating is unable to transport a variety of purely viscous materials. In addition, as the elasticity of mucus and mucus simulants is increased, there is a sharp increase in the clearance rate up to an optimal value, followed by a slow decrease in mucus transport [112]. Thus there is strong evidence of a mucus viscoelastic "sweet spot," yet such results have yet to be reproduced with high-fidelity models. We point the reader to Chaps. 7 and 8 and the article by Teran, Fauci, and Shelley [113] for issues of optimal transport in complex fluids.

From the clinical and observational perspectives, there are numerous interpretations that have guided treatments for compromised mucus clearance. If respiratory mucus becomes too runny, i.e., more viscous than elastic, gravity may dominate ciliary transport, as observed when mucus drains from the nasal and sinus cavities. Often, gravitational drainage therapy is preceded by inhalation of hypertonic saline solutions to "thin" the airway mucus. On the other hand, if mucus becomes too "hardened," i.e., more elastic than viscous, it hinders transport by cilia for potentially many reasons. For instance, the storage modulus may be so high that mucus resists the stress that a beating cilium is capable of generating in the power stroke. Thus, cilia are unable to penetrate the mucus layer. Often a drug or physical therapy is capable of softening mucus temporarily to reinstate MCC. The viscoelasticity of mucus is apparently regulated to obtain the best compromise between the elasticity needed to prevent gravitational drainage (which leaves airway epithelia exposed to pathogens) and to provide efficient ciliary transport [26]. Healthy mucus is a gel with relatively low viscosity and elasticity that is easily transported by ciliary action, whereas pathological mucus has higher viscosity and elasticity and is less easily cleared [28, 73, 114]. Although, the viscosity of mucus results in energy loss, these loses are necessary for mucus to be displaced and either expectorated or swallowed [26, 32]. The elasticity of mucus is potentially important to minimize energy, with little energy loss from physiological forcing. As mucus becomes more viscous, there is a tendency for the ciliary beat frequency to decrease [43, 115] and the length and coordination of the metachronal wavelength become less efficient. We refer to several reviews related to these interpretations associated with efficiency of clearance and mucus viscoelasticity [95, 103, 116–119]. Again, without accurate modeling and simulations, it is virtually impossible to quantify all of these competing effects.

To explore the interplay between phasic forcing conditions and the viscoelastic properties of a complex fluid such as mucus, we consider the model problem of a viscoelastic material between two plates, with the upper plate stationary and the lower plate oscillating with frequency ω and maximum speed U_0,

$$u_x(y=0) = U_0 \cos(\omega t). \tag{2.11}$$

In this way, the movement of the lower plate captures, in a geometrically simplified sense, the effect of ciliary beating. We seek to assess the work consumed in phasic boundary forcing of a viscoelastic layer, and how elasticity contributes to the workload. Assuming "viscometric conditions" typical of rheometers, the velocity in the gap is given by [120]

$$u_x(y) = U_0 \left(1 - \frac{y}{H} \cos(\omega t)\right), \tag{2.12}$$

so that the shear rate is spatially uniform throughout the plate gap:

$$\dot{\gamma}_{xy}(y) = \frac{\partial u_x}{\partial y} + \frac{\partial u_y}{\partial x} = \frac{U_0}{H} \cos(\omega t). \tag{2.13}$$

Finally, the rate of work is given by the dissipation function [21]:

$$\boldsymbol{\tau} : \nabla \mathbf{u} = \tau_{xy} \dot{\gamma}_{xy}. \tag{2.14}$$

In one period, $2\pi/\omega$, of the oscillating plate, the net work done is

$$W = \int_0^{2\pi/\omega} \tau_{xy}\dot{\gamma}_{xy}dt. \tag{2.15}$$

For a Hookean elastic solid $\tau_{xy} = G\gamma_{xy}$, so $W = 0$, whereas for a simple viscous fluid, $\tau_{xy} = \eta\dot{\gamma}_{xy}$, and

$$W^* = \int_0^{2\pi\omega} \tau_{xy}\dot{\gamma}_{xy}dt = \eta\left(\frac{U_0}{H}\right)^2 \int_0^{2\pi/\omega} \cos^2(\omega t)dt = \eta\left(\frac{U_0}{H}\right)^2 \frac{\pi}{\omega}. \tag{2.16}$$

Thus the network is minimal for elastic materials and maximal for viscous materials.

For a *linear* viscoelastic material of Maxwell type, the scalar shear stress constitutive equation is

$$\tau_{xy} + \lambda(d\tau_{xy})/dt = \eta\dot{\gamma}_{xy} = \eta\frac{U_0}{H}\cos(\omega t), \tag{2.17}$$

which yields

$$\tau_{xy} = \eta\left(\frac{U_0}{H}\right)^2 \frac{1}{1+(\omega\lambda)^2)}\left[e^{-t/\lambda} + \cos(\omega t) + \omega\lambda\sin(\omega t)\right], \tag{2.18}$$

so that

$$W = \eta\left(\frac{U_0}{H}\right)^2 \frac{\pi}{\omega}\left(\frac{1}{1+(\omega\lambda)^2}\right)^2 \left[\frac{\omega\lambda}{\pi}\left(e^{-2\pi/\omega\lambda} - 1\right) + 1 + (\omega\lambda)^2\right]. \tag{2.19}$$

Figure 2.8, shows values of W normalized by the viscous work per cycle W^*, given in Eq. (2.16), and illustrates the compromise between elastic and viscous characteristics of the fluid. This simple calculation involves far too restrictive assumptions on geometry and constitutive modeling, and it assumes that work per cycle is a relevant metric for biology. The point of this illustration is to show that one can formulate any number of metrics, e.g., mass transport for asymmetric forcing conditions, but any quantitative analysis requires a constitutive model, so that a heavy premium is placed on the accuracy of constitutive modeling for lung mucus.

Since the mucin network controls viscoelasticity, the rheological properties of mucus can be modified, potentially dramatically, by overproduction of mucins (common in COPD), by the presence of other molecules that disrupt the gel (such as inhaled or ingested drugs or environmental toxins), or by dehydration of the mucus layer (common in cystic fibrosis (CF)). It follows that the balance between secretion and clearance must be maintained so that mucus viscoelasticity remains within efficient ranges for the mechanism(s) that mucus performs in different parts of the human body. The evolution from healthy to pathological mucus occurs by multiple processes such as abnormal secretion of salts and water, increased

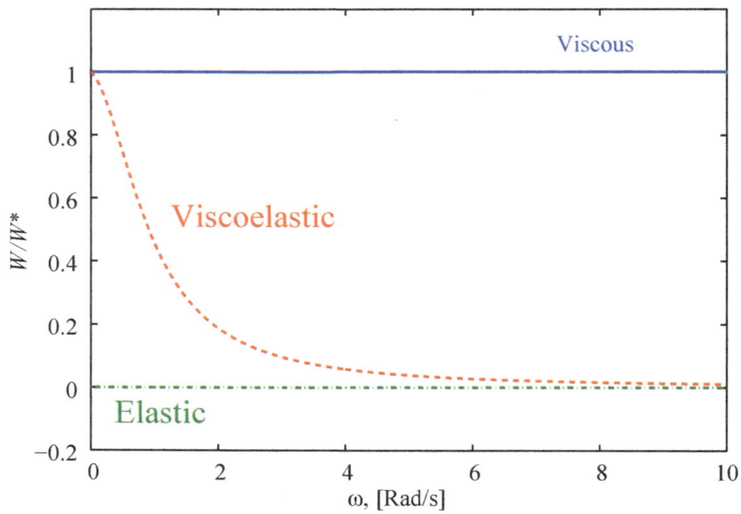

Fig. 2.8 Normalized rate of work for viscous, elastic, and viscoelastic materials as a function of driving frequency in a parallel plate shear cell under viscometric flow assumptions. Normalization is with respect to the viscous work per cycle W^*, given in Eq. (2.16)

production of mucins, infiltration of mucus with inflammatory cells, and heightened bronchia-vascular permeability [121]. Some inflammatory conditions induce mucin overproduction and hypersecretion. In cystic fibrosis (CF) and other chronic inflammatory airways diseases [57, 122–125], other large polymers predominate in the airway secretions; these biological macromolecules together with bacteria and other cell components are prevalent in the larger airways of the respiratory tract, constituting *sputum* that is recovered in the clinic. Breakdown in mucus clearance leads to a cascade of deleterious effects, including clogging of airways and safe harbor for infectious microbes [121]; see Fig. 2.9.

On the other hand, if the mucus mesh becomes more dilute or porous or chemically unresponsive so that bacteria and viruses diffuse more freely and rapidly, epithelial cells and tissue are under-protected and once again risk of infection increases [126, 127]. Similar conditions arise in the viscoelasticity of cervical mucus (CVM). For instance, in women with bacterial vaginosis, the viscosity of the CVM is lower than in those with normal flora, which may be responsible for the increased risk of infection by HIV and other sexually transmitted pathogens, as well as other adverse gynecological conditions [126]. An understanding of the diffusive barrier properties of mucus in the human body and the development of predictive models based on experimental rheological data are major areas of open research, with public health implications.

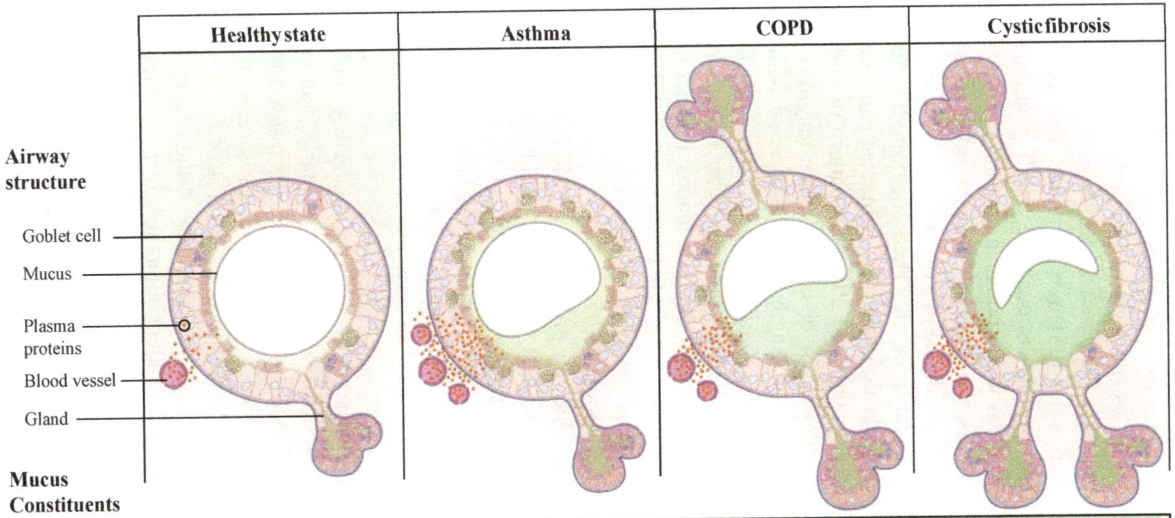

Mucus Constituents	Healthy state	Asthma	COPD	Cystic fibrosis
Mucin	X X	X X X X X	X X X X	X X X X
Plasma proteins	X	X X X X	X X	X X
Inflammatory cells	X	X X X	X X X	X X X X X
DNA		X	X X	X X X X X
Actin		X	X X	X X X X X
Bacteria			X X	X X X X X

Fig. 2.9 Contributions of mucosa disease to abnormal mucus in healthy state, asthma, COPD, and CF. Figure from: [121]

2.3.2 Modeling Mucociliary Clearance (MCC)

Mathematical models of mucociliary transport incorporate the effects of a large number of cilia beating in a coordinated manner with the two-layer (mucus and PCL) airway surface liquid (ASL). Low Reynolds numbers characterize the fluid mechanics of the ASL, since small velocities (microns per second), length scales (microns), and high viscosities (low-frequency shear viscosity of typical lung mucus is 2–3 orders of magnitude greater than water [33]) are prevalent. The reader is forewarned that low Reynolds number is a simple viscous fluid "guiding parameter" and only a small part of the story in viscoelastic fluid mechanics. The sentence just above makes perfect sense in viscous fluids, yet it does not say anything about the elasticity of the medium, the frequency-dependence of viscous and elastic moduli, nonlinearity of the microstructure during transport, etc. Theoretical models, to the extent that they encode the relevant viscoelastic properties of mucus, can provide information on the transport induced by an imposed cilia beat pattern or stresses from air drag. However, several length scales are involved: molecular length scales, relative to the biochemical structure of mucus; length scales associated with the cilium tip ($\sim$1 μm); lengths associated with the cilium length, cell size, and ciliary wavelength; length scales associated with the length and diameter of an airway (5–10 mm). Of particular importance with regard to cilia-mucus interaction is the scale of the cilium tip with respect to the entangled network of molecules constituting the mucus, since at these scales the cilium tip is comparable to the molecular characteristics of the mucin network [28, 107, 128].

Flow models, to be useful, should enable the prediction of MCC and, more importantly, suggest means for modifying the system. King et al. [129] proposed a planar two-layer fluid model to study the transport of mucus in the respiratory tract due to cilia beating and air motion. While their model was based on certain restrictive assumptions, such as zero mean PCL transport, they predicted that clearance increases as the elastic modulus decreases in agreement with experimental observations.

In contrast to the well-described axial transport of mucus along airway surfaces via ciliary action, theoretical analyses predict that the PCL is nearly stationary. However, experimental studies have concluded that the entire PCL is transported at approximately the same rate as mucus, 39.2 ± 4.7 and 39.8 ± 4.2 μm/s, respectively [117]. Removing the mucus layer reduced PCL transport by $> 80\,\%$, to 4.8 ± 0.6 μm /sec, a value close to that predicted from theoretical analyses of the ciliary beat cycle; hence, the rapid movement of PCL is dependent upon the transport of mucus [117]. In addition, movement of the PCL has proven important for mixing effects [117]. These seminal studies, and a revisitation of the conclusions drawn from tracer markers in the PCL and mucus layers, were the focus of initial conversations by Forest and the applied math group at UNC with faculty in the cystic Fibrosis Center at UNC and Richard Superfine from Physics at UNC. That was a decade ago, with interactions leading to the Virtual Lung Project at UNC.

Several theoretical mucociliary transport models are discussed by Sleigh et al. [93, 130]. We summarize some of them next and point the reader to Chaps. 7 and 8 for a related discussion of locomotion through complex fluids.

- *Envelope model* [131]. This model assumes that cilia are densely packed, so that the fluid effectively experiences an oscillating material "sheet".
- *Sublayer model* [132]. Initially proposed by Blake, in this model individual cilium is modeled as a flexing cylindrical body anchored at the cell surface. The action of a large number of cilia is modeled by a continuous distribution of force per unit volume within the cilia sublayer.
- *Discrete cilia model* [133]. The cilia are represented by a distribution on of Stokeslets with appropriate mirror images to satisfy the no-slip velocity boundary condition at the cell surface. The flow in the cilia sublayer is then determined by summing the individual cilium velocity fields, in the infinite plane approximation.
- *Traction layer model* [134, 135]. This model is the continuous version of the sublayer model. The action of a large number of cilia is modeled by a continuous distribution of force per unit volume within the cilia sublayer.

2.3.3 Modeling Mucus Transport in the Human Respiratory Tract

In healthy subjects, the layer of mucus lining the respiratory tract epithelia is only $\sim$5–10 μm thick and rests on an $\sim$7 μm sol phase. The diameter of an adult human trachea is $\sim$1 cm, while the diameters for subsequent branches are successively smaller [136]. The manipulations of physical therapists are known to be clinically efficient, but outcomes are mostly empirical since the biophysical mechanisms involved in these manipulations are not well understood. Mauroy and coworkers [137] developed a model of mucus clearance in idealized rigid human bronchial trees to study the interaction between tree geometry, mucus physical properties, and amplitude of flow rate in the tree. Their results showed that airflow rate and viscoelastic properties of mucus determine the maximal possible mucus thickness in each branch of the tree, resulting in a specific distribution of mucus thickness and properties along the tree. In general, most models of mucus transport in the respiratory tract include the following assumptions [105]:

- Mucus layer thickness is assumed constant within single airway bifurcations.
- Transport of the mucus layer has a constant net velocity.
- Mucus production rates in the terminal bronchioles are considered equal.
- Thickness of the mucus layer lining the wall of a given airway tube is negligibly small with respect to the airway diameter.

Recent work by Sorin Mitran [138, 139] attempts to bridge cilia geometry, cilia force generation, cilia density, a viscous PCL layer, a viscoelastic mucus layer, and an air phase toward a predictive tool for mucus hydrodynamics. Similar modeling and simulation tools are under development by (and partially with) our colleagues, including Ricardo Cortez, Lisa Fauci, Anita Layton, and Karin Leiderman.

2.3.4 Cough Clearance

A second mechanism for the expulsion of mucus from the airways is cough clearance, which becomes a primary clearance mechanism when MCC fails. The hydration of mucus dramatically affects its viscous and elastic properties, which in turn determines how effectively it is cleared by ciliary action and cough [121]. This may help explain why lung diseases caused by impaired ciliary function are less severe than those caused by dehydration, which impedes both clearance mechanisms [121].

King and coworkers [140] examined the relationship between mucus rheology, depth of the mucus layer, and clearance by simulated cough. Cough clearance and adhesion were explored in experiments on mucus transport. They found that high elasticity of the mucus, $G' > G''$, impedes clearance. These findings show an opposite relationship to that seen in ciliary clearance, suggesting that healthy mucus may exhibit intermediate levels of elasticity because it must be capable of responding to both forms of clearance, i.e., mucociliary and cough. It follows that medications that decrease viscosity, such as mucolytics, may benefit ciliary clearance but hamper cough clearance, while medications that decrease the adhesion of secretions to the epithelial surface are likely to improve airflow-dependent clearance [141].

Camassa et al. [90] studied, experimentally and theoretically, flows where an annular viscous liquid film lining the wall of a tube is forced upwards against gravity by turbulent airflow up the core of the tube. This core-annular flow configurations mimics mucus clearance in the trachea and was pursued to reproduce and extend seminal experiments by Kim et al. [91, 142, 143]. The authors derived a longwave, fully nonlinear asymptotic model to interpret experimental observations and data against model simulations. Traveling wave solutions of their model predict a transition between different mass transport regimes. Past a certain threshold that can be identified with surface tension of the liquid-air interface, sufficiently large-amplitude waves begin to trap boluses of fluid, which propagate upward disconnected from the wetting layer similar to vortex rings sliding between the viscous film and air. This theoretical result is then confirmed by a second set of experiments that show ring waves of annular fluid propagating over the underlying creeping flow. By tuning the parameters of the experiments, the strength of this phenomenon can be adjusted in a way that is predicted qualitatively by the model. Recent results based on a different turbulent airdrag closure have brought the theory and experiments much closer to quantitative accuracy [Camassa et al. 2013, preprint]. The extension of these experiments and models to viscoelastic fluids that mimic mucus is in progress. Furthermore, the modeling platform of Mitran [144] is proposed to explore fully resolved air–liquid pumping simulations. The goal of these studies is to assess the efficiency of turbulent air drag in transporting mucus layers in airways. The relative efficiency of air–liquid pumping versus coordinated cilia or chest cavity compressive pumping is unknown.

2.4 Diffusion in Mucus

The following discussion can be more profitably read after Chap. 3, since passive microrheology is based on diffusion of microbead probes in soft matter. The deposition and clearance of particulates in mucus layers have been under investigation for decades [145]. Generation of suitable therapies that effectively deposit particles in the mucus layer has long been thought to be the only relevant factor for successful drug absorption. However, efficient deposition of particles, although still important, is not always a sufficient condition for successful drug delivery. Particle diffusion through the mucus barrier, just like the flow of mucus, is not a simple process where particles obey simple diffusion laws. New experimental tracking methods have provided a wealth of data on particles with diameters down to tens of nm up to microns, where the particles not only range in size (hydrodynamic radius) but also in their surface interactions with the mucin gel network. Because of this heightened awareness, there is a compelling need to understand and control what happens after particles have landed in the mucus layer. We refer to our recent articles and references therein [73, 146–148].

In the lungs, objects trapped in the mucus gel are transported at rates up to 5–10 mm/min by beating cilia and are delivered to the GI tract for inactivation and digestion. The luminal gel layer of respiratory mucus is replaced as rapidly as every 10–20 min, resulting in extremely efficient clearance of inhaled particles. Trapping and rapid clearance is crucial to protect the airway epithelium from the onslaught of pathogens and environmental toxins we breathe everyday. However, the "mucociliary escalator" also serves as a major barrier to the delivery of therapeutic nanoparticles. Strategies to address this barrier and more efficiently deliver therapeutic nanoparticles to the lungs include mucoadhesive particles, mucus-penetration particles, and mucolytics [145]. The success of these strategies lies in the experimental and theoretical understanding of the diffusive properties of mucus and the underlying interactions between the deposited particulates and the mucus gel matrix.

The study of self-diffusion of nm and micron-scale particles in mucus is important to determine the movement of toxic agents within the layer, as well as for drug delivery and gene delivery therapies [96]. This study is twofold. On one hand, there is the need to understand to what extent particles, characterized by a given surface charge and hydrophobic or hydrophilic properties, are able to diffuse through the network structure. On the other hand, one needs to understand how the network hydrophobic and hydrophilic regions, negatively charged biopolymers [96], and topological constraints arising from the pores present in the network affect such diffusion. That is, if one controls the particle diameter and modifies surface chemistry, then physics of adhesion or repulsion of nearby polymer chains can likewise dramatically alter the diffusive paths. If one varies the particle diameter with controlled surface chemistry, then the observed diffusive scaling varies significantly. This is accentuated if the particle diameter is comparable to the pore/mesh scales in the polymeric network [149, 150].

The various mucin molecules and other proteins in mucus form a heterogeneous, three-dimensional network with a potentially fractal length-scale distribution. The incorporation of the complex mucus microstructure into estimates of particle diffusion in mucus has been modeled in various ways. One approach is to consider the microstructure as a physical obstruction to particle diffusion. This effect is then modeled as a reduced self-diffusion coefficient of the particles. In addition to reduced particle mobility (i.e., lower diffusion coefficients than in the pure saltwater solvent), these obstructive effects can result in a broad distribution of effective particle diffusivities within the gel matrix; clearly one particle may not sample the same length scales of the mucus network as another. It is an intriguing open problem to explore how these effective diffusivities (again, assuming particles obey simple diffusion but with reduced diffusivity relative to the solvent) reflect the heterogeneity of the mucus microstructure. One can, for example, use the mean diffusivity of multiple particle paths with a range of particle diameters to infer a mesh distribution of pore sizes. Carrying this out in a reasonably rigorous manner remains an open problem, yet it is highly relevant to understand the diffusion of diverse species in mucus, including viruses, bacteria, airborne particulates, and many other inhaled substances.

Studies of passage times through mucus layers are riddled with complications that arise because the particles not only have to overcome the gel matrix barrier but also it is not completely known how different particles will interact with mucus under pathological conditions. For instance, Sanders and coworkers [96] performed a study of passage times in CF and COPD sputum. They observed that a low percentage of nanospheres, with a diameter of 270 nm, moved through a 220-μm-thick CF sputum layer after 150 min. Whereas larger nanospheres (560 nm) were almost completely blocked by the sputum, smaller nanospheres (124 nm) were retarded only by a factor of 1.3 as compared with buffer. In addition, they found that sputum from a patient with COPD retarded the transport of nanospheres to the same extent as CF sputum [96]. Interestingly, the authors found that nanospheres diffused significantly more easily through the more viscoelastic sputum samples. These findings are in contrast to studies that showed that mucolytic agents, which decrease the viscoelasticity of biogels, enhance the transport of drugs and colloidal drug carriers [151,152]. Sanders and coworkers argued that this increase in the mean diffusion of the nanospheres in the more viscoelastic medium is due to the increased heterogeneity in the network. This claim is supported by studies with synthetic gels. Mallam et al. [153] observed that increasing the concentration of junctions in these gels changes the network structure from a homogeneous microporous matrix into a more heterogeneous macroporous network. These studies point to the importance of determining the mesh length-scale spectrum and identifying degrees of heterogeneity in the sample. Below we summarize several research efforts to determine mesh sizes in mucus samples. Characterization of heterogeneity is more challenging and relies on the formulation of mathematical models and tools capable of faithfully reproducing the observed diffusive behavior of particles in mucus. Several steps have been taken in this direction and we summarize them at the end of this section.

If we are able to screen two related effects: the binding-unbinding kinetics of the various diameter particles to the mucus mesh and the repulsion vs. attraction of the mesh to the particle surface chemistry, *then, in principle*, single particle microrheology should reveal sufficient information about the mesh length-scale distribution of a given mucus sample. One should note, however, that the inference of mesh length scales is not immediate. The particle increment process (the measurable data) is a result of all lengthscale fluctuations of the mucus network, not simply those associated with the particle diameter. Therefore, the challenge is to somehow learn how different length-scale particles behave across the entire relaxation spectrum of the mucus gel and then to be able to learn the mucus network structure from the cumulative results of particle paths across a range of particle diameters. This inverse problem is far from solved; each particle diameter reflects a different sampling of the colored noise spectrum from the mucus microstructure.

2.4.1 Modeling Diffusion in Mucus

A standard practice in the microrheology and drug delivery literature is to report mean-squared displacement (MSD) of particle position data on a log-log plot. MSD is calculated as

$$\left\langle \Delta \mathbf{r}^2(\tau) \right\rangle = \left\langle [\mathbf{r}(t+\tau) - \mathbf{r}(t)]^2 \right\rangle, \tag{2.20}$$

where $\mathbf{r}(t)$ is the position of the particle at time t, τ is the lag time between the two positions taken by the particle used to calculate the displacement $\Delta\mathbf{r}$, and the average $\langle \cdots \rangle$ designates a time average over t and/or an ensemble-average over several trajectories.

Reported values of MSD in mucus [73, 75, 154–156] show a sub-diffusive MSD scaling over an intermediate dynamic range,

$$\left\langle \Delta r^2(\tau) \right\rangle = D\tau^{\alpha}, \tag{2.21}$$

where the power-law exponent is $0 < \alpha < 1$ and often a transition to linear scaling ($\alpha = 1$) over sufficiently long lag times. The prefactor D is the effective diffusion coefficient with units $\mu\text{m}^2/\text{s}^{\alpha}$.

For direct modeling of tracer particles in a viscoelastic medium, a sufficiently robust family of stochastic processes is needed that reflects these fundamental MSD signatures of transient sub-diffusion. With a robust family of stochastic processes in hand, then one can build inference methods to fit model parameters to experimental data and give a means to characterize diffusive properties of a given particle in a viscoelastic medium [146, 148]. Below we summarize two models that have been successfully used to describe diffusion of particles in mucus, together with simple Brownian motion.

- **Brownian Motion**

  The velocity of a particle driven by Brownian motion is described by the Langevin equation

$$m\frac{d^2\mathbf{x}}{dt^2} = -\zeta\frac{d\mathbf{x}}{dt} + \mathbf{F}(t), \tag{2.22}$$

where $\mathbf{x}$ is the position of a Brownian particle with mass m and drag coefficient ζ. The force $\mathbf{F}$ comes from random fluctuations and is assumed to be white noise, i.e.,

$$\begin{aligned}\langle\mathbf{F}(t)\rangle &= 0\\ \langle\mathbf{F}(t)\mathbf{F}(s)\rangle &= 2\zeta k_B T\delta(t-s),\end{aligned}$$

here k_B is the Boltzman constant and T the absolute temperature. In the zero mass limit (inertialess), Brownian motion is described by the equation

$$\zeta d\mathbf{x} = \mathbf{F}(t)\,dt. \tag{2.23}$$

The MSD of a particle with a diffusion coefficient $D = k_B T/(6\pi\eta a)$ and undergoing Brownian motion is

$$\left\langle \Delta r^2(\tau)\right\rangle = D\tau.$$

- **Fractional Brownian Motion**

  Fractional Brownian motion [157] is a self-similar Gaussian process with stationary increments and a uniform MSD scaling behavior,

$$\left\langle \Delta r^2(\tau)\right\rangle = D_{\text{fBm}}\tau^{\alpha}, \tag{2.24}$$

where D_{fBm} is the generalized diffusion coefficient with dimensions L^2/t^{α}. The autocorrelation function for fBm is likewise known and given by

$$E\left[B_\alpha(t)B_\alpha(s)\right] = \frac{1}{2}\left(|t|^\alpha + |s|^\alpha - |t-s|^\alpha\right). \tag{2.25}$$

Furthermore, long-range correlations are given by

$$\langle\xi_\alpha(0)\xi_\alpha(t)\rangle \approx \alpha(\alpha-1)t^{\alpha-2}, \tag{2.26}$$

where ξ_α is the fractional Gaussian noise, so that the Langevin equation describing simple fBm is

$$\zeta d\mathbf{x} = \xi_\alpha(t)\,dt. \tag{2.27}$$

From Eq. (2.26) is easy to see that uncorrelated, regular Brownian motion corresponds to $\alpha = 1$. For $0 < \alpha < 1$ the prefactor is negative and the increments are negatively correlated, rendering the associated process sub-diffusive. Conversely, when $\alpha > 1$ the motion is persistent (positively correlated), resulting in superdiffusion in which successive steps tend to follow in the same direction. fBm has been used to model a variety of processes including diffusion of biopolymers inside cells [158], monomer diffusion in a polymer chain [159], bacteria chromosomal loci [160], polymer translocation [161], diffusion in crowded fluids [162], and diffusion of one micron particles in HBE mucus [73].

Hill and coworkers [73] found for different mucus concentrations and over the experimental time scales (1 min) that single particle and ensemble MSD data were remarkably well approximated by a uniform power law and therefore consistent with a scaling of the form (2.24). Furthermore, the power law and prefactor were well described by the following functions of wt% of solids in the mucus samples:

$$\alpha \approx -0.17\,\text{wt\%} + 1.1, \qquad \text{for } 1.5 \leq \text{wt\%} \leq 5, \tag{2.28}$$

$$D_{\text{fBm}} \approx 1.6\exp(-1.5\,\text{wt\%}), \qquad \text{for } 1.5 \leq \text{wt\%} \leq 5, \tag{2.29}$$

where the units of D_{fBm} are $\mu\text{m}^2/s^{\alpha}$. These excellent fits to fBm need to be validated for longer time scales in order to apply the fBm models to predictions of passage time distributions of particles through mucus barrier layers. It is an open problem of intense study in our research group to predict passage time distributions and their scaling with thickness of the mucus layer.

- **Generalized Langevin Equations**

  Inspection of Eqs. (2.23) and (2.27) shows a general form of Langevin equations describing the diffusion of particles. In fact, if $\mathbf{F}(t)$ in Eq. (2.22) is not white noise, the motion of the particle has been described by a GLE of the form

$$m\frac{d^2\mathbf{x}}{dt^2} = -\int_0^t \frac{d\mathbf{x}(s)}{ds} K(t-s)ds + \mathbf{F}(t), \tag{2.30}$$

and the FDT connects the memory kernel function, $K(t)$, with the random fluctuation force by

$$\langle \mathbf{F}(t)\mathbf{F}(s)\rangle = k_B T K(|t-s|).$$

In the Laplace transform space, the one-dimensional GLE becomes

$$mz^2\tilde{X}(z) = -\tilde{K}(z)z\tilde{X} + \tilde{F}. \tag{2.31}$$

This is similar to the simple Langevin equation for viscous diffusion, except with a frequency-dependent drag coefficient given by the transform of the memory kernel. In the viscous limit, $\tilde{K}$ reduces to the constant Stokes drag coefficient: $6\pi\eta a$, where η is the fluid viscosity and a is the particle radius.

Fricks et al. [146] explored GLEs with memory functions consisting of an arbitrary sum of exponentials (a Prony series) and showed that the GLE can be transformed to a large vector Langevin equation. This transformation allows the application of well-known tools for inference from noisy time-series data (e.g., maximum likelihood estimators and the Kalman filter) to GLEs and for numerical generation of paths for given Prony series kernels. These tools were implemented to illustrate the transient anomalous diffusive statistics for the two classic, solvable models of polymer chain dynamics, namely Rouse and Zimm models. These models exhibit transient sub-diffusive behavior with intermediate time scale MSD scaling of $t^{1/2}$ and $t^{2/3}$, respectively, which are reproduced accurately in [146]. However, the inference of parameters in the memory kernel was restricted to a small number of memory time scales, on the order of less than 10, whereas mucus and similar biological fluids have a broad spectrum with decades of memory time scales.

McKinley et al. [148] made an important advance for GLEs and their application to transient anomalous sub-diffusion, showing how to prescribe an arbitrarily specified sub-diffusive power-law scaling in MSD on intermediate time scales (between the shortest and longest time scales of memory in the fluid). To do so, they generalized the Rouse and Zimm models, showing (and proving) how the scaling in the memory spectrum can be dictated so that the MSD scales with a tunable power law between 0 and 1. Shortly afterward, Amitai et al. [163] published a related result on tunable MSD exponents. McKinley et al. [148] further showed that the zero mass limit of the GLE, for an arbitrary memory kernel, is given by a sum of exponentials and is exactly solvable, giving an explicit particle path formula [148],

$$X(t) = CB(t) + \sum_{k=1}^{N} C_k Z_k(t), \tag{2.32}$$

where $X(t)$ is the particle position at time t, $B(t)$ is a standard Brownian motion, and the second term in the right-hand side is a sum of Ornstein–Uhlenbeck (OU) processes. In this sum, each term represents one color of noise in the process and the full-colored noise spectrum is related by explicit polynomial interpolation to the kernel memory spectrum. Each Z_k satisfies the stochastic differential equation:

$$dZ_k(t) = -\lambda_k Z_k + dB_k(t). \tag{2.33}$$

These results now pose a three-parameter family of GLE memory kernels, essentially dictated by the shortest and longest time scale of memory and the intermediate power law exponent of MSD, with which to fit to experimental particle time series data. These kernels are candidates along with fractional

Brownian motion for best fit to particle paths in mucus and other biological fluids. The determination of which models fit the data more accurately is an open challenge currently being explored in collaboration with S. McKinley, J. Mellnik, N. Pillai, and M. Lysy [164].

2.4.2 Mesh Size Distribution

Since the distribution of mesh spacing (and attractive versus repulsive interactions) is determinant in the diffusive properties of particulates through mucus, several studies have focused on determining the mesh size of different types of mucus based on an obstruction scaling model. We note that these studies must assume a model for the mucus mesh, and all inferences are based on those model assumptions. A more fundamental approach than the one summarized next is a very important project.

2.4.3 Obstruction Scaling Model

The obstruction scaling model [165, 166] assumes that the reduction in diffusivity is due to the particle encountering polymer chain obstacles. However, this approach assumes an effective diffusivity (i.e., Brownian motion), even though it is now evident that particles above a few hundred nm have sublinear scaling of MSD with time. Said differently, if one fixes a time scale of observation of paths, then it is possible to associate a viscous diffusivity of the sample fluid for that fixed time scale. The downside of this approach is that one will get different results for the effective diffusivity for every time scale. Nonetheless, this modeling approach is a common standard so it is worth understanding before proposing alternatives.

The model assumes that the effective radius of the mesh spacing is greater than the hydrodynamic radius of the diffusing particle and there is no interaction between the solute and the polymer. From this model, the ratio of diffusion in a gel and diffusion in pure water is given by

$$\frac{D_g}{D_w} = \exp\left[-\frac{\pi}{4}\left(\frac{r_s + r_f}{r_g + r_f}\right)^2\right], \tag{2.34}$$

where D_g is the diffusion coefficient of the particle in the polymer gel, D_w is the diffusion coefficient in water, r_s is the particle radius, r_f is the gel fiber radius, and r_g is the effective radius of the pore. For an extended discussion about these models of diffusion through mucus, we refer the reader to Cu and Saltzman [167]. Some reported values of mucus mesh sizes inferred from these assumptions are shown in Table 2.3.

Note that the results from these studies imply that the mucus gel network has pores that are larger than the diameter of many known viruses [168]. It is clear that mucus employs methods other than obstruction to prevent viruses from infecting

Table 2.3 Mesh size in mucus samples

Mucus type	Mesh size (nm)
Cervicovaginal mucus (CVM)[a]	340 ± 70
Cystic fibrosis sputum[b]	145 ± 50
Chronic rhinosinusitis mucus[c]	150 ± 50
CVM treated with nonionic surfactant N9[d]	130 ± 50

References: [a][168], [b][150], [c][156], [d][169]

mucosal surfaces, indicating that mucus is not just a steric barrier to deposited particulates. Namely, mucus is also an effective "adhesive" that can immobilize particles by hydrophobic and electrostatic interactions and hydrogen bonding [145]. These binding affinities suggest at the very least a generalization of the obstruction scaling model to include waiting times for particle binding and unbinding, which leads to sub-diffusive scaling [170]. Note, it is precisely these binding affinities that surface treatments of synthetic particles are designed to screen, in order for passive microrheology to faithfully reflect the innate fluctuations of the material. In particular, antibodies found in mucosal secretions have been reported to immobilize viruses and bacteria [171–175]. While one region of an antibody is capable of forming low affinity bonds with mucus, the other region can specifically link to the surface of pathogens. Thus, even viruses that are smaller than the average mucus mesh spacing and that do not bind to mucin molecules can be trapped with the help of antibodies [28]. Our research group has teamed with Sam Lai in the School of Pharmacy at UNC to study the intricate kinetic and diffusive interactions between antibodies, viruses, and mucus gels [176, 177]. However, to circumvent the mucus barrier, some viruses have evolved to contain hydrophilic coatings that enhance their mobility through the barrier by minimizing interactions with the components of the mucus gel network [173].

One also has to keep in mind that the mucus gel is not a rigid structure. As discussed above, the orientation and spacing between the components of the matrix gel is maintained by a series of interactions between the various macromolecules and small molecules in the solvent. Mucus network fluctuations can be strongly affected by changes in the small molecules in the solvent, e.g., significant changes in GI mucus versus PH has been observed by several research groups [46, 49, 178–181]. Then the use of average mesh values to describe the gel network is a crude way to estimate mean passage time. Our focus has been on more accurate methods of estimating passage time distributions for particles in mucus versus particle diameter, which can vary dramatically with disease conditions [73]. To do so, it is necessary to first identify models to describe diffusion through mucus. Our group has spent a significant effort on modeling of the primitive particle time series afforded by the advanced microscopy and particle tracking tools. As noted earlier, we have focused on parametric methods [73, 146–148], based on assumed stochastic processes and models from which we infer parameters of the model from the experimental particle time series data, and on nonparametric methods [73], based on statistical analyses of the time series data without any assumptions on an underlying model for the particle paths. Continued advances are needed in the understanding of how particle size and surface chemistry affect the passage times through mucus layers. These studies also provide insights into the length-scale distribution of the mucus network, although it is an open and intriguing problem to infer details of the mucus network structure from particle fluctuations. Clearly, attractive and repulsive particle-mucus microstructure kinetics need to be filtered in order to isolate fluctuations arising purely from mucus microstructure. This goal is the reason why two-particle microrheology was developed, for a discussion see Chap. 3.

3 Modeling Structure and Dynamics Within a Single Cell: The Mitotic Yeast Spindle

A predictive simulation of the structure and dynamics within an individual living cell remains a fundamental modeling and computational challenge. Cells are highly complex structures and there are multitudes of organizational charts for a single cell, e.g., search for "eukaryotic cell component chart." Most research labs in cell biology focus on specific aspects of cells and cellular processes. We will restrict our discussion to mitosis in yeast. Yeast are a model system for eukaryotic cells, which are distinguished by a membrane-bound nucleus and nuclear chromosomes packaged into chromatin fibers. The lab of Kerry Bloom at UNC explores the intricate behavior of these structures during different phases of the yeast cycle, and our work in yeast mitosis has been with Kerry Bloom and his students, specifically Andrew Stephens and Jolien (Verdaasdonk) Tyler.

3.1 Modeling Mitosis in Yeast Cells

In cell biology, yeast provide model systems for the study of the cell cycle and regulatory mechanisms. Yeast are readily available, and while less complex than animal cells, the cell cycle in yeast is remarkably similar. In general, the cell cycle encompasses a series of events leading to the division and duplication of the cell. Within these processes, mitosis and its regulation play a key role. *Mitosis is the stage of the cell cycle where the cell focuses its energy toward a single goal: chromosome segregation.* Models of cell mitosis based on experimental evidence serve as in vitro labs, where different theories and mechanisms, not accessible experimentally, can be tested. Modeling of cell mitosis can be performed at different levels. On one hand, one can use generalized logical network models where the self-regulating cell division system is modeled as an intricate molecular network [182–185]. In addition to the different molecular players, the network includes a series of checkpoints that place cell division under external control and ensure that every single step

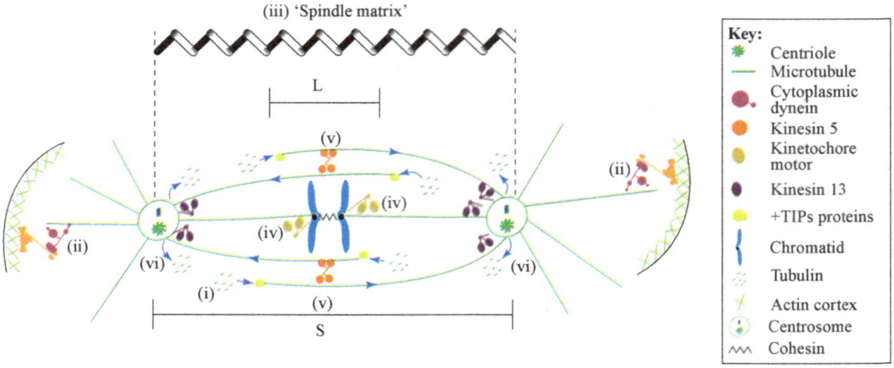

Fig. 2.10 Schematic of the mitotic spindle. Figure from [185]

has been completed before the next one begins. For example, sister chromatids are not separated until chromosomes are correctly aligned during the metaphase phase. Another approach consists on describing the mechanical and physical interaction of the main components of the cell via force balances [185–191]. The approach described next falls within this category. In this model the regulation of cell division results from a force balance applied to the mitotic spindle.

The mitotic spindle ensures the equal distribution of chromosomes during cell division. In the mitotic spindle, sister chromatids are bi-oriented and bound via the kinetochore to microtubules emanating from opposite spindle pole bodies; see Fig. 2.10. The kinetochore is a specialized protein/DNA structure built on centromere DNA that binds to the plus-end of dynamically growing and shortening kinetochore microtubules (kMTs). In yeast, each of the sixteen chromosomes is tethered to the spindle via a single kMT [192, 193]. Other microtubules extend inward from the spindle pole bodies and are not attached to replicated sister chromatids. These microtubules are known as interpolar microtubules. Interpolar microtubules (ipMT), from opposite spindle poles, overlap and are cross-linked by microtubule motor proteins and microtubule-associated proteins, represented in red in Fig. 2.10. These proteins (when double bound) exert outward forces as they slide pulling the ipMTs apart [194,195]. In addition, the replicated chromosomes form the sister chromatids (blue in Fig. 2.10) and exert inward forces through their connection to the kMTs. Two proteins play a fundamental role in the structure of the sister chromatids: cohesin and condensin [196–200]. These complexes, together with DNA, constitute the "chromatin spring" [201, 202]. The balance of microtubule-based extensional force and a chromatin spring contractile force is necessary to produce a steady-state spindle length and tension at the kinetochore that satisfies the spindle checkpoint [203]. The spindle checkpoint is a control mechanism that ensures that sister chromatids are attached and aligned before the two poles separate to form the two daughter cells.

The microtubules, microtubule-based motor proteins, and kinetochore components of the segregation apparatus have been explored with biophysical techniques, leading to a detailed understanding of their function [204–208]. However, the inherent complexity of the cell division process has made it challenging to understand the underlying mechanisms, even for a single phase such as metaphase. Mathematical and computational models are necessary to integrate experimental results with biochemical and biophysical cellular components, with the goal to understand the mechanochemical principles of the cell division process. From such a basic framework, one can then begin to understand cellular dysfunction and then apply that understanding in beneficial ways. Different models have been formulated to study individual components of the metaphase spindle. For instance, mathematical models of the spindle aim to account for the distribution and dynamics of spindle microtubules [188, 189, 209–212]. One class of models consists of stochastic equations describing kMT plus-end dynamics [188,211,213,214]. These models include spatial gradients in dynamic instability across the spindle, as well as tension-mediated regulation of kMT plus-end dynamics [188]. Another stochastic model that includes kinetochore attachment and detachment was formulated by Gay and coworkers [189]. Although microtubule dynamics were not explicitly modeled, this model incorporated a spatial gradient in kMT detachment rate that is analogous to the spatial gradients of Gardner and colleagues [188]. Overall, with appropriate tuning of the spatial gradients, these models were able to recapitulate experimentally observed features of microtubule plus-end dynamics and kinetochore separation. However, none of the models explicitly consider the physical properties of the chromatin spring (for a review see the article by Mogilner and Craig [206]) which is one focus of the Bloom lab.

Some models have coupled the microtubules and motor dynamics to the chromatin spring. For a review, the reader is referred to Mogilner et al. [185]. What these models have in common is that they assume the chromatin behaves as a Hookean spring, i.e., a linear force-extension relation. The chromatin spring is presumed to be derived via cohesion between sister chromatids [189], cohesin and condensin-based chromatin loops [202, 215], or an entropic worm-like chain [216]. However, Stephens et al. [191, 202] developed a series of models based on experimental observations in budding yeast cells and showed that the dynamics of the chromatin spring are not explained by a simple Hookean spring assumption. Using spindle length, chromatin dynamics, and stretching of individual and multiple chromosomes in the spindle, Stephens and coworkers probed the physical nature of the spring by comparison of measurable data with a predictive mathematical model, described next.

3.1.1 Force Balance Within the Budding Yeast Mitotic Spindle

The components contributing to force balance in the mitotic spindle are depicted in Fig. 2.11. We consider three main force-generating processes: (i) an extensional force arising from double-bound motors walking directionally along ipMTs, F_{ip};

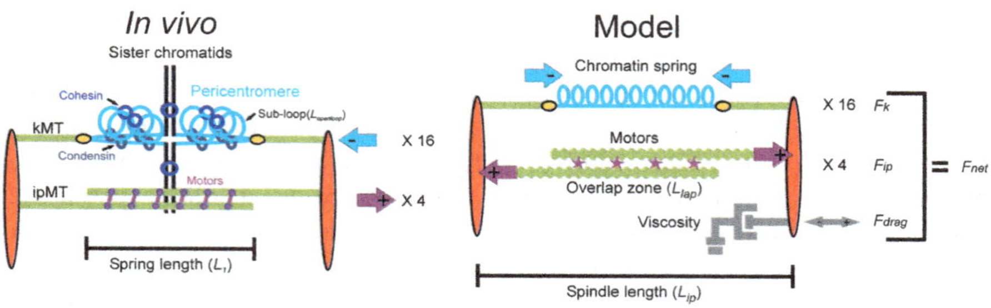

Fig. 2.11 Representation of the mitotic spindle in vivo and modeling assumptions. Figure adapted from [191]

$$[\mathrm{D}] \underset{k_{\mathrm{on},1}}{\overset{k_{\mathrm{off},1}}{\rightleftarrows}} [\mathrm{Sip}] \underset{k_{\mathrm{on},2}}{\overset{k_{\mathrm{off},2}}{\rightleftarrows}} [\mathrm{U}] \underset{k_{\mathrm{off},2}}{\overset{k_{\mathrm{on},3}}{\rightleftarrows}} [\mathrm{SkMt}]$$

Fig. 2.12 Schematic representation of the motors population balance. Values of the "off" and "on" rates are discussed in the text and in Table 2.4. Figure from [191]

(ii) an opposing contractile (restoring) force generated by the chromatin spring, F_k; and (iii) a viscous drag force, F_{drag}, that accounts for the cumulative viscous drag on the spindle. Holding one spindle pole fixed and summing all forces acting along the primary spindle axis, the net force on the spindle pole is

$$\sum F_{\mathrm{on\,spindle\,pole}} = F_{\mathrm{ip}} + F_k + F_{\mathrm{drag}} = F_{\mathrm{net}}. \tag{2.35}$$

The basis for a stable spindle length in the model is that a quasi-steady state is reached between inward and outward forces, with fluctuations about a mean spindle length arising from microtubule-based motor activity. Each force contribution is calculated as follows.

1. **Motor Force,** F_{ip}
   The outward force (F_{ip}) arises from the sliding of antiparallel ipMTs due to plus-end-directed motors bound in the overlap region (denoted double-bound motors, $D(t)$). The forces exerted by each motor, F_m, are additive, so F_{ip} is proportional to the total number of double-bound motors:

$$F_{\mathrm{ip}} = D(t) \cdot F_m. \tag{2.36}$$

Four types of motors are considered in the model: S_{ip} for motors single bound to ipMTs, S_{kMT} for motors single bound to the kMTs, U for unbound or free motors, and D for double-bound motors in the ipMT overlap zone. The total number of motors $D + S_{\mathrm{ip}} + U + S_{\mathrm{kMT}}$ is conserved. The population dynamics is summarized in Fig. 2.12.

Table 2.4 Transition rates of motors; see Fig. 2.12

Description	Dependence	Value	Notes
$\mathbf{k}_{\mathbf{on,1}}$: attachment rate of single-bound motors to ipMTs	Constant	$0.13\,s^{-1}$	(a)
$\mathbf{k}_{\mathbf{on,2}}$: attachment rate of unbound motors to ipMTs	$L_{ip}, L_{kMT}^{\text{left, right}}$	Eq. (2.37)	(b)
$\mathbf{k}_{\mathbf{on,3}}$: attachment rate of unbound motors to kMTs	$L_{ip}, L_{kMT}^{\text{left, right}}$	Eq. (2.38)	(b)
$\mathbf{k}_{\mathbf{off,1}}$: detachment rate of double-bound motors	Constant	$0.3\,s^{-1}$	(c)
$\mathbf{k}_{\mathbf{off,2}}$: detachment rate of single-bound motors	Constant	$0.3\,s^{-1}$	(c)

(a) The dynamics of attachment of single-bound motors in ipMTs to become double-bound motors is assumed to follow a binomial process,

$$S_{ip} \to D \sim B\left(S_{lap}, 0.12\right),$$

where S_{lap} is the number of motors in the overlap region, L_{lap}, of the ipMTs,

$$S_{lap} = S_{ip} \cdot \frac{2L_{lap}}{L_{ip} + L_{lap}}.$$

The probability of success is constant and equal to 12 %. Since the relation between the probability of attachment, p_{on}, and the rate of attachment, $k_{on,1}$, is

$$p_{on} = 1 - e^{k_{on,1}},$$

it follows that $k_{on,1} \approx 0.13$.

(b) The rate of attachment of free motors to ipMTs and kMTs is assumed to be proportional to the tubulin concentration (constant) and the percentage of the total length that is available for attachment:

$$k_{on,2} = [\text{tubulin}] \cdot \frac{L_{ip} + L_{lap}}{L_{ip} + L_{lap} + L_{kMT}^{\text{left}} + L_{kMT}^{\text{right}}}, \tag{2.37}$$

$$k_{on,3} = [\text{tubulin}] \cdot \frac{L_{kMT}^{\text{right or right}}}{L_{ip} + L_{lap} + L_{kMT}^{\text{left}} + L_{kMT}^{\text{right}}}. \tag{2.38}$$

(c) The rate of detachment of bound motors is assumed constant:

$$k_{\,off,1} = k_{off,2} = 0.3\,s^{-1}.$$

2. **Drag Force**, F_{drag}

At the low Reynolds numbers inside the cell, F_{drag} is proportional to the velocity of the spindle length (denoted L_{ip}) given by the Stokes drag law [217],

$$F_{drag} = -C_{drag} V_{ip}, \tag{2.39}$$

where $V_{ip} = dL_{ip}/dt$.

3. **Spring force**, F_k
   The length of the chromatin spring L_{sp} is total spindle length L_{ip} minus the length of each kinetochore microtubule $\left(L_{\text{kMT}}^{\text{left}}, L_{\text{kMT}}^{\text{right}}\right)$, where $L_{\text{kMT}}^{\text{left}}$ and $L_{\text{kMT}}^{\text{right}}$ are the length of the left and right kMT, respectively. Thus the chromatin length is $L_{\text{sp}} = L_{\text{ip}} - L_{\text{kMT}}^{\text{left}} - L_{\text{kMT}}^{\text{right}} - L_{\text{rest}}$, where L_{rest} is the rest length of the spring in the absence of force. To determine L_{sp}, the length of the left and right kMTs need to be defined. We follow the model of Gardner et al. [188] as explained next.

3.1.2 Kinetochore Microtubule Length Dynamics

The kMTs grow and shrink stochastically through polymerization and depolymerization, but the process is biased by the state of the kMTs, i.e., their length relative to a threshold length and the tension in the kMTs. The dynamics of this process was studied by Gardner et al. [188], giving the following relationships for the probabilities of rescue (growth), p_r, and catastrophe (shortening), p_c,

$$p_r = |0.21 - 9.5F_k|, \tag{2.40}$$

$$p_c = \left|0.38 - 0.65\left(L_{\text{kMT}} - 0.75\right)^2\right|. \tag{2.41}$$

To find if at a given time step a kinetochore microtubule is growing or shortening, the following procedure is implemented. Two random numbers, r_c and r_r, are drawn from a uniform distribution. These numbers are used to compute two logical values, a_1 and a_2, as

$$a_1 = \begin{cases} 0 & \text{if } p_c < r_c \\ 1 & \text{if } p_c > r_c \end{cases} \qquad a_2 = \begin{cases} 0 & \text{if } p_r < r_r \\ 1 & \text{if } p_r > r_r \end{cases}$$

Changes in kMT length are then determined by the following rules:

a_1	a_2	Process
0	0	Do nothing
0	1	Rescue (rate 17 nm/s)
1	0	Catastrophe (rate 25 nm/s)
1	1	Same as previous time step

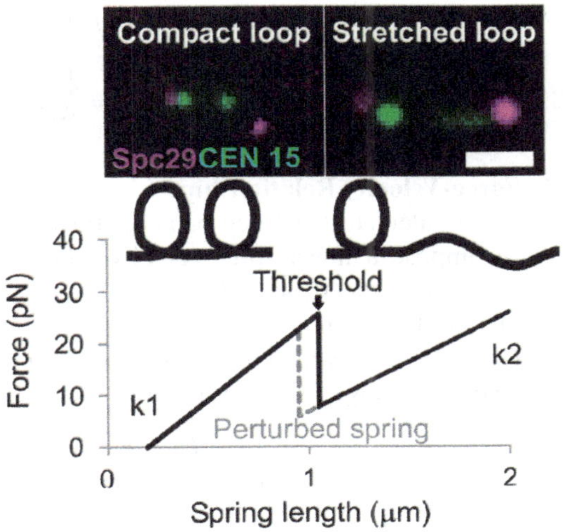

Fig. 2.13 Experimental evidence and graphic representation of piecewise continuous spring law. In the experimental data (*top*) the spindle pole bodies (Spc29) are labeled *red*, while the *green spots* correspond to labeled DNA arrays that are 1.8 kb from the centromere (CEN 15). Figure adapted from [191]

3.1.3 Functional Form of the Spring Force

The model assumes a *nonlinear spring force* that posits a threshold value for the spring extension. Spring lengths above the threshold value result in a decreased spring constant and an increase in rest length as depicted in Fig. 2.13.

The spring piecewise continuous spring force is given by

$$F_{k,i} = \begin{cases} -k_{\text{sp}}\left(L_{\text{sp}} - L_{r1}\right), & L_{\text{sp}} < X_{\text{thres}}, \\ -k_{\text{sp}}\left(\frac{L_1}{L_1 + x_{\text{loop}}}\right)\left(L_{\text{sp}} - L_{r2}\right), & L_{\text{sp}} \geq X_{\text{thres}}, \end{cases} \tag{2.42}$$

where x_{loop} and L_1 are constants and the "switching" between states is assumed instantaneous.

In addition, the 16 springs are arranged in parallel and linked to their two nearest neighbors. The links are assumed soft linear springs with spring constant $k_{\text{cross-link}} \ll k_{\text{sp}}$. Then the force law in each link is given by

$$\begin{aligned} F_{\text{cross-link}}\big|_{n+1} &= -k_{\text{cross-link}}\left[\left(L_{\text{sp}} - L_{\text{sp}}\big|_{n+1}\right)\cos\theta\right], \\ F_{\text{cross-link}}\big|_{n-1} &= -k_{\text{cross-link}}\left[\left(L_{\text{sp}}\big|_{n-1} - L_{\text{sp}}\right)\cos\theta\right], \end{aligned}$$

where θ is the angle between a cross-link and a spring, and we assumed that adjacent springs are close enough to each other so that $\theta \approx 0$ and $\cos\theta \approx 1$. The force exerted by the spring is then

$$\begin{aligned} F_{k,i}^{\text{cross-link}} &= F_{k,i} + F_{\text{cross-link}}\big|_{i+1} + F_{\text{cross-link}}\big|_{i-1}, \\ F_{k,i}^{\text{cross-link}} &= F_{k,i} - k_{\text{cross-link}}\left(L_{\text{sp}}\big|_{i-1} - L_{\text{sp}}\big|_{i+1}\right), \end{aligned}$$

where $F_{k,i}$ is calculated using Eq. (2.42). The total spring force is

$$F_k = \sum_{i=1}^{16} F_{k,i}. \tag{2.43}$$

4. **Force-Velocity Relationship**.
   To calculate the force balance, Eq. (2.35), in a physically meaningful way we impose a linear force-velocity relationship defined by two parameters: a maximum (stall) force, $F_M = 6\,\text{pN}$, and a maximum speed, $V_{\max} = 50\,\text{nm/s}$. The net force (sum of forces) felt by double-bound motors on ipMTs is $F_{\text{net}} = F_{\text{ip}} + F_k + F_{\text{drag}}$, which is then distributed evenly across double-bound motors and gives an average force per motor:

$$\frac{F_{\text{net}}}{D} = \frac{F_{\text{ip}} + F_k + F_{\text{drag}}}{D} = \frac{F_M D + F_k + F_{\text{drag}}}{D},$$
$$F_{\text{per motor}} = F_M + \frac{F_k + F_{\text{drag}}}{D}.$$

   The interpolar microtubule (therefore spindle) velocity can be determined as

$$V_{\text{ip}}(t) = \frac{F_{\text{net}}}{D} \cdot \frac{V_{\max}}{F_M} = V_{\max}\left(1 + \frac{F_k(t) + F_{\text{drag}}(t)}{F_M D(t)}\right). \tag{2.44}$$

   From Eq. (2.44) we see that:

   - If $F_k + F_{\text{drag}} = 0$, then $V_{\text{ip}} = V_{\max}$. In other words, if the forces from the chromatin spring and fluid drag cancel, the spindle will move at the maximum speed of one motor.
   - If the net force on the ipMTs is zero, $F_{\text{net}} = 0$, then $V_{\text{ip}} = 0$ and the spindle is stationary. In this quasi-equilibrium condition, each motor is at or near stall force (6 pN) and this force arises from the spring and drag forces acting on the motors.
   - The average force acting on a motor can be found as $F_{\text{per motor}} = F_M(1 - V_{\text{ip}}/V_{\max})$.

5. **Numerical Integration**
   Solving Eq. (2.44) for the spindle velocity gives

$$V_{\text{ip}}(t) = \left(\frac{U_{\max}}{F_M D(t) + U_{\max} C_{\text{drag}}}\right) \cdot \left(F_M D(t) + F_k(t)\right),$$

   so that

$$V_{\text{ip}}(t) = \left(\frac{U_{\max}}{F_M D(t) + U_{\max} C_{\text{drag}}}\right) \times$$
$$\left(F_M D(t) - k_{\text{sp}}\left(L_{\text{ip}}(t) - L_{\text{kMT}}^{\text{left}}(t) - L_{\text{kMT}}^{\text{right}}(t) - L_{\text{rest}}\right)\right).$$

To integrate this equation in time and find $L_{\text{ip}}(t+\Delta t)$, we perform a predictor-corrector scheme. The spindle length in the predictor step, $L_{\text{ip}}(t^*)$, is calculated as

$$L_{\text{ip}}(t^*) = L_{\text{ip}}(t) + \Delta t V_{\text{ip}}(t).$$

After this predictor step, we use $L_{\text{ip}}(t^*)$ to find $D(t^*)$ and $L_{\text{kMT}}^{\text{left,right}}(t^*)$ and perform a corrector step

$$L_{\text{ip}}(t+\Delta t) = L_{\text{ip}}(t) + \frac{\Delta t}{2}\left[V_{\text{ip}}(t) + V_{\text{ip}}(t^*)\right],$$

where

$$V_{\text{ip}}(t^*) = \left(\frac{U_{\max}}{F_M D(t^*) + U_{\max} C_{\text{drag}}}\right) \times \left(F_M D(t^*) - k_{\text{sp}}\left(L_{\text{ip}}(t^*) - L_{\text{kMT}}^{\text{left}}(t^*) - L_{\text{kMT}}^{\text{right}}(t^*) - L_{\text{rest}}\right)\right).$$

Solutions of the model have been shown to correctly recapitulate the observed experimental behavior and give insight into the dynamics of the chromatin network; for a detailed discussion see [191, 218].

Finally, we point out several aspects of the modeling amenable to further investigation. In this sense, the open research questions regarding this model include:

- *How microtubules attach to the chromosomes.* It is well known that as the kMTs grow and shorten, they "probe" space until they capture the chromosomes in a process rightly called search-and-capture [206,219,220]. The dynamic instability leading to growth/shrinkage of kMTs is included in the model; however the attachment/detachment dynamics are not included, i.e., the model assumes that the chromosomes are attached to the kMTs at all times.
- *Three-dimensional resolution.* The model assumes one-dimensional force balances. A generalization to a full 3-D geometry is a current research project.
- *Spring force law.* The piecewise continuous function used in the model correctly recapitulates the experimental observations. However, it assumes a single "unfolding" event and an *instantaneous* transition. New laws can be incorporated into the model based on a formal description of the transition states of the chromatin loops. For example, dynamics like those used to describe folding and unfolding of proteins [221] are being explored.
- *Astral microtubules.* In yeast cells, during metaphase there are approximately 2–3 astral microtubules extending from the spindle pole into the cytoplasm (see Fig. 2.10) versus the ~40 spindle microtubules. The astral microtubules are

critical for spindle orientation and are acted upon by cytoplasmic dynein [222]. While the spindle force balance proposed in the model guarantees that the sister chromatids are "centered" with respect to the spindle poles, the position of the chromatids with respect to the cell nucleus is governed by the astral microtubules. This coupling is another ongoing project.

- *Nuclear membrane forces.* The nuclear membrane has a heterogeneous morphology. As interactions of the spindle with the nucleus wall are introduced in the model, nuclear shape needs to be considered.
- *Inward motor forces.* An important component of the spindle machine is the minus-end motor Kar3 [194, 223, 224]. Kar3 is a nonessential gene that nonetheless contributes to the fidelity of chromosome segregation in mitosis. Kar3 is found in metaphase along the ipMTs as well as kMTs and microtubule plus-ends. The simplest model is that Kar3 opposes the outward motors Cin8 and Kip1. This more detailed resolution of motor activity is another research project.
- *Kinetochore forces.* Kinetochore forces are implicit as the mechanism that translates the spatial catastrophe gradient and tension-dependent rescue from the chromatin spring to kinetochore microtubule plus-ends. The rupture force to dissociate the kinetochore from a microtubule is approximately 9pN [225]. Note that this is comparable to the force of a single motor protein (6pN) so that its effects on the model might not be negligible.
- Variation among the chromatin springs is not incorporated in the model including: histone exchange in the pericentric chromatin [226], the likelihood that spring constants for different chromosomes are not identical nor are the switching thresholds, and variation in chromatin protein number (e.g., cohesin, condensin). These sources account for in vivo noise but are not believed to alter the overall trends or behavior of the model.

We end this discussion with a quote from Mogilner et al. [185] "A complete understanding of complex mitotic processes will inevitably require multidisciplinary efforts, of which modeling will undoubtedly be a major part. Three aspects of modeling will be crucial for success. First, the iterative character of the model-experiment loop will allow models to be adapted and improved. Although the initial models proposed will probably not survive experimental scrutiny, the development of first, even relatively crude, models is essential for the emergence of second generation models. Second, modeling will become more comprehensive and powerful through the combination of mathematical and computational approaches. Also, detailed mechanistic models will have to be combined with informatics-type models to deal with incomplete and sometimes noisy data of high-throughput studies. Third, simplistic models might have to become very detailed [...] Which way mitotic models will turn out is unclear, but the great challenge is to build adequate models of mitosis without making the models as complex as the mitotic spindle itself." Our philosophy is aligned with theirs!

4 Modeling Cell Motility

In the interests of space, we only briefly highlight the modeling approach of our group (Q. Wang, X. Yang, J. Zhao at U. So. Carolina, A. Chen, T. Wessler at UNC) toward the fundamental phenomenon of living cell motility. There are many reviews easily found in a search, such as [227, 228], that can lead a modeling approach. Our group has teamed with Ken Jacobson, Maryna Kapustina, Denis Tsyngkov, and Tim Elston to explore an intriguing cell oscillation phenotype and to search via modeling for the mechanochemical mechanisms to explain this remarkable oscillatory phenomenon. For details of our approach, we refer to our book chapter [229]; for details of the oscillatory cell phenotype, we refer to [230], and we recommend in particular the "biosights" video podcast for their paper. Our modeling approach assumes a phase-field formulation of the cell, based on a relatively coarse organizational structure that ignores molecular-scale complexity. This level of description is chosen because we want to understand the minimum mechanical and chemical species and processes necessary to reproduce the experimental observations. For an excellent overview of cell mechanics refer the reader to the recent article by Hoffman and Crocker [231].

In our approach, the cell is modeled as a composite of substructures (see Fig. 2.14): a lipid bilayer membrane, a thin cortex (cytoskeleton) that is the primary structural component, a nucleus, and a cytoplasm that fills the remainder of the cell's interior volume. Each substructure (phase) is governed by specific material properties and constitutive relations. In the phase field formalism, the boundary between adjacent phases is diffuse, modeled with a thin transition layer, and an energy functional prescribes the mass, momentum, and energy exchange across the transition layer. The cytoplasm contains various protein filaments, other organelles, and aqueous cytosol [232] and should properly be modeled as a viscoelastic fluid phase. Many laboratories focus specifically on the viscoelastic properties of the cytoplasm and in particular on fluid flow and transport within; we posit a

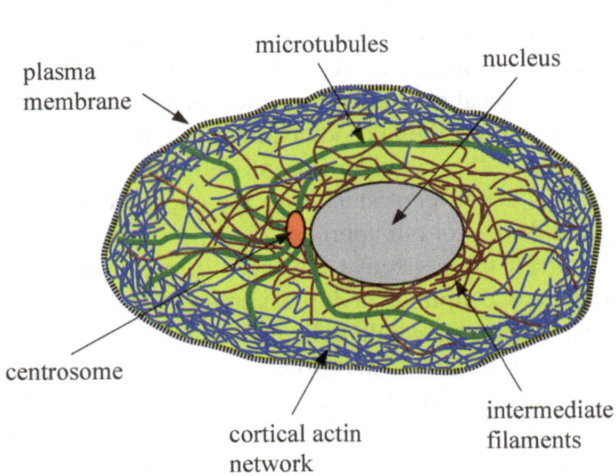

Fig. 2.14 Schematic of cell components. Figure from [232]

homogeneous viscoelastic fluid phase in our level of modeling, averaging overall the molecular complexity. The cortical layer not only provides the cell with mechanical integrity, but also provides a pathway for chemical activation and momentum transfer to the rest of the cell. The cytoskeleton is a network of protein filaments that spans the cell body and it is continuously remodeling. This constant remodeling makes the cytoskeleton highly adaptable, allowing the cell to change shape, to move, and to divide or merge. In our modeling, we adopt the description of the cortical layer as an active nematic gel [233].

Active polar nematic gel models have emerged as a new and popular topic in soft matter and complex fluids [233–235]. In an active material system, energy is continuously supplied by internal as well as external sources to drive the movement of the material system. The driving force behind these cell motility studies was the urge to understand the interaction between molecular motors and cytoskeletal filaments in cell motion and self-propelled motion of certain living cells. In a living cell, cross-linking proteins bind two or more self-assembled filaments (e.g., F-actin or F-actin and microtubules) to form a dynamical gel, in which motor proteins bind to filaments and hydrolyze nucleotide ATP. This process coupled to a corresponding conformational change of the binding protein turns stored energy into mechanical work, thereby leading to relative motion between bound filaments. In active nematic gel models, these molecular processes are upscaled to activation terms in the velocity field and extra stresses in the momentum balance.

Self-propelled gliding motion of certain bacterial species is another example of such an active material system, where molecular motors drive the cellular motion in a matrix of another material [233–245]. Both continuum mechanical models and kinetic theories have been proposed for active complex fluid systems [235, 240, 245, 246]. The mathematical framework incorporates the source of "active forcing" into an otherwise passive material system. The models are based on free energy considerations, both equilibrium and nonequilibrium, where one can keep track of dissipative and conservative principles, and the challenge for biological fidelity is to construct relevant energy potentials and chemical-mechanical activation functions. These potentials require detailed viscous and elastic properties of the fundamental cell components or phases, for which experimental techniques are now advanced enough to make progress. The energy formulation is likewise compatible with mathematical modeling, numerical algorithms, and simulation tools that have been developed for the hydrodynamics of multiphase complex fluids in evolving spatial domains. The simultaneous modeling of reaction and diffusion of biochemical species is self-consistent with the energetic formulation. These advances lay the groundwork for our approach.

Given the collective advances in membrane and cytoskeletal modeling, cell-substrate coupling, and biochemical kinetics, it is now feasible to develop a coarse-grained, whole cell model for migration on substrates or suspended in a saltwater solvent. This global cell-substrate model will enable us to investigate cell motility, dynamics of signaling proteins, cytoskeleton-substrate coupling, and contact cue guidance of motile cells. The model predictions will provide qualitative comparisons with cell experiments in the first proof-of-principle stage and potentially guide future experiments on detailed mechanisms associated with motility. As properties of each substructure become more quantified, the model will be able to make predictions to guide cell motility experiments. Given the complex nature of cell migration on topographically designed substrates, one should adopt a theoretical and computational platform that is capable of reproducing a variety of dynamical modalities.

Among the competing mathematical models for multiphase soft matter phenomena, a field phase approach is sufficiently versatile to handle the complexity of this challenge and to incorporate additional biological complexity. This is, admittedly, a top-down approach and is not to be expected to resolve detailed molecular or even supramolecular structure of the lipid bilayer, cortex, cytoplasm, and nucleus. These four phases are coarse-grained into constitutive relations and material parameters, with laws for interphase mass, momentum and energy exchange, and activation (what makes the cell alive) provided by a set of chemical species. These small molecule proteins responsible for activation are the focus of molecular biologists. There are many to choose from, and we begin with a handful of species that diffuse within the four phases with phase-specific diffusion coefficients that allows the model to enforce membrane-bound or cortex-bound constraints, for example. The concentrations of these kinetic species, their reactions with one another, and their activation energy within each phase are hypothesized and tested with simulations. Our current simulations are focused on the oscillatory phenotype. We posit one protein species that binds to filaments in the cortex, modeled as a contractile activation proportional to protein concentration; another protein species triggers unbinding by the kinetics of reaction and diffusion among the protein species.

The test of this modeling approach rests on whether we are able to establish fundamental mechanisms of cell motility that follow from a minimal set of coupled mechanochemical processes and structures within single cells, together with cues from their environment. We refer again to our review article [229] where we put the mechanical phases, chemical processes, and external environment into one formulation. The success of this framework remains to be determined. In particular, there are significant challenges to full three-dimensional simulations of these models, requiring the tuning of all material phases and chemical species. A current focus of the Jacobson lab and our modeling effort calls into question the geometry of cells suspended in a solvent. Essentially all modeling efforts assume a spherical equilibrium cell morphology. However, Jacobson and Kapustina have convincing evidence that the equilibrium morphology is highly folded if viewed from two-dimensional focal planes and indeed severely blistered if viewed as a three-dimensional structure. This evidence implies in our phase field formulation that cells have far more surface area available than that required to enclose the cell volume with a sphere. We are currently exploring excess surface area versus volume constraints, which are natural within the phase-field formulation. We likewise are exploring the current leading constitutive characterizations and material properties of cell membranes, cortex, cytoplasm, and nucleus, as well as the current leading candidates for chemical activation species and their mechanical activation rules within the cortex.

References

1. R.B. Bird, C.F. Curtiss, R.C. Armstrong, O. Hassager, *Dynamics of Polymeric Liquids, 2 Volume Set* (Wiley-Interscience, New York, 1996)
2. A.N. Beris, B.J. Edwards, *Thermodynamics of Flowing Systems: With Internal Microstructure* (Oxford University Press, New York, 2011)
3. R.G. Larson, *The Structure and Rheology of Complex Fluids*, vol. 2 (Oxford university press, New York, 1999)
4. M. Rubinstein, R.H. Colby, *Polymer Physics* (OUP, Oxford, 2003)
5. R.J. Braun, Annu. Rev. Fluid Mech. **44**, 267 (2012)
6. F. Guilak, L.G. Alexopoulos, M.L. Upton, I. Youn, J.B. Choi, L. Cao, L.A. Setton, M.A. Haider, Ann. NY Acad. Sci. **1068**(1), 498 (2006)
7. J.D. Ferry, *Viscoelastic Properties of Polymers*, vol. 3 (Wiley, New York, 1980)
8. C.W. Macosko, *Rheology: Principles, Measurements, and Applications* (VCH, New York, 1994)
9. T.A. Waigh, Rep. Progress Phys. **68**(3), 685 (2005)
10. T.M. Squires, T.G. Mason, Annu. Rev. Fluid Mech. **42**, 413 (2010)
11. D.T. Chen, Q. Wen, P.A. Janmey, J.C. Crocker, A.G. Yodh, Condens. Matter Phys. **1**, 301 (2010)
12. J.C. Crocker, B.D. Hoffman, Methods Cell Biol. **83**, 141 (2007)
13. A. Aufderhorst-Roberts, W.J. Frith, A.M. Donald, Eur. Phys. J. E **37**(5), 1 (2014)
14. E.J. Fong, Y. Sharma, B. Fallica, D.B. Tierney, S.M. Fortune, M.H. Zaman, Ann. Biomed. Eng. **41**(4), 837 (2013)
15. M. Tanaka, S. Wada, M. Nakamura, in *Computational Biomechanics* (Springer, Berlin, 2012), pp. 181–187
16. Y.C. Fung, *Biomechanics: Mechanical Properties of Living Tissues* (Springer, New York, 1993)
17. M.A. Meyers, P.Y. Chen, A.Y.M. Lin, Y. Seki, Progress Mater. Sci. **53**(1), 1 (2008)
18. A.S. Popel, P.C. Johnson, Annu. Rev. Fluid Mech. **37**, 43 (2005)
19. N.W. Tschoegl, *The Phenomenological Theory of Linear Viscoelastic Behavior: An Introduction* (Springer, Berlin, 1989)
20. P. Kollmannsberger, B. Fabry, Annu. Rev. Mater. Res. **41**, 75 (2011)
21. R.G. Larson, *Constitutive Equations for Polymer Melts and Solutions* (Butterworths, Boston, 1988)
22. H.C. Öttinger, *Stochastic Processes in Polymeric Fluids: Tools and Examples for Developing Simulation Algorithms* (Springer, Berlin, 1996)
23. R. Keunings, Micro-Macro methods for the multiscale simulation of viscoelastic flow using molecular models of kinetic theory, Rheol. Rev., British Society of Rheology 67–98 (2004)
24. R.G. Owens, T.N. Phillips, *Computational Rheology*, vol. 2 (World Scientific, Singapore, 2002)
25. A. Allen, J.P. Pearson, Eur. J. Gastroenterol. Hepatol. **5**(4), 193 (1993)
26. A. Silberberg, Cell Motility **2**(S1), 25 (1982)
27. D.F. Katz, Am. J. Obstet. Gynecol. **165**, 1984 (1991)
28. R.A. Cone, Adv. Drug Deliv. Rev. **61**(2), 75 (2009)
29. G. Garcia, R.C. Boucher, T.C. Elston, Biophys. J. **104**(3), 716 (2013)
30. G. Herschlag, G. Garcia, B. Button, R. Tarran, B. Lindley, B. Reinhardt, T.C. Elston, G. Forest, J. Theor. Biol. **325**, 42 (2013)
31. E. Fröhlich, E. Roblegg, in *Intracellular Delivery II* (Springer, New York, 2014), pp. 139–163
32. M.R. Knowles, R.C. Boucher, J. Clin. Invest. **109**(5), 571 (2002)
33. S.K. Lai, Y.Y. Wang, D. Wirtz, J. Hanes, Adv. Drug Deliv. Rev. **61**(2), 86 (2009)
34. D.J. Thornton, J.K. Sheehan, Proc. Am. Thoracic Soc. **1**(1), 54 (2004)
35. L.A. Sellers, A. Allen, E.R. Morris, S.B. Ross-Murphy, Carbohydrate Res. **178**(1), 93 (1988)
36. C. Taylor, A. Allen, P.W. Dettmar, J.P. Pearson, Biomacromolecules **4**(4), 922 (2003)

37. P. Verdugo, Cold Spring Harb. Perspect. Med. **2**(11), 1 (2012)
38. R. Bansil, E. Stanley, J.T. LaMont, Annu. Rev. Physiol. **57**(1), 635 (1995)
39. R. Bansil, B.S. Turner, Curr. Opin. Colloid Interface Sci. **11**(2), 164 (2006)
40. J.L. Desseyn, J.P. Aubert, N. Porchet, A. Laine, Mol. Biol. Evol. **17**(8), 1175 (2000)
41. C.M. Evans, J.S. Koo, Pharmacol. Therapeutics **121**(3), 332 (2009)
42. D.J. Thornton, N. Khan, J.K. Sheehan, in *Glycoprotein Methods and Protocols* (Springer, Berlin, 2000), pp. 77–85
43. M.S. Quraishi, N.S. Jones, J. Mason, Clin. Otolaryngol. Allied Sci. **23**(5), 403 (1998)
44. D.J. Thornton, M. Howard, N. Khan, J.K. Sheehan, J. Biol. Chem. **272**(14), 9561 (1997)
45. X. Cao, R. Bansil, K.R. Bhaskar, B.S. Turner, J.T. LaMont, N. Niu, N.H. Afdhal, Biophys. J. **76**(3), 1250 (1999)
46. R. Bhaskar, D. Gong, R. Bansil, S. Pajevic, J.A. Hamilton, B.S. Turner, J.T. LaMont, Am. J. Physiol. Gastrointest. Liver Physiol. **261**(5), G827 (1991)
47. B.D. Raynal, T.E. Hardingham, J.K. Sheehan, D.J. Thornton, J. Biol. Chem. **278**(31), 28703 (2003)
48. I. Carlstedt, J.K. Sheehan, A.P. Corfield, J.T. Gallagher, Essays Biochem. **20**, 40 (1985)
49. J.P. Celli, B.S. Turner, N.H. Afdhal, R.H. Ewoldt, G.H. McKinley, R. Bansil, S. Erramilli, Biomacromolecules **8**(5), 1580 (2007)
50. P. Gniewek, A. Kolinski, Biophys. J. **99**(11), 3507 (2010)
51. J.K. Sheehan, C. Hanski, A.P. Corfield, C. Paraskeva, D.J. Thornton, Biochem. Soc. Trans. **23**(4), 819 (1995)
52. G.E. Yakubov, A. Papagiannopoulos, E. Rat, T.A. Waigh, Biomacromolecules **8**(12), 3791 (2007)
53. C. Taylor, A. Allen, P.W. Dettmar, J.P. Pearson, Biochim. Biophys. Acta (BBA)-Gen. Subjects **1674**(2), 131 (2004)
54. I. Carlstedt, J.K. Sheehan, *Mucus and Mucosa* (Pitman, London, 1984), pp. 157–172
55. B. Button, L.H. Cai, C. Ehre, M. Kesimer, D.B. Hill, J.K. Sheehan, R.C. Boucher, M. Rubinstein, Science **337**(6097), 937 (2012)
56. H. Matsui, M.W. Verghese, M. Kesimer, U.E. Schwab, S.H. Randell, J.K. Sheehan, B.R. Grubb, R.C. Boucher, The journal of immunology **175**(2), 1090 (2005)
57. S. Girod, J.M. Zahm, C. Plotkowski, G.C. Beck, E. Puchelle, Eur. Respir. J. **5**(4), 477 (1992)
58. S.M. Gouveia, J.M. Tiffany, Biochim. Biophys. Acta (BBA)-Proteins Proteomics **1753**(2), 155 (2005)
59. A.L. Innes, S.D. Carrington, D.J. Thornton, S. Kirkham, K. Rousseau, R.H. Dougherty, W.W. Raymond, G.H. Caughey, S.J. Muller, J.V. Fahy, Am. J. Respir. Crit. Care Med. **180**(3), 203 (2009)
60. A.E. Bell, L.A. Sellers, A. Allen, W.J. Cunliffe, E.R. Morris, S.B. Ross-Murphy, Gastroenterology **88**(1), 269 (1985)
61. A.R. Talbot, T.M. Herr, D.S. Parsons, The Laryngoscope **107**(4), 500 (1997)
62. M. Robinson, A.L. Hemming, J.A. Regnis, A.G. Wong, D.L. Bailey, G.J. Bautovich, M. King, P.T. Bye, Thorax **52**(10), 900 (1997)
63. S.H. Donaldson, W.D. Bennett, K.L. Zeman, M.R. Knowles, R. Tarran, R.C. Boucher, New Engl. J. Med. **354**(3), 241 (2006)
64. L. Gheber, A. Korngreen, Z. Priel, Cell Motility Cytoskeleton **39**(1), 9 (1998)
65. A. Kater, M.O. Henke, B.K. Rubin, Ann. NY Acad. Sci. **1112**(1), 140 (2007)
66. M. King, Biorheology **24**(6), 589 (1987)
67. A.B. Lansley, Adv. Drug Deliv. Rev. **11**(3), 299 (1993)
68. D.M. Saito, A.L. Innes, S.D. Pletcher, Am. J. Rhinol. Allergy **24**(1), 1 (2010)
69. V.J. Broughton-Head, J. Shur, M.P. Carroll, J.R. Smith, J.K. Shute, Am. J. Physiol. Lung Cell. Mol. Physiol. **293**(5), L1240 (2007)
70. C.H. Lee, W. Yicheng, S.C. Shin, Y.W. Chien, Contraception **65**(6), 435 (2002)
71. S. Shak, D.J. Capon, R. Hellmiss, S.A. Marsters, C.L. Baker, Proc. Natl. Acad. Sci. **87**(23), 9188 (1990)

72. C.A. Vasconcellos, P.G. Allen, M.E. Wohl, J.M. Drazen, P.A. Janmey, T.P. Stossel, Science **263**(5149), 969 (1994)
73. D.B. Hill, P.A. Vasquez, J. Mellnik, S. McKinley, A. Vose, F. Mu, R.C. Boucher, M.G. Forest, PloS One **9**(2), e87681 (2014)
74. P.G. Bhat, D.R. Flanagan, M.D. Donovan, J. Pharm. Sci. **85**(6), 624 (1996)
75. A. Macierzanka, N.M. Rigby, A.P. Corfield, N. Wellner, F. Böttger, E.C. Mills, A.R. Mackie, Soft Matter **7**(18), 8077 (2011)
76. C.S. Albertini-Yagi, R.C. Oliveira, J.E. Vieira, E.M. Negri, L.R. de Oliveira, P.H. Saldiva, G. Lorenzi-Filho, Respir. Physiol. Neurobiol. **145**(1), 101 (2005)
77. M. King, J.M. Zahm, D. Pierrot, S. Vaquez-Girod, E. Puchelle, Biorheology **26**(4), 737 (1989)
78. B.K. Rubin, O. Ramirez, M. King, J. Appl. Physiol. **69**(2), 424 (1990)
79. W.D. Bennett, W.F. Chapman, T.R. Gerrity, CHEST J. **102**(2), 412 (1992)
80. B.K. Rubin, B. Finegan, O. Ramirez, M. King, CHEST J. **98**(1), 101 (1990)
81. B. Dasgupta, R.P. Tomkiewicz, W.A. Boyd, N.E. Brown, M. King, Pediatric Pulmonol. **20**(2), 78 (1995)
82. E. Daviskas, S.D. Anderson, K. Gomes, P. Briffa, B. Cochrane, K. Chan, I.H. Young, B.K. Rubin, Respirology **10**(1), 46 (2005)
83. D.B. Hill, B. Button, *Mucins* (Springer, New York, 2012), pp. 245–258
84. H. Matsui, V.E. Wagner, D.B. Hill, U.E. Schwab, T.D. Rogers, B. Button, R.M. Taylor, R. Superfine, M. Rubinstein, B.H. Iglewski, R.C. Boucher, Proc. Natl. Acad. Sci. **103**(48), 18131 (2006)
85. B.K. Rubin, Respir. Care **47**(7), 761 (2002)
86. L. Morgan, M.A. Pearson, R. de Iongh, D.W. Mackey, H. van der Wall, M. Peters, J. Rutland, Eur. Respir. J. **23**(4), 518 (2004)
87. A. Wanner, M. Salathé, T.G. O'Riordan, Am. J. Respir. Crit. Care Med. **154**(6), 1868 (1996)
88. M. Lippmann, D.B. Yeates, R.E. Albert, Brit. J. Ind. Med. **37**(4), 337 (1980)
89. H. Matsui, B.R. Grubb, R. Tarran, S.H. Randell, J.T. Gatzy, C.W. Davis, R.C. Boucher, Cell **95**(7), 1005 (1998)
90. R. Camassa, M.G. Forest, L. Lee, H.R. Ogrosky, J. Olander, Phys. Rev. E **86**(6), 066305 (2012)
91. C.S. Kim, M.A. Greene, S. Sankaran, M.A. Sackner, J. Appl. Physiol. **60**(3), 908 (1986)
92. R.C. Boucher, Annu. Rev. Med. **58**, 157 (2007)
93. M.A. Sleigh, J.R. Blake, N. Liron, Am. J. Respir. Crit. Care Med. **137**(3), 726 (1988)
94. J.H. Widdicombe, S.J. Bastacky, D.X. Wu, C.Y. Lee, Eur. Respir. J. **10**(12), 2892 (1997)
95. M. Salathe, Annu. Rev. Physiol. **69**, 401 (2007)
96. N.N. Sanders, S.C. De Smedt, J. Demeester, J. Pharm. Sci. **89**(7), 835 (2000)
97. E. Aiello, M.A. Sleigh, J. Cell Biol. **54**(3), 493 (1972)
98. M.J. Sanderson, M.A. Sleigh, J. Cell Sci. **47**(1), 331 (1981)
99. J.L. Devalia, R.J. Sapsford, C.W. Wells, P. Richman, R.J. Davies, Respir. Med. **84**(4), 303 (1990)
100. P.L. Leopold, M.J. O'Mahony, X.J. Lian, A.E. Tilley, B.G. Harvey, R.G. Crystal, PloS One **4**(12), e8157 (2009)
101. J.H. Widdicombe, J. Anatomy **201**(4), 313 (2002)
102. M.A. Sleigh, Cell Motility **2**(S1), 19 (1982)
103. M.A. Sleigh, Eur. J. Respir. Dis. Suppl. **128**, 287 (1983)
104. R.C. Boucher, J. Physiol. **516**(3), 631 (1999)
105. R. Sturm, J. Thoracic Dis. **4**(1), 48 (2012)
106. R. Toczyłowska-Mamińska, K. Dołowy, J. Cell. Biochem. **113**(2), 426 (2012)
107. J.R. Blake, J. Biomech. **8**(3), 179 (1975)
108. C. Marriott, Adv. Drug Deliv. Rev. **5**(1), 19 (1990)
109. http://www.nhlbi.nih.gov/health/health-topics/topics/pcd/
110. C. Marriott, S.S. Davis, Thorax **33**(5), 545 (1978)
111. B.J. Romanczuk, W.P. Potsic, A.L. McCall, M. Khan, C.K. Shih, M. Litt, Cleft Palate J. **15**, 93 (1978)

112. C.K. Shih, M. Litt, M.A. Khan, D.P. Wolf, Am. Rev. Respir. Dis. **115**(6), 989 (1977)
113. J. Teran, L. Fauci, M. Shelley, Phys. Rev. Lett. **104**(3), 038101 (2010)
114. A.L. Innes, P.G. Woodruff, R.E. Ferrando, S. Donnelly, G.M. Dolganov, S.C. Lazarus, J.V. Fahy, CHEST J. **130**(4), 1102 (2006)
115. C.K. Luk, M.J. Dulfano, Clin. Sci. **64**(Pt 4), 449 (1983)
116. E. Houtmeyers, R. Gosselink, G. Gayan-Ramirez, M. Decramer, Eur. Respir. J. **13**(5), 1177 (1999)
117. H. Matsui, S.H. Randell, S.W. Peretti, C.W. Davis, R.C. Boucher, J. Clin. Invest. **102**(6), 1125 (1998)
118. P. Satir, S.T. Christensen, Annu. Rev. Physiol. **69**, 377 (2007)
119. P. Satir, M.A. Sleigh, Annu. Rev. Physiol. **52**(1), 137 (1990)
120. B. Lindley, E.L. Howell, B.D. Smith, G.J. Rubinstein, M.G. Forest, S.M. Mitran, D.B. Hill, R. Superfine, J. Non-newtonian Fluid Mech. **156**(1), 112 (2009)
121. J.V. Fahy, B.F. Dickey, New Engl. J. Med. **363**(23), 2233 (2010)
122. M. Dawson, D. Wirtz, J. Hanes, J. Biol. Chem. **278**(50), 50393 (2003)
123. C. Galabert, J. Jacquot, J.M. Zahm, E. Puchelle, Clin. Chim. Acta **164**(2), 139 (1987)
124. S.H. Randell, R.C. Boucher, Am. J. Respir. Cell Mol. Biol. **35**(1), 20 (2006)
125. N.N. Sanders, S.C. De Smedt, E. Van Rompaey, P. Simoens, F. De Baets, J. Demeester, Am. J. Respir. Crit. Care Med. **162**(5), 1905 (2000)
126. S.S. Olmsted, L.A. Meyn, L.C. Rohan, S.L. Hillier, Sexually Transmitted Dis. **30**(3), 257 (2003)
127. R. Chatelin, P. Poncet, M. Tokman, *Proceedings of the 2nd European Conference on Microfluidics-Microfluidics*, Toulouse, 2010, pp. 8–10
128. M.B. Institute, Mathematical and computational challenges in cilia- and flagella-induced fluid dynamics. http://mbi.osu.edu/2012/mccdescription.html (2012)
129. M. King, M. Agarwal, J. Shukla, Biorheology **30**(1), 49 (1993)
130. D.J. Smith, E.A. Gaffney, J.R. Blake, Respir. Physiol. Neurobiol. **163**(1), 178 (2008)
131. J.R. Blake, J. Fluid Mech. **46**(1), 199 (1971)
132. J.R. Blake, M.A. Sleigh, Biol. Rev. **49**(1), 85 (1974)
133. D.J. Smith, E.A. Gaffney, J.R. Blake, Bull. Math. Biol. **69**(5), 1477 (2007)
134. D. Lubkin, E.A. Gaffney, J.R. Blake, Bull. Math. Biol. **69**(1), 289 (2007)
135. S.M. Ross, A wavy wall analytical model of muco-ciliary pumping. Ph.D. Thesis, Johns Hopkins University, 1971
136. Y.C. Fung, *Biomechanics: Motion, Flow, Stress, and Growth*, vol. 990 (Springer, New York, 1990)
137. B. Mauroy, C. Fausser, D. Pelca, J. Merckx, P. Flaud, Phys. Biol. **8**(5), 056006 (2011)
138. S. Mitran, J. Comput. Phys. **244**, 1–336 (2013)
139. S. Mitran, Comput. Struct. **85**(11), 763 (2007)
140. M. King, G. Brock, C. Lundell, J. Appl. Physiol. **58**(6), 1776 (1985)
141. J.A. Voynow, B.K. Rubin, CHEST J. **135**(2), 505 (2009)
142. C.S. Kim, A.J. Iglesias, M.A. Sackner, J. Appl. Physiol. **62**(3), 959 (1987)
143. C.S. Kim, C.R. Rodriguez, M.A. Eldridge, M.A. Sackner, J. Appl. Physiol. **60**(3), 901 (1986)
144. S. Mitran, Bearclawsoftware. http://mitran.web.unc.edu/codes
145. J. Kirch, C.A. Ruge, C. Schneider, J. Hanes, C.M. Lehr, *Nanostructured Biomaterials for Overcoming Biological Barriers* (Royal Society of Chemistry, UK, 2012), pp. 239–272
146. J. Fricks, L. Yao, T.C. Elston, M.G. Forest, SIAM J. Appl. Math. **69**(5), 1277 (2009)
147. C. Hohenegger, M.G. Forest, Phys. Rev. E **78**(3), 031501 (2008)
148. S.A. McKinley, L. Yao, M.G. Forest, J. Rheol. **53**, 1487 (2009)
149. S.K. Lai, Y.Y. Wang, R. Cone, D. Wirtz, J. Hanes, PloS One **4**(1), e4294 (2009)
150. J.S. Suk, S.K. Lai, Y.Y. Wang, L.M. Ensign, P.L. Zeitlin, M.P. Boyle, J. Hanes, Biomaterials **30**(13), 2591 (2009)
151. J. Khan, Y. Iiboshi, L. Cui, M. Wasa, A. Okada, J. Parenter. Enteral Nutr. **23**(1), 19 (1999)
152. M. Stern, N.J. Caplen, J.E. Browning, U. Griesenbach, F. Sorgi, L. Huang, D.C. Gruenert, C. Marriot, R.G. Crystal, D.M. Geddes, E. Alton, Gene Therapy **5**(1), 91 (1998)

153. S. Mallam, F. Horkay, A.M. Hecht, E. Geissler, Macromolecules **22**(8), 3356 (1989)
154. J. Fiegel, T. Brenza, R. Hamed, *Controlled Pulmonary Drug Delivery* (Springer, Berlin, 2011), pp. 143–163
155. J. Kirch, A. Schneider, B. Abou, A. Hopf, U.F. Schaefer, M. Schneider, C. Schall, C. Wagner, C.M. Lehr, Proc. Natl. Acad. Sci. **109**(45), 18355 (2012)
156. S.K. Lai, J.S. Suk, A. Pace, Y.Y. Wang, M. Yang, O. Mert, J. Chen, J. Kim, J. Hanes, Biomaterials **32**(26), 6285 (2011)
157. B.B. Mandelbrot, J.W. Van Ness, SIAM Rev. **10**(4), 422 (1968)
158. G. Guigas, M. Weiss, Biophys. J. **94**(1), 90 (2008)
159. D. Panja, J. Stat. Mech. Theory Exp. **2010**(2), L02001 (2010)
160. S.C. Weber, A.J. Spakowitz, J.A. Theriot, Phys. Rev. Lett. **104**(23), 238102 (2010)
161. J.L. Dubbeldam, V.G. Rostiashvili, A. Milchev, T.A. Vilgis, Phys. Rev. E **83**(1), 011802 (2011)
162. D. Ernst, M. Hellmann, J. Köhler, M. Weiss, Soft Matter **8**(18), 4886 (2012)
163. A. Amitai, Y. Kantor, M. Kardar, Phys. Rev. E **81**(1), 011107 (2010)
164. M. Lysy, N. Pillai, D.B. Hill, M.G. Forest, J. Mellnik, P. Vasquez, S.A. McKinley, Comparison for single particle tracking in biological fluids. J. Am. Stat. Assoc., submitted (2014)
165. B. Amsden, Macromolecules **31**(23), 8382 (1998)
166. B. Amsden, Macromolecules **32**(3), 874 (1999)
167. Y. Cu, W.M. Saltzman, Adv. Drug Deliv. Rev. **61**(2), 101 (2009)
168. S.K. Lai, Y.Y. Wang, K. Hida, R. Cone, J. Hanes, Proc. Natl. Acad. Sci. **107**(2), 598 (2010)
169. L.M. Ensign, B.C. Tang, Y.Y. Wang, A.T. Terence, T. Hoen, R. Cone, J. Hanes, Sci. Trans. Med. **4**(138), 138ra79 (2012)
170. http://cancerres.aacrjournals.org/content/48/14/4032
171. R. Diskin, J.F. Scheid, P.M. Marcovecchio, A.P. West, F. Klein, H. Gao, P.N. Gnanapragasam, A. Abadir, M.S. Seaman, M.C. Nussenzweig, P.J. Bjorkman, Science **334**(6060), 1289 (2011)
172. J. Mestecky, M.W. Russell, S. Jackson, T.A. Brown, Clin. Immunol. Immunopathol. **40**(1), 105 (1986)
173. S.S. Olmsted, J.L. Padgett, A.I. Yudin, K.J. Whaley, T.R. Moench, R.A. Cone, Biophys. J. **81**(4), 1930 (2001)
174. L.M. Walker, S.K. Phogat, P.Y. Chan-Hui, D. Wagner, P. Phung, J.L. Goss, T. Wrin, M.D. Simek, S. Fling, J.L. Mitcham, J.K. Lehrman, F.H. Priddy, O.A. Olsen, S.M. Frey, P.W. Hammond, S. Kaminsky, T. Zamb, M. Moyle, W.C. Koff, P. Poignard, D.R. Burton, Science **326**(5950), 285 (2009)
175. X. Wu, Z.Y. Yang, Y. Li, C.M. Hogerkorp, W.R. Schief, M.S. Seaman, T. Zhou, S.D. Schmidt, L. Wu, L. Xu, N.S. Longo, K. McKee, S. OÕDell, M.K. Louder, D.L. Wycuff, Y. Feng, M. Nason, N. Doria-Rose, M. Connors, P.D. Kwong, M. Roederer, R.T. Wyatt, G.J. Nabel, J.R. Mascola, Science **329**(5993), 856 (2010)
176. A. Chen, S.A. McKinley, S. Wang, F. Shi, P.J. Mucha, M.G. Forest, S.K. Lai, Biophys. J. **106**(9), 2028 (2014)
177. S.A. McKinley, A. Chen, F. Shi, S. Wang, P.J. Mucha, M.G. Forest, S.K. Lai, PloS One **9**, e100598 (2014)
178. S.S. Davis, S. Scobie, A. Inglis, Biorheology **12**(3–4), 225 (1975)
179. A. Garner, G. Flemström, A. Allen, J.R. Heylings, S. McQueen, Scand. J. Gastroenterol. Suppl. **101**, 79 (1983)
180. Z. Hong, B. Chasan, R. Bansil, B.S. Turner, K.R. Bhaskar, N.H. Afdhal, Biomacromolecules **6**(6), 3458 (2005)
181. L.M. Lichtenberger, Annu. Rev. Physiol. **57**(1), 565 (1995)
182. A. Csikász-Nagy, Brief. Bioinform. **10**(4), 424 (2009)
183. A. Fauré, A. Naldi, F. Lopez, C. Chaouiya, A. Ciliberto, D. Thieffry, Mol. Biosyst. **5**(12), 1787 (2009)
184. D.J. Irons, J. Theor. Biol. **257**(4), 543 (2009)
185. A. Mogilner, R. Wollman, G. Civelekoglu-Scholey, J. Scholey, Trends Cell Biol. **16**(2), 88 (2006)

186. G. Civelekoglu-Scholey, J.M. Scholey, Cell. Mol. Life Sci. **67**(13), 2231 (2010)
187. E.N. Cytrynbaum, J.M. Scholey, A. Mogilner, Biophys. J. **84**(2), 757 (2003)
188. M.K. Gardner, C.G. Pearson, B.L. Sprague, T.R. Zarzar, K. Bloom, E.D. Salmon, D.J. Odde, Mol. Biol. Cell **16**(8), 3764 (2005)
189. G. Gay, T. Courtheoux, C. Reyes, S. Tournier, Y. Gachet, J. Cell Biol. **196**(6), 757 (2012)
190. D.J. Odde, Curr. Biol. **15**(23), R956 (2005)
191. A.D. Stephens, R.A. Haggerty, P.A. Vasquez, L. Vicci, C.E. Snider, F. Shi, C. Quammen, C. Mullins, J. Haase, R.M. Taylor, J.S. Verdaasdonk, M.R. Falvo, Y.J. Jin, M.G. Forest, K. Bloom, J. Cell Biol. **200**(6), 757 (2013)
192. E.T. O'Toole, M. Winey, J.R. McIntosh, Mol. Biol. Cell **10**(6), 2017 (1999)
193. J.B. Peterson, H. Ris, J. Cell Sci. **22**(2), 219 (1976)
194. W.S. Saunders, M.A. Hoyt, Cell **70**(3), 451 (1992)
195. W.S. Saunders, V. Lengyel, M.A. Hoyt, Mol. Biol. Cell **8**(6), 1025 (1997)
196. Y. Blat, N. Kleckner, Cell **98**(2), 249 (1999)
197. K.A. Hagstrom, V.F. Holmes, N.R. Cozzarelli, B.J. Meyer, Genes Dev. **16**(6), 729 (2002)
198. P.C. Megee, C. Mistrot, V. Guacci, D. Koshland, Mol. Cell **4**(3), 445 (1999)
199. T. Tanaka, M.P. Cosma, K. Wirth, K. Nasmyth, Cell **98**(6), 847 (1999)
200. B.D. Wang, D. Eyre, M. Basrai, M. Lichten, A. Strunnikov, Mol. Cell. Biol. **25**(16), 7216 (2005)
201. D.C. Bouck, K. Bloom, Curr. Biol. **17**(9), 741 (2007)
202. A.D. Stephens, J. Haase, L. Vicci, R.M. Taylor, K. Bloom, J. Cell Biol. **193**(7), 1167 (2011)
203. K. Bloom, E. Yeh, Curr. Biol. **20**(23), R1040 (2010)
204. G. Alushin, E. Nogales, Curr. Opin. Struct. Biol. **21**(5), 661 (2011)
205. J.R. McIntosh, M.I. Molodtsov, F.I. Ataullakhanov, Q. Rev. Biophys. **45**(2), 147 (2012)
206. A. Mogilner, E. Craig, J. Cell Sci. **123**(20), 3435 (2010)
207. N.T. Umbreit, T.N. Davis, Exp. Cell Res. **318**(12), 1361 (2012)
208. Y. Watanabe, Nat. Rev. Mol. Cell Biol. **13**(6), 370 (2012)
209. G. Civelekoglu-Scholey, D.J. Sharp, A. Mogilner, J.M. Scholey, Biophys. J. **90**(11), 3966 (2006)
210. E.N. Cytrynbaum, P. Sommi, I. Brust-Mascher, J.M. Scholey, A. Mogilner, Mol. Biol. Cell **16**(10), 4967 (2005)
211. B.L. Sprague, C.G. Pearson, P.S. Maddox, K.S. Bloom, E.D. Salmon, D.J. Odde, Biophys. J. **84**(6), 3529 (2003)
212. R. Wollman, E.N. Cytrynbaum, J.T. Jones, T. Meyer, J.M. Scholey, A. Mogilner, Curr. Biol. **15**(9), 828 (2005)
213. M.K. Gardner, D.J. Odde, K. Bloom, Methods **41**(2), 232 (2007)
214. M.K. Gardner, B.L. Sprague, C.G. Pearson, B.D. Cosgrove, A.D. Bicek, K. Bloom, E.D. Salmon, D.J. Odde, Cell. Mol. Bioeng. **3**(2), 163 (2010)
215. S.A. Ribeiro, J.C. Gatlin, Y. Dong, A. Joglekar, L. Cameron, D.F. Hudson, C.J. Farr, B.F. McEwen, E.D. Salmon, W.C. Earnshaw, P. Vagnarelli, Mol. Biol. Cell **20**(9), 2371 (2009)
216. K.O. Greulich, E. Wachtel, J. Ausio, D. Seger, H. Eisenberg, J. Mol. Biol. **193**(4), 709 (1987)
217. G.K. Batchelor, *An Introduction to Fluid Mechanics* (Cambridge University Press, Cambridge, 1967)
218. A.D. Stephens, C.E. Snider, J. Haase, R.A. Haggerty, P.A. Vasquez, M.G. Forest, K. Bloom, J. Cell Biol. **203**(3), 407 (2013)
219. T.L. Hill, Proc. Natl. Acad. Sci. **82**(13), 4404 (1985)
220. T.E. Holy, S. Leibler, Proc. Natl. Acad. Sci. **91**(12), 5682 (1994)
221. G.I. Bell, Science **200**(4342), 618 (1978)
222. C.G. Pearson, K. Bloom, Nat. Rev. Mol. Cell Biol. **5**(6), 481 (2004)
223. S.A. Endow, S.J. Kang, L.L. Satterwhite, M.D. Rose, V.P. Skeen, E.D. Salmon, EMBO J. **13**(11), 2708 (1994)
224. S. Inoué, E.D. Salmon, Mol. Biol. Cell **6**(12), 1619 (1995)
225. B. Akiyoshi, K.K. Sarangapani, A.F. Powers, C.R. Nelson, S.L. Reichow, H. Arellano-Santoyo, T. Gonen, J.A. Ranish, C.L. Asbury, S. Biggins, Nature **468**(7323), 576 (2010)

226. J.S. Verdaasdonk, R. Gardner, A.D. Stephens, E. Yeh, K. Bloom, Mol. Biol. Cell **23**(13), 2560 (2012)
227. S. Li, J.L. Guan, S. Chien, Annu. Rev. Biomed. Eng. **7**, 105 (2005)
228. S. Li, N.F. Huang, S. Hsu, J. Cell. Biochem. **96**(6), 1110 (2005)
229. Q. Wang, X. Yang, D. Adalsteinsson, T.C. Elston, K. Jacobson, M. Kapustina, M.G. Forest, *Computational Modeling of Biological Systems* (Springer, New York, 2012), pp. 257–296
230. M. Kapustina, T.C. Elston, K. Jacobson, J. Cell Biol. **200**(1), 95 (2013)
231. B.D. Hoffman, J.C. Crocker, Annu. Rev. Biomed. Eng. **11**, 259 (2009)
232. P.A. Pullarkat, P.A. Fernández, A. Ott, Phys. Rep. **449**(1), 29 (2007)
233. R. Voituriez, J.F. Joanny, J. Prost, Phys. Rev. Lett. **96**(2), 028102 (2006)
234. L. Giomi, L. Mahadevan, B. Chakraborty, M.F. Hagan, Phys. Rev. Lett. **106**(21), 218101 (2011)
235. L. Giomi, M.C. Marchetti, Soft Matter **8**(1), 129 (2012)
236. A. Baskaran, M.C. Marchetti, Phys. Rev. Lett. **101**(26), 268101 (2008)
237. A. Baskaran, M.C. Marchetti, Proc. Natl. Acad. Sci. **106**(37), 15567 (2009)
238. S. Fürthauer, M. Neef, S.W. Grill, K. Kruse, F. Jülicher, New J. Phys. **14**(2), 023001 (2012)
239. L. Giomi, M.C. Marchetti, T.B. Liverpool, Phys. Rev. Lett. **101**(19), 198101 (2008)
240. B.M. Haines, I.S. Aranson, L. Berlyand, D.A. Karpeev, Comm. Pure Appl. Anal. **11**(1), 19 (2012)
241. B.M. Haines, A. Sokolov, I.S. Aranson, L. Berlyand, D.A. Karpeev, Phys. Rev. E **80**(4), 041922 (2009)
242. J.P. Hernandez-Ortiz, P.T. Underhill, M.D. Graham, J. Phys. Condens. Matter **21**(20), 204107 (2009)
243. D.L. Koch, G. Subramanian, Annu. Rev. Fluid Mech. **43**, 637 (2011)
244. S.D. Ryan, L. Berlyand, B.M. Haines, D.A. Karpeev, Multiscale Model. Simul. **11**(4), 1176 (2013)
245. D. Saintillan, M.J. Shelley, Phys. Rev. Lett. **100**(17), 178103 (2008)
246. X. Yang, M.G. Forest, W. Mullins, Q. Wang, Soft Matter **6**(6), 1138 (2010)

Part II
Rheology of Complex Biological Fluids

Chapter 3
Theoretical Microrheology

Roseanna N. Zia and John F. Brady

Abstract The importance of microrheology in the study of biological systems has a long and rich history, tracing its roots to the work of the botanist Robert Brown in the early nineteenth century. Indeed, passive microrheology and Brownian motion are one and the same. Brown's observation of microscopic pollen grains dancing about in water was initially thought to reveal some sort of "fundamental life force." However, upon further investigation, it turned out the motion depended only on the microscopically small size of the particles. The mysterious phenomenon went unexplained until the turn of the next century when Einstein and Perrin utilized Brownian motion to prove the atomic nature of matter. In addition to this profound result, the foundation of modern-day passive microrheology had been laid. Einstein combined the theory of diffusion with the Stokes' solutions for creeping flow to yield the Stokes–Einstein relation connecting observable particle motion—diffusion—to material properties: the viscosity. Perrin's experiments confirmed the theory. But Einstein's arguments and the Stokes–Einstein relation rely on the existence of equilibrium and other narrow criteria. New approaches have extended the idea of tracking the motion of a Brownian particle to understand material properties far beyond this limited model. These advancements are critical to the study of many biological systems which conduct much of their function in a nonequilibrium condition. In this chapter we will see how one can study biological systems from a rheological perspective, showing the unique role played by microrheology in understanding such systems. In a sense, Brown's initial hypothesis was not too far off the mark: rather than being driven by life, Brownian motion plays the role of the invisible hand that drives many of the processes required for life to proceed and indeed may have played a role in the very origin of the life process.

R.N. Zia (✉)
Cornell University, Ithaca, NY, USA
e-mail: zia@cbe.cornell.edu

J.F. Brady
California Institute of Technology, Pasadena, CA, USA
e-mail: jfbrady@caltech.edu

S.E. Spagnolie (ed.), *Complex Fluids in Biological Systems*, Biological and Medical Physics, Biomedical Engineering,
DOI 10.1007/978-1-4939-2065-5_3

1 Introduction

Complex biological fluids span a remarkable range of systems, from blood to biofilms to the crowded, watery interior of eukaryotic cells. Uniting these materials is the rich structure their name suggests: a macroscopic medium comprising a collection of microscopically small domains—a microstructure—which fluctuates and relaxes over observable time scales. Imposing a flow on such a material deforms the embedded microstructure, giving rise to dramatic changes in material behavior, including now-familiar non-Newtonian behaviors such as increases or decreases in viscosity (shear thickening and thinning) and viscoelastic behavior. Such changes in behavior are linked to biological functionality such as the shear thinning that allows mucus to be coughed out of the lungs [1–4]. This important bio-rheological fluid is studied in detail in Chap. 2. Rheology—the study of material flow behavior—is traditionally carried out by imposing bulk displacements on a macroscopic sample of material in a viscometer, e.g. via a shearing or extensional flow. When displacements are steady in time, a range of flow strengths may be imposed to study flow rate-dependent behaviors. Oscillatory and other time-varying motions may be applied in order to study viscoelastic and transient responses. Constitutive relations between the imposed flow and the stress may then be developed and, from these, material properties such as the viscosity inferred. Barnes et al. [5] provide a thorough review of traditional rheological techniques.

Theoretical rheology has the additional goal of connecting these non-Newtonian macroscopic behaviors with their underlying microstructural origins. For example, as flow strength is increased, the role played by Brownian motion in structural evolution decreases. For weak flows, this leads to a reduction in viscosity (shear thinning) [6, 7]. Hydrodynamic interactions between particles become increasingly important as flow strength increases, however, and can lead to increases in viscosity (shear thickening) [7–11]. Transient flows give further insight into the micromechanics of rate-dependent processes in steady-state flow: sudden removal of external forcing demonstrates that the microstructure relaxes over multiple time scales, each associated with distinct physical processes [12–17]. Such time-dependent behavior is a hallmark of non-Newtonian fluids; they can display both liquid-like (viscous) and solid-like (elastic) behavior, depending on the rate with which they are perturbed relative to the relaxation time scale(s) of the microstructure. Linear viscoelastic properties are typically studied by imposing a small-amplitude oscillatory shearing motion on a bulk sample of material [18]. In this linear-response regime, the resultant shear stress is linear in the imposed strain and strain rate, with coefficients that form the real and imaginary parts of a complex modulus. The phase shift and amplitude change with respect to strain rate or strain are thus related in a simple way to the storage (elastic modulus) and loss (viscous modulus) of energy in the sheared material. Viscoelastic behavior is found even in systems as simple as a colloidal dispersion of hard spheres [16, 17, 19–22]. Importantly, theoretical approaches to colloidal dispersions permit separation of the suspension stress (and viscosity) into solvent and particle phases, and the particle-phase stress can be further resolved into hydrodynamic, interparticle-force, and entropic contributions [20, 23–25].

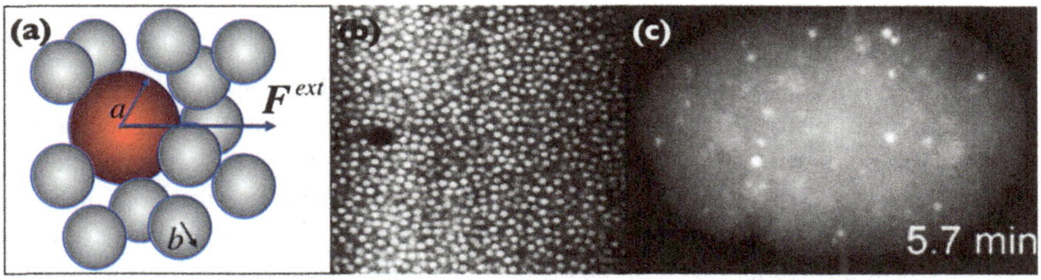

Fig. 3.1 (**a**) Microrheology model system. (**b**) Confocal microscopy image of microrheology of a concentrated colloidal dispersion of PMMA beads in Newtonian solvent (P. Habdas and E. Weeks, with permission). (**c**) Spatiotemporal changes in P granule size in the 1-cell *C. elegans* embryo. From [35]. Reprinted with permission from AAAS

Such contributions can become more (or less) important under variations in, e.g., ionic strength; thus, biologically relevant changes in pH, for example, can lead to profound changes in material properties such as viscosity. Recent work in the direct observation of particle microstructure, in contrast to indirect methods such as light and neutron scattering, has successfully merged the micromechanical approach of theory with experiments, allowing computation of the suspension stress via the statistical mechanics theory [26, 27]. But in these traditional rheological approaches, material perturbations are applied over macroscopic length scales and may therefore exclude important materials of interest. These include, for example, materials whose response varies over microscopic distances, or materials not available in sufficient quantities for bulk interrogation, e.g., rare biological fluids. Understanding the in vivo viscoelastic behavior of actin and determination of the diffusive speed of DNA, proteins, and nano-pharmaceuticals devices inside cells are a just few examples [28–31].

An alternative approach is to interrogate the material over microscopic length scales—a microscale version of rheometry. This type of microscale probing, known as "microrheology," traces its roots to the turn of the twentieth century and the work of Einstein [32] and Perrin [33]. Microrheology comprises a theoretical framework and experimental technique in which the motion of a Brownian particle (or set of particles) is tracked and its motion studied to infer the properties of the surrounding medium [34] [see Fig. 3.1a]. Advances in microscopy and other techniques (see, e.g., Crocker and Grier [36]) have prompted considerable new study in recent years, both experimental and theoretical [17, 34, 37–43] [see Fig. 3.1b, c].

Much work has since followed, including the extension of the Stokes–Einstein relation to viscoelastic materials [44] and studies of the effect of probe size [45] and shape [21]. Microrheological techniques have been used to study a diverse set of systems: cells [46–48], actin networks [49, 50], gelatin [51], DNA and polyethylene oxide solutions [52], and the behavior of colloids near the glass transition [37], as well as fundamental interactions between pairs of colloidal spheres [53–55] and entropic forces in binary colloids [56]. Microrheology has also been proposed as a tool for studying basic physics in atomic or molecular systems and for high-throughput material screening [57, 58].

Both equilibrium (linear-response) properties and the nonlinear response of a material can be probed, via *passive* and *active* microrheology, respectively. In the former, thermal fluctuations of a particle cause it to undergo a random-walk process; equilibrium and linear viscoelastic properties are then obtained by correlating the random thermally driven displacements of tracers to the complex modulus through a generalized Stokes–Einstein relation. In order to obtain nonlinear response properties, however, the material must be driven out of equilibrium. In this active (or nonlinear) microrheology regime, tracer particles undergo displacements due not only to random thermal fluctuations but also due to the application of an external force applied directly to the tracer, or "probe," by applying a constant or oscillatory external force to the particles, for example, by using optical tweezers or magnetic fields [37, 38, 41, 59, 60]. As with macrorheology, dynamic-response properties such as the viscosity can then be measured. Since the tracer interrogates the material at the microscopic length scale, much smaller samples are required compared to traditional macrorheology, and localized material heterogeneity can be explored. Recently a Stokes–Einstein-like relation has been developed which connects nonequilibrium particle motion and material properties [43]. Microrheology thus holds a particular benefit for rare biological materials and small systems such as cells. The theory that predicts the microviscosity and microdiffusivity of dilute systems of colloids and defines the relationship between micro- and macrorheology has recently been established [17, 21, 39, 40, 42, 43, 61]. Recent experiments confirm the theory [38, 41].

In the majority of theoretical approaches in microrheology, focus has been placed on the model of a colloidal dispersion of hard spheres [17, 39, 40, 42, 43, 62], with some effort focused on spheroidal probes [63] and suspensions of ellipsoids [64]. This simple model system recovers much of the rich physical behavior of more complicated systems (e.g., the viscoelastic behavior of entangled polymers, the shear thinning and thickening of particles of many shapes and polydisperse composition) [65]. While biological fluids are inevitably more complex than a simplified model system, e.g., with particles that may be irregularly shaped or polydisperse, often such details make quantitative rather than qualitative contributions to behavior. But inclusion of such details in a model can quickly render its analysis intractable, hindering the study of some of the most basic questions regarding their material behavior. Toward achieving fundamental insight into such systems, in this chapter we shall focus on the model system of a colloidal dispersion, about which much is known, allowing us to address many questions of material behavior rigorously and precisely.

The remainder of this chapter is organized as follows: in Sect. 2, passive microrheology is discussed, including historical perspectives, the Stokes–Einstein relation, and its extension to non-Newtonian materials via the generalized Stokes–Einstein relation. Limitations on its use are described, with a focus on the breakdown of the continuum assumption and alternative approaches. We then turn our attention to nonequilibrium material properties and active, nonlinear microrheology

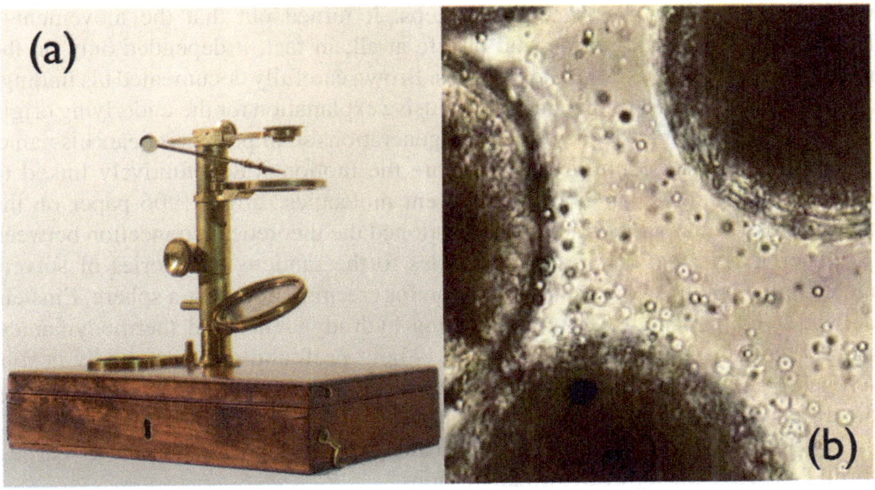

Fig. 3.2 (**a**) Robert Brown's microscope (by permission of the Linnean Society of London). (**b**) Reproduction of Brown's experiment: pollen grains released from pollen sacs (image courtesy P. Jones)

in Sect. 3. We review the microrheology model system, the microviscosity, the force-induced diffusion, and a recently developed nonequilibrium Stokes–Einstein relation giving the suspension stress in terms of diffusion and viscosity. In Sect. 3.5 an overview of time-dependent flows in microrheology is presented, followed by a brief discussion of the Brownian dynamics simulation method in Sect. 3.6. An important connection between active microrheology and a newly proposed "nonequilibrium equation of state" is the focus of Sect. 4. An effort is made throughout to highlight major results and to connect them to the study of relevant complex biological fluids. In Sect. 5, guidelines for implementation in laboratory experiments are suggested. The final section, Sect. 6, offers a summary and discussion.

2 Passive Microrheology: Brownian Motion

In 1827 Robert Brown peered through his microscope into a collection of tiny pollen grains suspended in water [66] (see Fig. 3.2). He watched them dance vigorously through the fluid, as if propelled by some invisible force. Contemporaneous scientific thought included the notion that living matter was composed of microscopic "fundamental life-force" blocks. Brown hypothesized that the particles were *self*-propelled and that the motion evidenced this fundamental life force. In an effort to test this provocative hypothesis, he killed the pollen particles, expecting the motion to cease. His excitement gave way to surprise and puzzlement when the lifeless particles continued their unrelenting dance. In the ensuing weeks Brown stared through his microscope at suspensions of coal dust, arsenic, and pulverized stone—all executing the same lively steps. It turned out that the movement—Brownian motion—did not depend on life at all; in fact, it depended only on the microscopically small size of the particles. Brown carefully documented his findings and published them in 1828 with no conclusive explanation for the underlying origin of the motion, leaving a legacy for future generations so important it bears his name. Three quarters of a century passed before the motion was definitively linked to its origin: the molecular motion of solvent molecules. In his 1906 paper on the theory of Brownian motion, Einstein developed the theoretical connection between the macroscopic motion of small particles to the random trajectories of solvent molecules. From this, and Stokes' solution for creeping flow past a sphere, Einstein derived a fundamental relation connecting hydrodynamics and thermodynamics: the Stokes–Einstein relation. The result is both profound and profoundly useful: Avogadro's number and a method to determine the viscosity of a fluid by tracking the diffusive motion of a single-probe particle.

We thus commence our discussion of theoretical microrheology in its earliest form: Einstein's theory of Brownian motion.

2.1 Single-Particle Diffusion and the Viscosity of Newtonian Solvents

In his theory of Brownian motion Einstein presented two approaches to connect thermal motion and diffusion to mechanical motion and hydrodynamics [32]. A central assumption in both approaches is thermodynamic equilibrium between the solvent and the suspended particles. In one approach, one imagines a dilute suspension of colloids settling under gravity toward the bottom of a container, thereby generating a weak spatial concentration gradient from bottom to top. The gradient in the colloid number density $n = N/V$ drives a diffusive flux upward. Because the system is at equilibrium in the external gravitational potential V_g, the diffusive flux, $\mathbf{j}_D$, is exactly balanced by the advective flux owing to sedimentation, $\mathbf{j}_{\text{sed}}$. Assuming a Fickian diffusion process, $\mathbf{j}_D = -\mathbf{D} \cdot \nabla n$, and an advective flux given by hydrodynamic mobility, $\mathbf{M}$, times the driving force of gravity, $\mathbf{j}_{\text{sed}} = -n\mathbf{M} \cdot \nabla V_g$, and recognizing that the equilibrium Boltzmann distribution, $n \sim \exp(-V_g/k_B T)$, applies, Einstein deduced that

$$\mathbf{D} = k_B T \mathbf{M}, \tag{3.1}$$

the Einstein–Smoluchowski equation relating the translational diffusivity $\mathbf{D}$ to the thermal energy $k_B T$ and the hydrodynamic mobility $\mathbf{M}$. Implicit in this result is the notion that the fluid behavior produced by the tracer's motion obeys the Stokes equations; that is, the fluid is a homogeneous, isotropic, incompressible continuum and the particle is small in some sense, such that the inertial terms in the Navier–Stokes equations may be neglected, and Stokes' equations govern fluid motion. To connect particle motion to fluid motion, one can solve Stokes' equations for the fluid velocity, which in turn yields the hydrodynamic force exerted on the particle by the fluid. The linearity of Stokes flow demands a linear relationship between particle motion $\mathbf{U}$ and this hydrodynamic force, $\mathbf{F}^H$, as $\mathbf{U} = -\mathbf{M} \cdot \mathbf{F}^H$, where the hydrodynamic mobility $\mathbf{M}$ couples force to motion, thus obtaining the probe mobility $\mathbf{M}$. For a translating rigid sphere of radius a, the Stokes mobility $\mathbf{M} = \mathbf{I}/6\pi\eta a$, giving the Stokes–Einstein relation for the translational diffusivity of a spherical particle in a solvent

$$\mathbf{D} = \frac{k_B T}{6\pi\eta a}\mathbf{I}, \tag{3.2}$$

where k_B, as in equation (3.2) k is Boltzmann's constant, T is the absolute temperature, η is the viscosity of the solvent, and $\mathbf{I}$ is the isotropic tensor. The Stokes–Einstein relation asserts that a small particle of size a placed in a fluid of viscosity η and temperature T will fluctuate due to random impacts from solvent molecules. These fluctuations cause the particle to undergo a random-walk process—it diffuses. Its diffusivity $\mathbf{D}$ is driven by the thermal energy $k_B T$, which is then dissipated back into the solvent by viscous drag $\sim 1/\eta$. This direct connection between the thermal energy of the solvent and the hydrodynamic drag on a microscopic particle is a statement of the fluctuation-dissipation theorem. It is easy to show the corresponding relationship between particle rotation and diffusion to obtain the rotational diffusivity of spheres:

$$\mathbf{D}_{\text{rotation}} = \frac{k_B T}{8\pi\eta a^3}\mathbf{I}. \tag{3.3}$$

Without much trouble the solution can be extended to particles of other shapes immersed in a simple continuum Newtonian solvent.

The first microrheology experiment was carried out by Jean-Baptiste Perrin in an effort to prove Einstein's theory [33, 66]. Projecting light through a suspension of gamboge spherules, Perrin tracked the motion of individual particles by tracing their movements on paper (the precursor to modern particle-tracking microscopy!) As shown in the traces excerpted from Perrin's laboratory notebook (reproduced in Fig. 3.3), the particles executed a random walk through the fluid. An ensemble average showed a linear growth of the mean-square displacement over time, confirming the hypothesis of diffusion due to random impacts from solvent molecules. Perrin published his results in his Ph.D. thesis and in the work *Brownian Movement and the Molecular Reality* [33]. It is generally accepted as a central finding in the proof of the existence of the atom and earned Perrin a Nobel Prize in 1926. The result is profound in its simplicity: just watching a particle move revealed the existence of the atom. It is as though a particle is a microscope, a lens through which we were able to "see" atoms and molecules for the first time. Einstein's theory and Perrin's experiments thus formed the basis of modern-day passive microrheology.

While his underlying goal in this work was to obtain Avogadro's number, N_A, an important outcome of Einstein's study was the leading-order correction to the viscosity of dilute suspensions which accounted for the presence of particles.

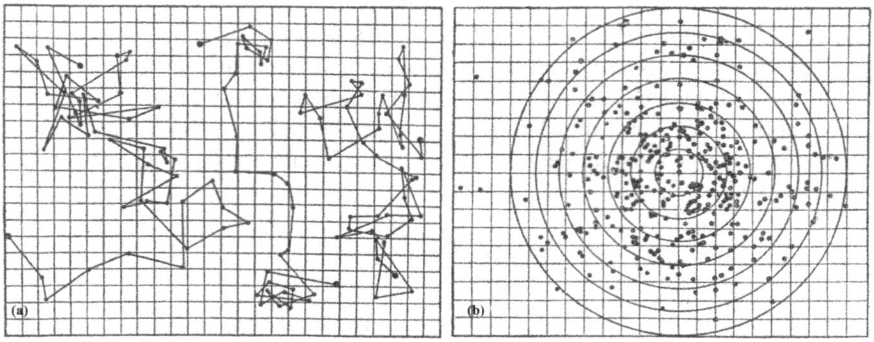

Fig. 3.3 Reproduced from the book of Jean-Baptiste Perrin, *Les Atomes* [67]. Reprinted with permission. (**a**) Three tracings of the motion of colloidal particles of radius 0.53 μm, as seen under the microscope, are displayed. Successive positions every 30 s are joined by *straight line* segments. The mesh size is 3.2 μm. (**b**) Normalized distribution of the displacements

He first utilized the diffusive motion of sugar molecules of unknown size a to determine the quantity $N_A a = \mathscr{R}T/6\pi\eta D$, where $\mathscr{R}$ is the universal gas constant. To show that the sugar itself was composed of atoms (or molecules) and would dissolve in a fluid, he asserted that the viscosity of the fluid must now be that of a *suspension* rather than a pure solvent. He derived an expression for the effective viscosity of a dilute suspension, $\eta^{\text{eff}} = (1 + 5\phi/2)\eta$, where the volume fraction of particles $\phi = (4\pi a^3/3)(N_A \rho/\mathscr{M})$, ρ is the particle density, and $\mathscr{M}$ the molar mass. The two expressions were combined to reveal Avogadro's number and the size of the diffusing sugar particles. *But by its very nature, Einstein's derivation shows that a tracer diffusing in a suspension is not connected in a simple way to the bulk suspension viscosity.* That is,

$$D \neq \frac{k_B T}{6\pi\eta^{\text{eff}} a}, \tag{3.4}$$

a fact made clear some 70 years later by the work of Batchelor [6, 68, 69] .

The Stokes–Einstein relation thus holds for a rigid particle in a Newtonian solvent and is thence limited to a rather narrow class of simple fluids. But many biological fluids are far from simple: they frequently comprise suspensions of many micro- or nanoscale particles, short- or long-chain polymeric structures which may be entangled or give rise to elastic networks, among others. In such materials, key assumptions of the Stokes–Einstein relation may no longer hold. In the next sections we examine how the Stokes–Einstein relation can be modified or its experimental application adjusted to accommodate a wider class of complex fluids. Two important cases in biological fluids include homogenous viscoelastic fluids and particle suspensions, treated next.

2.2 Extension to Viscoelastic Fluids: The Generalized Stokes–Einstein Relation

The non-Newtonian behavior of many fluids, in particular biological fluids, renders questionable the use of the Stokes–Einstein relation to infer bulk rheology [44, 54]. Examples include entangled polymer melts, solutions of flexible polymers, and gels which include mucus, saliva, synovial fluid, and the interior of eukaryotic cells [1, 4, 29, 46, 48, 70–72]. Such soft materials exhibit a time-dependent viscosity, with both liquid-like and solid-like responses to flow. For example, the mucus that lines the lungs and airway provides optimal healthy function with both viscous and elastic behavior. Strong flows, e.g., coughing, ideally produce viscous flow. But when the elastic modulus dominates the behavior of airway mucus, serious (sometimes life-threatening) pathology results, for example, in the case of cystic fibrosis [1–4]. When such a material forms a homogenous and isotropic continuum relative to the length scale of a probe particle, one may propose to model it as "Newtonian-like." Suppose such a material undergoes low-Reynolds number flow; one can write down a generalized form of the Stokes equations to model its behavior:

$$\nabla p = \int_{-\infty}^{t} \eta(t-t')\nabla^2 \mathbf{u}\, dt', \tag{3.5}$$

where p is the dynamic pressure, $\eta(t-t')$ is the time-dependent viscosity, $\mathbf{u}$ is the fluid velocity and is divergence free, and t is time.[1] A simple Newtonian fluid has no memory, and so $\eta(t-t') = \eta\delta(t-t')$, where δ is the Dirac delta distribution, η is the Newtonian viscosity, and Stokes' equations are recovered. To explore frequency-dependent response, one can take the Laplace transform of Eq. (3.5) and solve for the frequency-dependent fluid velocity. This is in turn a function of the frequency-dependent viscosity, $\hat{\eta}(s)$, where the hat symbol signifies the Laplace-transformed quantity, and the frequency s is the transform of the time variable. The hydrodynamic force exerted by the fluid on the particle can then be computed directly. The properties of Stokes flow demand a linear coupling between the hydrodynamic force and particle velocity; this simple relation connects the frequency-dependent and frequency-independent hydrodynamic mobilities, $\hat{\mathbf{M}}(s)$ and $\mathbf{M}$, respectively:

$$\hat{\mathbf{M}}(s) = \frac{\eta}{\hat{\eta}(s)}\mathbf{M}. \tag{3.6}$$

[1]Equation (3.5) is appropriate for consideration of fluid relaxation in the sense of a steady flow. It should not be confused with the unsteady Stokes equations which, used in conjunction with a Langevin equation, are appropriate for the study of vorticity diffusion in molecular liquids and the time-dependent Stokes drag. Further discussion of the topic may be found in e.g. Russel (1981) [73].

The material viscosity is thus obtained by solving the fluid mechanics problem subject to the boundary condition determined by *average* probe motion. To formulate a Stokes–Einstein-like connection for such a fluid, one must now connect *fluctuating* probe motion—passive diffusion—to material properties. This is done via the Langevin equation governing particle motion. The generalized Langevin equation describes the motion of a particle in a medium with memory [74]:

$$m\frac{d\mathbf{U}(t)}{dt} = \int_0^t \mathbf{R}(t-t')\cdot\mathbf{U}(t')dt' + \mathbf{f}^B(t), \tag{3.7}$$

where m is the mass of the probe, $\mathbf{U}$ is its velocity, $\mathbf{R} = \mathbf{M}^{-1}$ is the hydrodynamic resistance tensor, and $\mathbf{f}^B(t)$ is the random fluctuating (Brownian) force acting on the probe due to solvent collisions. A Laplace transform yields the frequency-dependent probe motion (which is overdamped in this viscous system). The velocity autocorrelation is thus given by $\langle\hat{\mathbf{U}}(s)\cdot\mathbf{U}(0)\rangle = m(U(0))^2\hat{M}$, where $U(0)$ is the initial probe speed and is uncorrelated with the Brownian force. For simplicity, the probe has been assumed to be isotropic, $\hat{\mathbf{M}}(s) = \hat{M}(s)\mathbf{I}$. Following Einstein's approach, Mason and Weitz [44] note that when one assumes the probe to be in thermal equilibrium with the suspending fluid, the equipartition theorem demands that the kinetic energy of the probe be given by one-half k_BT per degree of freedom, and $\langle\hat{\mathbf{U}}(s)\cdot\mathbf{U}(0)\rangle = k_BT\hat{M}$ for a translating probe. Mason and Weitz further explored the connection between diffusion, the velocity autocorrelation, $D = \int_0^t\langle\mathbf{U}(t')\cdot\mathbf{U}(0)\rangle dt'$, and the definition of the diffusivity as one-half the time rate of change of the mean-square displacement, $\langle\Delta r^2(t)\rangle$, of the probe. From it they obtained the generalized Stokes–Einstein relation, $\hat{D}(s) = k_BT\hat{M}(s)$ [44].

The frequency-dependent viscosity is a response function and is causal; it is thus straightforward to show via the Kramers–Kronig relations [75, 76] and analytic continuation to a Fourier transform that it is equivalent to a complex viscosity which comprises real and imaginary parts, $\eta^*(\omega) = \eta'(\omega) - i\eta''(\omega)$, where i is the imaginary unit and ω is the frequency. Causality demands restriction of the Fourier transform to the upper half of the complex plane [77], and thus $i\omega$ corresponds to the Laplace variable s. While the latter notation, $\hat{\eta}(s)$, is more compact, the former notation, $\eta^*(\omega)$, is more revealing: physically, the real and imaginary parts correspond to the viscous (loss) and elastic (storage) moduli, respectively. The frequency-dependent mobility may then be rewritten by simple substitution, $\mathbf{M}^*(\omega) = (\eta/\eta^*)\mathbf{M}$, and in turn, the generalized Stokes–Einstein relation (for a rigid sphere) reads:

$$\mathbf{D}(\omega) = \frac{k_BT}{6\pi\eta^*(\omega)a}\mathbf{I}. \tag{3.8}$$

An equally important definition of the material modulus owes its origins to the study of solid mechanics and thus is defined with reference to an imposed strain, rather than an imposed strain *rate*. The so-called bulk modulus, $G^*(\omega) = G'(\omega) + iG''(\omega)$, where G' and G'' are the storage and loss moduli, respectively.

Its connection to the complex viscosity η^* is $G^*(\omega) = i\omega\eta^*(\omega)$. As shown by Mason and Weitz [44], the bulk rheology of the material may be inferred by tracking the mean-square displacements (or their frequency transform), as

$$\hat{G}(s) = \frac{k_B T}{3\pi a s \langle \Delta \hat{r}^2(s) \rangle} = |G^*(\omega)|, \tag{3.9}$$

recovering the connection between single-particle tracking and bulk material rheology.

As noted above, however, expressions (3.8) and (3.9) apply only to isotropic, incompressible, homogenous materials whose internal structure is small compared to the microrheology probe—that is, the probe sees the structure as a continuum. While this extends the Stokes–Einstein relation to a much larger class of materials, many important materials (such as colloidal dispersions) are still excluded. This is the subject of the next section in passive microrheology. We begin with a discussion of the continuum limit.

2.3 *Validity of the Stokes–Einstein Relation?*

The diffusion of a microscopic particle in a homogenous, isotropic continuum fluid is a consequence of random impacts the colloid suffers from constituent molecules of the surrounding fluid. These impacts exert a fluctuating force on the colloid. Many thousands of impacts on a single colloid may occur in the time τ_p over which the momentum of the colloid can change; a simple scaling analysis shows the number of collisions scales as $\sim (a_p/a_s)^4$, where a_p is the colloid size and a_s is the solvent molecule size [61, 78]. Thus, for a particle-to-solvent-molecule size ratio of only 10:1, a particle will suffer 10,000 random impacts before it can experience any change in its momentum. Over this particle momentum-relaxation time scale, the colloid executes a random walk due to the solvent impacts: it diffuses. Correspondingly, any momentum acquired by the particle due to application of, say, an external force decays nearly instantaneously under the continuous barrage of solvent impacts. The colloid thus interacts with the solvent as a continuum and experiences the impacts as both a fluctuating stochastic force and a mean (steady) viscous drag.[2] This so-called continuum limit, in which microscopic particles interact with a solvent as a continuum fluid of viscosity η and density ρ, centers on a separation in length scales. The particle is very large compared to the fluid molecules. In the present discussion the continuum approximation is an appropriate model for the interaction between a single colloid

[2]Colloid motion and the associated Stokes drag reach steady state over a short but finite time. However, over times relevant to study of diffusion coefficients, the unsteadiness in particle motion is superfluous. Further discussion of the topic may be found in, e.g., Russel (1981) [73].

and a solvent. The particle dynamics are in turn modeled by the Langevin equation, whereby particle inertia is balanced by the (mean) viscous drag and the (fluctuating) Brownian force. Over times long compared to the particle's momentum-relaxation time scale, the Langevin equation recovers the Stokes–Einstein relation connecting particle diffusion and material viscosity (see, e.g., Brady [78] or Zia [61]).

But a probe particle embedded in a *suspension* interacts with both solvent and other colloidal particles, which occurs over several length scales and gives rise to multiple relaxation processes [12, 13, 15–17, 19, 21, 22, 79]. A probe particle moving through such a "microstructure" thus probes multiple time scales, and the connection between its diffusive motion and "suspension" viscosity requires a closer look. To do so, we consider the regimes of diffusion in colloidal dispersions alongside the microstructural contributions to suspension viscosity.

Diffusive Regimes In colloidal dispersions particles undergo three well-defined diffusive processes: short-time self-diffusion, long-time self-diffusion, and gradient or collective diffusion, each corresponding to a distinct physical process [80]. In the infinite-dilution limit they are identical and given by a Stokes–Einstein diffusivity (3.2). The short-time self-diffusivity corresponds to the motion of a colloidal (probe) particle inside a local "cage" of nearest-neighbor colloids. Such motion occurs over length scales much smaller than colloid size or equivalently, over times much shorter than that required for the tracer to move its size, $t \ll \tau_D = a^2/D$ but long compared to the momentum relaxation time of the tracer $t \gg \tau_p = m/6\pi\eta a$. As the probe wiggles around it must drag fluid. The no-slip surfaces of nearby particles resist this fluid motion, thus slowing probe motion. That is, the probe diffuses more slowly than were it alone in a solvent. This so-called short-time self-diffusion, denoted $\mathbf{D}_0^s$, occurs over time scales too short to cause changes to the microstructure, however, and is thus a strictly hydrodynamic phenomenon. At longer times, the probe is able to exchange places with its neighbors, and if one waits sufficiently long, $t \gg \tau_D$, the probe will execute a random walk throughout the suspension. This regime corresponds to the so-called long-time self-diffusivity, $\mathbf{D}_\infty^s$. Finally, distinct from both regimes of self-diffusion is the collective diffusivity, $\mathbf{D}^c$, of particles down a concentration gradient.

The well-known expressions for gradient, short-, and long-time self-diffusion in dilute dispersions of hydrodynamically interacting particles were developed by Batchelor in the 1970s and 1980s in a series of seminal papers [6,23,69]. To account for particle interactions, Batchelor generalized Einstein's argument by invoking a broader condition for thermodynamic equilibrium in a multicomponent system: uniformity of both the temperature *and* of the chemical potential. In this approach, the driving force for particle diffusion is a gradient in chemical potential. Batchelor separated the diffusion problem into two parts: a virial expansion of the chemical potential to obtain the concentration dependence of the driving force and a virial expansion of the hydrodynamic mobility functions describing the viscous response of the particles to applied forces. This general approach allowed the treatment of two key problems: the gradient diffusion of a single species and simultaneous gradient diffusion of multiple particle species. In the former regime, he noted the equivalence of gradient diffusion to sedimentation and found that gradient diffusion in dilute dispersions increases with the volume fraction $\phi = 4\pi a^3 n/3$ of particles of number density n: $\mathbf{D}^c = \mathbf{D}_0(1 + 1.45\phi)$, where $\mathbf{D}_0 = (k_B T/6\pi\eta a)\mathbf{I}$ is the diffusion of a particle of size a alone in a solvent of viscosity η [6] .

In his treatment of multicomponent gradient diffusion in dilute dispersions, Batchelor extracted an important case: when one species, i, say, is present in only trace amounts compared to the concentration of another species, j, the former undergoes tracer diffusion through the suspension. That is, it is formally equivalent to self-diffusion. The expression for the multicomponent gradient diffusivity $\mathbf{D}^c$ then yields the self-diffusion $\mathbf{D}_i^s$ of a tracer species i through a suspension of species j:

$$\mathbf{D}_i^s = \mathbf{D}_i \left(1 + \sum_{j \neq i} K_{ij}\phi_j\right), \tag{3.10}$$

where $\mathbf{D}_i = (k_B T/6\pi\eta a_i)\mathbf{I}$ is the self-diffusivity of an isolated particle of species i and size a_i. Here, K_{ij} is a bulk mobility coefficient comprising two contributions for a system of hydrodynamically interacting particles; when both species are of the same size, $K_{ij} = -C + S_{ij}$.[3] The first contribution, C, pertains to diffusive particle motion within a local cage of nearest neighbors. That is, over times much shorter than that required for the tracer to move its size, $t \ll a_i^2/D$, (i.e., before it can deform the microstructure) but long compared to the momentum relaxation time of the tracer $t \gg \tau_p = m/6\pi\eta a_i$. In this limit, $C = -1.83$ and $S_{ij} = 0$ and Eq. (3.10) gives the short-time self-diffusivity in dilute, monodisperse hard-sphere dispersions [6, 69]:

$$\mathbf{D}_0^s = \mathbf{D}(1 - 1.83\phi), \quad \phi \ll 1. \tag{3.11}$$

Because the species i can be a small subset of species j, the subscripts have been removed. In contrast to the linear growth of the gradient diffusivity $\mathbf{D}^c$ with volume fraction, the self-diffusion $\mathbf{D}_0^s$ of a tracer species through a suspension *diminishes* linearly with concentration of the background species.

Over long times, $t \gg a^2/D$, the tracer can escape its local cage. After many such displacements its motion becomes a random walk from cage to cage, and $S_{ij} = 0.27$, arising from both structural deformation of the microstructure and from hydrodynamic interactions between the particles. Equation (3.10) then gives the long-time self-diffusivity, $\mathbf{D}_\infty^s$, in a dilute suspension of hydrodynamically interacting hard spheres [69]:

---

[3] Batchelor combined statistical mechanics theory and the hydrodynamics of Stokes flow to derive expressions and some values for the coefficients K_{ij}, C, and S_{ij} by modeling three systems: particle velocity during sedimentation, gradient diffusion, and tracer diffusion in a polydisperse suspension. For each an average of the appropriate hydrodynamic functions for particle motion is weighted by the distribution of positions of particle pairs [6, 69].

Table 3.1 Comparison of diffusion coefficients, with $D_0 = k_B T/6\pi\eta a$ [6, 69, 80, 81]

	Dilute		Concentrated	
Mode	HI	No HI	HI	No HI
D_0^S	$D_0(1-1.83\phi)$	D_0	$D_0^S(\phi)^a$	D_0
D_∞^S	$D_0(1-2.1\phi)$	$D_0(1-2.0\phi)$	$D_0^S/[1+2.0\phi\, g(2;\phi)]$	$D_0/[1+2.0\phi\, g(2;\phi)]$

[a]As determined by dynamic simulation

$$\mathbf{D}_\infty^s = \mathbf{D}(1-2.1\phi), \quad \phi \ll 1. \tag{3.12}$$

These results have been extended by others for concentrated suspensions. For example, Brady obtained the short- and long-time self-diffusion coefficients in concentrated suspensions of colloids in the presence of strong and weak hydrodynamic interactions. In the latter case, an extension to higher concentrations is given by [69]:

$$\mathbf{D}_\infty^s(\phi) = \mathbf{D}_0^S(\phi)[1+2.0\phi\, g(2;\phi)]^{-1}, \tag{3.13}$$

where $g(2,\phi)$ is the pair distribution of particles at contact. A summary of self-diffusion coefficients is given in Table 3.1. For the collective diffusivity, a generalized form of the Stokes–Einstein equation [82],

$$D^c(\phi) = D_0 K(\phi)\frac{d[\phi Z(\phi)]}{d\phi} \tag{3.14}$$

reflects the contribution of hydrodynamic and thermodynamic forces in the sedimentation coefficient, $K(\phi)$, and the gradient of the compressibility factor, $Z(\phi)$, respectively.

The plot shown in Fig. 3.4 summarizes the three diffusive processes. The dependence of each on colloid volume fraction reflects the effect of neighboring particles on average particle motion. For example, collective diffusion increases linearly with ϕ at dilute concentrations. As seen in Eq. (3.14), however, the decreasing hydrodynamic contribution nearly cancels the increasing thermodynamic contribution, producing a near insensitivity to volume fraction at higher concentrations [6, 82]. Alternatively, passive self-diffusion is always hindered as the concentration of nearby particles grows. The short-time self-diffusion decreases with volume fraction as the number of nearby no-slip surfaces increases. The long-time diffusion is always slower than short-time self-diffusion, owing to the distortion of the microstructure that must occur in order for the probe to exchange places with its neighbors. The transition from short- to long-time self-diffusion is not a random process; the intermediate regime is characterized by correlated motion as the probe interacts with its neighbors. Careful identification of such a transition in the mean-square displacement is critical to understanding the size dependence of diffusion coefficients measured in experiments, for understanding measurements of so-called "ballistic" motion, and for revealing the fundamental connection between the frequency-dependent diffusivity $D(\omega)$ and the complex

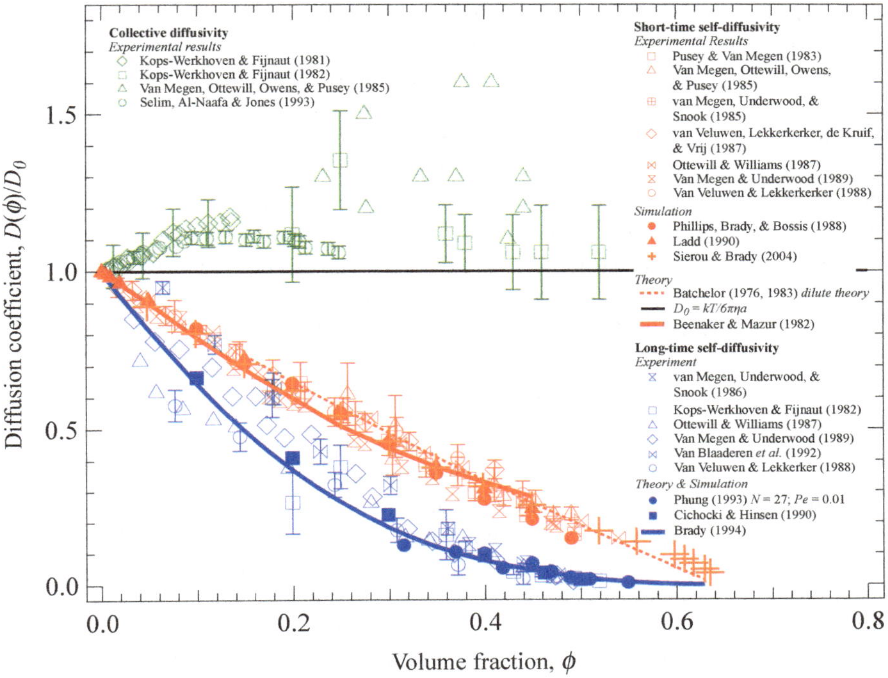

Fig. 3.4 Diffusion processes are set by the mechanism of interaction with neighboring particles. *Green symbols*, collective or gradient diffusion, D^c; *red symbols*, short-time self-diffusion; *blue symbols*, long-time self-diffusion. Data shown are for monodisperse suspensions of hard spheres

viscosity $\eta^*(\omega)$. To distinguish short- from long-time behavior, for example, one can plot the diffusion coefficient over long times. When the temporal slope of the diffusion coefficient ceases to change, the long-time limit has been reached. The intimate connection between the short- and long-time diffusivities is discussed further in the active microrheology section.

Suspension Viscosity It has been established in theory and experiment that the thermal motion of a colloidal tracer through a viscoelastic material reveals its linear viscoelastic character [34, 36, 44, 52, 83]. That is, the short- and long-time self-diffusion in suspensions reflect temporal relaxation processes that can be connected to rates of energy dissipation and storage [21, 34, 36, 44, 52, 80, 83]. In such passive microrheology experiments, the mean-square displacement of a diffusing probe particle $\langle|\Delta\mathbf{x}(t)|^2\rangle$ is proportional to the linear viscoelastic creep compliance of the material in which it is immersed, $J(t)$, as $\langle|\Delta\mathbf{x}(t)|^2\rangle = (k_B T/\pi a)J(t)$ for a spherical particle of size a [83]. The mean-square displacement can thence be related to the complex mobility. In theory all relaxation modes are excited by the random fluctuating motion, which is measured by dynamic light scattering in reciprocal time (frequency). But in practice it can be difficult to resolve these measurements over very fast or slow frequencies [36, 44]. Alternatively, in traditional macroscale shear rheology, small-amplitude oscillatory displacements can be applied to a macroscopic sample of material over a range of frequencies ω to reveal its viscoelastic properties, e.g., the complex viscosity $\eta^*(\omega) = \eta'(\omega) - i\eta''(\omega)$. Such displacements are small compared to the length scale of the microstructure, but every particle on average experiences the same perturbation. Thus the measured properties represent a suspension average. In the high-frequency limit, $\omega \to \infty$, displacements occur much more quickly than the fastest relaxation process of the microstructure, precluding any appreciable microstructural rearrangement. The suspension viscosity thus probed is purely hydrodynamic in origin, $\eta^*(\omega \to \infty) = \eta'(\omega \to \infty) \equiv \eta'_\infty$ [20]. In the dilute limit, to $O(\phi^2)$ this is given by [84]:

$$\frac{\eta'_\infty}{\eta} = 1 + \frac{5}{2}\phi + 5.0\phi^2, \quad \phi \ll 1. \tag{3.15}$$

In more concentrated suspensions, other relations have been proposed [19, 85–87], with limited validity at concentrations near close packing. The high-frequency limit thus probes the change in viscosity that is due simply to the presence of the particles: the surfaces of the particles obey a no-slip condition that gives rise to a distortion in fluid streamlines, leading to an $O(\phi)$ increase in viscous dissipation.

Because the short-time diffusion of a tracer and macroscopic high-frequency oscillation both probe the suspension without deforming its structure and are hydrodynamic phenomena, it is natural to seek a comparison between the two. That is, one can ask whether they relate directly via a Stokes–Einstein relation,

$$D_0^s(\phi) \stackrel{?}{=} \frac{k_B T}{6\pi\eta'_\infty(\phi)a}. \tag{3.16}$$

For dilute suspensions this relation would give $\eta'_\infty/\eta = 1 + 1.83\phi$, which is clearly not equal to Einstein's result, $\eta^{\text{eff}}/\eta = 1 + 2.5\phi$. Thus it is not possible to recover exactly the value of the bulk *shear* viscosity from the motion of a single diffusing particle. Differences arise when the probe particle selectively interacts with the immersing material or is not large enough relative to the characteristic microstructural length scale of the immersing fluid. Indeed, in the dilute limit it interacts with only one other colloid at a time. In such cases, however, the diffusive motion of the probe is still indicative of the linear viscoelastic character of the immersing medium but on a microscopic rather than macroscopic scale. In fact, qualitative agreement between the high-frequency viscosity, $\eta'_\infty(\phi)$, and the inverse of the short-time self-diffusivity, $1/D_0^S(\phi)$, is excellent. As seen in Fig. 3.5, quantitative agreement is also quite good—but not exact. In fact, the inverse of the short-time self-diffusivity is systematically lower than the high-frequency viscosity, $1/D_0^S(\phi) < \eta'_\infty(\phi)$, a disparity which arises precisely because the characteristic length of the probe particle is comparable to that of the suspending microstructure. Indeed, Almog and Brenner noted such effects in their study of a non-Brownian probe driven through a non-Brownian suspension [88]. Comparison

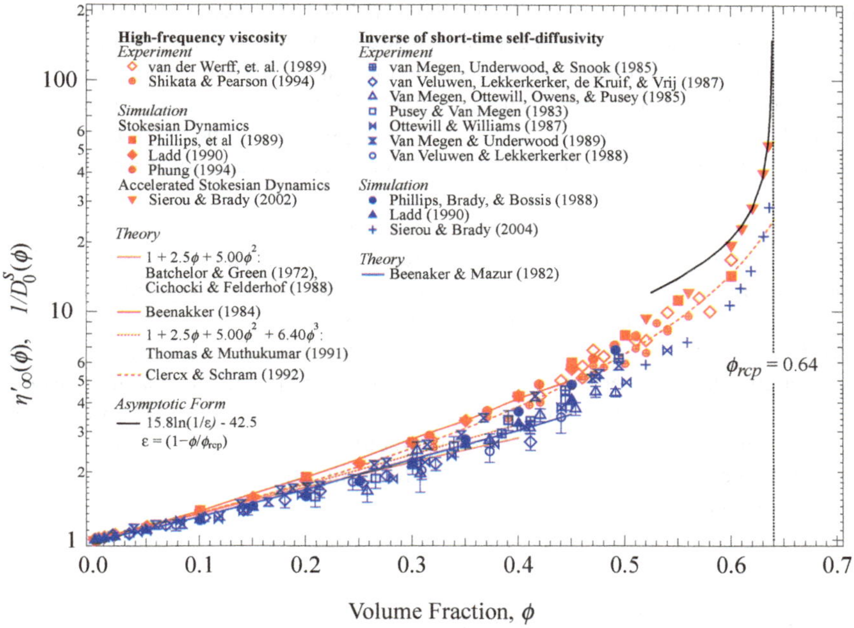

Fig. 3.5 Comparison of the infinite-frequency viscosity (*red symbols*) and the short-time self-diffusivity in colloidal dispersions (*blue symbols*), from dilute to close packing. Asymptotic theories as shown. Data shown are for monodisperse suspensions of hard spheres

of the zero-shear-rate viscosity, $\eta_0(\phi)$, to the inverse long-time self-diffusivity, $1/D^S_\infty(\phi)$, reveals similar qualitative agreement (and quantitative differences) as that seen in Fig. 3.4 [89]. In fact, Brady proposes that this inverse relationship between the self-diffusivity and the suspension viscosity carries over to arbitrary frequencies, differing only by a scale factor, that the appropriately normalized frequency-dependent self-diffusivity should be a universal function of the frequency for all volume fractions and that this frequency dependence should be virtually identical to that of the effective viscosity [80]. This universal scaling for the frequency-dependent viscosity has been verified experimentally [90]. How this view can be reconciled with the correlated motion during the transition from short- to long-time diffusion, however, remains an open question.

Nonetheless, if the goal is to recover *exactly* the "bulk" equilibrium rheology, the probe must interrogate the material over a length scale that is large compared to the microstructure. Passive single-probe microrheology works well when the probe is large in some sense, for example, in water, whose constituent molecules are of size about an angstrom, or a polymer melt, whose constituent monomer size is also small. It is also known from the work of Batchelor that for dilute suspensions a very large probe recovers precisely Einstein's viscosity correction. The extent to which this holds at higher volume fractions is not known. As many important biological fluids are composed of solvents plus particles on the order of nanometers to microns in size, understanding the interplay between probe and microstructure is an important aspect of the rheology of biological systems. For example, a probe inserted in a networked structure may diffuse locally inside a network pore; naïve application of the Stokes–Einstein relation for short-time measurements will produce a viscosity more indicative of the interstitial fluid than of the energy required to deform the network, often not the desired result. Even the application of the generalized Stokes–Einstein relation may obtain only pore-solvent viscosity and local network elastic strength.

At least two alternative approaches in microrheology can be taken to circumvent the continuum issue. In the first, pioneered by Crocker and coworkers [54], one tracks the motion of a *pair* of probes diffusing passively through the medium. In this so-called two-point microrheology, the motion of two tracers is tracked and correlations in their motion analyzed, enabling the inference of material properties over length scales larger than the probe radius [53–55]. This approach is discussed next. In addition, in Sect. 3 we will see that the motion of a single probe *driven actively* through a colloidal dispersion—active microrheology—can be related closely to the suspension viscosity and in turn, directly to the short- and long-time self-diffusivity of the probe.

2.4 Dual-Probe Microrheology

Tracking the motion of a single diffusing probe particle through a material has produced results consistent with the theoretical predictions of the Stokes–Einstein and generalized Stokes–Einstein relations for a wide range of fluids. The strongest agreement between theory and experiment is seen in the probing of simple Newtonian fluids and in some complex fluids whose internal microstructure interacts with the probe as a continuum. Examples of the latter include solutions of polymers and polymer gels [91–97]. However, in other studies of e.g., entangled polymer melts [98] and the intracellular medium [50, 99] the agreement is poor. For example, recent studies of intracellular diffusion report unexpectedly large diffusion coefficients [99, 100], superlinear diffusion [48, 100], and multiple diffusive regimes [70]. Results can also be probe-size dependent: in F-actin networks, small probes undergo sterically hindered diffusion, but larger probes must deform the network in order to diffuse [49, 101]. Interpretation of probe statistics thus requires consideration of the length and time scales present in the structure, which lead to structural and dynamical heterogeneities, respectively. Valentine et al. showed that careful statistical analysis of probe motion can distinguish between dynamical and structural heterogeneity in a material [92]. In addition, chemical as well as mechanical factors may contribute to difficulty in interpreting probe statistics. For example, common in biological systems are attractive interactions between probe and medium or chemical interactions between probe and structure. These fundamentally alter the dynamics of probe motion. A review of the experimental literature on the effects of chemical interactions between probe and structure can be found in Squires and Mason [97]. However, a structural feature common among such systems is the presence of multiple length and relaxation time scales which are also comparable to that of the diffusing probe—invalidating the continuum hypothesis and calling into question the validity of Stokes–Einstein relations. One successful approach to resolving issues of non-continuum interactions between probe and structure is dual-probe (two-point) microrheology [54, 55].

In an effort to extend the connection between tracer motion and the viscoelastic properties of the embedding material, Crocker and coworkers [54] introduced the model of dual-probe microrheology. They noted that a single probe inserted into a soft, homogeneous matrix may sample only the fluid within the pore, may interact locally with the matrix itself, or may move from pore to pore by deforming its local cage, yielding values for the viscoelastic moduli that are quite different from the viscoelastic modulus one would measure by a bulk shearing motion. Their model merges, in a sense, the approach of microscale and bulk interrogation: they proposed that two *widely* separated probes will interact with each other hydrodynamically—across the bulk material—with an interaction strength that scales, to leading order, as a Stokeslet, $\sim 1/r$. That is, the motion of one particle entrains the other through the complex fluid, giving rise to correlated fluctuations in particle position. Such correlations may then be, in turn, connected to a mean-square displacement tensor. The Laplace transform of this tensor is scaled on the ratio of the particle size to separation distance, a/r, giving a form of the generalized Stokes–Einstein relation (3.8):

$$\hat{D}(s) = \frac{k_B T}{2\pi r s \hat{\eta}(s)}, \tag{3.17}$$

where, as before, s is the transformed time variable and $\hat{\eta}(s)$ is the frequency-dependent viscosity. Frequency modes corresponding to length scales on the order of particle size do not produce in-phase motion of the probes. Ideally then, the only frequency modes encoded into the complex modulus $\hat{\eta}(s)$ are those corresponding material response (modulus) over length scales on the order of the separation distance. That is, the issue of a non-continuum length scale is circumvented by interrogating the material over a long length scale. Rather than using a single large probe to interact with many structural features simultaneously, two small tracers probe the bulk material. Crocker and coworkers then conducted experiments in several materials to test the connection between viscous moduli obtained via traditional shear rheology and that obtained single- and dual-probe microrheology with promising results. Despite its success in recovering bulk rheology for a wide class of complex materials, however, this technique is hindered by its intrinsic requirement for prohibitively large statistical sampling.

Recent years have seen an emergence of an alternative approach, active microrheology, which has shown success in resolving the limitations of continuum and equilibrium materials. This approach is discussed next.

3 Nonequilibrium Systems: Active Microrheology

In the previous sections we saw that the random thermal displacements of a Brownian tracer particle can reveal equilibrium, linear-response properties of an embedding material. The regimes of validity of such passive microrheology were shown to include the interrogation of homogeneous, incompressible, isotropic, continuum Newtonian and non-Newtonian fluids via the Stokes–Einstein and frequency-dependent Stokes–Einstein relations, respectively, and that the latter relation could be extended to a dual-probe model to study materials with more complex structure and relaxation behaviors. But several shortcomings remained. First and most obvious is the exclusion of nonequilibrium material behaviors. By their very nature, thermal fluctuations of the probe cannot drive a material from equilibrium and hence cannot probe nonlinear response properties. Secondly, in all but the simplest continuum fluids, displacements of a single probe cannot be related in a simple way to recover the *exact* bulk viscoelastic or other material properties as measured via traditional macroscopic shear rheology. While dual-probe passive microrheology has been shown to recover reasonably well the bulk rheology of a wide class of complex materials, this technique is hindered by its intrinsic requirement for prohibitively large statistical sampling. The model of nonlinear or "active" microrheology can overcome many of these limitations and obtain accurate predictions of both non-continuum and nonequilibrium material properties.

To interrogate the behavior of materials driven out of equilibrium, the colloidal probe is actively driven through the medium by an externally applied force. The force may be steady or oscillate in time (e.g., via magnetic fields or laser tweezers) imparting to it a nonzero average velocity. If driven by a *fixed* external force, the embedding material will slow the mean motion of the probe. This reduction may be analyzed via application of Stokes' drag law [39] and the mean motion connected to material viscosity [39, 40]. Even when the external force is fixed, probe velocity will fluctuate, leading to a diffusive spread of its trajectory and force-induced diffusion, the so-called microdiffusivity [42]. These results can also be combined to obtain the full tensorial expression for the suspension stress, which leads ultimately to a nonequilibrium Stokes–Einstein relation [43]. An oscillatory motion applied to the probe can interrogate the linear response viscoelastic properties [21] and nonlinear oscillatory behavior as well [22]. As in the previous sections for passive microrheology, we will make use of the workhorse model for complex fluids: a dispersion of colloidal spheres in a Newtonian solvent. Because the particles are Brownian, an understanding of the statistics of the particle distribution is required in order to determine average material properties.

We begin the discussion with an outline of the microrheology model system.

3.1 Model System

The microrheology model system comprises a Brownian probe particle of size a driven through a dispersion of neutrally buoyant colloidal particles, all of size b, immersed in a Newtonian solvent of viscosity η and density ρ. The relative strength of fluid inertia to viscous dissipation defines a Reynolds number, $\mathrm{Re} = \rho U a/\eta$, where U is the characteristic probe velocity. Because the probe and bath particles are small, $\mathrm{Re} \ll 1$; on the particle time scale, inertia can therefore be neglected and the fluid mechanics are governed by Stokes' equations. As the probe moves through the bath, it drives the suspension from equilibrium. Simultaneously, the Brownian motion of the bath particles acts to recover their equilibrium configuration, giving rise to an entropic restoring force of order $k_B T/b$, where $k_B T$ is the thermal energy of the bath. The degree to which the suspension is driven from equilibrium, and hence its effect on probe motion, is determined by the strength of external probe forcing F^{ext} compared to thermal restoring force $k_B T/b$, defining a Péclet number: $\mathrm{Pe} = F^{ext}/(k_B T/b)$. This interplay between probe motion and microstructural response gives rise to changes in probe velocity, which can be used to interrogate suspension properties [17, 38–43].

In general, the particles interact through hydrodynamic and interparticle forces which may be both short and long-ranged; the simplest model for the interactive force, which shall be adopted here, is the hard-sphere potential. Thusly defined, the colloids exert no force on each other until their surfaces touch, i.e., when their separation $r = a + b$. At contact, an infinite potential prevents their overlap.

The radii a and b at which the particles exert the hard-sphere force may or may not be the same as their physical or hydrodynamic radii, a_h and b_h, where the no-slip boundary condition is met. Various physical conditions of the colloid or solvent can extend the effective size of the particle beyond the hydrodynamic radius, e.g., steric repulsion or an ionic screening layer. The particles then repel each other at their extended or "thermodynamic" radii, a and b. This approach forms the foundation of the excluded annulus model of Morris and Brady [25], in which the ratio of the two radii ($\lambda_a = a/a_h$ and $\lambda_b = b/b_h$) can be modulated to account for the relative importance of hydrodynamic-to-interparticle forces. When $\lambda_a, \lambda_b \gtrsim O(1)$, the particles are able to approach each other closely enough to experience (long-range) hydrodynamic interactions. For $\lambda_a, \lambda_b \to 1$, lubrication interactions also become important. At the opposite extreme of $\lambda_a, \lambda_b \to \infty$, long-range interparticle repulsion keeps the particles sufficiently separated that hydrodynamic interactions are negligibly weak. The configuration of particles, resulting from imposed forces, flows, and interactions, is given by an N-particle probability distribution governed by a Smoluchowski equation; the shape of this structure can be described by the pair-distribution function $g(\mathbf{r})$ [39, 40]. For example, a suspension of microorganisms in suspension may be Boltzmann-distributed, $g \sim \exp[-V(r)/k_B T]$, where $V(r)$ is the potential of their interaction. For a suspension of algae particles this interaction

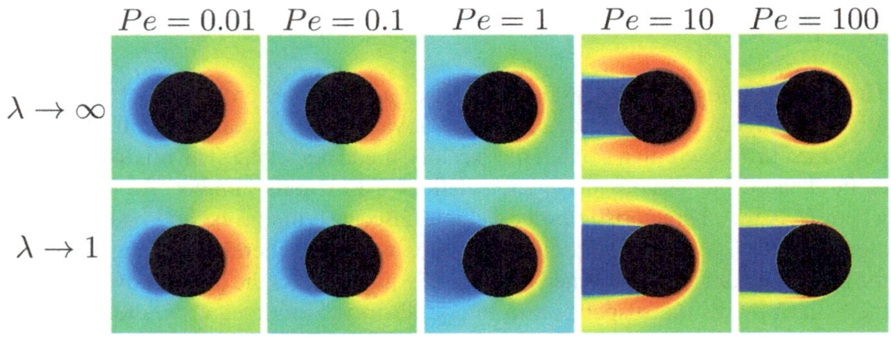

Fig. 3.6 Contour plots of the deformed microstructure around an externally forced probe driven to the right. The Péclet number increases in each from *left* to *right* as shown. *Top row*: negligibly weak hydrodynamic interactions, $\lambda \to \infty$. *Bottom row*: strong hydrodynamic interactions, $\lambda = 1$. *Black center* is the volume excluded to the probe. *Red regions* indicate areas of high pair density, *blue* indicates strong depletion, and *green* indicates undisturbed structure. From [40, 62], used in accordance with the Creative Commons Attribution 3.0 Unported License

may comprise a strong attraction at longer ranges that causes them to aggregate into colonies, along with a soft repulsion that permits fluid interpenetration and the activity of surface groups in the colony.

As the probe particle moves through the suspension it must push neighboring particles out of its way; a buildup of background-particle concentration forms in front of the advancing probe and a deficit or wake trails it. Brownian diffusion of the bath particles acts to restore isotropy, but as Pe is increased advection wins the competition and the microstructural deformation becomes highly anisotropic. Overall, the microstructure hinders the probe, slowing it down. Figure 3.6 gives an illustration of the deformed microstructure (the total structure $g(\mathbf{r})$ minus the equilibrium structure $g^{eq}(\mathbf{r})$ gives the deformed shape) at several values of the Péclet number (where Pe increases from left to right in the frames) and for the limits of weak and strong hydrodynamic interactions ($\lambda_a, \lambda_b = 1$ and $\lambda_a, \lambda_b \to \infty$, respectively). The evolution of microstructural shape can clearly be seen. The microstructural differences between the two rows are discussed in detail in Sect. 3.2.

3.2 *Microviscosity*

There are two primary modes of probe motion in microrheology: one in which the probe is driven by a fixed imposed force and the other in which its velocity is prescribed. The fixed-force case is the more interesting of the two and more relevant to the study of biological complex fluids, where a translating probe is free to fluctuate and undergo diffusive motion as well as steady translation. The velocity $\mathbf{U}$ of a probe particle translating under a fixed force, $\mathbf{F}^{ext}$, alone through an otherwise quiescent solvent is given by Stokes' drag law. The probe velocity, $\mathbf{U} = \mathbf{U}^{\text{Stokes}} = \mathbf{F}^{\text{ext}}/(6\pi\eta a_h)$, is proportional to the applied force and inversely proportional to its hydrodynamic radius a_h and the solvent viscosity η. The probe velocity is thus a measure of the solvent viscosity. When the probe is driven through a suspension of other particles, however, the microstructure slows the probe's motion. Squires and Brady [39] and Khair and Brady [40] interpreted the mean speed reduction as the viscous drag of the bath and defined an effective viscosity η^{eff} via application of Stokes' drag law to the average velocity of the probe. In the fixed-force case, this reads:

$$\mathbf{F}^{\text{ext}} = 6\pi\eta a \frac{\eta^{\text{eff}}}{\eta} \langle \mathbf{U} \rangle. \tag{3.18}$$

The effective viscosity is then given by

$$\frac{\eta^{\text{eff}}}{\eta} = \frac{F^{\text{ext}}}{6\pi\eta a \langle U \rangle}, \tag{3.19}$$

where $\langle U \rangle = -\langle \mathbf{U} \rangle \cdot \mathbf{F}^{\text{ext}}/F^{\text{ext}}$. The effective viscosity may also be written as

$$\frac{\eta^{\text{eff}}}{\eta} = 1 + \frac{\eta^{\text{micro}}}{\eta}, \tag{3.20}$$

where the microviscosity η^{micro} is the viscous drag of the particle microstructure—above and beyond the solvent viscosity. In general it includes contributions due to hydrodynamic, interparticle, and Brownian drag: $\eta^{\text{micro}} = \eta^H + \eta^P + \eta^B$. In the case of negligibly weak hydrodynamic interactions, $\lambda_a, \lambda_b \to \infty$, only the interparticle contribution is present [39]:

$$\frac{\eta^P}{\eta} = \frac{3}{4\pi} \text{Pe}^{-1} \phi_b \left(1 + \frac{a}{b}\right)^2 \hat{\mathbf{u}} \cdot \int \mathbf{n} g(2) d\Omega. \tag{3.21}$$

Here, ϕ_b is the volume fraction of bath particles, $\hat{\mathbf{u}}$ is a unit vector parallel to the applied probe force, $\mathbf{n}$ is a unit vector along the line of centers of a probe and bath particle, $g(2)$ is the contact value of the pair-distribution function of the bath relative to the probe (the "microstructure"), and $d\Omega$ is the element of solid angle. When hydrodynamic interactions cannot be neglected, the statistical mechanics theory can be applied in a straightforward way to define the corresponding integrals for the hydrodynamic, interparticle, and Brownian contributions to the microviscosity, η^H, η^P, and η^B. Here we shall discuss the physical relevance of the three contributions to the microviscosity without derivation of the corresponding formulae; these are given in detail by Khair and Brady [40].

Physically, there are two microstructural contributions to the effective viscosity: the rigidity of individual particles and the distribution of those particles relative to each other. The former, the non-deformability of the particles and the no-slip

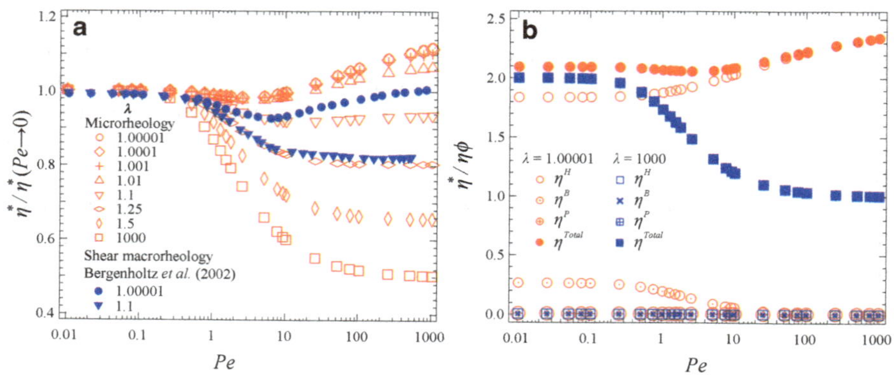

Fig. 3.7 (**a**) The effective viscosity normalized by its zero-Pe value, plotted versus Pe for a range of values of λ, the ratio of the hard-sphere to the hydrodynamic radius of the particles. *Open symbols*: microviscosity, with $\eta^* = \eta^{\text{micro}}/\phi$ and $\text{Pe} = F^{\text{ext}}b/k_B T$. A transition occurs near $\lambda = 1.1$ after which the suspension no longer thickens with increasing Pe [39, 40]. *Filled symbols*: macroviscosity, with $\eta^* = (\eta^{\text{eff}} - 1 - 5\phi/2)/\phi^2$. (**b**) Hydrodynamic, Brownian, and interparticle contributions to the microviscosity, η^H, η^B, and η^P, respectively, along with their combined total, $\eta^{\text{Total}} = \eta^{\text{micro}}$. The system is monodisperse ($a/b = 1$, $\lambda_a = \lambda_b \equiv \lambda$). Data same as [40], reprinted with permission

condition at their surfaces, leads to distorted fluid streamlines and increasing viscous dissipation—a hydrodynamic effect. However, the overall *arrangement* of particles is also deformable. This has interesting consequences for the viscosity, which depends both on the strength and on the rate of the structural distortion. Such flow-dependent viscosity is a hallmark of non-Newtonian behaviors such as shear thinning, shear thickening, and viscoelasticity.

In Fig. 3.7a the dependence of the microviscosity on the strength of hydrodynamic interactions has been highlighted; several familiar trends emerge. A Newtonian plateau at low Pe exists regardless of the importance of hydrodynamic interactions (i.e., for all values of λ). When the Péclet number increases, the plateau gives way to the thinning behavior characteristic of colloidal dispersions and many other complex fluids. As expected, the onset of this thinning behavior occurs when the Brownian motion becomes weaker than advection and can no longer effectively dissipate the work done by the probe, near $\text{Pe} = 1$. In the absence of hydrodynamic interactions, the thinning continues until a Newtonian plateau is reached at high Pe. With stronger hydrodynamic interactions, the thinning gives way to thickening near $\text{Pe} = 5$. The transition from thinning to thickening shifts to slightly lower Péclet numbers as λ decreases, reflecting the interplay between entropic and hydrodynamic forces. These trends are in close qualitative agreement with the corresponding values obtained via dilute theory for sheared dilute colloidal dispersions [7], which are shown as the filled symbols in Fig. 3.7a, and with Brownian dynamics simulation [102] and experiment [38].

Figure 3.7b highlights each of the individual contributions—hydrodynamic, Brownian, and interparticle—to the microviscosity. Two limiting cases are shown: weak (square symbols) and strong (circular symbols) hydrodynamic interactions. In both cases, a transition to force thinning occurs near Pe = 1 as the relative strength of Brownian motion weakens. For strong hydrodynamic interactions, the transition from thinning to thickening behavior near Pe = 5 coincides with the rapid decay of the Brownian contribution η^B and growth of the hydrodynamic contribution η^H, demonstrating that in fact it is only the hydrodynamic contribution that gives rise to the thickening behavior. This can be understood in terms of hydrodynamic mobility: in the presence of strong hydrodynamic interactions, the relative mobility between particles is zero at contact. This lubrication interaction prevents contact, thus permitting no contribution by the interparticle viscosity η^P.

The micromechanical origin of this behavior is encoded in the statistical distribution of colloidal particles, which is governed by a Smoluchowski equation. In the dilute limit, the pair Smoluchowski equation governs their distribution [39]:

$$\nabla \cdot \left[\mathbf{D} \cdot \left(\frac{1}{2k_B T} \mathbf{F}^{\text{ext}} g + \nabla g \right) \right] = 0. \tag{3.22}$$

The boundary conditions associated with this equation are zero relative flux at particle contact and no long-range order. Equation (3.22) gives a generic form, applicable for either a fixed-force or fixed-velocity probe. In the latter case, one must simply solve the appropriate force balance and linearity relation between hydrodynamic force and particle velocity to obtain the external force in terms of the imposed velocity [39, 62].

The contour plots in Fig. 3.6 give the microstructural solutions to Eq. (3.22) for several values of the Péclet number, with Pe increasing from left to right in the figure. The top row gives the microstructural perturbation in the absence of hydrodynamic interactions, $\lambda \to \infty$. The second row reflects strong hydrodynamic interactions, $\lambda = 1$. The volume excluded by the probe is shown in black, and the probe moves from left to right. Red regions indicate a concentration of bath particles above the equilibrium number density; blue indicates a depletion; and green represents the equilibrium microstructure. In the linear-response regime, Pe $\ll$ 1, the perturbation is a diffusive dipole. As the forcing strength increases, particles are swept downstream and the symmetry is broken. The downstream (blue) depletion region shrinks, becoming a defined particle-poor wake that trails the probe. Particles accumulate on the upstream face of the probe (shown in red), forming a particle-rich boundary layer that thins as forcing strength grows. In both the absence and presence of hydrodynamic interactions, the boundary layer thins as Pe increases. In the case of strong hydrodynamics, when Pe becomes large the reduction in boundary-layer thickness allows the probe and bath particles to approach one another closely enough to experience lubrication interactions. Not only does this inhibit their relative approach, separation is also difficult, effectively causing the probe to drag particles along—a phenomenon absent at long times when hydrodynamic interactions are weak. It is this combination of squeeze flow and suction that leads to force (shear) thickening [7, 40].

Hydrodynamic interactions thus have a pronounced effect on distortions in the nonequilibrium microstructure. Since hydrodynamic lubrication reduces both the relative rates of advection and diffusion to zero near interparticle contact, the boundary layer that forms at high Péclet number is noticeably more diffuse for $\lambda = 1$ than $\lambda \to \infty$. It has been shown [39] (for negligible hydrodynamic interactions) that the thickness of the boundary layer scales as Pe^{-1} while the concentration of bath particles within scales as Pe. But in the presence of strong hydrodynamic interactions, $\lambda = 1$, the concentration of particles inside the boundary layer is smaller, which mirrors the weaker gradients near contact [40, 62]. Wake structure is also affected by hydrodynamic interactions. When they are strong (bottom row in Fig. 3.6), the boundary layer separates from the probe later than when they are weak (top row). This has an important consequence for the rheology, as the hydrodynamic contribution to the effective viscosity depends foremost on the total number of particles near the probe and not directly on microstructural asymmetry. When forcing becomes stronger, the boundary layer remains attached longer, wrapping around the probe into the downstream region.

We return briefly to our earlier discussion about passive microrheology and the connection (or lack thereof) between the high- and low-frequency viscosity and the short- and long-time self-diffusivity. For dilute dispersions of hard spheres, the so-called high-frequency viscosity for microrheology is given by $\eta^H(\omega \to \infty) \equiv \eta'_\infty = (1 + 1.83\phi)\eta$ in the limit of full hydrodynamic interactions [40] and arises simply due to the presence of the particles and is independent of deformation of their configuration; it is a strictly hydrodynamic phenomenon. In this linear-response regime ($\mathrm{Pe} \ll 1$), and in the opposite limit of steady probe motion, $\eta^H(\omega \to 0) = (1 + 2.1\phi)\eta$ (for a constant-force probe) [21, 40]. In the absence of hydrodynamics, the pertinent values are $\eta^P(\omega \to \infty) = (1 + 2.0\phi)\eta$ and $\eta^P(\omega \to 0) = (1 + 1.0\phi)\eta$. *The correspondence between the short- and long-time self-diffusivity and the infinite- and zero-frequency microviscosity is exact.* This important result can be understood in the following way. Recall from Sect. 2.3 that in passive microrheology, the mean-squared displacement of a diffusing probe particle is proportional to the linear viscoelastic creep compliance of the material in which it is immersed, $J(t) \sim t/\eta^{\mathrm{eff}}$ [83], and thus to the complex mobility (Fig. 3.8 [103]). The creep compliance is a memory kernel that encodes the temporal dependence of relaxation for a material. At short times, when no relaxation can occur, it corresponds to the infinite-frequency viscosity (cf. Fig. 3.8a). A probe diffusing over long distances, however, must await the relaxation of structure over longer times, corresponding to the zero-frequency, or steady, limit. This is precisely what one expects from the linear-response theory: the thermal motion giving rise to diffusion is the same as the mean motion due to weak external force.

Constant External Force Versus Constant External Velocity Thus far we have discussed the motion of a probe driven by a fixed external force and the corresponding deformation of the microstructure through which it moves. Alternatively, one could drive the probe at a fixed velocity, **U**. Fixed-force experiments are typically carried out by driving the probe with magnetic tweezers [37]. In fixed-velocity

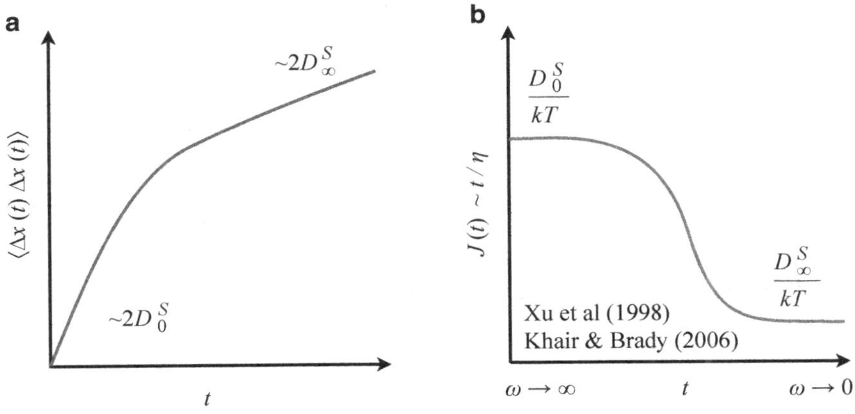

Fig. 3.8 (**a**) Transition from short- to long-time diffusive behavior. (**b**) Correspondence of the short- and long-time self-diffusion to the infinite- and zero-frequency microviscosity [40, 103]

experiments, the probe can be held in a stiff optical trap [38] while the bath is moved past it at a fixed rate.[4] In the fixed-velocity case, the probe's motion is prescribed and so it cannot diffuse, which has interesting dynamical consequences [39, 62]. It was recently shown that hydrodynamic interactions cause dissimilarities in the microstructural evolution and hence quantitative differences in the microviscosity [62]. In the fixed-force case, however, the probe velocity fluctuates as it moves through the bath. As a result, the probe experiences a collision-induced diffusive spread of its trajectory [42]. The constant-force probe is the more biologically relevant mode, e.g., for gravitational settling or propulsion via a motor protein, and gives rise to diffusive probe motion—a subject to which we now turn our attention.

3.3 *Force-Induced Diffusion: Microdiffusivity*

The same collisions between probe and bath particles that reduce mean probe speed also cause fluctuations in probe velocity, similar in many ways to shear-induced diffusion in non-colloidal [104, 105] and colloidal [25, 106] suspensions. Zia and Brady [42] found that collisions between probe and bath particles cause the probe to undergo a random-walk process: the long-time mean-square fluctuational motion of the probe is diffusive. They derived an analytical expression for the force-induced

[4] The idea of a fixed force or fixed velocity is an approximation, the accuracy of which is dictated by the uniformity of the applied field or the stiffness of the optical trap. Such approximations can be made quite accurate [37, 38].

diffusion, or "microdiffusivity" for a dilute suspension of colloids, in the absence of hydrodynamic interactions:

$$\frac{\mathbf{D}^{\text{eff}}}{D_0} = \left[\mathbf{I} - \frac{3}{4\pi}\phi_b \left(1 + \frac{a}{b}\right)^2 \oint_{r=a+b} \mathbf{n}\mathbf{d}\, d\Omega \right]. \tag{3.23}$$

In the limit Pe $\to 0$, the entropically hindered diffusion of a particle in a dilute suspension without hydrodynamic interactions is recovered, $D^{\text{eff}} = D^s_\infty = D_0(1 - 2\phi_b)$ [6, 69], for $a/b = 1$. Here, D_0 is the diffusivity of an isolated probe alone in a solvent. To separate passive diffusion from the force-induced diffusion, the hindered diffusion $-2\phi_b D_0$ is extracted from the effective diffusivity

$$\frac{\mathbf{D}^{\text{eff}}}{D_0} = (1 - 2\phi_b)\,\mathbf{I} + \mathbf{D}^{\text{micro}}, \tag{3.24}$$

where the diffusive motion of the probe due to external forcing, the *microdiffusivity*, is given by

$$\frac{\mathbf{D}^{\text{micro}}}{D_0} = \frac{3}{4\pi}\phi_b \left(1 + \frac{a}{b}\right)^2 \oint \mathbf{n}\mathbf{d}'\, d\Omega, \tag{3.25}$$

where $\mathbf{d}'$ is the fluctuation vector describing the collision-induced fluctuations in probe motion [42]. The effective diffusivity of a tracer particle is its bare diffusivity, $D_0\mathbf{I}$, minus the entropic hindrance of the bath, $2\phi_b D_0 \mathbf{I}$, plus an enhancement due to mechanical scattering by the other bath particles, $\mathbf{D}^{\text{micro}}$. The force-induced microdiffusivity is proportional to the number density of bath particles, the isolated probe self-diffusivity, and to the first moment of the hard-sphere deflections. They showed that the deflection field obeys a Smoluchowski equation forced by gradients in the microstructure:

$$\nabla^2 \mathbf{d}' - \text{Pe}\,\hat{\mathbf{u}} \cdot \nabla \mathbf{d}' = \frac{2}{(1 + a/b)} \nabla g, \tag{3.26}$$

with a no-flux condition at contact and no fluctuations occurring at infinite separations. The fluctuation field $\mathbf{d}'$ is a probability-weighted displacement. While the steady microstructure shown in Fig. 3.6 describes the likelihood of a collision, the fluctuation field describes the likelihood that such a collision will produces a deflection of a certain magnitude in a given direction. That is, the former describes the likelihood of a diffusive "kick," while the latter describes the size and direction of that kick. Contour plots of the fluctuation field governed by Eq. (3.26) are given in Fig. 3.9, which illustrate the fluctuations in microstructure that give rise to probe diffusion. The top row represents fluctuations that cause diffusion along the direction of external forcing, while the bottom row gives fluctuations that give rise to transverse diffusion. The fluctuation field is shown for several values of the Péclet number, with the probe (excluded volume) as the black center moving from

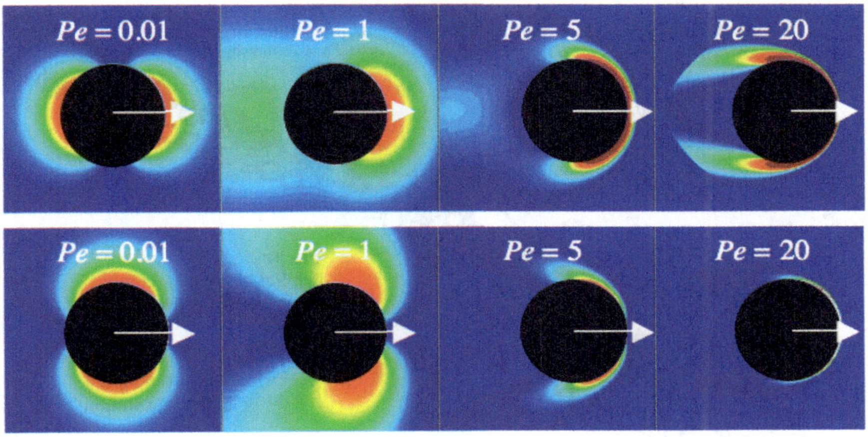

Fig. 3.9 The fluctuation field longitudinal (*top row*) and transverse (*bottom row*) to probe forcing. *Blue areas* indicate regions of weak or no deflection; *red areas* indicate probability of strong deflection. Figure from [42], reprinted with permission

left to right. Areas in red indicate likelihood of strong deflections, and areas in blue indicate low probability of deflections.

The diffusion tensor can also be resolved into orthogonal components parallel and transverse to the direction of forcing, $D_{\parallel}^{\text{micro}}$ and $D_{\perp}^{\text{micro}}$, respectively. At small Pe Brownian motion dominates and the diffusive behavior of the probe characteristic of passive microrheology is recovered but with an incremental flow-induced microdiffusivity that scales as $D^{\text{micro}} \sim D_0 \text{Pe}^2 \phi_b$:

$$D_{\parallel}^{\text{micro}} = \frac{79}{180}\left(1+\frac{a}{b}\right)^2 D_0\,\text{Pe}^2\phi_b, \tag{3.27}$$

$$D_{\perp}^{\text{micro}} = \frac{11}{60}\left(1+\frac{a}{b}\right)^2 D_0\,\text{Pe}^2\phi_b. \tag{3.28}$$

The force induced microdiffusivity is anisotropic, with diffusion longitudinal to the direction of forcing larger in both limits compared to transverse diffusion but more strongly so in the high-Pe limit. At the other extreme of high Péclet number the fluctuational motion is still diffusive, and the diffusivity becomes primarily "force induced", scaling as $D^{\text{micro}} \sim D_0 \text{Pe} \phi_b$:

$$D_{\parallel}^{\text{micro}} = \frac{1}{4}\left(1+\frac{a}{b}\right)^2\left(\ln 2-\frac{1}{4}\right) D_0\,\text{Pe}\,\phi_b + O(1), \tag{3.29}$$

$$D_{\perp}^{\text{micro}} = \frac{1}{32}\left(1+\frac{a}{b}\right)^2 D_0\,\text{Pe}\,\phi_b + O(1). \tag{3.30}$$

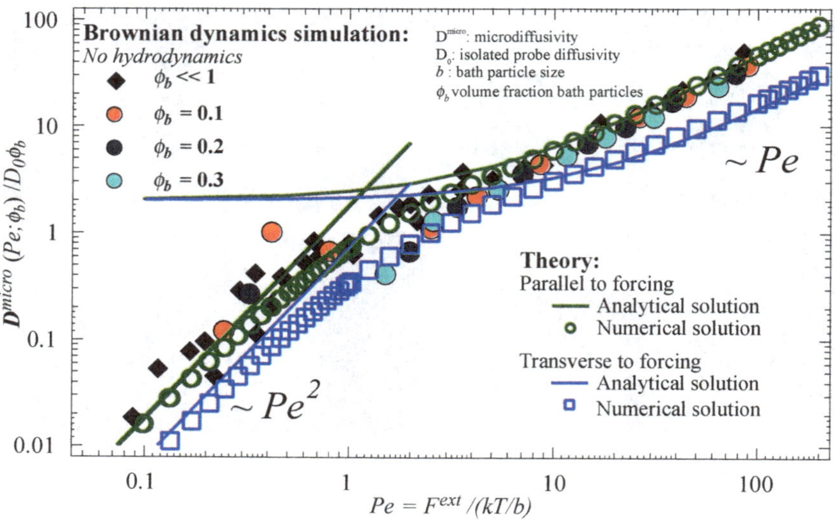

Fig. 3.10 Microdiffusivity in the absence of hydrodynamic interactions. *Solid lines*: asymptotically weak and strong forcing [Eqs. (3.27)–(3.30)]. *Open symbols*: numerical solution of (3.25) and (3.26). *Filled symbols*: Brownian dynamics simulations. Data from [42], reprinted with permission

As in the low-Pe limit, the large-Pe microdiffusivity is also transversely anisotropic, with a longitudinal-to-transverse preference of approximately $7/2$. The effect of the hard-sphere collisions is a Pe-large diffusive scattering of the probe's trajectory.

A plot of the microdiffusivity components is shown in Fig. 3.10. For weak probe forcing, Brownian motion dominates the behavior, and the probe essentially goes nowhere on average. In the other extreme of high Péclet number the fluctuational motion is still diffusive, and the diffusivity becomes primarily force induced, scaling as $(F^{ext}/\eta)\phi_b$, where η is the viscosity of the solvent; its scattering off the mean path is Pe-large. For microscopic particles in biological systems undergoing such motion, the ideal range of Pe for coherent motion is thus Pe ~ 1, as shown by the numerical solutions to (3.25) and (3.26) (the open symbols in the figure), which agree closely with the asymptotic limits.

The dilute theory results can also be scaled up to the higher concentrations relevant in biological systems, as shown by the filled circles in the figure, representing the microdiffusivity obtained via Brownian dynamics simulation [42], in which the Péclet number was scaled on the long-time self-diffusivity. The simple rescaling gives excellent agreement between the dilute theory and the results shown for concentrated suspensions. This important result extends the applicability of the dilute theory to concentrated systems, which is critical to its use in biological complex fluids, which are, more often than not, crowded and concentrated. These latter results highlight the benefit of including dynamic simulations in theoretical approaches to microrheology. They provide an ideal "experimental" system in which all particle-level interactions—hydrodynamic as well as thermodynamic—can be rigorously and systematically controlled. Brownian dynamics simulations are discussed further in Sect. 3.6.

3.4 A Complete Picture: Microviscosity, Microdiffusivity, and Normal Stresses

To give a physical picture of the importance of normal stresses, one can imagine a soft particle, such as an organelle inside the cell or such as a bubble, placed in a suspension. If the particle is compressible, then a change in the osmotic pressure will cause it to shrink or expand—and possibly burst—due to the particle pressure. If the particle is soft, normal stress differences will elongate it—possibly causing rupture. These effects have important implications in the consideration of nanoparticle drug delivery and the escape of viruses from the endosomal pathway, for example.

The microviscosity and microdiffusivity, as noted above, can be obtained by two simple quantities: mean and mean-square displacement of a probe driven through a complex fluid. The notion that diffusive flux is driven by stress gradients leads to the idea that the stress can be related directly to the microdiffusivity and suggests that the anisotropy of the diffusion tensor reflects the presence of normal stress differences in nonlinear microrheology [43].

The work of Einstein and Batchelor discussed in the passive-microrheology section of this chapter connected gradients in osmotic pressure and chemical potential to collective diffusion [23, 32]; these treatments applied to equilibrium Brownian suspensions. An extension of the connection between gradients in the pressure or more generally in the stress Σ to diffusion in nonequilibrium systems holds appeal in that the simple measurement of mean-square displacement would then provide a measurement of suspension stress and overcome a long-standing shortcoming of passive microrheology. A naïve first approach is simply to apply the relation $\partial\Sigma/\partial n \propto \mathbf{D}$ to nonequilibrium suspensions. While much of the correct qualitative behavior is recovered, it falls short of predictive theory. Zia and Brady assert that the primary feature of the equilibrium model limiting its generalization to nonequilibrium systems is that it was derived from an equation of state, while for nonequilibrium systems an equation of motion is required. Cauchy's momentum balance is the appropriate equation of motion of the suspension. Zia and Brady derived this connection, finding that stress gradients drive both diffusive and advective flux of the probe, corresponding to the microdiffusivity and the microviscosity, respectively [43]:

$$-\frac{\Sigma^{\mathrm{neq}}}{n_a k_{\mathrm{B}} T} = \left(2 - \frac{\eta^{\mathrm{micro}}}{\eta\phi_b}\right)\frac{\mathbf{D}^{\mathrm{micro}}}{D_0} + \frac{1}{3}\mathrm{tr}\left(\frac{\mathbf{D}^{\mathrm{micro}}}{D_0}\right)\mathbf{I}, \tag{3.31}$$

where the superscript neq indicates the nonequilibrium value and n_a is the number density of probes. They rearranged this expression to assert a *nonequilibrium Stokes–Einstein relation* [43]:

$$\mathbf{D}^{\mathrm{micro}} = -\left(\frac{\Sigma^{\mathrm{neq}}}{n_a} + \mathbf{P}\right) \cdot \mathbf{M}^{\mathrm{neq}}, \tag{3.32}$$

where the isotropic pressure term with $\mathbf{P} \equiv \mathbf{R} \cdot \mathrm{tr}(\mathbf{D}^{\mathrm{micro}})\mathbf{I}/3$. The nonequilibrium mobility tensor is given by

$$\mathbf{M}^{\mathrm{neq}} = \left(2 - \frac{\eta^{\mathrm{micro}}}{\eta \phi_b}\right)^{-1} \mathbf{M} \tag{3.33}$$

and $\mathbf{M} = \mathbf{I}/6\pi\eta a$. Like the equilibrium Stokes–Einstein relation (3.1), the nonequilibrium Stokes–Einstein relation (3.32) gives a relationship between fluctuation and dissipation. Fluctuations, $\mathbf{D}^{\mathrm{micro}}$, driven by the energy, $\Sigma^{\mathrm{neq}}/n_a$ and $\mathbf{P}$, are dissipated back to the solvent by viscous drag $(\mathbf{M}^{\mathrm{neq}})^{-1}$.

This is an important result in the quest to fully characterize complex fluids by relating tracer motion to material properties. *In order to obtain viscosity, diffusion, viscoelastic behavior, and all elements of the stress tensor, one need only track the displacements of a single-probe particle* [43].

Normal stress differences can be important in the context of single-particle forcing in many physical systems; they may cause soft particles, e.g., subcellular organelles, to elongate or even rupture. The normal stress differences are defined by $N_1 \equiv \Sigma_{zz} - \Sigma_{yy}$ and $N_2 \equiv \Sigma_{yy} - \Sigma_{xx}$, where z is the direction of probe forcing and x and y are the two orthogonal axes. According to (3.31), the first normal stress difference is then

$$-\frac{\langle N_1^{\mathrm{neq}} \rangle}{n_a k_B T \phi_b} = \frac{D_{\parallel}^{\mathrm{micro}} - D_{\perp}^{\mathrm{micro}}}{D_0 \phi_b} \left[2 - \frac{\eta^{\mathrm{micro}}}{\eta \phi_b}\right]. \tag{3.34}$$

Due to the axisymmetric structure around the probe, the second normal stress difference is zero. While normal stress differences indicate how a soft object might elongate, the osmotic pressure indicates how a compressible object might shrink or expand. The osmotic pressure is the trace of the probe-phase stress tensor; the corresponding nonequilibrium contribution is

$$\frac{\langle \Pi^{\mathrm{neq}} \rangle}{n_a k_B T \phi_b} = \frac{1}{3} \frac{D_{\parallel}^{\mathrm{micro}} + 2D_{\perp}^{\mathrm{micro}}}{D_0 \phi_b} \left[3 - \frac{\eta^{\mathrm{micro}}}{\eta \phi_b}\right]. \tag{3.35}$$

The strength of this approach is the ability to predict all of the rheologically interesting information about a material: shear thinning, shear thickening, normal stress differences, viscoelastic behavior, and force-induced diffusion—all by tracking the motion of a *single colloidal probe* [43].

One can ask how well this phenomenological theory compares to the known results for the stress obtained via the statistical mechanics theory [24, 107]:

$$-\frac{\langle\Sigma\rangle}{n_a k_B T} = 1 + \frac{3}{8\pi}\phi_b\left(1+\frac{a}{b}\right)^3 \int_{r=a+b} \mathbf{nn}\, g(\mathbf{r})d\Omega. \tag{3.36}$$

Equation (3.36) is an exact formula (for hard spheres) of the stress contribution from the probe particles no matter what the concentration of bath (or probe) particles and for any type of forcing, e.g., shearing motion or an external force for microrheology.

The predictive theory presented above in Eqs. (3.34) and (3.35) is shown in Fig. 3.11, alongside the micromechanical approach (3.36) and with a measurement of the stress via Brownian dynamics simulations [43]. The normal stress differences were compared side by side among the three approaches, showing excellent agreement (Fig. 3.11a). For large Pe, the first normal stress difference scales linearly in the forcing and in the volume fraction of bath particles, $\langle N_1\rangle/n_a k_B T \sim \text{Pe}\,\phi_b$. For very weak forcing, Pe $\ll 1$, $\langle N_1\rangle$ vanishes as $\sim O(\text{Pe}^4)$—because the Brownian motion of the bath particles easily restores deformation caused by the probe's motion, and the structure is nearly isotropic. One could expect instead that $\langle N_1\rangle$ should vanish as Pe^2, as might be predicted from the near-equilibrium, linear-response theory. But this is where the importance of normal viscous stresses plays a special role in microrheology. Without these, the decay in $\langle N_1\rangle$ for very weak forcing would indeed scale quadratically in Pe. Although viscosity plays a role in the stress due to fluctuations, the viscous drag due to mean motion always acts to slow the probe; at small Pe, the effect is a stronger suppression of the advective motion, which results in a stronger suppression of microstructural asymmetry. Preliminary results for the microdiffusivity in the presence of hydrodynamic interactions are also shown in the figure (filled triangles), obtained via Stokesian dynamics simulation [61]. Hydrodynamics appear to contribute a primarily quantitative effect. The first normal stress difference is more pronounced, as one would expect: close-range longitudinal encounters between a probe and bath particle in the lubrication limit scale as the inverse of the separation distance, while the transverse encounter produces a logarithmically weak force.

The osmotic pressure computed via the theory is shown in Fig. 3.11b, where it is also compared to the statistical mechanics and Brownian dynamics results, with excellent agreement. For large Pe, the osmotic pressure scales linearly in the forcing and in the volume fraction of bath particles. For very weak forcing, Pe $\ll 1$, the total nonequilibrium osmotic pressure asymptotes to the value predicted by the Carnahan–Starling equation of state [108]. A comparison of the osmotic pressure to the dilute theory result for macroscopically sheared suspensions shows excellent agreement [7, 43]. Overall, the similarity between micro and macro is very strong: the same scaling in Pe is evident for the full range of Pe and the quantitative agreement is good. However, fundamental differences manifest in the second normal stress difference, which is zero in microrheology, and $O(\text{Pe})$ for strong forcing in macrorheology [43, 106].

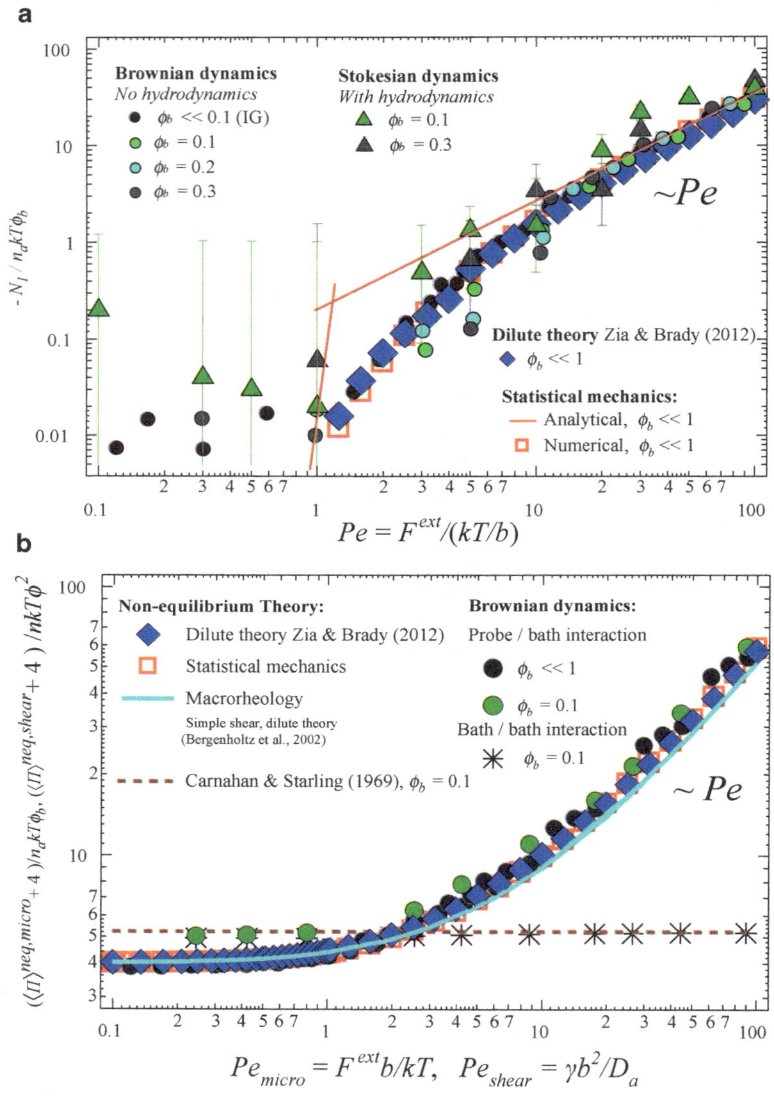

Fig. 3.11 (**a**) First normal stress difference via dilute theory, Eq. (3.34) (*filled diamonds*); statistical mechanics theory, Eq. (3.36) (*open squares*); Brownian dynamics simulation (*filled circles*); Stokesian dynamics simulations with hydrodynamic interactions (*filled triangles*). (**b**) Osmotic pressure via dilute theory, Eq. (3.35) (*filled diamonds*); statistical mechanics theory, Eq. (3.36) (*open squares*); Brownian dynamics simulation (*filled circles*) [43]. To highlight the strictly equilibrium contribution of the bath particles, the osmotic pressure of bath particles via dynamic simulation is given by the *asterisk symbols*, which are indistinguishable from the Carnahan–Starling equation of state (*dashed lines*) [108]. Data from [43], reprinted with permission by the Society of Rheology

Although the proposed theory was constructed for and compared with dilute dispersions, the physical ideas underlying the theory should hold more generally, and indeed, it can be seen in Fig. 3.11 that the proposed relation (3.31) also applies to much more concentrated colloidal dispersions, when Pe is rescaled on the long-time self-diffusivity [42, 43]. This important result holds both practical and fundamental appeal. From a practical standpoint the applicability of the theory to concentrated systems is critical to its use in many systems of interest, e.g., watery biophysical systems such as the crowded interior of the cell.

3.5 *Time-Dependent Flows*

Most work thus far in nonlinear microrheology has focused on steady dynamic behavior, to establish the relationship between steady-state microstructural mechanics and transport properties such as the diffusivity, viscosity, and stress [17, 39, 40, 42, 43]. Transient behavior has been studied in the near-equilibrium, linear-response regime—in the microrheological context via its connection to low-amplitude oscillatory probe forcing and the complex modulus [21, 80]. Such studies show that for very weak forcing, the microstructural response that drives stress (and relaxation) is indistinguishable from equilibrium fluctuations. Previous macrorheological studies of transient behavior in colloidal dispersions also reveal a range of interesting time-dependent phenomena, including temporary stress overshoot behavior, viscoelastic and memory effects such as strain recovery, and the existence of multiple relaxation modes [16].

Stress formation and relaxation in colloidal dispersions accompany the storage and loss of microstructural memory. In equilibrium colloidal dispersions, the thermal fluctuation of a single particle in a solvent is dissipated back to the solvent via viscous drag [32, 33]. The decay of viscous particle momentum occurs on the order of 10^{-7} s (for a 0.5-μm particle in water); from a particle perspective, memory loss is nearly instantaneous (for a discussion of relaxation on the solvent time scale, see e.g., Russel [73] or Hocquart and Hinch [109]). But a particle diffusing through a dispersion of other particles deforms the suspension, giving rise to stresses that relax on time scales much longer than the individual-particle momentum relaxation time. The temporal decay of stress (and velocity) fluctuations can be understood from the perspective of linear-response theory [110, 111]: while the shear stress in an equilibrium system is zero on average, thermal fluctuations produce small perturbations in microstructural isotropy, which give rise to temporary shear stresses, which are characterized by the shear stress autocorrelation function, $C_s(t) = \langle \sigma'_{xy}(t)\sigma'_{xy}(0)\rangle$ [79]. That is, stress relaxation accompanies memory decay.

A material may have multiple relaxation modes which can be probed via low-amplitude oscillatory displacements, for example. The high-frequency limit corresponds to the shortest relaxation time scale of the material; for hard-sphere colloidal dispersions the stress decays as $t^{-1/2}$ at very short times (in the absence of hydrodynamic interactions) [19–21, 80]. Nonequilibrium transient behavior has also been studied experimentally for sheared dispersions, where it has been shown that multiple mechanisms play a role in suspension stress and viscosity e.g., hydrodynamic, interparticle, and Brownian forces—and give rise to distinct relaxation processes. For example, Mackay, Kaffashi, and coworkers [12, 15] studied the decay of stress immediately after the cessation of imposed strain rate on a sheared suspension; they found that the hydrodynamic stress decays instantaneously, as it must—the hydrodynamic stress is proportional to the imposed strain rate, and thus must vanish in the absence of flow. Watanabe and coworkers [13, 14] analyzed stress development and relaxation in sudden start-up and cessation of shearing flow and found both short- and long-time relaxation modes. And although at long times after flow has been initiated a colloidal dispersion behaves as a viscous fluid, at early times during the startup of the flow, an "overshoot" in the material stress or viscosity may be observed [16], mirroring the microstructural rearrangements responsible for the transition from elastic to viscous behavior. When the flow is shut off, one would expect the elastic behavior to manifest as creep recovery: removal of the force should allow the microstructure to return to a previous state. These studies show that the macroscopic stress relaxes via distinct transport processes, but the microstructural evolution that accompanies this relaxation was not as thoroughly studied until recently.

The formation and relaxation of nonequilibrium stress in colloidal dispersions were recently studied by Zia and Brady using the framework of nonlinear microrheology [17]. The comparison between transient nonlinear macrorheology and microrheology shows that in both cases, the early stress evolution scales with the square root of time and shows the same shear (force) thinning over a range of Pe [16]. In both micro- and macrorheology, the long-time relaxation behavior collapses onto a single curve, indicating that memory of the initial condition is lost. After the forcing is removed, the motion of the probe and suspension do not cease instantaneously. The suspension relaxation occurs over at least two time scales: first, an initially rapid decay in stress that accompanies diffusive relaxation of the boundary layer. Stronger departures from equilibrium recover faster: the stress relaxation time depends on the boundary-layer thickness, which it must, because the primary contribution to suspension stress is due to near-contact particle interactions. The dilute microrheology theory, in the absence of hydrodynamic interactions, captures the physics of the relaxation process: the pair-level equation obtained for the relaxation of the viscosity also found agreement with the concentrated suspension in accompanying simulations [17]. That is, the relaxation rate of the boundary layer is evidently independent of volume fraction. These results are consistent with similar findings by Foss [16].

Creep recovery after shutoff provides insight into the connection between stress, free-energy storage, and entropic memory. During start-up and at steady state, the probe's motion compresses the particle distribution, restricting its entropy; the free-energy change required to do so is released by the microstructure as it relaxes and regains access to more configuration states. As shown by Zia and Brady [17], when the force is abruptly removed, not all of the strain can be recovered; the remaining energy is dissipated viscously to the solvent by the probe and the bath particles. The storage is entropic in origin (not enthalpic), and this is the origin of nonequilibrium suspension stress. They showed that the connection between the entropic force and the osmotic compressibility supports this view of the stress as energy storage.

Recent work in large-amplitude oscillatory flows is useful for understanding the relative importance of different forces, conservative and nonconservative, in determining unsteady and nonlinear flow properties. Recent work includes large-amplitude oscillatory forcing (LAOF) microrheology, in which a colloidal dispersion is deformed by a probe particle driven with an oscillatory force of arbitrary magnitude and frequency [22]. This is the microrheological analogue to the large amplitude oscillatory shear (LAOS) experiment.

3.6 Brownian Dynamics Simulations

An alternative approach to the Smoluchowski theory emphasized thus far is Brownian dynamics, in which one examines the detailed motion of the individual particles; no assumption of diluteness is then required. Here, the dynamics are governed by the Langevin equation, a stochastic force balance that includes Brownian, external, hydrodynamic, and other interparticle forces:

$$\mathbf{M}\cdot\frac{d\mathbf{U}}{dt}=\mathbf{F}^{H}+\mathbf{F}^{\text{ext}}+\mathbf{F}^{B}+\mathbf{F}^{P}, \tag{3.37}$$

where $\mathbf{M}$ is the mass (or moment of inertia) tensor and $\mathbf{U}$ is the particle velocity. The left-hand side is zero because inertia is not important for colloidal dispersions. Periodic replication of the cell simulates an infinite domain. On the right-hand side are the forces that act on a particle, which include the hydrodynamic drag $\mathbf{F}^{H}$, along with external, Brownian, and interparticle forces, $\mathbf{F}^{\text{ext}}$, $\mathbf{F}^{B}$, and $\mathbf{F}^{P}$, respectively. The external force (or probe velocity) is prescribed, and the Brownian force obeys the appropriate statistics. The velocity of the particle $\mathbf{U}=(\mathbf{F}^{\text{ext}}+\mathbf{F}^{B}+\mathbf{F}^{P})/6\pi\eta\, a$ can then be integrated forward in time to obtain a particle's displacement. To begin, a probe of size a is placed among a randomly distributed bath of particles of size b in the simulation cell. At each time step every particle is given a randomly directed Brownian displacement, simulating a continuum Newtonian solvent of viscosity η. The probe is also displaced at each time step in the direction of the external force $\hat{\mathbf{F}}^{\text{ext}}$. When two particles contact one another, the hard-sphere collision is treated via a "potential-free" algorithm [102, 112], where the overlap between two particles is prevented by separating the colliding pair along their line of centers until they are no longer in contact. The collision contributes to the particle's displacement and velocity and to the average stress in the suspension. For a complete description of Brownian dynamics of active microrheology, see Carpen and Brady [102].

To obtain the microviscosity, the probe velocity $\langle\mathbf{U}\rangle$ and applied force $\langle\mathbf{F}^{\text{ext}}\rangle$ are the quantities that are directly computed in simulation and theory and measured in experiment: in the constant-force case one measures the reduction in probe speed due to collisions with the background bath particles [17, 102]. As before (cf. Sect. 3.2) we interpret the reduced mean speed as an effective viscosity of the dispersion. Recall the definition of the effective viscosity for constant external force:

$$\frac{\eta^{\text{eff}}}{\eta} = \frac{F^{\text{ext}}}{6\pi\eta a\langle U\rangle}.$$

Here, $\langle U\rangle = U^0 + \langle U^P\rangle$ and $U^0 = F^{\text{ext}}/6\pi\eta a$, where the bath particles slow the probe's mean speed, $\langle U^P\rangle < 0$. Thus the mean speed of the probe provides a measure of the effective viscosity [17, 102]:

$$\frac{\eta_F^{\text{eff}}}{\eta} = \left[1 + \frac{\langle U^P\rangle}{F^{\text{ext}}/6\pi\eta\, a}\right]^{-1}, \tag{3.38}$$

and the *reduction* in mean speed gives a measure of the particle contribution to the viscosity, η_F^P:

$$\frac{\eta_F^P}{\eta} \equiv -\frac{\langle U^P\rangle}{F^{\text{ext}}/6\pi\eta\, a}. \tag{3.39}$$

The subscript F indicates constant-force mode. Finally, in the case of a dilute bath, the interparticle contribution to the viscosity measured in simulation is identical to the microviscosity defined by theory,

$$\frac{\eta_F^P}{\eta} = \frac{\eta^{\text{micro}}}{\eta}, \quad \phi_b \ll 1. \tag{3.40}$$

The average noted by the angle brackets indicates a time average over the duration of the simulation, given a sufficient duration for the probe to encounter a statistically large number of microstructural arrangements. Equivalently, the ensemble average of an instantaneous velocity can be computed across simulations to achieve the same result.

For the microdiffusivity, the trajectory of the probe can be measured directly, and the ensemble average across many simulations of its mean, $\langle x\rangle$, and mean-square displacement $\langle \Delta x^2\rangle$ can be computed to obtain the total or effective probe diffusivity:

$$\mathbf{D}^{\text{eff}}(\text{Pe}) = \frac{1}{2}\frac{d\langle \Delta\mathbf{x}\Delta\mathbf{x}\rangle}{dt}. \tag{3.41}$$

The passive diffusion of a tracer is recovered when $\text{Pe} = 0$; this value may then be subtracted from the effective diffusivity in Eq. (3.41) to obtain the force-induced portion of probe diffusion, $\mathbf{D}^{\text{micro}}$. The procedure is entirely analogous to that used in an experiment. The results of Brownian dynamics simulation conducted by Zia and Brady [42, 61] are shown in Fig. 3.10. As is seen in the plot, no assumption of diluteness is required in such simulations, allowing one to test the scaling-up of the dilute theory to concentrated suspensions. As noted above, defining the Péclet number in terms of the long-time (passive) self-diffusivity, rather than the bare diffusion, collapses the data for concentrated dispersions onto the dilute theory.

The nonequilibrium contribution to the particle stress due to a collision between the probe and a bath particle is $\langle \Sigma \rangle = -n_a \langle \mathbf{r}\mathbf{F}^P \rangle$ where the angle brackets $\langle \cdot \rangle$ indicate an average over the duration of the simulation. Here, $\mathbf{r}$ is the collisional displacement along the line of centers of the colliding particles. Simulations conducted by Zia and Brady, shown in Fig. 3.11, were conducted for a range of Pe and at concentrations up to a volume fraction of bath particles $\phi_b = 0.45$ [43]. The scaling in volume fraction ϕ_b appears to hold for much more concentrated systems as is seen in the plot of the first normal stress difference (Fig. 3.11a) and in the plot of the osmotic pressure (Fig. 3.11b). The good agreement adds weight to the supposition that Eq. (3.31) is general and not restricted to dilute colloidal dispersions. This important outcome has both practical and fundamental implications. In practice the theory predicts that the relation (3.31) holds for a large range of concentrations which opens a wide class of natural and engineered complex media for investigation.

It is important to ask whether collisions between probe and bath provide a measure of stress that corresponds to values measured via bulk rheology; that is, do collisions between the bath particles themselves matter? In a dilute bath, interactions between the bath particles make an $O(\phi_b^2)$ contribution to the nonequilibrium osmotic pressure, which is small for a dilute bath; hence, the probe-phase pressure dominates the pressure of all the particles (probe plus background bath particles) [43]. This simple scaling prediction is borne out by the Brownian dynamics simulations, in which the bath particles interact via excluded-volume interactions. The contribution to the osmotic pressure of both probe and bath particles can be monitored during simulations. The stress measured by probe/bath-particle collisions, $\langle \Sigma \rangle / n_a k_B T \phi_b$, and that for bath/bath-particle collisions, denoted $\langle \Sigma \rangle_{bb} / n_b k_B T \phi_b$, was recorded. The osmotic pressure thus obtained, as a function of Pe, is shown alongside those for the dilute bath in Fig. 3.11b. While the osmotic pressure measured by the probe, $\langle \Pi \rangle / n_a k_B T$, increases linearly with Pe, the osmotic pressure due to the bath/bath-particle interactions $\langle \Pi \rangle_{bb} / n_b k_B T$ remains at the equilibrium value regardless of Pe and is indistinguishable from the equilibrium osmotic pressure predicted by the Carnahan–Starling equation of state [108]. Zia and Brady also found this to hold at higher concentrations. The behavior seen in the figure reflects the fact that most of the bath particles interact with each other due to Brownian motion only, whereas the probe and bath particles interact due to the external forcing. In order to understand if this matches what one would expect from traditional macrorheology experiments, the dilute theory results for a bath undergoing simple shear [7] are plotted alongside the microrheology results in figure; the two agree.

4 A "Non-equilibrium Equation of State"

The work reviewed in Sect. 3.4 suggests that nonlinear microrheology may provide more than a microscale technique for interrogating complex fluids. As discussed previously in the introduction to this chapter, it provides a window through which one can view the atomic (microstructural) world. In 1906, Einstein invented the "passive microrheology" theory to prove the atomic nature of matter. He proposed that if matter were indeed composed of atoms, then the motion of a small particle suspended in a fluid would move with the same random trajectories as the solvent particles. Combining the theory from kinetics, diffusion, and thermodynamics he gave the relationship between fluctuation and dissipation at equilibrium, which showed that the diffusive motion of a small particle is indeed evidence of the existence of the atom. Perrin confirmed the theory with measurement in 1909. This is a profound conclusion, drawn by simply watching the motion of a single particle in a fluid. Together, theory and experiment produced the numerical value for Avogadro's number by relating the bulk temperature at the macroscale to the energy of individual particles at the microscale. Thus the ideal gas equation of state $PV = \bar{n}\mathscr{R}T$ (where $\bar{n}$ is the number of moles) became $P = nk_BT$ (where n is the number density of particles). Inspired by this, others have watched a particle move in a complex fluid—but now for a system that is not at equilibrium. By studying fluctuations away from equilibrium, Zia and Brady have proposed an analogous non-equilibrium relation between fluctuation and dissipation, shown in Eq. (3.32) [43]. It can be understood as follows.

Kinematically, the diffusivity is given by the sum of correlations in velocity fluctuations over time,

$$D \sim \int U'U'dt \sim (U')^2\tau, \tag{3.42}$$

where τ is the characteristic time scale for the decay in the velocity correlations. For Brownian motion, $(U')^2 \sim k_BT/m$ and $\tau \sim m/6\pi\eta a$, where m is the particle mass, and the usual Stokes–Einstein diffusivity is recovered. But more generally, the magnitude of the velocity fluctuations defines the "temperature," $T \sim U'U'$, whence [43]:

$$\Sigma \sim n_a C T^{\text{eff}}. \tag{3.43}$$

They define the proportionality factor $C \equiv (M^{\text{neq}})^{-1}\tau$ with the "effective temperature" of the dispersion given by particle velocity fluctuations, $T^{\text{eff}} \equiv \langle U'U' \rangle$. The proportionality of the stress to $\langle U'U' \rangle$ prompted them to interpret the stress, or osmotic pressure, as the energy density of the particle phase. Zia and Brady thus interpreted Eq. (3.43) can thus also be interpreted as a "non-equilibrium equation of state" for the particle phase, with C the factor of proportionality that connects the microscopic energy density to the macroscale effective temperature. They also proposed that one may also view the time dependence τ of the factor C as the time scale over which the energy stored by the microstructure is dissipated [17]. The proposed generality of this relation may be important in unifying the analysis, understanding, and treatment of a wide class of soft materials.

5 Experimental Measurement

We have reviewed a theoretical framework for microrheology that relates the viscosity, diffusion, and particle-phase stress of a colloidal dispersion to the mean and mean-square displacement of a probe driven through the material, which can be of practical use in making physical measurements. As noted below, detailed knowledge of the embedding material is not required. In order to obtain the stress tensor, the only quantity which must be measured is the total displacement of the probe over time. From this, the average speed $\langle \mathbf{U} \rangle = d\langle \mathbf{x} \rangle / dt$ can be computed to obtain the microviscosity via the relation

$$\frac{\eta_F^{\mathrm{micro}}}{\eta} = \frac{6\pi\eta a}{\langle F^{\mathrm{ext}} \rangle_F} \langle \mathbf{U} \rangle, \tag{3.44}$$

where $\langle F^{\mathrm{ext}} \rangle_F$ is a constant external force applied to the probe and η_F^{micro} is the microviscosity. From the total and mean displacement one may then compute the microdiffusivity as

$$\mathbf{D}^{\mathrm{micro}} = \frac{1}{2} \frac{d}{dt} \langle \mathbf{x}'(t) \mathbf{x}'(t) \rangle, \tag{3.45}$$

where $\mathbf{x}'(t) \equiv \mathbf{x}(t) - \langle \mathbf{x} \rangle (t)$ is the displacement from the mean as function of time and the angle brackets signify an average over realizations. The development of the relation (3.31) is sufficiently general as to apply to a very wide class of complex media, ranging from colloidal dispersions to gels to solids with suspended dislocations. Combining (3.31) with (3.44) and (3.45), Zia and Brady proposed that the stress in a general material can be obtained by measuring the mean and mean-square motion of a probe driven through the medium by a constant external force:

$$-\frac{\Sigma^{\mathrm{neq}}}{n_a k_{\mathrm{B}} T} = \left(2 - \frac{\eta_F^{\mathrm{micro}}}{\eta \phi_b} \right) \frac{\mathbf{D}^{\mathrm{micro}}}{D_0} + \frac{1}{3} \mathrm{tr} \left(\frac{\mathbf{D}^{\mathrm{micro}}}{D_0} \right) \mathbf{I}. \tag{3.46}$$

It should be noted that for $\mathrm{Pe} \lesssim 1$, the strong influence of Brownian motion can make it difficult to detect the small differences between the longitudinal and transverse microdiffusivity and hence between the corresponding normal stresses. It is thus recommended that the regime most suitable for experimental measurement using this analysis is $\mathrm{Pe} \gtrsim 1$.

6 Summary

We have presented a broad overview of theoretical microrheology as an approach for interrogating and understanding the behavior and material properties of complex fluids, with a view toward application to complex biological fluids. To the extent that biological fluids span an enormous range of complexity, we have attempted to elucidate the applicability (and limitations) of both the passive and active forms of microrheology.

In passive microrheology, we saw that the Stokes–Einstein relation is a powerful technique for probing continuum materials and that its applicability can be extended via a frequency-dependent version, due to Mason and Weitz, to interrogate a wider range of biologically relevant materials. When structural and dynamical heterogeneity in a material preclude the use of either of these relations, the dual-probe model of Crocker and co-workers has been shown to effectively correlate the motion of a pair of tracers to the bulk rheology the some very complex materials. In all of these models, however, one major shortcoming remains: the ability to understand nonlinear material properties. Microrheology provides powerful complement to more traditional models for the study of biological membranes, for example that presented next, in Chap. 4.

To this end, Squires and Brady introduced the theory of active, nonlinear microrheology and showed that the "microviscosity" measured via single-particle motion and an interpretation of Stokes' drag law showed, in the absence of hydrodynamic interactions, close qualitative and near-quantitative agreement with the pair-level particle-phase contribution to material viscosity obtained via traditional macroscale shear rheology. Khair and Brady extended this model to study linear viscoelasticity, and later to include hydrodynamic interactions, recovering the shear-thickening behavior that plays a prominent role in complex fluid rheology. However, neither study had yet addressed the effect of probe fluctuations. More importantly, these studies were able to obtain only scalar expressions for the viscosity—precluding any interpretation of the material stress.

Zia and Brady took on this challenge, formulating an expression for the stress tensor measured via microrheology given in terms of two contributions: fluctuation (the microdiffusivity) and dissipation (the viscosity). They proposed that this non-equilibrium fluctuation-dissipation relation constitutes a "nonequilibrium equation of state," thus providing the desired relation between particle tracking and rheology in *nonequilibrium* colloidal dispersions. The dilute theory accurately predicts behavior in much more concentrated colloidal dispersions, increasing its appeal and practical value as a tool for interrogating the typically concentrated and crowded environments found in complex biological fluids.

References

1. E.M. App, R. Kieselmann, D. Reinhardt, H. Lindemann, B. Dasgupta, M. King, P. Brand, Chest **114**(1), 171 (1998)
2. R. Banerjee, J.R. Bellare, R.R. Puniyani, Biochem. Eng. J. **7**(3), 195 (2001)
3. R. Wüstneck, N. Wüstneck, B. Moser, U. Pison, Langmuir **18**(4), 1125 (2002)
4. S.K. Lai, Y.Y. Wang, D. Wirtz, J. Hanes, Adv. Drug Deliv. Rev. **61**(2), 86 (2009)
5. H.A. Barnes, H.F. Hutton, K. Walters, *An Introduction to Rheology* (Elsevier, Amsterdam, 1989)
6. G.K. Batchelor, J. Fluid Mech. **74**, 1 (1976)
7. J. Bergenholtz, J.F. Brady, M. Vicic, J. Fluid Mech. **456**, 239 (2002)
8. H.A. Barnes, J. Rheol. **33**(2), 329 (1989)
9. P. D'Haene, J. Mewis, G. Fuller, J. Colloid Interface Sci. **156**(2), 350 (1993)
10. J.W. Bender, N.J. Wagner, J. Colloid Interface Sci. **172**(1), 171 (1995)
11. B.J. Maranzano, N.J. Wagner, J. Chem. Phys. **117**(22), 10291 (2002)
12. M.E. Mackay, B. Kaffashi, J. Colloid Interface Sci. **174**(1), 117 (1995)
13. H. Watanabe, T. Sato, K. Osaki, Macromolecules **29**, 3890 (1996)
14. H. Watanabe, M..L. Yao, Y. A., K. Osaki, T. Shikata, H. Niwa, Y. Morishima, Rheo. Acta **35**(11), 443 (1996)
15. B. Kaffashi, V.T. O'Brien, M.E. Mackay, S.M. Underwood, J. Colloid Interf. Sci. **187**(1), 22 (1997)
16. D.R. Foss, Rheological behavior of colloidal suspensions: the effects of hydrodynamic interactions. Ph.D. thesis, California Institute of Technology, 1999
17. R.N. Zia, J.F. Brady, J. Rheol. **57**(2), 457 (2013)
18. A.C. Pipkin, *Lectures on Viscoelasticity Theory* (Springer, New York, 1986)
19. R.A. Lionberger, W.B. Russel, J. Rheol. **38**(6), 1885 (1994)
20. J.F. Brady, J. Chem. Phys. **99**(1), 567 (1993)
21. A.S. Khair, J.F. Brady, J. Rheol. **49**(6), 1449 (2005)
22. J.W. Swan, R.N. Zia, J.F. Brady, J. Rheol. **58**(1), 1 (2014)
23. G.K. Batchelor, J. Fluid Mech. **52**(2), 245 (1972)
24. J.F. Brady, J. Chem. Phys. **98**(4), 3335 (1993)
25. J.F. Morris, J.F. Brady, J. Fluid Mech. **312**, 223 (1996)
26. C. Gao, S.D. Kulkarni, J.F. Morris, J.F. Gilchrist, Phys. Rev. E. **81**(4), 041403 (2010)
27. X. Cheng, J.H. McCoy, J.N. Israelachvili, I. Cohen, Science **333**(6047), 1276 (2011)
28. G.L. Lukacs, P. Haggie, O. Seksek, D. Lechardeur, N. Freedman, A.S. Verkman, J. Biol. Chem. **275**(3), 1625 (1999)
29. A.S. Verkman, Trends Biochem. Sci. **27**(1), 27 (2002)
30. J. Suh, D. Wirtz, J. Hanes, PNAS **100**(7), 3738 (2003)
31. J.R. Heath, M.E. Davis, L. Hood, Sci. Am. **300**, 44 (2009)
32. A. Einstein, Ann. der Physik **19**(4), 371 (1906)
33. J.B. Perrin, Ann. de Chimie et de Physique (VIII) **18**, 5 (1909)
34. F.C. MacKintosh, C.F. Schmidt, Curr. Op. Colloid Interf. Sci. **4**(1), 300 (1999)
35. C.P. Brangwynne, C.R. Eckmann, D.S. Courson, A. Rybarska, C. Hoege, J. Gharakhani, F. Jülicher, A.A. Hyman, Science **324**(5935), 1729 (2009)
36. J.C. Crocker, D.G. Grier, J. Colloid Interf. Sci. **179**(1), 298 (1996)
37. P. Habdas, D. Schaar, A.C. Levitt, E.R. Weeks, Europhys. Lett. **67**(3), 477 (2004)
38. A. Meyer, A. Marshall, B.G. Bush, E.M. Furst, J. Rheol. **50**(1), 77 (2006)
39. T.M. Squires, J.F. Brady, Phys. Fluids **17**(7), 073101 (2005)
40. A.S. Khair, J.F. Brady, J. Fluid Mech. **557**, 73 (2006)
41. L.G. Wilson, A.W. Harrison, A.B. Schofield, J. Arlt, W.C.K. Poon, J. Phys. Chem.B **113**(12), 3806 (2009)
42. R.N. Zia, J.F. Brady, J. Fluid Mech. **658**, 188 (2010)
43. R.N. Zia, J.F. Brady, J. Rheol. **56**(5), 1175 (2012)

44. T.G. Mason, D.A. Weitz, Phys. Rev. Lett. **74**(7), 1250 (1995)
45. Q. Lu, M.J. Solomon, Phys. Rev. E **66**(6), 061504 (2002)
46. A.R. Bausch, F. Ziemann, A.A. Boulbitch, K. Jacobson, E. Sackmann, Biophys J. **75**(4), 2038 (1998)
47. W.H. Guilford, R.C. Lantz, R.W. Gore, Am. J. Physiol-Cell. Ph. **268**(5), C1308 (1995)
48. A.W.C. Lau, B.D. Hoffman, A. Davies, J.C. Crocker, T.C. Lubensky, Phys. Rev. Lett. **91**(19), 198101 (2003)
49. T. Gisler, D.A. Weitz, Phys. Rev. Lett. **82**, 1606 (1999)
50. F. Ziemann, J. Rädler, E. Sackmann, Biophys. J. **66**(6), 2210 (1994)
51. H. Freundlich, W. Seifriz, Z. Phys. Chem. **104**, 233 (1923)
52. T. Mason, K. Ganesan, J.H. Van Zanten, D. Wirtz, S.C. Kuo, Phys. Rev. Lett. **79**(17), 3282 (1997)
53. J.C. Crocker, J. Chem. Phys. **106**(8), 2837 (1997)
54. J.C. Crocker, M.T. Valentine, E.R. Weeks, T. Gisler, P.D. Kaplan, A.G. Yodh, D.A. Weitz, Phys. Rev. Lett. **85**(4), 888 (2000)
55. A.J. Levine, T.C. Lubensky, Phys. Rev. Lett. **85**, 1774 (2000)
56. J.C. Crocker, J.A. Matteo, A.D. Dinsmore, A.G. Yodh, Phys. Rev. Lett. **82**(21), 4352 (1999)
57. V. Breedveld, D.J. Pine, J. Mat. Sci. **38**(22), 4461 (2003)
58. K.M. Schultz, E.M. Furst, Lab on a Chip **11**, 3802 (2011)
59. E.M. Furst, Curr. Opin. Coll. Surf. Sci. **10**, 79 (2005)
60. I. Sriram, R.J. DePuit, T.M. Squires, E.M. Furst, J. Rheol. **53**(2), 357 (2009)
61. R.N. Zia, Individual particle motion in colloids: microviscosity, microdiffusivity, and normal stresses. Ph.D. thesis, California Institute of Technology, 2011
62. J.W. Swan, R.N. Zia, Phys. Fluids. **25**(8), 083303 (2013)
63. A.S. Khair, J.F. Brady, J. Rheol. **52**(1), 165 (2008)
64. R.J. DePuit, T.M. Squires, J. Phys. Condens. Matter **24**(46), 464106(1 (2012)
65. A.P. Minton, J. Pharm. Sci. **96**(12), 3466 (2007)
66. M. Haw, *Middle World: The Restless Heart of Matter and Life* (Palgrave Macmillan, New York, 2006)
67. J.B. Perrin, *Les Atomes* (Librairies Félix Alcan, Paris, 1914)
68. G.K. Batchelor, J.T. Green, J. Fluid Mech. **56**(2), 375 (1972)
69. G.K. Batchelor, J. Fluid Mech. **131**, 155 (1983)
70. A. Caspi, R. Granek, M. Elbaum, Phys. Rev. Lett. **85**(26), 5655 (2000)
71. K. Luby-Phelps, Int. Rev. Cytol. **192**, 189 (2000)
72. B.R. Daniels, B.C. Masi, D. Wirtz, Biophys. J. **90**(12), 4712 (2006)
73. W.B. Russel, Ann. Rev. Fluid Mech. **13**, 425 (1981)
74. R. Zwanzig, J. Stat. Phys. **9**(3), 215 (1973)
75. R.d.L. Kronig, JOSA **12**(6), 547 (1926)
76. H. H. Dramers, In *La diffusion de la lumiere par les atoms*, Atti del Congresso Internazionale dei Fisici, Vol. 2 (Zanichelli, Bologna, 1927), pp. 545–557
77. T. Pritz, J. Sound Vib. **279**(3–5), 687 (2005)
78. J.F. Brady, *Introduction to Suspension Dynamics* (2007, unpublished)
79. G. Nägele, J. Bergenholtz, J. Chem. Phys. **108**(23), 9893 (1998)
80. J.F. Brady, J. Fluid. Mech. **272**, 109 (1994)
81. J.M. Rallison, E.J. Hinch, J. Fluid Mech. **167**, 131 (1986)
82. W.B. Russel, D.A. Saville, W.R. Schowalter, *Colloidal Dispersions* (Cambridge University Press, Cambridge, 1989)
83. R. Zwanzig, J. Chem. Phys. **43**(2), 714 (1965)
84. G.K. Batchelor, J. Fluid Mech. **83**, 97 (1977)
85. Z. Cheng, J. Zhu, P. Chaikin, S.E. Phan, W.B. Russel, Phys. Rev. E **65**(4), 041405 (2002)
86. A. Sierou, J.F. Brady, J. Fluid Mech. **448**, 115 (2001)
87. J. Mewis, N.J. Wagner, *Colloidal Suspension Rheology* (Cambridge University Press, New York, 2012)
88. Y. Almog, H. Brenner, Phys. Fluids **9**(1), 16 (1997)

89. D.R. Foss, J.F. Brady, J. Fluid Mech. **407**, 167 (2000)
90. J.C. van der Werff, C.G. de Kruif, C. Blom, J. Mellema, Phys. Rev. A **39**(2), 795 (1989)
91. F. Ziemann, J. Rädler, E. Sackmann, Biophys. J. **66**(6), 2210 (1994)
92. M.T. Valentine, P.D. Kaplan, D. Thota, J.C. Crocker, T. Gisler, R.K. Prud'homme, M. Beck, D.A. Weitz, Phys. Rev. E **64**(6), 061506 (2001)
93. B.R. Dasgupta, S.Y. Tee, J.C. Crocker, B.J. Frisken, D.A. Weitz, Phys. Rev. E **65**(5), 051505 (2002)
94. J.H. van Zanten, S. Amin, A.A. Abdala, Macromolecules **37**(10), 3874 (2004)
95. B.R. Dasgupta, Microrheology and dynamic light scattering studies of polymer solutions. Ph.D. thesis, Harvard University, 2004
96. B.R. Dasgupta, D.A. Weitz, Phys. Rev. E **71**(2), 021504 (2005)
97. T.M. Squires, T.G. Mason, Ann. Rev. Fluid Mech. **42**, 413 (2010)
98. A. Tuteja, M.E. Mackay, S. Narayanan, S. Asokan, M.S. Wong, Nano letters **7**(5), 1276 (2007)
99. C.P. Brangwynne, G.H. Koenderink, F.C. MacKintosh, D.A. Weitz, Trends Cell Bio. **19**(9), 423 (2009)
100. C.P. Brangwynne, G.H. Koenderink, F.C. Mackintosh, D.A. Weitz, J. Cell Bio. **183**(4), 583 (2008)
101. F. Gittes, B. Schnurr, P.D. Olmsted, F.C. MacKintosh, C.F. Schmidt, Phys. Rev. Lett. **79**(17), 3286 (1997)
102. I.C. Carpen, J.F. Brady, J. Rheol. **49**(6), 1483 (2005)
103. J. Xu, V. Viasnoff, D. Wirtz, Rheol. Acta **37**(4), 387 (1998)
104. F. Gadala-Maria, A. Acrivos, J. Rheol. **24**(6), 799 (1980)
105. D. Leighton, A. Acrivos, J. Fluid Mech. **181**, 415 (1987)
106. J.F. Brady, J.F. Morris, J. Fluid. Mech. **348**, 103 (1997)
107. D.A. McQuarrie, *Statistical Mechanics* (Harper and Row, London, 1976)
108. N.F. Carnahan, K.E. Starling, J. Chem. Phys. **51**(2), 635 (1969)
109. R. Hocquart, E.J. Hinch, J. Fluid Mech. **137**, 217 (1983)
110. M.S. Green, J. Chem. Phys. **22**(3), 398 (1954)
111. R. Kubo, Rep. Prog. Phys. **29**(1), 255 (1966)
112. D.M. Heyes, J.R. Melrose, J. Non-Newtonian Fluid Mech. **46**(1), 1 (1993)

Chapter 4
Membrane Rheology

Arthur A. Evans and Alex J. Levine

Abstract Surfactant monolayers and lipid bilayers are intrinsically two-dimensional structures with viscoelastic mechanical properties. Monolayers display a plethora of complex broken symmetry phases, each with its own rheological signature, while bilayers are of fundamental biological importance in forming the cell membrane and the principal internal partitions of the cell. Understanding the low-energy excitations and mechanical response of these materials is thus an important probe of novel two-dimensional phases and essential to biomechanics at the cellular level, cell recognition, and transport across membranes; as such, a number of macroscopic and microscopic techniques have been developed to explore the rheological properties of monolayers and membranes. In this chapter we review the fundamental physics and rheology of molecularly thin membranes, paying particular attention to the fact that these systems are necessarily bounded on one or both sides by an aqueous fluid. We develop the basic theory of both the in- and out-of-plane viscoelastic response of membranes and monolayers and apply this theory to the study of particle transport at the surface. Such transport measurements form the basis of typical rheological experiments. We also report on more recent investigations regarding the role of nontrivial membrane geometry on particle transport and examine a novel approach to monolayer and membrane microrheology using the thermal fluctuations of particles submerged beneath the membrane. We conclude with a discussion of open questions in the field and some speculations on future research directions.

A.A. Evans (✉)
University of Massachusetts Amherst, 666 North Pleasant Street, Amherst, MA 01003, USA
e-mail: artio.evans@gmail.com

A.J. Levine
University of California Los Angeles, 405 Hilgard Ave, Los Angeles, CA 90095, USA

S.E. Spagnolie (ed.), *Complex Fluids in Biological Systems*, Biological and Medical Physics, Biomedical Engineering,
DOI 10.1007/978-1-4939-2065-5_4

1 Overview of Membranes and Langmuir Monolayers

There is a class of molecules that, when introduced to water, spontaneously assemble at the air/water interface into a layer only one molecule thick. This nanoscale film is termed a Langmuir monolayer and its dynamics or rheology is the focus of the present chapter. While the ability to self-assemble into nanometer-thick monolayers sounds like a dispatch from the frontiers of modern nanoscience, these systems are ubiquitous in nature and their study ancient in origin. In fact, such quotidian materials as olive oil and soap form these remarkable nanoscale structures. The historical record is rife with observations providing hints about these systems. Pliny the Elder remarked on the effect of oil on a body of water in the first century of the common era [1]:

> ... all sea water is made smooth by oil, and so divers sprinkle oil on their face because it calms the rough element and carries light down with them.

He was noting what was a common practice among divers in the Mediterranean who, by releasing a small quantity of oil that goes to the surface of the water above them and forms a thin film, flattened the waves and allowed for the better penetration of light into the depths. The spreading of oil on water was noted repeatedly over the next 2,000 years by others including Ben Franklin, who observed it from shipboard on a trip to England to complain about taxes (on behalf of the Pennsylvania Assembly). He later performed more careful experiments to determine how much oil was necessary to affect a certain area of the air/water interface [2]. It took over another century for these experiments to be repeated more precisely by Lord Rayleigh [3] and then Agnes Pockels [4] and to receive their modern interpretation: these oil layers, as calculated by Lord Rayleigh, were actually one molecule thick [5].

In 1917 Irving Langmuir presented the modern picture of what causes these monolayers to form in terms of the *amphiphilic* nature of the molecules in question. Each molecule of this type consists of a charged or polar "head group" that is hydrophilic and a nonpolar part, typically a short carbon chain, that is hydrophobic. At low concentrations such molecules become localized at the air/water interface where their head groups remain in contact with water while their hydrophobic tails are exposed to the air. In recognition of his understanding of the localization and orientational order of these molecules at the air/water interface, Langmuir received the Nobel Prize in Chemistry in 1932. Today the standard experimental approach to studying the phase behavior of such surface-active agents or *surfactant* molecular systems is in a *Langmuir trough.*

Despite the antiquity of these earliest observations, Langmuir monolayers continue to present important questions in physics, chemistry, biology, and the applied sciences. The main reason for the interest in the basic physical sciences is that Langmuir monolayers present a rare opportunity to study complex ordered phases in effectively two-dimensions. As briefly reviewed below, surfactant systems at the air/water interface present a plethora of complex broken symmetry phases including those that have no three-dimensional counterpart, such as the hexatic phase that exists between the solid and liquid states only in two dimensions.

In addition, Langmuir monolayers present novel dynamical questions associated with flow in a two-dimensional fluid (or perhaps in a viscoelastic solid) coupled to the underlying aqueous layer. From the point of view of the applied sciences, there are intriguing issues associated with the ability to form and then deposit monolayers of particular molecules onto another material. In this way, one can control the structure of complex heterostructures at the nanoscale, at least in the one direction normal to these monolayers.

Langmuir monolayers also serve as an important biomimetic model for in vitro experiments relevant to the life sciences. There are well-known examples of surfactant monolayers in biology. The one receiving the most attention is that found coating the lung. These pulmonary surfactants are essential for normal breathing as they dramatically lower the surface tension of the air/water interface in the lung [6, 7]. Without this effect, the work required to inflate the lung against surface tension would be prohibitive. Moreover, biology abounds with surfactant-based membranes, typically having the structure of a lipid bilayer [8]. Such structures bound the cell and its internal organelles, such as mitochondria or the nucleus. In fact, some organelles, such as the smooth and rough endoplasmic reticulum (ER), are essentially a complex network of folded and highly invaginated membranes; the "smooth" ER is the site of lipid synthesis while the "rough" ER appears rough due to the large number of attached ribosomes which are the main engines of protein synthesis in the cell. Understanding the mechanics of and phase separation within these bilayers in the nonequilibrium environment of the cell is a forefront problem in the life sciences and biological physics. Giant unilamellar vesicles (GUVs) are a useful in vitro testing ground for exploring the molecular basis of phase separation in cellular mimics. These large vesicles display surfactant-mediated phase separation, curvature-induced localization, and dynamical "blebbing," all of which relate to processes that effect cellular function. Furthermore, the hydrodynamics associated with the transport (either actively or via passive diffusion) of transmembrane proteins embedded in these bilayers is essential for developing a complete picture for complex biological issues such as cell–cell signaling. The basic physics, however, remains obscure and is a subject of some debate. It has been suggested that one can learn relevant dynamical/rheological information about these biological membranes as well as lung surfactant monolayers using the Langmuir trough, which allows for precise chemical control of the system and experimental access with a variety of probes, as discussed later in this chapter.

Since Langmuir monolayers sit at the confluence of a number of research traditions in a variety of related fields, it is not practical to survey the subject too broadly. We do not attempt to give an exhaustive overview of the field. In this chapter, we concentrate on the dynamics or rheology of Langmuir monolayers giving special emphasis to microrheological investigations. To give sufficient background to appreciate the continuing interest in monolayer rheology, we briefly review (and refer the reader to more thorough reviews regarding) the equilibrium phase behavior of monolayers. We also discuss a number of other probes used to explore membrane/monolayer rheology. We propose that by developing new experimental techniques and new theories of transport in membranes/monolayers or the fluid hydrodynamics near these membranes, one can address a number of questions relevant to their physics, chemistry, and application in the life sciences. For example, the microrheological measurements discussed here may be able to address the (visco-)elastic response of the various complex broken symmetry phases of monolayers. Such measurements can test theories of their equilibrium phase behavior as some phase transitions have dynamical signatures in new hydrodynamic modes. Given the historical difficulty in identifying phase transitions in these systems (discussed below), it is possible that having a fluctuation-based or microrheological approach to observing the new generalized elasticity emerging from the broken continuous symmetries of the various phases might be useful in addressing the equilibrium phase diagram of Langmuir monolayers.

In addition, these studies are also relevant to understanding phase separation in multicomponent monolayers, particulate transport within the layers, and the complex folding and buckling of monolayers in response to area pressure. Finally, we will show that the dynamics of particles in a the fluid near the monolayer or a membrane is modified by the viscoelasticity of that boundary. As a result one can perform a type of noncontact microrheology using particles that do not touch (and thus do not disrupt the structure of) the monolayer. This noncontact approach is also important for intracellular microrheology. Understanding in a truly quantitative manner how to interpret the Brownian fluctuations of particles in a fluid near a monolayer or membrane is essential in this field since the cell is bounded by a viscous membrane and its interior is quite densely packed with a variety of other membrane-bound structures. As a result, nearly all tracer particles in cells are necessarily near one or more membranes.

The remainder of this chapter is organized as follows: after our brief review of the equilibrium phase behavior of Langmuir monolayers (Sect. 2) we develop the basic theory of linear response for a 2D membrane embedded in a 3D viscous fluid (Sect. 3). Following this we discuss macro- and microrheological methods that have been developed for measuring the viscoelastic moduli of these membranes (Sect. 4), and finally close with a discussion of open questions (Sect. 5).

Phase Diagram of Langmuir Monolayers The equilibrium phase behavior of surfactant systems is remarkably complex and a complete review of their fascinating phenomenology is beyond the scope of this chapter. It is nevertheless worthwhile to comment briefly on their equilibrium phase behavior inasmuch as some features are fundamental to understanding those dynamical properties to be discussed in more detail here. Fortunately, the reader interested in learning more about the various equilibrium phases of surfactant monolayers, the experimental probes of their structure, and the theoretical explorations of their various phase transitions may consult one of a few excellent reviews of the subject [10, 11].

The complexity of monolayer phase behavior arises in part from the fact that the head groups of the surfactants can break translational symmetry in a number of ways, leading to, e.g., various periodic (in fact, quasiperiodic—more on that below) arrangements and the hydrophobic tail groups can be disordered or aligned in various ways with respect to the air/water surface normal and with respect to the crystallographic axis defining the order of the head groups. The traditional method of investigation of their phase behavior has been to examine two-dimensional

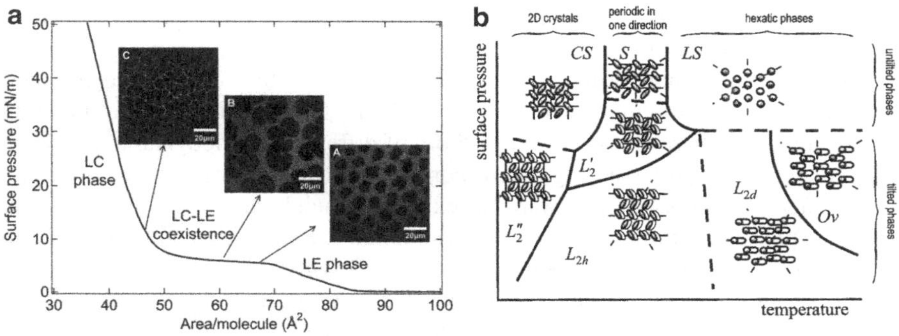

Fig. 4.1 (**a**) Isotherm for a Langmuir monolayer composed of dipalmitoylphosphatidylcholine (DPPC), the primary component of pulmonary surfactant. As the trough area is decreased, the surface pressure increases and the monolayer of DPPC undergoes phase transitions from the liquid-expanded (LE) phase to the liquid-condensed (LC) phase, with a region of coexistence in between. Reproduced from [9] with permission. (**b**) An example of how complicated the phase transitions in monolayers can become. The vast array of chain and tilt order associated with the molecules means that there exists the possibility that each crystallographic symmetry leads to a specific thermodynamic phase. Adapted from [10] with permission

pressure Π versus area A isotherms, i.e., the variation of the pressure as a function of the area of the surfactant layer at fixed temperature. Having access to the independent thermodynamic variables surface pressure (henceforth just "pressure") and temperature, one can develop a phase diagram spanned by those variables.

At a fixed temperature the generic behavior of the system can be roughly summarized as follows (see also Fig. 4.1a for a specific example). At largest areas corresponding to lowest pressures, there is a translationally disordered phase that is the two-dimensional analog of a nearly ideal gas. In this phase the free energy is dominated by the translational entropy of the surfactants so that the pressure vs. area follows the usual ideal gas law $\Pi = nk_{\mathrm{B}}T$, where $n = N/A$ is the two-dimensional number density of molecules on the surface of area A. For historical reasons this ideal gas phase is labeled *liquid-expanded phase*, which is also denoted in the literature as *LE* or L_1. Upon increasing pressure at fixed surfactant number, the hyperbolic pressure vs. area isotherm terminates at a critical pressure $\Pi_c(T)$ at which one encounters two-phase coexistence between the aforementioned liquid-expanded phase and a denser *liquid-condensed phase* (L_2). The pressure remains on a plateau $\Pi = \Pi_c(T)$ throughout the coexistence region; this behavior is indicative of a canonical first-order phase transition. In some systems, deviations from this Gibbs phase rule have been observed, calling the first-order nature of the transition into question, but direct observations of two-phase coexistence in the 1980s [12] strongly support this former interpretation. Since studies have observed (quasi-)long-range order in spatial arrangement of the head groups, the term liquid condensed appears to be a misnomer. Modern reviews suggest that the term *condensed phase* or simply the designation L_2 be used hereafter. We follow this suggestion in the current chapter.

The L_2 phase has a significantly higher area modulus (less compressibility) than the L_1 phase. In the more densely packed L_2 phase the hydrophobic tails of the surfactants are also now ordered and generically tilted with respect to the air/water normal. Given the simultaneous order between the head groups and the direction of the chain tilt, there are, in fact, a plethora of different thermodynamic phases in this condensed regime (see Fig. 4.1b). For example, the chains can tilt towards a nearest neighbor molecule or towards a next nearest neighbor molecule. In addition, there are phase transitions which involve a simultaneous distortion of the head group lattice and a change in the direction of the tail tilt with respect to the now lower symmetry unit cell of the heads. Finally, there are untilted tail phases typically obtained at even higher area pressures where closer head group packing presumably forces the tails to align with the local interface normal.

Phase behavior measurements can rely in part on the pressure-area isotherms, where small signals of kinks indicate second-order transitions and short coexistence plateaus are associated with first-order transitions. But direct evidence of the crystallographic symmetry of the head group packing in the condensed phases comes from scattering experiments. As early as 1938, electron-scattering studies by Germer and Storks demonstrated the triangular packing of the head groups. The specular reflection signal contains information about the electron-density distribution in the direction normal to the interface and thus informs one regarding tail tilt [13, 14], while diffuse scattering has been used to study the height-height correlation spectrum on the surface, from which the bending modulus of the surface may be extracted. The key scattering technique for studying the packing of the surfactants, however, is grazing incidence x-ray diffraction, where the intensity vs. in-plane momentum of the scattered photons provides a direct measurement of the Brillouin zone of the head group lattice [15].

There are other experimental techniques used to probe the chain order on the interface over longer length scales using optical probes. These include polarized fluorescence microscopy (PFM) and Brewster angle microscopy (BAM) [16]. In the former, one introduces small molecular fluorescent dyes at low concentration. These preferentially absorb in the higher density phases of the monolayer and thus identify islands of condensed phase(s) in the condensed-liquid-expanded coexistence regime. In the latter technique, laser illumination is directed at the monolayer at the Brewster angle in a polarization state such that there should be no reflected light from a pure air/water surface. The observed reflections are solely due to the local change in index of refraction due to the surfactants. As this index depends sensitively on the orientation of the tails, subtle changes in tail orientation and tilt are observable. This technique has, in fact, observed new ordered phases of the monolayer that produced no observable signal in the pressure vs. area isotherms.

Two-dimensional order also opens up a new phase of matter in between the solid and liquid states, characterized by the presence of quasi-long-range orientational order but short-range translational order. This new phase, deemed the hexatic, is predicted to arise from the unbinding of thermally generated dislocations in the crystal lattice [17]. The hexatic phase has been observed in colloidal suspensions confined into a single layer by two flat plates [18, 19]. Hexatic phases also display a universal combination of elastic constants and are only terminated by another unbinding transition of dislocation dipoles (or disclinations); the existence and values of these predictions require a fundamental understanding of the mechanical response of the two-dimensional system.

2 Membrane Mechanics

There is no such thing as a typical membrane, and because the molecular constituents that compose Langmuir monolayers and lipid bilayers can interact via chemical reactions, in general a large number of possible models are available to describe the equilibrium configurations of such materials; herein we discuss the simplest models that allow for a continuum elasticity approach and refer the interested reader to other sources for a review on the treatment of lipid membranes and vesicles in other contexts [10, 11, 20, 21].

While actually composed of a number of constitutive molecules, if we consider deformations on length scales much larger than the size of these components then continuum elasticity may be used to good effect. Furthermore, if the internal structure of the membrane is not drastically changed by these deformations such that the position of any point on the material may be described by a suitably defined "middle surface," then we may use the language of differential geometry and elasticity to write the energy of deformation associated with such a surface. By the location of the surface with a position vector $\mathbf{r}$ and an orthonormal vector field $\{\hat{\mathbf{n}}, \hat{\mathbf{t}}_1, \hat{\mathbf{t}}_2\}$, we fully determine the middle surface of the undeformed membrane. From this we may construct the two important tensor quantities that will enable the quantitative description of deformation of the surface: the metric tensor $g_{\alpha\beta} = \partial_\alpha \mathbf{r} \cdot \partial_\beta \mathbf{r}$ and the curvature tensor $d_{\alpha\beta} = \partial_\alpha \hat{\mathbf{n}} \cdot \partial_\beta \hat{\mathbf{t}}$.

When the surface deforms, new metric and curvature tensors $g^*_{\alpha\beta}$ and $d^*_{\alpha\beta}$ must be defined. To quantify these deformations we define the strain tensor $E_{\alpha\beta} = g^*_{\alpha\beta} - g_{\alpha\beta}$ and the bending tensor $K_{\alpha\beta} = d^*_{\alpha\beta} - d_{\alpha\beta}$. Then, in the framework of linear elasticity, these modes of deformation may be used to construct the simplest elastic energy for a membrane:

$$\mathscr{F} = \frac{1}{2} \int \left[2\mu E^\alpha_\beta E^\beta_\alpha + \lambda (E^\alpha_\alpha)^2 + \kappa (K^\alpha_\alpha)^2 \right] dS, \tag{4.1}$$

where we used the metric tensor to raise the tensorial indices, i.e., $E^\alpha_\alpha = g^{\alpha\gamma} E_{\gamma\alpha}$ (see, e.g., [22, 23]). The "Lamé coefficients" μ, λ are energy penalties per unit area associated with deformations of the membrane. Since the parametrization of our surface cannot matter, the functional above depends only on the tensor invariants of deformation, i.e., the trace and determinant of the strain tensor, and the trace of the bending tensor, so the material described is isotropic in space. Physically, the off-diagonal terms in the strain tensor relate to shear modes, while the trace component incorporates the ability of the membrane to compress or stretch. Similarly, the trace of the bending tensor is related to the mean curvature of the surface, and κ is the bending rigidity that penalizes changes of this mean curvature from the flat reference state.

For an initially flat membrane, deformations of the surface may be parametrized by the vector $\zeta\hat{\mathbf{n}} + \xi_\alpha\hat{\mathbf{t}}^\alpha$, and the strain and bending tensors are simplified to $E_{\alpha\beta} \approx \frac{1}{2}(D_\alpha\xi_\beta + D_\beta\xi_\alpha), K_{\alpha\beta} \approx D_\alpha D_\beta\zeta$, with D_α the covariant derivative. The Euler-Lagrange equations for the surface, in the presence of external normal forces p and in-plane forces f_α can then be written as

$$\kappa\partial^4\zeta = p, \tag{4.2}$$

$$\mu\partial^2\xi_\alpha + (\mu+\lambda)\partial_\alpha\partial_\beta\xi_\beta = f_\alpha, \tag{4.3}$$

where $\partial^2 = \partial_x^2 + \partial_y^2$ is the linearized 2D Laplacian. Note that in this linear regime the normal and tangential modes decouple, and thus we may treat them separately. This situation is complicated by curvature of the shell, and we discuss this below. In-plane elasticity and flow, coupled to external fluids from a three-dimensional subphase, will be the primary topic of this chapter, but the undulatory (or flexural) deformations of a membrane (or shell) are important for biological membranes that are heavily cross-linked and thus display elastic behavior. As a result the mechanical response of viral capsids [24–27], red blood cells [28], and other soft materials (both in and out of fluids) [29–32] have been studied using this formulation.

We should state that this is the simplest elastic membrane model: an isotropic, linearly elastic material. Many biologically relevant materials will obey different constitutive laws. For example, Langmuir monolayers existing in different broken symmetry phases will not necessarily behave isotropically, and thus additional terms from the strain tensor may be required to describe their elastic response, while lipid bilayers behave like a fluid in plane, so only the bending response is elastic. Furthermore, many soft materials obey viscoelastic constitutive relations, an issue that we discuss below in the framework of dynamic linear response.

3 Dynamical Linear Response

With the wealth of data-gathering techniques for the equilibrium phases of Langmuir monolayers and armed with the knowledge that the elastic moduli of the membrane can be affected by subtle changes in symmetry associated with the molecular constituents, we now turn to the study of dynamics.

Understanding the mechanical response of Langmuir monolayers requires knowledge of how the two-dimensional membrane couples to the three-dimensional solvent. Ultimately, if one can calculate the response function for the combined membrane/subphase system, then the behavior of the monolayer can be probed. The calculations presented below are equally valid for membranes (bilayers) as they are for monolayers[1]. Hereafter we refer to the two-dimensional viscoelastic medium as *the membrane* for convenience, but the results are valid for monolayers as well.

Consider a rigid particle embedded in the membrane which reports on the displacement field at that point. One may extract the relevant dynamical/rheological information about the membrane from that particle's linear response to an applied force. To characterize that response we introduce a susceptibility tensor $\chi_{ij}(\omega)$ defined by

$$U_j = \frac{1}{-i\omega} V_j = \chi_{jk}(\omega) F_k, \tag{4.4}$$

where $V(\omega)$ is the particle's velocity, which is simply related to the membrane deformation U_j in the Fourier (frequency) domain, and F_j is the force applied. We consider here and throughout a coordinate system in which the z axis is normal to the interface and points outwards from the fluid subphase. The interface occupies the $z = 0$ plane. We treat the fluid subphase as having infinite depth and lateral extent, although calculations have been performed that include a subphase of finite depth [33].

The membrane at the air/water interface breaks the full rotational symmetry of the problem; as a consequence, the susceptibility tensor is no longer proportional to the identity. It remains diagonal, however, and has only two independent, non-vanishing components. This may be inferred from the following argument. On one hand, if the force were applied in the z direction (i.e., normal to the interface), there can be no motion induced in the xy plane by rotation invariance about the z axis. On the other hand, if a force F applied in the $\hat{\mathbf{x}}$ direction were to generate motion in the $\hat{\mathbf{z}}$ direction, then the sign of that displacement would be reversed under the application of an equal and opposite force $-F$. However, these applied forces are related by a symmetry of the system—a rotation of π about the z axis– and thus must produce the same displacement in the vertical direction. We conclude that the vertical displacement in response to a force in the xy plane must vanish. Finally, the same rotational symmetry requires $\chi_{xx} = \chi_{yy}$. Consequently, there are only two distinct components of the mobility tensor to calculate, $\chi_{xx} = \chi_{yy}$ and χ_{zz}. The nontrivial components of the susceptibility tensor can be calculated from linear modes of deformation of the coupled membrane and subphase system.

3.1 Flat Membranes

For in-plane deformations, two linear modes exist [34] corresponding to longitudinal and transverse (shear) excitations [35]. These modes are linearly independent for a flat (visco-)elastic membrane bounding a fluid subphase. Thus we may calculate the response of the subphase due to either shear or compression waves independently.

[1] We neglect slip between the two leaflets of the bilayer here.

Working in the limit of zero Reynolds number, we solve the Stokes equation $\eta\nabla^2\mathbf{u} = \nabla p$ in the subphase subject to the condition of incompressibility, $\nabla\cdot\mathbf{u} = 0$. We must apply boundary conditions to this bulk fluid flow, and one of the most commonly used (and generally justifiable) assumptions is to impose no slip between the bulk fluid and the membrane, i.e., $\mathbf{u}|_{z=0} = \partial\xi/\partial t$. To exploit the translational invariance of the problem in the plane of the membrane it is useful to decompose these deformations into Fourier modes such that $\mathbf{u} = \int \mathbf{u}(\mathbf{q},\omega)e^{i\mathbf{q}\cdot\mathbf{x}}d^2\mathbf{x}$ so that the shear and compression response of the bulk fluid generated by membrane waves are given by [35]:

$$\mathbf{u}^T = -i\omega\xi^T e^{-|\mathbf{q}|z}e^{i\mathbf{q}\cdot\mathbf{x}-i\omega t}, \tag{4.5}$$

$$\mathbf{u}^L = -i\omega(1-|\mathbf{q}|z)\xi^L e^{-|\mathbf{q}|z}e^{i\mathbf{q}\cdot\mathbf{x}-i\omega t}, \tag{4.6}$$

where we have suppressed the $(\mathbf{q},\omega)$ dependence for notational convenience. Here the transverse u^T and longitudinal u^L fluid flows are acquired from $u^T_\alpha = P^{(T)}_{\alpha\beta}u_\beta$ and $u^L_\alpha = P^{(L)}_{\alpha\beta}u_\beta$, with the projection operators $P^{(T)}_{\alpha\beta} = \delta_{\alpha\beta} - P^{(L)}_{\alpha\beta} = \delta_{\alpha\beta} - \hat{q}_\alpha\hat{q}_\beta$. Physically, this bulk fluid flow generates stresses in the membrane. We now consider the problem of stress balance across that interface:

$$\mu\partial^2\xi_\alpha + (\mu+\lambda)\partial_\alpha\partial_\beta\xi_\beta - \eta w_\alpha = f_\alpha, \tag{4.7}$$

where $\eta w_\alpha = \eta\partial_z u_\alpha|_{z=0}$ is the shear stress generated by the subphase fluid acting on the membrane surface. If the membrane is viscoelastic, these Lamé constants are complex and frequency dependent. In that case, the above equation must be rewritten as a convolution integral over time, but this leads to an equation analogous to the one above when it is recast in the frequency domain. There the real elastic constants are replaced by their complex frequency-dependent ones. Transforming to Fourier space $(\mathbf{q},\omega)$ once again, and dividing this equation into transverse T and longitudinal L components, we find that

$$(2\mu+\lambda)|\mathbf{q}|^2\xi^L_\alpha - \eta w^L_\alpha = f^L_\alpha, \tag{4.8}$$

$$\mu|\mathbf{q}|^2\xi^T_\alpha - \eta w^T_\alpha = f^T_\alpha, \tag{4.9}$$

$$w^L_\alpha = 2i\omega|\mathbf{q}|\xi^L, \tag{4.10}$$

$$w^T_\alpha = i\omega|\mathbf{q}|\xi^T, \tag{4.11}$$

where we have used the results from Eq. (4.5) to find the transverse and longitudinal components of the bulk fluid stress. From these results it is straightforward to calculate the Green's function defined by $\xi_\alpha = \mathscr{G}_{\alpha\beta}f_\beta$ for the longitudinal and transverse modes:

$$\mathscr{G}_{\alpha\beta} = \mathscr{G}^L \frac{q_\alpha q_\beta}{|\mathbf{q}|^2} + \left(\delta_{\alpha\beta} - \frac{q_\alpha q_\beta}{|\mathbf{q}|^2}\right)\mathscr{G}^T, \tag{4.12}$$

$$\mathscr{G}^L = \frac{1}{(2\mu+\lambda)|\mathbf{q}|^2 - i\omega\eta|\mathbf{q}|}, \tag{4.13}$$

$$\mathscr{G}^T = \frac{1}{\mu|\mathbf{q}|^2 - i\omega\eta|\mathbf{q}|}. \tag{4.14}$$

The response tensor for a force applied to a (point) particle embedded in the membrane can be calculated using these Green's functions and inverted into real space if so desired.

Out-of-plane deformations can be calculated similarly. In this case there is one mode associated with the deformation, and the resistance to these deformations comes from surface tension or bending rigidity. The response function is

$$\mathscr{G}_\perp = \frac{1}{\kappa|\mathbf{q}|^4 + \gamma|\mathbf{q}|^2 - 2i\omega\eta|\mathbf{q}|}, \tag{4.15}$$

where κ is the bending rigidity of the membrane and γ is the surface tension. Undulatory waves in membranes have been used to probe the passive thermal fluctuations of biological and other soft surfaces that display a bending rigidity [36], while thermally activated capillary waves in micro- and nanoscopic fluids have been probed using grazing incidence x-ray scattering to extract rheological data [37, 38]. In active membrane systems, such as plasma membranes or other protein-embedded surfaces, the chemistry of the inclusion coupled with hydrostatic or osmotic pressure leads to material and transport properties that can be measured and modified [39–44].

We now discuss the physical implications of these results generally. The subtle issues of measuring the flow properties of membranes lies in the fact that these membranes never occur without a viscous fluid in contact with them, and the competing physics between membrane-based 2D viscoelasticity and 3D hydrodynamics changes the mechanical response of the membrane depending on the length scale on which the flow response is probed. It should be borne in mind that length scales introduced by the curvature of the reference state of the membrane further complicate this analysis. We comment on this in the next section.

For these flat membranes there is one natural length scale that arises in calculating the response function of the membrane—the Saffman–Delbrück (SD) length, named in homage to the original hydrodynamic calculation that was performed to determine the mobility of a rigid cylindrical inclusion moving in a liquid membrane [45]. This calculation generally applies to the diffusion of proteins in cellular membranes and the motion of lipid rafts at the interface of other biomembranes. The SD length emerges from a simple dimensional analysis of a lipid membrane with two-dimensional viscosity η_m resting on an infinite Newtonian fluid of viscosity η: the ratio of these two parameters $\eta_m/\eta = \ell$ introduces a length scale ℓ which serves as a natural cutoff between the purely 2D physics of the membrane and the 3D hydrodynamics of the subphase. The most natural interpretation of the SD length is that it is the distance in the plane of the membrane over which in-plane momentum bleeds off into the third dimension, the subphase. At lengths smaller compared to the SD length, $L \ll \ell$, the membrane dynamics is dominated by in-plane stress balance and the subphase plays a small role. At larger length scales, $L \gg \ell$, however, in-plane momentum is converted into flows in the subphase, which now dominates the dynamics.

The SD length cuts off the log divergence of the point-force response function in two-dimensional elasticity theory or hydrodynamics, where this divergence is known as Stokes' paradox [46]. For a membrane with no coupling to a three-dimensional fluid phase, the logarithmic divergence associated with 2D Stokes flow leads to the result that boundary conditions at the edge of the membrane will always be important in determining the point-force response. For Langmuir monolayers, however, the loss of in-plane momentum to the subphase on the scale of the SD length screens this effect and cuts off the log divergence discussed by Stokes.

3.2 Curved Surfaces

So far we have explored the linear response of nearly flat membranes, where deformation modes can be successfully decomposed into shear, compression, and undulatory deformations, but when a surface is curved intrinsically then the surface geometrically couples the in-plane and out-of-plane responses. In fact, the radius of curvature of the reference state results in corrections to the membrane hydrodynamics analogous to those associated with the SD length. One may say that geometry acts as another type of viscosity in such circumstances. It is not merely of academic importance to study curved membranes, since lipids in solution naturally self-assembly into micelles, vesicles, and other shapes that have one (or more) natural radii of curvature. Again, the literature in the cellular biology, biological physics, and biochemistry community is extensive [20, 47, 48], and here we shall mention only the rheological consequences (i.e., dynamical mechanical response) for introducing curvature into the problem. Once this natural curvature is introduced, the strain and bending tensors couple in- and out-of-plane deformations at leading order:

$$E_{\alpha\beta} = \frac{1}{2}(D_\alpha \xi_\beta + D_\beta \xi_\alpha) - d_{\alpha\beta}\zeta, \tag{4.16}$$

$$K_{\alpha\beta} = D_\alpha D_\beta \zeta - d_{\gamma\alpha} d^\gamma_\beta \zeta. \tag{4.17}$$

Note that for $d_{\alpha\beta} \equiv 0$ these reduce to the form that will yield the 2D force-balance equations discussed above. Except in very simple situations the Euler-Lagrange equations that result from balancing the free energy above with hydrodynamic forces are very difficult to solve, owing to the deformation out of plane changing the reference surface that the embedding fluid acts on.

To isolate the rheological effects of curvature one may consider in-plane flows in a membrane with nontrivial geometry, but which cannot deform normal to the surface. This constraint fixes the geometry of the system. Without this constraint, the flows could in principle change the local geometry which would then act back on the flows making the hydrodynamic problem nonlinear. To our knowledge, this more complex problem has not been adequately analyzed. Specializing to purely viscous membranes, the covariant 2D Stokes equations may be formulated and then balanced at the membrane surface with the 3D hydrodynamics. The coupling between geometry and fluid mechanics can then be seen clearly in the modification of the Stokes flow on a curved surface [49]:

$$\eta_m[D^\beta D_\beta u_\alpha(\mathbf{r}) + K(\mathbf{r})u_\alpha(\mathbf{r})] = D_\alpha p(\mathbf{r}). \tag{4.18}$$

Here $K(\mathbf{r}) = \mathrm{Det}(\mathbf{d})$ is the local Gaussian curvature on the membrane surface parametrized by the vector $\mathbf{r}$. Obtaining the Gaussian curvature in the above equation is simplified if one uses the fact that the commutator of the covariant derivatives acting on a vector field $w_\alpha(r)$ is just the product of that field with the local Gaussian curvature: $[D^\beta, D_\alpha]w_\beta(r) = K(r)w_\alpha(r)$.

Understanding how curvature affects the rheological properties of an incompressible 2D fluid has implications for the flow patterns observed in cylindrical and spherical vesicles. For the case of a flat monolayer, there is only one natural length scale arising from the balance of 2D and 3D viscosities, but with the inclusion of a radius of curvature R there is now an additional regime that must be considered [49–51]. Ultimately, the accurate calculation and measurement of flow within lipid vesicles [52, 53] requires an understanding of this length scale. Furthermore, as discussed in the first section, the lipid properties themselves can lead to important morphological changes. In general, the bending energy discussed above requires an additional term such that

$$E_{bend} = \int \frac{\kappa}{2}(K_\alpha^\alpha - C_0)^2 dS, \tag{4.19}$$

where C_0 is a spontaneous curvature preferred by the lipid constituents. Since some lipids have conical or even chiral shapes, this contribution to the mechanical stability can lead to important shape changes in a membrane. For example, Fig. 4.2 shows the consequences of changing the lipid constituents of a biphasic membrane. The top row of vesicles is composed of an inverse-conical lipid, and through a cyclic change of temperature, the first pattern-forming phase separation occurs, followed by membrane tabulation similar to that seen in biological systems [54]. When this inverse-cone lipid is replaced with a cylindrical lipid, no such tubule formation occurs. These effects, among others, indicate that curvature-mediated behavior is important in transport and aggregation phenomena that occur in biological membranes [55–57].

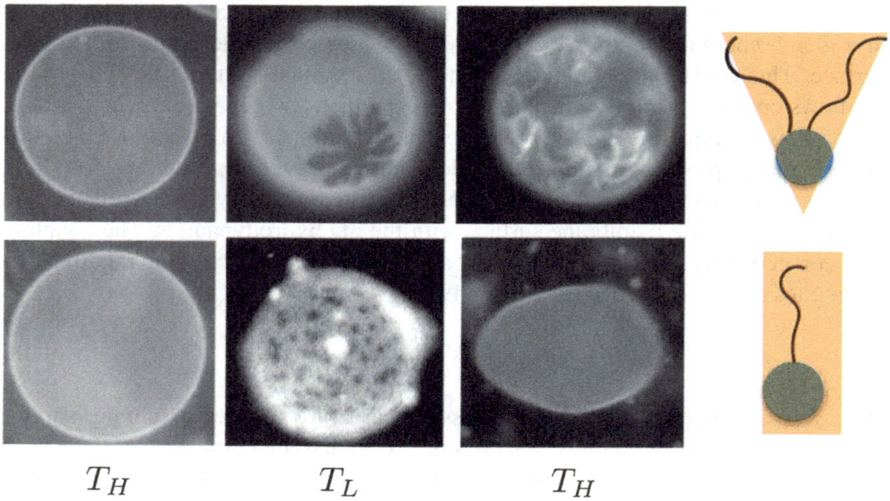

Fig. 4.2 *Top row*: A binary-mixture GUV (giant unilamellar vesicle), composed of DPPC and DMPE (*an inverse-cone lipid*). As the temperature is changed from hot (T_H) to cold (T_L) and back again, phase separation and tubulation occurs. *Bottom row*: A binary mixture of DPPC and DLPC (*a cylinder-shaped lipid*). No tubules are formed throughout the temperature cycling process, indicating that the shape of the lipid is key to forming these structures. Figure adapted from [54] with permission

4 Monolayer Rheology Experiments

There are a number of experimental approaches to monolayer rheology. These have been well reviewed in the literature [58] and although we have made no attempt to be complete in our discussion of this active field of experimental research, the interested reader will find some representative references below. The measurement approaches may be divided into two broad groups: (1) macroscopic methods, which examine macroscopic flows in response to applied stresses, such as in a two-dimensional analog of Couette flow, forcing the monolayer to flow between two rigid barriers, or examining the mobility of large (i.e., non-Brownian) driven particles in the monolayer; and (2) microscopic methods, which consider the (usually equilibrium) fluctuations of small particles embedded in the monolayer. This distinction is well known in other contexts and typically referred to as rheology or microrheology, respectively. We draw attention to the differences in these methods mainly because they have been in poor agreement and may not even reflect the same sorts of dynamics within the monolayers.

4.1 Macroscopic Methods

Macroscopic measurements of interfacial flow must be carefully constructed due both to the fragility of the system under examination and to prevent the effects of unexpected physical phenomena such as Laplace or Marangoni stresses. If dilatational interface measurements are taken then the area expansion of the surface can lead to gradients in surface tension and drive Marangoni flows. Shear rheology is generally performed at constant surface area, however, and usually involves the controlled translation or rotation of a solid boundary in contact with the membrane. Traditional interface shear rheology has been performed using a channel flow apparatus with characteristic length scale L, in the limit of small Boussinesq number: $\mathrm{Bo} = \eta_m/(\eta L) \ll 1$; this dimensionless group, long used in interfacial rheological studies, is the ratio defined above relating the Saffman–Delbrück length to the size of the system. For larger values of Bo, an oscillatory needle confined to the interface provides a probe of rheological properties in the frequency domain. From the point of view of this chapter, each class of experiments measures some linear response function that is associated with the rheological properties of the monolayer.

Ghaskadvi and Dennin [59] developed a surface Couette apparatus that is capable of simultaneously measuring the shear response of a monolayer and visualizing the flows within it under applied shear stress. The basic idea is to reproduce the Couette flow used to study the viscosity of three-dimensional fluids by trapping them between two coaxial cylinders, which are then rotated relative to one other [60, 61]. To adapt this to the monolayer, one replaces the inner cylinder by a Teflon knife-edge disk hung from a wire. The outer cylinder is a circular barrier that can be rotated to generate the shear flow, while the angular displacement of the inner cylinder measures the shear stress.

Another approach to monolayer rheology relies on modifying the archetypal Poiseuille flow between parallel plates to the monolayer geometry. By using floating barriers to take the place of the walls in the three-dimensional experiment, one may drive monolayer flow between the plates by an area pressure gradient [62, 63]. This is rather complex to analyze since the plates enforce stick boundary conditions (e.g., the velocity field goes to zero) on the monolayer but not on the subphase which is partially entrained by the flowing monolayer. When the drag of the subphase dominates, the monolayer flow profile between the plates is elliptic, rather than the usual parabolic profile, which would be found if the monolayer's viscosity were the dominant source of dissipation. In general when dissipative stresses in both the monolayer and the bulk have to be accounted for, a difficult set of dual integral equations must be solved [63, 64] to obtain the flow profile and from that relate the flow data to a monolayer rheology.

Finally, dragging a macroscopic particle through the monolayer provides yet another class of monolayer rheological experiments [65, 66]. Typically one uses a needle, which may be dragged in the plane of the monolayer in the direction of its tip or broadside on. Understanding the drag force acting on the particle also provides a fairly complex hydrodynamic problem [35, 67, 68], but one that has been analyzed in great detail—see Ref. [65].

4.2 Microrheology of Membranes

Another approach to rheology is to use the inherent thermal fluctuations of soft materials as a probe of its mechanical/rheological properties. This basic idea of using strain fluctuations in a frequency-resolved manner to obtain rheological data is now quite common and generally termed *microrheology* due to the use of a microscopic tracer particle to obtain the strain fluctuation data. The details of this form of passive microrheology through particle tracking is well detailed in the literature. The reader interested in an overview of the techniques and some of the more simple extensions of the basic idea is referred to Refs. [9, 69–78].

Most of the principal benefits of microrheology are due to its passive nature. One does not need to be concerned with accessing nonlinearities in the mechanical response of highly compliant materials or damaging (i.e., structurally modifying) extremely fragile ones. Microrheology is well adapted to probing a very wide frequency range reaching quite high frequencies since one is not required to drive the system but only to be able to obtain passive frequency-resolved fluctuation data. The current state of the art allows for at least six orders of magnitude of frequency range up to 10^5 Hz [79]. In addition, microrheology is clearly helpful for systems that are inherently small (e.g., living cells) where there are significant experimental challenges associated with applying a well-controlled shear strain. The same point may be raised for monolayer and membrane rheology, where there are difficulties associated with quantitatively controlling the applied shear strain (although some of the techniques mentioned above address these issues). In addition, monolayer structures are notoriously fragile so that structural modification in response to macroscopic shear-based measurements are a source of concern. Finally, microrheology allows for rheological probes in spatially heterogeneous materials, such as in many phases of Langmuir monolayers. There is a reasonable expectation that microscopic fluctuation data in such materials should be dominated by their local rheology, opening the possibility of a type of scanning probe rheometer based on particle-tracking data from many sites within the material.

Because monolayer structures are commonly fragile and because it is difficult to drive them with high-frequency oscillatory shear, it would appear that microrheology is ideally suited to their exploration. This may be true, but there are new issues associated with the interpretation of microrheological data. In monolayers and membranes there appears to be generic and remarkably large discrepancies between the rheological data obtained using macro- and microrheology. As we outline below, recent work suggests that these discrepancies result from a poor understanding of the probe's response function and resulting inaccuracies with the interpretation of the fluctuation-based data. As such, it is worthwhile to briefly review how microrheology can go wrong. First, the probe's fluctuations reflect all thermally excited modes of the systems that couple to the probe's position. These may include hydrodynamic modes unrelated to shear. Second, the introduction of the exogenous probes may, in fact, locally perturb the structure of particularly soft and fragile materials. In fact, it is well known that in at least some polymer solutions the introduction of a rigid colloidal particle can produce a local depletion zone in which that probe fluctuates [72, 74]. This issue of probe-induced structural perturbations can be handled in three-dimensional materials by focussing on the correlated motion of two well-separated probes [72, 73, 80], and the combination of one- and two-particle microrheology has been used to probe the structure of the depletion zone [74]. For monolayers and membranes, we propose another way—using a noncontact approach in which the tracer particle is in the fluid near the monolayer but not in direct contact with it. This noncontact approach avoids the issue of the probe altering the monolayer's local structure and greatly simplifies the analysis of the (purely hydrodynamic) coupling of the probe to the monolayer. We believe that this non-contact approach puts interfacial and membrane microrheology on a stronger footing and opens up new possibilities for the quantitative study of particular fragile and even heterogeneous monolayers.

4.3 The Case of the Missing Modulus

Figure 4.3 shows there is a significant discrepancy between the monolayer modulus obtained by macroscopic methods and by particle-tracking microrheology [81]. This discrepancy appears to be quite common, appearing in a variety of surfactant systems. The modulus obtained via macroscopic methods exceeds that obtained by microrheology by a factor of $\sim 10^3 - 10^4$ over the accessible range of frequencies. The dependence of the modulus as a function of surfactant type (*e.g.*, chain length) or area pressure obtained from these two classes of measurements, however, appears to be similar. It is as though the microrheological method uniformly underreports the modulus (or, conversely the macroscopic methods overreport it) by a multiplicative factor that is only weakly dependent on surfactant chemistry or monolayer phase.

We term this large discrepancy, the "missing modulus problem," and discuss below possible explanations for eliminating this issue in both microrheology and rheology.

The resolution of this discrepancy is more than a simple problem of instrument calibration, but speaks to basic questions of the mechanical coupling of the probe particle to the monolayer or membrane. There are several possible ways to account for this inconsistency having to do with an imperfect knowledge of the interaction of the probe with the monolayer at the microscopic scale. As shown schematically in Fig. 4.4a, one would like to imagine that the probe is well embedded in an otherwise undisturbed monolayer and thus simply reports on the strain field at that point in space. Since the structure of the monolayer is generally quite fragile, the probe may produce a local structural distortion as shown in Fig. 4.4b resulting in the probe particle sitting in a less ordered and perhaps more compliant region of the monolayer. The nature of how the lipid fluid interacts with the bead is poorly understood, and the reactivity of the molecules with the bead surface certainly have the ability to affect the mobility of the particles. If the surface pressure of the membrane is high enough, the bead may be shoved entirely out of the membrane and diffuse along the surface without being embedded in it.

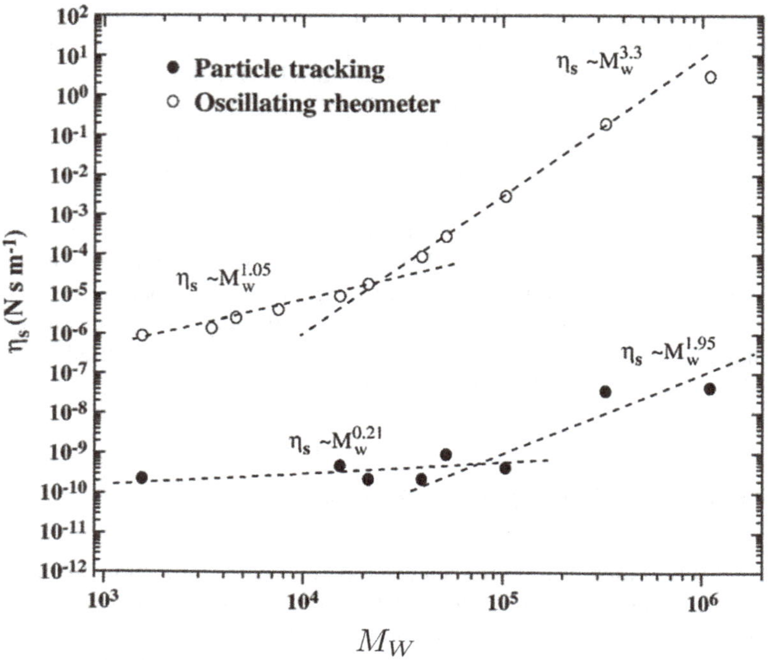

Fig. 4.3 Surface shear viscosity of poly(t-butyl acrylate) as a function of molecular weight at constant surface pressure, measured by macroscopic methods and microscopic particle tracking. In some regions as much as four orders of magnitude separate the two measurements, but the general trend in the moduli appears to be qualitatively similar. Adapted from Ortega et al. [81] with permission

Fig. 4.4 Schematic of particle-tracking microrheology and potential problems with the implementation. (**a**) The assumed set-up for particle tracking: a tracer particle is embedded in the membrane and it's thermal fluctuations are monitored. (**b**) The interaction between the molecules and the bead surface are not well understood and at this stage very difficult to measure, so the position and coupling of the bead may be uncertain due to this. (**c**) If the surface pressure is large enough the bead may be pushed entirely out of the interface. It is rare to have a microscope that has submicron resolution, and thus verifying that the bead is actually embedded in the monolayer is difficult

Such effects related to the perturbation of the material by the tracer particle have been observed in three-dimensional microrheology. Their single particle-tracking measurements and two-particle measurements give different rheological measures of the material. In that case, observing the correlated fluctuations of pairs of particles [72, 73] is necessary to determine the rheology of the undistorted material, and a combination of one- and two-particle microrheology measurements allows one to examine the local rheological perturbation surrounding the probes [80]. A third possibility is that the probe particle is actually not well embedded or even directly coupled to the monolayer at all, as shown in Fig. 4.4c. This is possible since the resolution of the microscope along the optical axis (normal to the plane of the monolayer) is significantly poorer than that in the plane perpendicular to that axis.

We have pointed out above that many assumptions upon which the interpretation of microrheology hinges are difficult to verify. Of course, it is also possible that microrheological techniques are correctly measuring the monolayer's rheology. The macroscopic methods, which rely on straining the monolayer over longer length scales, may be corrupted by structural heterogeneities in the material. Additionally, Langmuir monolayers are notoriously fragile systems, and traditional viscometry methods may access the nonlinear regime. Below we turn to a new approach to resolving this issue using noncontact microrheology.

4.4 Submerged Particle Microrheology

One approach to eliminating these difficulties is to modify microrheology by locating the probe particle away from the membrane. Specifically, if one places the probe particle at a known distance from the monolayer and within the fluid subphase, the questions arising about the detailed nature of the coupling of the probe to the monolayer are completely avoided. Issues associated with the local disruption of the monolayer by the probe are removed as well. In this noncontact form of microrheology the probe never directly interacts with the monolayer/membrane but is only hydrodynamically coupled to it. Although the coupling between particle and membrane is simplified, the interaction is substantially weaker. However, by examining the change in the fluctuations of the probe as a function of distance from the monolayer, one is able to extract the mechanical contribution of the monolayer to the observed fluctuations and thereby determine the frequency-dependent rheology of the monolayer in this noncontact approach. This technique using submerged particles or SPIM (submerged particle interfacial microrheology) requires a new theoretical understanding of the Green's function of the particle in a fluid bounded by a viscoelastic monolayer. On top of this, new experimental developments are necessary to measure the particle's position with sufficient spatial and temporal resolution to observe the change in the fluctuation spectrum due to the presence of the viscoelastic monolayer above the particle. The details of the measurement technique have been presented elsewhere [82]; here we expand on the calculations necessary to understand that fluctuation spectrum by presenting a more detailed account of the calculation of the necessary Green's function.

We determine the velocity response function for a spherical particle of radius a in a fluid with viscosity η, submerged a depth d beneath a monolayer (see Fig. 4.5a). The ultimate goal is to determine the velocity response $\mathbf{V}$ of the particle due to a force $\mathbf{F}$ as a function of the rheological properties of the membrane and

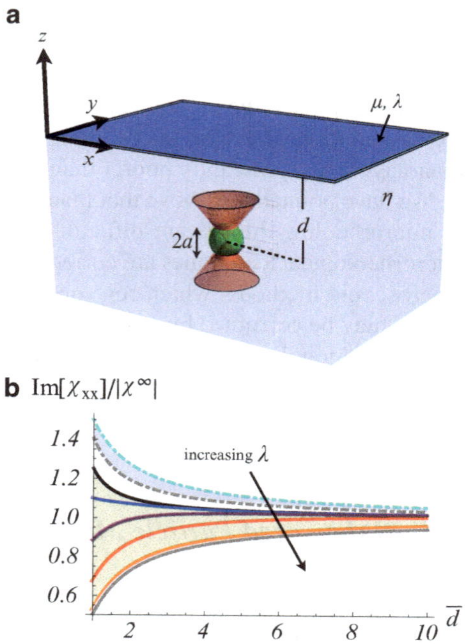

Fig. 4.5 (**a**) Schematic of submerged particle microrheology. A tracer bead of radius a is placed a distance $d = \bar{d}a$ beneath, the surface is contained using an optical trap, and its thermal fluctuations are measured. (**b**) Response functions calculated using the formalism described in the text. For a free surface the susceptibility measured is enhanced near the surface, while for increasing values of the in-plane compression modulus λ the susceptibility decreases

distance from the surface. To do this we will need to know the flow in both the monolayer and the bulk subphase. Given the linearity of the equations of motion, it is permissible to divide the solution of the Stokes equations in the bulk into two parts, i.e., $\mathbf{u} = \mathbf{u}^{(1)} + \mathbf{u}^{(2)}$. The first part $\mathbf{u}^{(1)}$ is the flow field resulting from a point force at the position of the probe, but with the monolayer replaced by a *perfect* slip surface; this field satisfies the force balance in the bulk but does not satisfy the stress balance condition on the monolayer. To correct this, we add the second solution $\mathbf{u}^{(2)}$, which is the fluid velocity field induced by the surface stresses arising from $\mathbf{u}^{(1)}$ and satisfying the homogeneous equations. The physical solution, $\mathbf{u}^{(1)}$ + $\mathbf{u}^{(2)}$, now satisfies all the necessary stress balance conditions in the bulk and in the interface. Associated with each bulk velocity field $\mathbf{u}^{(1,2)}$ there is an in-plane membrane displacement $\xi^{(1,2)}$ determined by the no-slip matching condition, and from which the interfacial displacement field can be computed.

The perfect slip solution $\mathbf{u}^{(1)}$ can be obtained using image solutions (see, for example, [83–85]). By assuming that the interface at $z = 0$ can exert no shear stresses we find the perfect slip solution in the spatial Fourier domain to be

$$u_j^{(1)} = \frac{F_x}{\eta k^2}\left(\delta_{xj} - \hat{k}_x\hat{k}_j\right)\cos(k_z d), \tag{4.20}$$

$$u_\beta^{(1)} = \frac{F_x}{2\eta\omega i}e^{-k_\perp d}\left[\frac{-2\delta_{x\beta}}{k_\perp} + \frac{k_x k_\beta}{k_\perp^3}(1 + k_\perp d)\right], \tag{4.21}$$

where k and $k_\perp$ are the three- and two-dimensional (in the plane of the interface) wavevectors, respectively. By symmetry, the normal velocity $u_z^{(1)}$ vanishes at the surface ($z = 0$), so that the associated normal displacement is zero, $\xi_z^{(1)} = 0$. Projecting the in-plane components of the interfacial velocity into longitudinal and transverse channels, $\xi_\beta = \xi_\beta^{(L)} + \xi_\beta^{(T)}$, we find

$$\xi_\beta^{(L1)} = \mathscr{L}_{\beta\alpha}\xi_\alpha^{(1)} = -\frac{F_\alpha}{2\eta\omega i}e^{-k_\perp d}\left(\frac{k_\alpha k_\beta}{k_\perp^3}(1-k_\perp d)\right), \tag{4.22}$$

$$\xi_\beta^{(T1)} = \mathscr{T}_{\beta\alpha}\xi_\alpha^{(1)} = -\frac{F_\alpha}{\eta\omega i}e^{-k_\perp d}\left(\frac{\delta_{\alpha\beta}}{k_\perp} - \frac{k_\alpha k_\beta}{k_\perp^3}\right). \tag{4.23}$$

From Eq. (4.7) and noting that the fluid stresses on the interface due to $v^{(1)}$ vanish by construction (as they must for a perfect slip surface), we find that the $v^{(1)}$ solution generates unbalanced interfacial stresses

$$S_\alpha^{(L)} = -k_\perp^2(2\mu+\lambda)\xi_\alpha^{(L1)}, \tag{4.24}$$

$$S_\alpha^{(T)} = -k_\perp^2\mu\xi_\alpha^{(T1)}. \tag{4.25}$$

The presence of unbalanced stresses at the interface are due to our imposing a boundary condition that is not correct. We know that the in-plane stresses must satisfy force balance throughout the membrane, which is a more complicated condition than perfect slip or no slip. However, due to the linearity of Stokes equations, we may add additional flow solutions to cancel any component of the flow that results from a nonzero stress; this is the essence of the SPIM calculation.

The correct "counter stresses" required to cancel these unbalanced stresses coming from the perfect slip solution $v^{(1)}$ are calculated using the fundamental solutions for longitudinal and transverse flow derived in a previous section:

$$u_\alpha^{(L2)} = -i\omega\int S_\alpha^{(L)}\frac{(1+k_\perp z)\,e^{ik_\perp\cdot x_\perp + k_\perp z}}{(2\mu+\lambda)k_\perp^2 - 2i\omega\eta k_\perp}\frac{\mathrm{d}^2k_\perp}{(2\pi)^2} \tag{4.26}$$

$$u_\alpha^{(T2)} = -i\omega\int S_\alpha^{(T)}\frac{e^{ik_\perp\cdot x_\perp + k_\perp z}}{\mu k_\perp^2 - i\omega\eta k_\perp}\frac{\mathrm{d}^2k_\perp}{(2\pi)^2}. \tag{4.27}$$

The integrals over the in-plane wavevector return the solution to the real spatial domain. This result may be thought of as the effect of two-dimensional rheology of the membrane—specifically how it differs from that of a perfect slip interface used in the calculation of $\mathbf{u}^{(1)}$. In principle there is nothing stopping us from imposing a no-slip interface and then calculating the rheological difference from this simplified case, but the image solutions for a no-slip interface are more complicated than those corresponding to the perfect slip.

Performing the integrals in Eqs. (4.26) and (4.27), we determine the velocity field correction from the surface rheology $\mathbf{u}^{(2)}$ at the position of the tracer. Now it is a simple matter of using Faxén's law [83, 84] to find the change in particle velocity due to the surface rheology. This contribution to the particle's velocity needs to be added to that first contribution giving the velocity of the particle in the subphase below a rigid wall with perfect slip boundary conditions: $\mathbf{u}^{(1)}$. From this sum, we write the susceptibility matrix χ_{ij} as defined earlier. For convenience, we take the case of an incompressible membrane at the interface. In that case, the in-plane part of the response function is given by

$$\frac{\chi_{xx}}{\chi^{\infty}} = 1 - \frac{9a}{16d} + \frac{a^3}{16d^3} - \frac{3a}{2\ell}\left[\mathrm{Ei}\left(\frac{-2d}{\ell}\right) - i\pi\right] e^{\left(\frac{2d}{\ell}\right)}, \tag{4.28}$$

with $\mathrm{Ei}(x)$ as the exponential integral function and $\chi^{\infty} = i/6\pi a\eta\omega$ the Stokes susceptibility of a spherical tracer infinitely far from interface. $\ell = \mu/(-i\eta\omega)$ is, once again, the Saffman–Delbrück (SD) length [86], which is complex for viscoelastic membranes. The phase of the complex SD length is determined by whether interfacial stresses are dissipative or reactive. In Fig. 4.5b we plot the response function for a membrane with varying degrees of elastic response. The first three terms in this expression correspond to the mobility of a sphere near a no-slip wall, and thus have no rheological significance.

The susceptibility approaches given by the simple Stokes drag on a sphere for any SD length for deep particles, i.e., those far from the membrane. This is to be expected and it is clear that distant tracers are useless as rheological probes of the membrane. Key microrheological data is obtained from probes near the membrane. To improve signal-to-noise ratios in measurements, we have found it advantageous to observe the change in the fluctuation spectrum of the probe as a function of depth near to the membrane. More details can be found in Ref. [82].

The first tests of this method involved studies of the fluctuation spectrum of tracers at various depths below monolayers of arachidic acid (AA), eicosanol, and dipalmitoylphosphatidylcholine (DPPC). Previous work had demonstrated three different dependencies of viscosity on surface pressure [9, 77, 78, 87–90]. AA is viscoelastic but with a large elastic modulus in the frequency range studied and thus should be indistinguishable from a rigid boundary. Each of these three types of monolayers displays distinct rheological characteristics that serve as benchmarks for testing the SPIM system.

As discussed in Ref. [82], a tracer particle was trapped at various depths beneath the monolayer using an optical tweezer. Its fluctuations in the plane parallel to the monolayer were then measured at those depths. Using the fluctuation-dissipation theorem [34] in the usual way, the imaginary part of the tracer's response function is measured from the power spectrum of those fluctuations in thermal equilibrium. To simplify the analysis, it was assumed that the three monolayers could be treated as being purely viscous, making ℓ real. The results of these measurements for tracers submerged below AA (red circles), eicosanol (blue squares), and DPPC (black triangles) monolayers as a function of surface pressure in Fig. 4.6. Fluctuations of

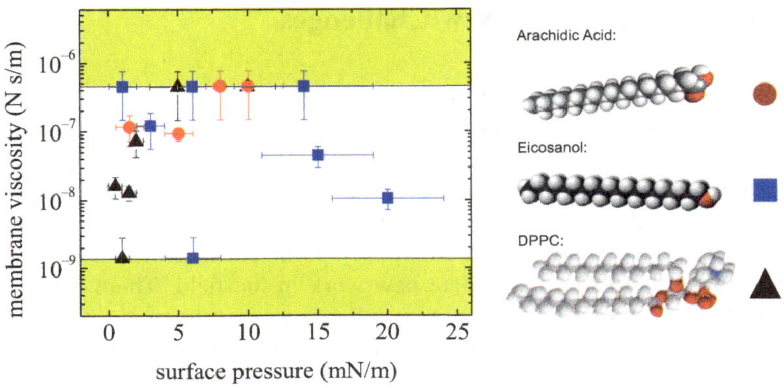

Fig. 4.6 Data obtained using noncontact microrheology. The *yellow* regions indicate areas in parameter space where the experimental technique is insensitive to changes in the viscoelastic moduli of the membrane. This sensitivity can be increased through various means (see text). The *left panel* of the figure is from Ref. [82] and used with permission

the immersed bead are measured and converted into the imaginary part of the tracer response function, normalized by the Stokes result. This is, in turn, converted into surface viscosities using Eq. (4.28).

Eicosanol has a phase transition to a fluid phase above a pressure threshold, and the rheological signature of this structural transition is indeed observed using SPIM in the expected regime (15 mN/m)—see Refs. [82, 90]—as well as the expected decreasing viscosity with increasing surface pressure. The surfaces viscosities, however, are about one order of magnitude less than that previously measured macroscopically [90]. Frequency dependence of modulus [61], errors in the macroscopic measurements, or the assumption of a purely viscous response in the fluid phase are all valid explanations for the persistence of the difference in modulus.

SPIM finds DPPC viscosities that are consistent with previous macroscopic measurements at higher pressures, indicating that the missing modulus from other microrheological techniques has been eliminated, at least in this case. A range of viscosities $(1–4) \times 10^{-7}$ Ns/m is observed at surface pressures between (2–5) mN/m. Previous active rheology measurements (using a driven rotating disk) gave 4×10^{-7} Ns/m [9] at similar pressures and temperatures, where the monolayer was in the same phase. Given the fact that SPIM reproduces rheological measurements obtained from macroscopic measurements, we believe that the correct resolution to the missing modulus problem is to be found in a correction to interfacial microrheology. Using a noncontact approach appears to be successful at providing this correction.

5 Open Questions and New Challenges

There remain a number of open questions related to membrane and monolayer microrheology. Some of these involve straightforward generalizations of the work presented above or mentioned in the references. These include finding a tractable approach to the complex mathematics associated with these sorts of hydrodynamic questions. Other theoretical questions are more open-ended; no paths towards their resolutions are obvious, and we simply point out that their solution would be interesting and likely to stimulate new work in the field. There are questions fundamentally of an experimental nature, where new instrument development may be necessary. Ultimately, the most interesting questions involve some combination of the abovementioned difficulties: new theories and new experiments are needed.

As pointed out in the introduction, biology presents a number of interesting membrane or monolayer systems composed of a mixture of lipids and protein surfactants. Generally speaking, one expects to find partial phase separation of these constituents in at least some ranges of temperature, area pressure, and composition. The cell membrane (although a nonequilibrium structure) appears to be phase separated [91]. Even single component systems present regions of phase coexistence where the membrane is inhomogeneous. With the introduction of multiple domains, translational invariance is broken in the membrane. If the characteristic length scale of the heterogeneity is small then a homogenization procedure may be employed [64]. In essence one may develop a type of mean field model or effective medium theory that incorporates the generalized elasticity of the various domains and their area fraction on the surface. The more interesting limit, however, is found in systems where the length scale of the membrane/monolayer heterogeneity is sufficiently large as to be observable. In that case, is it possible to push the submerged particle microrheology technique into a *scanning probe* mode, whereby moving the particle parallel to the interface, one obtains local rheological information about the monolayer or membrane above it? Since the principal surface stresses associated with the motion of the probe particle are essentially local, it is reasonable to suppose that such a local rheological probe is possible and that the lateral resolution of the rheological probe should scale inversely with the depth of the particle. Still, new theoretical work is necessary to learn how to interpret the fluctuation data below a heterogeneous membrane/monolayer and experimental proofs of principle are required to calibrate scanning or S-SPIM.

SPIM has the advantage of rendering irrelevant our ignorance of the direct mechanochemical coupling between the tracer probe and the monolayer. Nevertheless, the understanding of this coupling remains an open question. In some cases at least the boundary conditions assumed by Saffman and Delbrück for point-like particles [92] or Hughes Pailthorpe and White for particles with a finite size [93] appear to provide the experimentally observed mobilities [94, 95]. Other experiments, such as those by Gambin et al. [96] on engineered proteins, cannot be reconciled with these classic results for membrane/monolayer hydrodynamics. These experimental results, if supported by further work, may point to an important and fundamental problem in membrane hydrodynamics since the observed protein mobilities cannot be explained by treating the monolayer as a 2D Newtonian fluid coupled to the surrounding aqueous fluid via no-slip boundary conditions. One proposal to resolve this conflict by Naji and collaborators [97] suggested that proteins engineered to embed in the membrane cause local distortions of the lipid tilt order or chain stretch. In either case, if the dominant mode of dissipation occurs in a boundary layer around the protein due to its coupling to these other degrees of freedom, mobilities similar to those observed experimentally by Gambin *et al.* can be obtained. Coarse-grained molecular dynamic simulations for these protein/membrane interactions by Stevens and collaborators [98] are far from conclusive and the experimental results have not, to our knowledge, been corroborated by other groups so the situation remains murky. However, if the mobility of proteins in a membrane depends on the details of the local interaction between that protein and the local shell of surrounding lipids, many of our assumptions regarding the mobility of transmembrane proteins in a biological context will have to be revisited. This problem has important biological as well as physical implications and deserves further scrutiny.

Finally, a full theory of the dynamics of membranes and monolayers should account for dramatic changes in the monolayer's geometry. It is now well known that highly compressed monolayers can form giant folds [99–103]. Under some circumstances these folds are reversible, opening up to reform the flat monolayer with little to no loss of surfactant (to micelle formation) under reexpansion. This property is physiologically essential for lung surfactant. While the static structure of the folds is now understood, the associated in-plane flows during the folding process remain less well studied. Such flows play an important role in controlling the transport of particles within the monolayer and across the monolayer during cyclic compression that generates repeated folding and unfolding the monolayer. Recent unpublished experiments by M. Dennin and collaborators suggest that cyclic compression does enhance particulate transport across monolayers in a size-selective manner and that probes coupled to the monolayer become concentrated in the folds produced by the compression. We are not aware of any theoretical understanding of this underlying processes leading to these observations.

Other motions in the third dimension may be caused by active processes such as the inclusion of transmembrane/monolayer pumps, which necessarily generate active (nonthermal) forces in the direction normal to the membrane. This problem brings the rheology/dynamics of membranes and monolayers into contact with the broader field of elastic continua driven out of equilibrium by active processes. Common examples are found in biology including that of molecular motors acting on the cytoskeletal filaments. There one observes that active processes dramatically change the elastic/rheological properties of the material. It is unknown if there are similar examples in the world of membranes and monolayers.

References

1. G.P. Secundus, *Naturalis Historia: libri II*, vol. 1 Gaius Plinius Secondus, Natural History, Volume 1, Book 2, Chapter 106
2. B. Franklin, Phil. Trans. **64**, 445 (1774)
3. L. Rayleigh, Proc. Roy. Soc. Lond. **47**(286-291), 364 (1889)
4. A. Pockels, Nature **43**(1115), 437 (1891)
5. L. Rayleigh, Phil. Mag. **48**, 321 (1899)
6. B. Piknova, V. Schram, S. Hall, Curr. Opin. Struct. Biol. **12**(4), 487 (2002)
7. C. Alonso, T. Alig, J. Yoon, F. Bringezu, H. Warriner, J.A. Zasadzinski, Biophys. J. **87**(6), 4188 (2004)
8. K.E. Kasza, A.C. Rowat, J. Liu, T.E. Angelini, C.P. Brangwynne, G.H. Koenderink, D.A. Weitz, Curr. Opin. Cell Biol. **19**(1), 101 (2007)
9. K. Kim, S.Q. Choi, J.A. Zasadzinski, T.M. Squires, Soft Matter **7**, 7782 (2011)
10. V. Kaganer, H. Möhwald, P. Dutta, Rev. Mod. Phys. **71**(3), 779 (1999)
11. C.M. Knobler, R.C. Desai, Annu. Rev. Phys. Chem. **43**(1), 207 (1992)
12. A. Fischer, M. Lösche, H. Möhwald, E. Sackmann, J. Phys. Lattr. **45**(16), 785 (1984)
13. K. Kjaer, J. Als-Nielsen, C. Helm, L. Laxhuber, H. Möhwald, Phys. Rev. Lett. **58**(21), 2224 (1987)
14. C.A. Helm, H. Möhwald, K.K. r, J. Als-Nielsen, Europhys. Lett. **4**(6), 697 (1987)
15. R. Kenn, C. Böhm, A. Bibo, I. Peterson, H. Möhwald, J. Als-Nielsen, K. Kjaer, J. Phys. Chem. **95**(5), 2092 (1991)
16. S. Riviere, S. Hénon, J. Meunier, D. Schwartz, M.W. Tsao, C. Knobler, J. Chem. Phys. **101**, 10045 (1994)
17. D.R. Nelson, B.I. Halperin, Phys. Rev. B **19**, 2457 (1979)
18. C.A. Murray, *Bond-Orientational Order in Condensed Matter Systems* (Springer, New York, 1992), pp. 137–215
19. P. Bladon, D. Frenkel, Phys. Rev. Lett. **74**(13), 2519 (1995)
20. U. Seifert, Adv. Phys. **46**(1), 13 (1997)
21. M. Edidin, Nat. Rev. Mol. Cell Biol. **4**(5), 414 (2003)
22. F.I. Niordson, *Shell Theory* (Elsevier, North-Holland, 1985)
23. T.R. Powers, Rev. Mod. Phys. **82**, 1607 (2010)
24. J. Lidmar, L. Mirny, D.R. Nelson, Phys. Rev. E **68**(5), 051910 (2003)
25. R. Zandi, D. Reguera, R.F. Bruinsma, W.M. Gelbart, J. Rudnick, Proc. Natl. Acad. Sci. U.S.A. **101**(44), 15556 (2004)
26. W.S. Klug, R.F. Bruinsma, J.P. Michel, C.M. Knobler, I.L. Ivanovska, C.F. Schmidt, G.J. Wuite, Phys. Rev. Lett. **97**(22), 228101 (2006)
27. T. Kuriabova, A.J. Levine, Phys. Rev. E **77**, 031921 (2008)
28. Y. Park, C.A. Best, K. Badizadegan, R.R. Dasari, M.S. Feld, T. Kuriabova, M.L. Henle, A.J. Levine, G. Popescu, Proc. Natl. Acad. Sci. **107**(15), 6731 (2010)
29. A. Vaziri, L. Mahadevan, Proc. Natl. Acad. Sci. **105**(23), 7913 (2008)
30. B. Li, Y.P. Cao, X.Q. Feng, H. Gao, Soft Matter **8**(21), 5728 (2012)
31. J. Paulose, G.A. Vliegenthart, G. Gompper, D.R. Nelson, Proc. Natl. Acad. Sci. **109**(48), 19551 (2012)
32. A.A. Evans, A.J. Levine, Phys. Rev. Lett. **111**, 038101 (2013)
33. H. Stone, A. Ajdari, J. Fluid Mech. **369**, 151 (1998)
34. P.M. Chaikin, T.C. Lubensky, *Principles of Condensed Matter Physics* (Cambridge University Press, Cambridge, 1995)
35. A.J. Levine, F.C. MacKintosh, Phys. Rev. E **66**, 061606 (2002)
36. F. Brochard, J. Lennon, J. Phys. **36**(11), 1035 (1975)
37. S. Sinha, E. Sirota, S. Garoff, H. Stanley, Phys. Rev. B **38**(4), 2297 (1988)
38. J. Harden, H. Pleiner, P. Pincus, J. Chem. Phys. **94**, 5208 (1991)

39. J. Manneville, P. Bassereau, S. Ramaswamy, J. Prost, Phys. Rev. E Stat. Nonlin. Soft Matter Phys. **64**(2), 021908_1 (2001)
40. J. Prost, R. Bruinsma, Europhys. Lett. **33**(4), 321 (1996)
41. E. Helfer, S. Harlepp, L. Bourdieu, J. Robert, F. MacKintosh, D. Chatenay, Phys. Rev. E **63**(2), 021904 (2001)
42. R. Shlomovitz, N.S. Gov, Europhys. Lett. **84**(5), 58008 (2008)
43. R. Shlomovitz, N.S. Gov, Phys. Rev. Lett. **98**, 168103 (2007)
44. A.A. Evans, E. Lauga, Phys. Rev. E **84**, 031924 (2011)
45. P. Saffman, M. Delbrück, Proc. Natl. Acad. Sci. **72**(8), 3111 (1975)
46. H. Lamb, *Hydrodynamics*, 6th edn. (Cambridge University Press, Cambridge, 1932)
47. H. Döbereiner, J. Käs, D. Noppl, I. Sprenger, E. Sackmann, Biophys. J. **65**(4), 1396 (1993)
48. U. Seifert, K. Berndl, R. Lipowsky, Phys. Rev. A **44**(2), 1182 (1991)
49. M.L. Henle, A.J. Levine, Phys. Rev. E **81**(1), 011905 (2010)
50. M. Henle, R. McGorty, A. Schofield, A. Dinsmore, A. Levine, Europhys. Lett. **84**(4), 48001 (2008)
51. S. Aliaskarisohi, P. Tierno, P. Dhar, Z. Khattari, M. Blaszczynski, T.M. Fischer, J. Fluid Mech. **654**, 417 (2010)
52. F.G. Woodhouse, R.E. Goldstein, J. Fluid Mech. **1**(1), 1 (2012)
53. A.R. Honerkamp-Smith, F.G. Woodhouse, V. Kantsler, R.E. Goldstein, Phys. Rev. Lett. **111**, 038103 (2013)
54. Y. Sakuma, T. Taniguchi, T. Kawakatsu, M. Imai, Biophys. J. **105**(9), 2074 (2013)
55. K.C. Huang, R. Mukhopadhyay, N.S. Wingreen, PLoS Comput. Biol. **2**(11), e151 (2006)
56. R. Mukhopadhyay, K.C. Huang, N.S. Wingreen, Biophys. J. **95**(3), 1034 (2008)
57. B.J. Reynwar, G. Illya, V.A. Harmandaris, M.M. Müller, K. Kremer, M. Deserno, Nature **447**(7143), 461 (2007)
58. G.G. Fuller, J. Vermant, Annu. Rev. Chem. Biomol. Eng. **3**, 519 (2012)
59. R. Ghaskadvi, M. Dennin, Rev. Sci. Instrum. **69**(10), 3568 (1998)
60. C.F. Brooks, G.G. Fuller, C.W. Frank, C.R. Robertson, Langmuir **15**(7), 2450 (1999)
61. M. Twardos, M. Dennin, Langmuir **19**(8), 3542 (2003)
62. D. Schwartz, C. Knobler, R. Bruinsma, Phys. Rev. Lett. **73**(21), 2841 (1994)
63. H. Stone, Phys. Fluids **7**, 2931 (1995)
64. M.L. Henle, A.J. Levine, Phys. Fluids **21**, 033106 (2009)
65. T. Fischer, J. Fluid Mech. **498**, 123 (2004)
66. M.H. Lee, C.P. Lapointe, D.H. Reich, K.J. Stebe, R.L. Leheny, Langmuir **25**(14), 7976 (2009)
67. A.J. Levine, T.B. Liverpool, F.C. MacKintosh, Phys. Rev. E **69**, 021503 (2004)
68. A.J. Levine, T.B. Liverpool, F.C. MacKintosh, Phys. Rev. Lett. **93**, 038102 (2004)
69. T.G. Mason, K. Ganesan, J.H. van Zanten, D. Wirtz, S.C. Kuo, Phys. Rev. Lett. **79**, 3282 (1997)
70. T.M. Squires, T.G. Mason, Annu. Rev. Fluid Mech. **42**(1), 413 (2010)
71. M.H. Lee, D.H. Reich, K.J. Stebe, R.L. Leheny, Langmuir **26**(4), 2650 (2009)
72. J.C. Crocker, M.T. Valentine, E.R. Weeks, T. Gisler, P.D. Kaplan, A.G. Yodh, D.A. Weitz, Phys. Rev. Lett. **85**, 888 (2000)
73. A.J. Levine, T.C. Lubensky, Phys. Rev. Lett. **85**, 1774 (2000)
74. D.T. Chen, E.R. Weeks, J.C. Crocker, M.F. Islam, R. Verma, J. Gruber, A.J. Levine, T.C. Lubensky, A.G. Yodh, Phys. Rev. Lett. **90**, 108301 (2003)
75. V. Prasad, E.R. Weeks, Phys. Rev. Lett. **102**(17), 178302+ (2009)
76. V. Prasad, S.A. Koehler, E.R. Weeks, Phys. Rev. Lett. **97**(17), 176001+ (2006)
77. M. Sickert, F. Rondelez, H.A. Stone, Europhys. Lett. **79**(6), 66005+ (2007)
78. S.Q. Choi, S. Steltenkamp, J.A. Zasadzinski, T.M. Squires, Nat. Commun. **2**, 312+ (2011)
79. M. Buchanan, M. Atakhorrami, J. Palierne, F. MacKintosh, C. Schmidt, Phys. Rev. E **72**(1), 011504 (2005)
80. A.J. Levine, T.C. Lubensky, Phys. Rev. E **63**, 041510 (2001)
81. F. Ortega, H. Ritácco, R.G. Rubio, Curr. Opin. Colloid Interface Sci. **15**(4), 237 (2010)

82. R. Shlomovitz, A.A. Evans, T. Boatwright, M. Dennin, A.J. Levine, Phys. Rev. Lett. **110**, 137802 (2013)
83. J. Happel, H. Brenner, *Low Reynolds Number Hydrodynamics: With Specific Applications to Particulate Media* (Springer, Kluwer Boston, 1983)
84. S. Kim, S.J. Karrila, *Microhydrodynamics: Principles and Selected Applications* (Dover Publications, New York, 2005)
85. C. Pozrikidis, *Boundary Integral and Singularity Methods for Linearized Viscous Flow* (Cambridge University Press, Cambridge, UK, 1992)
86. P.G. Saffman, M. Delbrück, Proc. Natl. Acad. Sci. **72**(8), 3111 (1975)
87. J. Krägel, J.B. Li, R. Miller, M. Bree, G. Kretzschmar, H. Möhwald, Colloid Polymer Sci. **274**, 1183 (1996)
88. R.S. Ghaskadvi, J.B. Ketterson, P. Dutta, Langmuir **13**(19), 5137 (1997)
89. R.S. Ghaskadvi, S. Carr, M. Dennin, J. Chem. Phys. **111**, 3675 (1999)
90. C.F. Brooks, G.G. Fuller, C.W. Frank, C.R. Robertson, Langmuir **15**, 2450 (1999)
91. K. Simons, W.L. Vaz, Annu. Rev. Biophys. Biomol. Struct. **33**, 269 (2004)
92. N. Oppenheimer, H. Diamant, Biophys. J. **96**(8), 3041 (2009)
93. B. Hughes, B. Pailthorpe, L. White, J. Fluid Mech. **110**(1), 349 (1981)
94. S.L. Veatch, S.L. Keller, Phys. Rev. Lett. **94**(14), 148101 (2005)
95. P. Cicuta, S.L. Keller, S.L. Veatch, J. Phys. Chem. B **111**(13), 3328 (2007)
96. Y. Gambin, R. Lopez-Esparza, M. Reffay, E. Sierecki, N. Gov, M. Genest, R. Hodges, W. Urbach, Proc. Natl. Acad. Sci. U.S.A. **103**(7), 2098 (2006)
97. A. Naji, A.J. Levine, P.A. Pincus, Biophys. J. **93**(11), L49 (2007)
98. M.J. Stevens, J. Chem. Phys. **121**(23), 11942 (2004)
99. W. Lu, C.M. Knobler, R.F. Bruinsma, M. Twardos, M. Dennin, Phys. Rev. Lett. **89**(14), 146107 (2002)
100. L. Pocivavsek, S.L. Frey, K. Krishan, K. Gavrilov, P. Ruchala, A.J. Waring, F.J. Walther, M. Dennin, T.A. Witten, K.Y.C. Lee, Soft Matter **4**(10), 2019 (2008)
101. J. Fang, M. Dennin, C.M. Knobler, Y.K. Godovsky, N. Makarova, H. Yokoyama, J. Phys. Chem. B **101**(16), 3147 (1997)
102. A. Gopal, K.Y.C. Lee, J. Phys. Chem. B **105**(42), 10348 (2001)
103. K.Y.C. Lee, Annu. Rev. Phys. Chem. **59**, 771 (2008)

Chapter 5
Rheology and Mechanics of the Cytoskeleton

Hamed Hatami-Marbini and Mohammad R.K. Mofrad

Abstract The cytoskeleton is a series of intertwined proteins with the primary function of providing the cell with structure and shape. The distinctive rheology and mechanical properties of the cytoskeleton are detrimental for cell's ability to perform its different and often critical functions in health and disease. From the mechanics point of view, the cytoskeleton is a dynamic and constantly changing structure which can be considered as a randomly crosslinked semiflexible polymer network. This chapter briefly reviews salient features of the cytoskeletal biopolymer network and discusses various computational and numerical methods that have been proposed for investigating its rheology and mechanical properties. In particular, an overview of the physical and mechanical behavior of individual constituents of the three-dimensional cytoskeletal network, i.e., F-actin, microtubules, intermediate filaments, binding proteins, and cross-link filaments, is first given. Then recent advancements on characterizing the response of cells to external stimuli are discussed. For this purpose, a summary of common experimental and computational studies on cytoskeletal rheology and mechanics are presented and critically compared against each other.

1 Introduction: Intracellular Structure and Composition

Cells, complex systems of self-assembled polymers, are basic building blocks of all living things. They are composed of numerous subunits with distinctive rheology and structural properties. Cells respond and interact with their surrounding environment by changing their morphology and biological signaling; therefore,

---

H. Hatami-Marbini (✉)
School of Mechanical and Aerospace Engineering, Oklahoma State University, Stillwater, OK 74078, USA
e-mail: hhatami@okstate.edu

M.R.K. Mofrad
Molecular Cell Biomechanics Laboratory, Departments of Bioengineering and Mechanical Engineering, University of California Berkeley, Berkeley, CA 94720, USA
e-mail: mofrad@berkeley.edu

S.E. Spagnolie (ed.), *Complex Fluids in Biological Systems*, Biological and Medical Physics, Biomedical Engineering,
DOI 10.1007/978-1-4939-2065-5_5

many aspects of cellular physiology such as motility and force generation directly rely upon their shape and structure. Cellular force generation often leads to remodeling of the extracellular matrix and/or pathological conditions. The particular shape and integrity of cells are determined by their internal structure, cytoskeleton, which lies between the nucleus and membrane. The cytoskeleton plays numerous roles and can be studied from different perspectives. For example, many cells spread and migrate during the development as well as at maturity for wound healing and cancer metastasis among others. Moreover, cells sense mechanical forces and convert them into a cascade of biochemical signaling events influencing their function and properties. The possibility of examining cells from a variety of perspective is why there exists such a diverse conceptual opinion about the mechanical properties of cytoskeleton. While biophysicists often model the cytoskeleton as a series of randomly cross-linked polymer chains, structural engineers see the cytoskeleton as a continuum over a certain length scale. It is noted that these very different models for cytoskeletal mechanics have often been backed by compelling experimental measurements; therefore, future studies are required for developing a better and possibly unified description for all of the observed phenomena.

The primary objective of this chapter is to present a concise overview of various key experimental and computational models regarding cytoskeletal rheology and mechanics. The cytoskeleton is a three-dimensional composite network of actin filaments, microtubules, and intermediate filaments through which water, solutes, and small organelles diffuse. Depending on the observation length scale, network fiber density, and degree of filament cross-linking, the cytoskeleton and subsequently the cells can have wide-range material properties from a viscous fluid to an elastic gel. For instance, while cells behave as a viscous fluid in slow deformation, they show elastic behavior in fast deformations. The main microstructural constituents of the cytoskeleton are actin filaments, microtubules, intermediate filaments, binding proteins, and cross-link filaments.

Actin filaments are considered as the primary structural component of cells; they play a key role in cell migration and respond rapidly to external forces. Actin filaments are twisted strands of filaments with a diameter of 7–9 nm and structural polarity, Fig. 5.1a. They constitute about 1 to 10 percent of all the proteins in most cells and are formed by polymerization of the globular monomeric actin (G-actin). The F-actin is composed of about 375 amino acids with a molecular weight of about 43 KDa. The growth of the filament occurs when adenosine-5'-triphosphate (ATP) binds to the barbed end and is hydrolyzed to ADP. Depolymerization occurs as ADP bound actin drops off the pointed ends. The polymerization and depolymerization processes are regulated by many factors such as ionic concentration of the surrounding environment, capping, binding, branching, and severing proteins. Actin filaments are the primary structural component of most cells and fall under the category of stiff structures [1–7]. These filaments have a persistence length of several micrometers and an effective Young's modulus of about 1–3 MPa. To put this value into perspective, note that the Young's moduli of polystyrene and bone are approximately 3 MPa and 9 MPa, respectively.

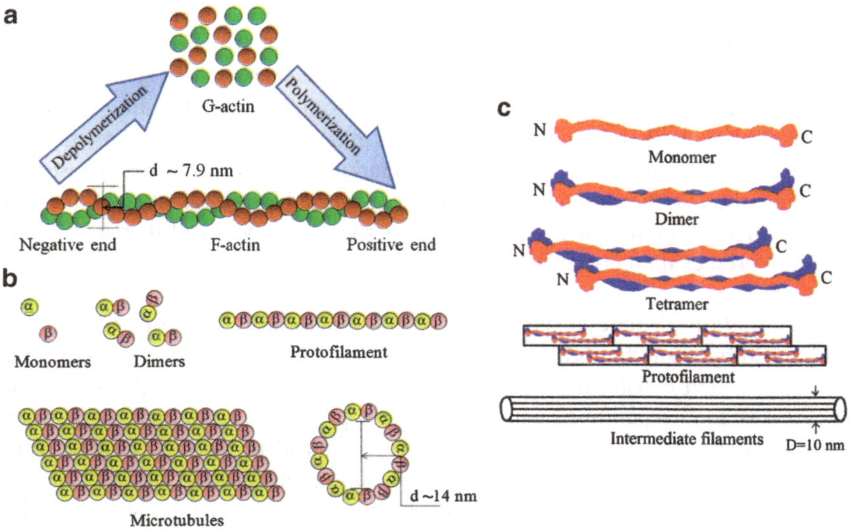

Fig. 5.1 Three main constituents of the cytoskeleton: (**a**) actin filaments, (**b**) microtubules, and (**c**) intermediate filaments. (**a**) Actin filaments are formed by polymerization of actin monomers (G-actin). The filaments have a distinct polarity because actin monomers are oriented in the same direction. Actin polymerization is a reversible process and G-actins can associate and dissociate from the filaments. (**b**) The microtubules are hollow rods with an inner diameter of about 14 nm and are composed of thirteen protofilaments. The protofilaments are built by assembly of dimers of α-tubulin and β-tubulin. (**c**) Intermediate filaments are composed of approximately eight protofilaments; each protofilament is formed from the end-to-end association of tetramers. The building blocks of tetramers are dimers which are two polypeptides wound around each other in a coiled–coiled structure

Microtubules are another component of the cytoskeleton; they are polymerized filaments constructed from tubulin, a heterodimer of α-tubulin and β-tubulin, and are arranged in a helical arrangement with a molecular weight of about 55 KDa, Fig. 5.1b. These filaments are in the form of small hollow cylinders with an outer diameter of about 25 nm. The bending stiffness of the microtubules is higher than that of the actin filaments. These filaments have a large persistence length of about 6 mm which is three orders of magnitudes larger than the average length scale of the cell [8]. Microtubules are highly dynamic with constant polymerization and depolymerization and a half-life of only a few minutes. They are involved in a number of cellular processes such as vesicle transport and cell division [9, 10].

Intermediate filaments are the third set of cytoskeletal filaments which are found nearly in all eukaryotic cells. The major function of intermediate filaments is to stabilize the cytoskeletal structure and integrity. They are smaller in size than microtubules but larger than microfilaments. The proteins forming intermediate filaments, Fig. 5.1c, are divided into five groups based on gene structure, primary structure, assembly properties, and development-regulated tissue-specific expression patterns [11]. Unlike F-actin and microtubules, intermediate filaments are composed of filamentous proteins with no enzymatic activity and are in the form of coiled-coil bundles of protofilaments. They are nonpolar structures with long-term stability, little solubility in salts, and no vectorial transport role. The persistence length of intermediate filaments is about few micrometers with bending modulus of about 300 MPa [12, 13]. They are very flexible in bending and can be stretched up to three-folds before rupture [14, 15]. Their remarkable extensibility is because, upon stretching, the α-helical coiled-coil dimer is converted into β-sheet-type structures. Because of their nonlinear mechanical properties, it is believed that intermediate filaments protect the cells from large deformation by acting as mechanical stress absorbers.

The structure of the cytoskeleton (e.g., the length and density of its filaments) is regulated by many binding proteins and cross-link filaments such as filamin and α-actinin, which constitutes a pathway for transmitting mechanical force through the cytoskeleton. Depending on the length and degree of cross-linkings of cytoskeletal fibers, cells can exhibit a wide range of material properties ranging from a viscous fluid to an elastic gel. The cytoskeletal components are constantly changing in order to accommodate both slow and fast cell dynamic processes such as cellular movement and division. In general, a cell behaves elastically (or viscoelastically) in response to quick (or slow) deformations through arrangement and rearrangement of microfilaments. For instance, treadmilling is a dynamic process where the addition and removal of actin subunits from two ends of a microfilament occur at the same rate.

A variety of actin-binding proteins (ABPs) regulates the formation of actin cytoskeleton network. These actin-binding proteins are an essential part of the cell and play important roles ranging from regulating assembly and disassembly of F-actins to controlling actin network dynamics and structure, Fig. 5.2. A plethora of actin monomers and a large number of actin-monomer-binding proteins are required for the rapid growth of actins in motile cells and also for their sudden reorganization in response to intra- and extracellular stimuli. These proteins are also required for the maintenance of cell integrity by interconnecting different cytoskeletal elements together. For instance, they connect actin filaments to microtubules and intermediate filaments. Another group of actin-binding proteins is molecular motor protein myosin which is involved in cytoskeleton dynamic behavior. Myosin binds to actin filaments, moves along them, and generates tensile forces. It converts the chemical energy of the ATP into mechanical energy in order to move along the cytoskeletal substrate. For example, two-headed myosin II generates tension and contraction inside the cytoskeleton.

Myosin is a motor protein which has a central role in force generation required for cellular locomotive and protrusive activities among others. Myosin is composed of two identical heavy chains and two pairs of light chains. There are three distinguishable domains in each heavy chain: the head, neck, and tail. The head region is located near the end of the two heavy chains and is where forces can be generated. The globular head domain includes binding sites for actin and is formed by folding one half of a single heavy chain. The other halves of the heavy chains twist around each other and form a helical fibrous structure. In the neck

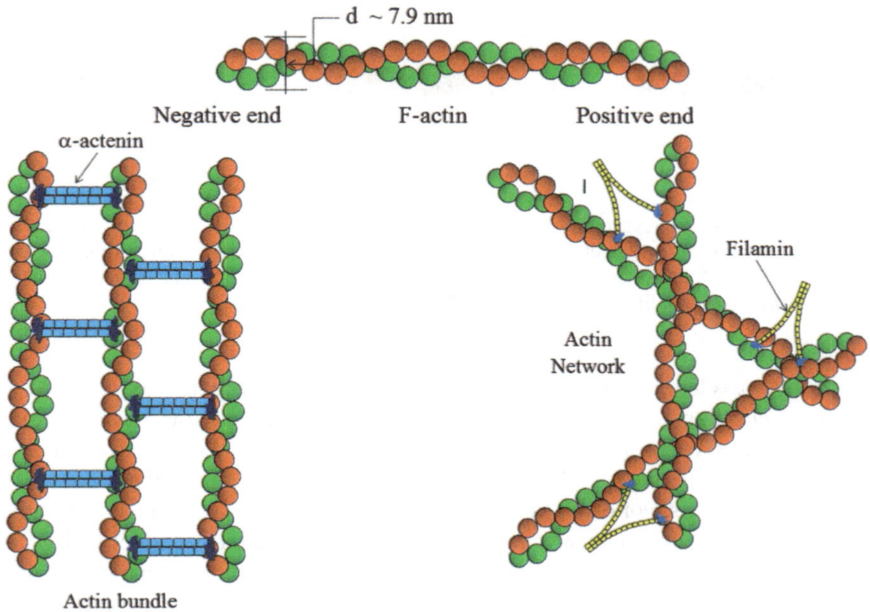

Fig. 5.2 Schematic representation of the role of actin-binding proteins (ABPs) in forming actin bundle and actin network. The smaller actin-binding proteins such as α-actinin often have different actin-binding sites and form actin bundles. On the other hand, the larger actin-binding proteins such as filamin usually push actin filaments apart and form actin networks

domain (the connection between the head and the tail), a pair of light chains stiffen and stabilize the structure. The energy required for myosin movement along actin filaments is obtained from ATP hydrolysis. The nucleotide-binding sites are at a distance of about 3.5 nm from actin-binding sites. The gamma phosphate sensor and the relay helix are essential within the myosin. Myosin uses the gamma phosphate protein sensor to distinguish between ATP- and ADP- bound states. In muscle cells, the myosin head moves 5–10 nm with respect to the sarcomere due to the lever arm rotation of about 70 degrees in each power stroke (i.e., the large sweeping movement of the myosin head), Fig. 5.3. This process starts by myosin head being released from the actin because of the ATP binding to myosin. The ATP is hydrolyzed quickly to ADP plus inorganic phosphate. The myosin head rotates from the sarcomere midline and attaches to the actin filament through ADP at a further location. Upon attachment, the ADP and inorganic phosphate are released. This causes the myosin lever arm to rotate and slide the myosin head toward the sarcomere midline. Chapter 9 discusses particle simulations and continuum kinetic theories which can be used to investigate the relation between the mechanics and interactions of solutions of motor proteins and actin filaments. These model could provide insight into the relation between the mechanics and interactions on the scale of individual particles.

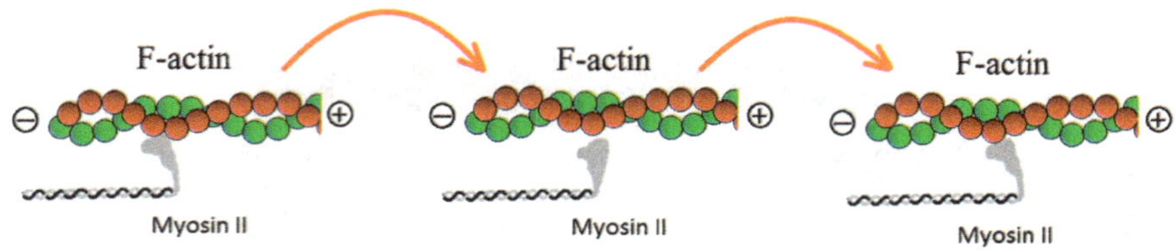

Fig. 5.3 Schematic representation of actin-myosin force generation. The interaction of myosin motor proteins with actin filaments results in the relative movement of actin filaments and myosins. This movement is ATP-dependent and results from the attachment, bending, and detachment of the myosin head to an actin filament

The cytoskeleton plays important roles in cell mechanics as well as in mechanosensation which itself regulates the cellular response to mechanical loadings. Cells can sense and respond to physical forces through a number of biological pathways such as changes in membrane channel activity, gene expression, or protein synthesis. For example, hair cells, located in the sensory cells of the inner ear, transduce the mechanical vibration of inner ear fluid into an electrical signal which propagates to the brain [16–18]. The details of force transmission into the ion channel during hair cell excitation are not fully known. Nevertheless, similar mechanosensitive ion channels such as the mechanosensitive channel of large conductance (MscL) have been investigated extensively [16, 19]. For instance, molecular dynamics simulations have shown how mechanical stress in the cell membrane acts directly on ion channels and may possibly change their conductance [20].

To date, a large number of computational and experimental methods have been developed in order to characterize the rheology and mechanics of the cytoskeleton. Although the importance of the cytoskeleton is well established, the field still lacks a complete understanding of the biophysical properties of the cytoskeletal network and there is no general theoretic model. In particular, the relationship between microstructural details and the macroscopic rheological behavior of the cytoskeleton remains elusive. The remaining of this chapter discusses some of the common theoretical and experimental methods that have been proposed to model and measure cytoskeletal rheology.

2 Cytoskeletal Rheology and Mechanics

The mechanical properties of the cytoskeleton have been studied using a number of different experimental and computational techniques. The primary cytoskeletal ingredients (F-actin, microtubules, and intermediate filaments) bundle together and form a dynamic network with unique properties. This network has an essential role in a variety of cellular functions such as adhesion, migration, and mechano-transduction. This section reviews some of the computational and experimental models that have been proposed for investigating the cytoskeletal rheology and mechanics.

Fiber networks are divided into different categories depending on the mechanical properties of the filaments and their network structure. The cytoskeleton is a semiflexible fiber network. The deformation of a filament in thermal networks is dominated by a strong entropic component which is often modeled as an entropic spring model,

$$f = -T\frac{\partial S}{\partial \ell}, \tag{5.1}$$

where f is the thermodynamic force, S is the entropy, ℓ is the filament contour length, and T is the absolute temperature [21]. The deformation of athermal semiflexible fibers is mainly enthalpic. Semiflexible networks include filamentous aggregates that are heavily cross-linked on the length scale proportional to their thermal persistence length, i.e., the length at which the thermal bending fluctuations become apparent. In particular, the persistence length,

$$\ell_p = \kappa / K_B T, \tag{5.2}$$

characterizes the competition between the bending energy and the thermal fluctuations. Here, K_B is the Boltzmann constant, and κ is the bending stiffness of the filament [22, 23]. In semiflexible fibers, the persistence length exceeds other relevant length scales of the polymer chain; therefore, there exists strong correlation on length scales larger than the average segment length. The average segment length ℓ_c equals the average filament length divided by the average number of segments. Many examples of structures composed of semiflexible polymers are found in biology. In cytoskeleton, the filamentous actin (F-actin) and microtubules have respective persistence lengths of $\sim$10 μm and $\sim$7 mm while their contour length is between 10 and 50 microns; therefore, $\ell_p > \ell_c$, [1, 8, 24, 25]. In semiflexible polymers, the energy is stored in bending, axial, torsion, and shear modes of deformation. The response of a filament in two dimensions can be modeled according to the classical linear elastic continuum beam theory which gives the elastic strain energy (Hamiltonian) as

$$U = \frac{1}{2}\int_{\ell}\left(\kappa(\nabla\psi(s))^2 + \eta\,(d\ell(s)/ds)^2\right) ds \quad \text{in 2D}, \tag{5.3}$$

where $\psi(s)$ is the rotation of the cross-sectional plane of the beam, $\ell(s)$ is the location of the beam cross section, κ and η are the bending and axial stiffness of the beam, and s is the beam arc-length parameter [26–28]. In general, the macroscopic behavior of a semiflexible network and its relation to the properties of its constituents are quite different than that of a flexible network. Thus, none of the many models developed for the behavior of flexible networks, such as rubbers, can accurately predict the mechanics of semiflexible systems [26, 29–34]. The mechanics of semiflexible fiber networks exhibit very interesting (and sometimes peculiar) behaviors. The cytoskeleton is highly heterogeneous and consists of semiflexible filaments with interconnection lengths on the same order as the fiber persistence length. Although many studies have been focused on elucidating the biophysical properties of cytoskeleton, the underlying physical origins of its complex and heterogeneous microstructure and mechanics are still unknown. Cytoskeleton is an active and dynamic structure undergoing constant remodeling in response to external and internal stimulants. For example, it has experimentally been observed that F-actin networks subjected to oscillatory shear deformation show negative normal stresses as large as shear stresses [35]. This phenomenon does not appear in the behavior of flexible polymer networks. In flexible fiber networks, similar to most materials, the tendency to expand in the direction normal to the applied shear forces yields positive normal stresses. Both stiffening and softening are among other unique properties of actin networks. Unlike common polymer gels, networks of semiflexible filaments nonlinearly stiffen in order to resist large deformations and maintain the network integrity [36, 37]. Furthermore, a reversible compressive stress-softening response in actin networks has been observed; this behavior is deemed to be necessary for preventing catastrophic fracture of the structure. In contrast to semiflexible networks, flexible and rigid polymers only exhibit monotonic stiffening when subjected to compression. The origins of above macroscopic behaviors are under study.

2.1 Experimental Studies

F-actin networks behave nonlinearly and can exhibit very different mechanical properties depending on the length scale of observations and experimental conditions. Therefore, it is necessary to pay extra attention to implications of length scales when analyzing and/or developing models for the mechanics of cytoskeleton [38]. The cytoskeleton has viscoelastic properties and can show characteristics of both elastic solids and viscous fluids. The effective elastic modulus and the amount of stored/dissipated mechanical energy in the cytoskeletal network can be studied by applying oscillatory external stresses. As it is discussed in Chap. 1, in these frequency-dependent viscoelasticity measurements, the reaction stress $\sigma_0 \sin(\omega t + \delta)$ due to the application of a small-amplitude oscillatory strain $\varepsilon_0 \sin(\omega t)$ is measured. Here σ_0 is the stress amplitude, ε_0 is the strain amplitude, δ is the phase shift between stress and strain, ω is the frequency of the oscillation, and t is time. The storage elastic modulus which is a measure of the stored mechanical energy is given by

$$G'(\omega) = \frac{\sigma_0}{\varepsilon_0} \cos \delta(\omega), \tag{5.4}$$

and the loss elastic modulus which is a measure of the energy dissipated as heat is given by

$$G''(\omega) = \frac{\sigma_0}{\varepsilon_0} \sin \delta(\omega). \tag{5.5}$$

It is seen that loss and storage moduli are frequency-dependent. In other words, depending on the applied frequency, a material may show solid-like or liquid-like behavior, see Chap. 4. Complex variables are used to write the above moduli as

$$G^*(\omega) = G'(\omega) + iG''(\omega) \tag{5.6}$$

The experimental measurements of the cell microrheology often characterize the complex modulus $G^*(\omega)$ over a wide range of frequencies. The real and imaginary parts of the complex modulus are used to represent the respective elastic energy stored and the frictional energy dissipated within the cell as a function of oscillatory frequency.

Different types of microrheology experiments have been proposed to measure and characterize the rheology of the cytoskeleton. Chapter 3 of this book gives an overview of theoretical microrheology for investigating the material properties of complex biological fluids. These techniques can be divided into two broad categories of active techniques and passive techniques. In passive techniques such as passive microrheology, fluorescence correlation spectroscopy, and dynamic light scattering, the behavior of the cytoskeleton is determined by monitoring the Brownian movements of inherent or external particles due to thermal fluctuations. The thermal fluctuations of a microscopic probe in a soft medium are measurable and representative of viscoelastic parameters of the medium surrounding the probe. For example, micron-sized beads are embedded into the cytoskeletal network in passive microrheology and their displacement is measured from video recordings or laser beam interferometry [39–44]. One- and two-particle methods of this technique are commonly used. In the one-particle technique, the positions of individual particles are recorded to calculate the mean square displacement of the Brownian motion. Then the fluctuation dissipation theorem is employed to determine the complex shear modulus of the environment [45]. One of the drawbacks of this method is that active movement of the probe particles may affect viscoelastic properties of the embedding domain and cause errors in the measurements. In order to overcome this problem, two-particle microrheology has been proposed in which the cross correlation of the displacement fluctuations of two particles located at a known distance from each other is measured. While short-length fluctuations are isolated in one-particle microrheology, the bulk rheology of the cytoskeletal network is estimated using the two-particle microrheology method [42, 44, 46].

The active microrheology methods involve the application of mechanical forces and characterization of the localized properties by applying an external force at the site of the interrogation. These methods include atomic force microscope, optical traps, microplates, micropipette aspiration, and magnetic traps. For example, the

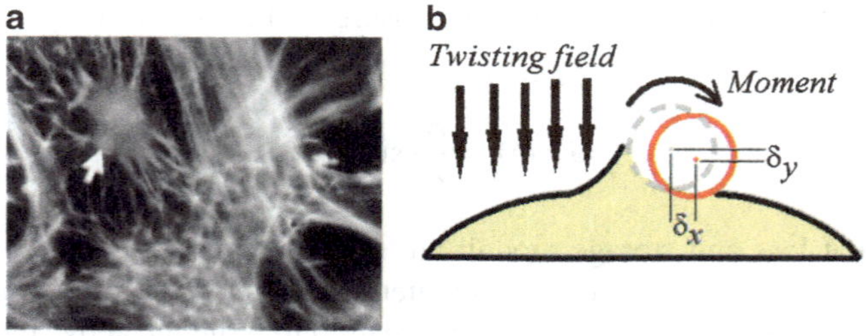

Fig. 5.4 (**a**) Ferrimagnetic beads *(arrow)* bind to the actin cytoskeleton (stained with phalloidin) of HASM cells via cell adhesion molecules (integrins). (**b**) The moment induced by a magnetic field causes the ferrimagnetic bead to rotate and to displace. The figure is adapted and obtained from Fabry et al. 2001

magnetic trap method uses the ferromagnetic microbeads coated with a protein which can bind to the intracellular proteins. These microbeads can be used to apply large forces either on the surface of cells or in the intracellular environment. The elastic modulus of cell types such as white blood cells and fibroblasts is about 1–10 KPa; therefore, large forces on the order of nano-Newtons are required to investigate the deformation of these cells. A high magnetic field is first used to magnetize the beads and create parallel magnetic moments in order to apply a torque, Fig. 5.4. Both ferromagnetic and paramagnetic particles are used to apply torques of several pico-newtons and forces up to 10 nN, respectively [47–54]. Furthermore, glass microneedles are used to apply nano-Newton or smaller forces on neurons and to initiate neurite extension [55]. Glass needles are calibrated to determine the bending constant of cells by applying small and meaningful forces. The calibration process uses a precalibrated large rod to calibrate a rod slightly smaller than itself. Each rod is then used to sequentially calibrate smaller rods until the given thin microneedle rod is calibrated. In the experiment, two needles are often mounted in a micromanipulator; one needle is calibrated for measuring the bending constant and the other needle is used as an unloaded reference point for bending of the calibrated needle and for possible drift of the micromanipulator system [56, 57].

2.2 Computational Studies: Continuum and Discrete Descriptions

The relationship between microstructural details and macroscopic rheological behavior of the cytoskeleton can be investigated using phenomenological and/or micromechanics models. Phenomenological models are based on macroscopic observations and ignore the discrete nature of the cytoskeleton. The cytoskeletal network is modeled as a continuum domain for which a macroscale constitutive relation such as a simple elastic, viscoelastic, and poroelastic model is chosen based on experimental observations and properties of interest. Chapter 1 of this book presents the mathematical framework of continuum mechanics and its application in modeling complex fluid mechanics. Continuum-level models are suitable if the length scale of interest is much larger than the characteristic length scale of the cell microstructure [48, 50, 58–60]. In other words, when considering whole-cell deformation, a cell behaves at the macroscopic length scale which is at least two to three orders of magnitudes larger than length scale of microstructural constituents. Examples of such cases are the behavior of erythrocytes/neutrophils in micropipette aspiration and magnetocytometry-induced deformations [61–64]. The micropipette aspiration method measures the cellular elasticity and viscosity by pulling on cells such as monocytes, erythrocytes (red blood cells), and leukocytes (white blood cells). In magnetocytometry, magnetic microbeads tightly bound to the cell are used to apply controlled forces to a single cell. In these situations, the length scale of the observed deformation is such that a continuum model can often capture the mechanical properties with a sufficient level of accuracy.

The continuum-level models often consider cells as single-phase materials and neglect their multiphasic nature. The biomechanical and biochemical properties of cells are influenced by ionic water solution and charged macromolecules forming the cell environment. For example, osmotic loads caused by variations in the ionic concentrations may change the cell volume. A number of multiphasic models have been developed to model possible structural roles of interactions among solid, fluid, and free ions in the cytoskeleton [65]. The first multiphasic model and most of the related work in this area have been developed in order to study the articular cartilage and chondrocyte cells. In the biphasic theory, the mixture is assumed to be intrinsically incompressible and the balance laws for solid and fluid phases are written in order to obtain the governing partial differential equations. It is known that viscoelastic behavior in cells can arise from flow-dependent and flow-independent mechanisms. The traditional biphasic model only considers the flow dependent viscoelastic behavior in order to describe the transient cell response. Nevertheless, there are instances in which the biphasic model is unsuccessful in describing the viscoelastic behavior of the cells. For example, an elastic biphasic model is unable to accurately describe the creep and time-dependent response of chondrocytes during micropipette aspiration [66, 67]. The triphasic continuum mixture model enhances the capability of the biphasic model by its ability to better describe mechanochemical couplings between different phases [65, 68]. In continuum-level modeling of the cell behavior, the small but important Brownian motions of the cytoskeleton due to thermal fluctuations are neglected. Moreover, the infinite number of time scales, a fundamental property of the power-law rheology of the cell response, is replaced with a limited number of time constants. Finally, although inhomogeneous anisotropic properties of cells can be modeled using appropriate constitutive laws, these models often cannot be related to and derived from the details of the cell microstructure. Therefore, the continuum-level methods are highly specific to experimental observations and usually cannot be generalized.

In the methods of micromechanics, the main microstructural features of the system, which are usually characterized separately, are included in the model. One of the fundamental features missing from continuum-level models is the intrinsic fibrous and discrete structure of cytoskeleton. The long fibers in the F-actin cytoskeleton are cross-linked with average segment length of about 10–100 nm [69, 70]. In addition to the required discretization of the spatial domain, many small time steps are necessary for numerical stability and a reasonable accuracy. Therefore, a fully discretized numerical model of the fiber network is computationally expensive and very time-consuming. In discrete models, the cytoskeleton microstructure is represented by a network of randomly cross-linked filaments. A complete review of these models has been presented in [29]. Since actin filaments are heavily cross-linked on the scale of their thermal persistence length, bending, twisting, and stretching are common modes of deformation. Therefore, the cytoskeleton is classified as a semiflexible network with strongly non-affine behavior. As it was stated earlier, the behavior of a semiflexible fiber network and its relationship to mechanical properties of the network constituents are more complicated than those of a flexible fiber network. The commonly used models for flexible networks are not appropriate to characterize the mechanics and rheology of semiflexible networks. Despite various research efforts, a unified model to describe the elasticity of semiflexible gels such as large shear moduli, nonlinear response, and power-law scaling of the cytoskeletal rheology does not exist. The cytoskeletal filaments resist bending deformation and have thermally induced fluctuations due to Brownian forces; therefore, they can be modeled as elastic rods with bending and stretching rigidities due to both enthalpic and entropic elasticity. The Langevin equation describes the hydrodynamics drag force of the filaments through the solvent.

The presence of cross-linkers in semiflexible gels significantly influences network properties. There are two distinct single-filament responses, i.e., bending and stretching modes, which may exist in the semiflexible networks. If individual filaments are only allowed to rotate and uniaxially deform under uniform loadings and the macroscopic strain distributes uniformly throughout the medium (affine deformation assumption), affine network models are appropriate to represent the network behavior [71]. The accuracy of these models reduces as the behavior of the network becomes non-affine, i.e., bending motions of the filaments become important. The reader is referred to our recent papers for a detailed discussion on affine versus non-affine behavior [26, 28, 70]. Hatami-Marbini and Picu recently developed a new methodology to solve boundary value problems on dense random fiber networks. This novel stochastic finite element based computational technique yields the statistics of the network mechanical behavior at a desired length scale by accounting for all possible configurations [38]. In this methodology, the total potential energy of the fiber network subjected to far-field loading is minimized in order to compute the nodal displacements. A regular mesh of square elements is then overlaid on the network in order to calculate the probability distribution of average stress and strain inside the domain. The elasticity of the discrete fiber network is mapped to a homogeneous continuum domain having correlated elastic moduli. The equivalent continuum domain has the main features and correlations of the network microstructure. The efficiency of this new technique has been analyzed by comparing its predictions with those obtained from Monte Carlo simulations [29, 38, 72, 73].

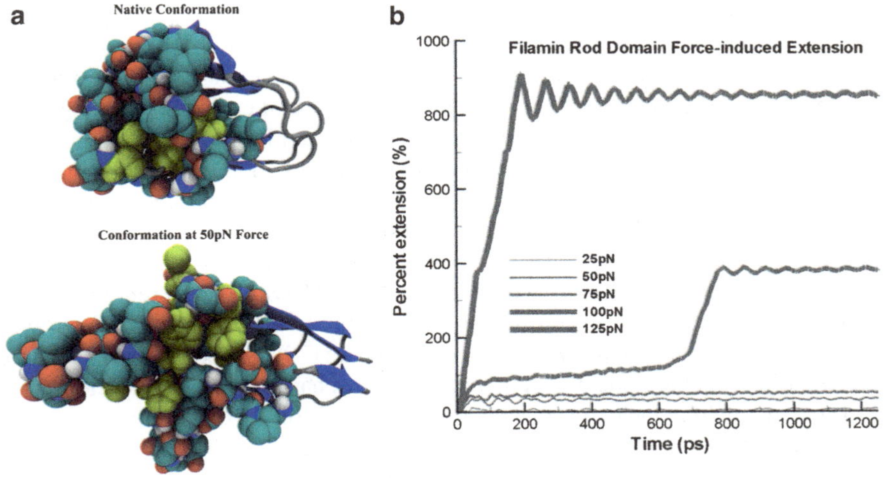

Fig. 5.5 (**a**) Molecular mechanics of filamin's rod domain; force along the rod domain to the right induces a conformational change in repeat 4 within 300 ps at $F = 50\,pN$. (**b**) The percentage extension of the whole rod domain is calculated versus time during the application of a constant force. These figures are obtained from Kolahi and Mofrad, 2008

In micromechanical models for F-actin network, the filaments are often assumed to be permanently hinged or rigidly cross-linked. Nevertheless, these cross-links undergo constant disassociation in the cytoskeleton and have variable strength depending on the ABPs' structures and properties. For instance, filamins are approximately 150 nm long V-shaped ABP which cross-link actin fibers preferentially. The filamin cross-link is composed of β-sheets which unfold under external loads and depict a sawtooth force displacement curve [74–78], Fig. 5.5. Therefore, it may not be appropriate to model the cross-links between filaments as simple rigid structural elements in discrete models of cytoskeleton. This issue is conveniently ignored in many of the current network models.

Moreover, the cytoskeleton is an active biological system which undergoes constant arrangement and rearrangement in response to various kinds of signals resulting from cell locomotion, division, and extension. The interplay of multiple phenomena including myosin contraction, actin network elasticity, and internal and external constraints imposed on the cytoskeleton creates the stress distribution inside the cell. For instance, Arp2/3 complex forms 70° branching networks, fimbrin and α-actinin put F-actins in parallel bundles, and filamin and spectrin form three-dimensional actin networks. Filamin and spectrin are protein complexes composed of multiple actin-binding domains which are responsible to arrange actin filaments into dense meshworks. Other types of ABPs are those which are used as a physical support or scaffold. They are less "directly" involved in regulating actin structure. Myosins are the most important member of this category of ABPs.

There are over fifteen different types of myosins which move their specific cargo inside the cell using actin filaments. Myosins are molecular motors which produce movement through ATP hydrolysis and can generate forces on order of pico-Newtons. The mechanical properties and structure of cytoskeleton are greatly influenced through the contribution of forces generated by myosin so their effect in cell mechanics cannot be overlooked. The effect of motor molecules may be included in the filamentous network models for the cytoskeleton. In these models, internal forces exerted by myosin motor molecules induce a state of prestress condition inside the system. These networks, whose elasticity is controlled by myosin, are often called active networks and their behavior is different than those of passive networks. It has been shown experimentally that active myosin stresses stiffen the network by about two orders of magnitude [79]. Nevertheless, a complete mathematical and theoretic model of active network has not yet been developed.

The stress field in the cytoskeletal network depends on the contractile machinery called stress fibers. They are formed by the actin-myosin interactions and are characterized by repeating units of myosin proteins. The movement of myosin motors in the stress fibers causes actin filaments to contract and slide past one another. The dynamics of stress fibers has not yet been well understood and is currently under study. However, it is known that their elasticity is a function of myosin spacings and changes over time. Cellular functions such as wound healing, proliferation, shape stability, and apoptosis depend strongly on how the force is acted upon the cell. Forces transmit into and out of the cytoskeleton through these bundles at the basal surface, where it interacts with the surrounding extracellular matrix, and where cytoskeletal contractility is resisted. At these interaction sites (called the focal adhesion), stress fibers often form [80, 81]. They also form along the direction of external forces, e.g., it is found that stress fibers orient with the direction of fluid flow in endothelial cells under shear stress [82,83]. The mechanical properties of stress fibers can be nonuniform because of the variations in myosin spacings along the stress fiber length. Moreover, the direction and strength of stress fibers depend on stress field within the cell. Forces are transferred and propagated directly and in a band-like structure from stress fibers to discrete sites on the nuclear envelope [84, 85]. As stress fibers enable the cell to focus myosin contractility along a specific direction of resistance, the cell can stiffen directionally to protect itself against excess stress and strains in this particular direction. Moreover, the cell explores the matrix rigidity and migrates in the direction of increasing rigidity through stress fibers [86]. Cell locomotion is believed to depend on and even be controlled by changes in substrate rigidity. In particular, cell response on culture surfaces is dictated by substrate rigidity: actin stress fibers are oriented along the stiffest direction of the microfabricated substrates [87].

In the literature, stress fibers have often been simulated by continuum models. These models predict the diffusion of stress away from the points of matrix attachment, as opposed to the directed and focused stress propagation observed in experiments. A recent study, however, showed that in translating the discrete microscale actin-myosin interaction to the continuum scale, these models are missing the perpendicular component of the myosin force acting on the actin filaments

[88, 89]. Inclusion of this correction in the model recovers the directed and focused nature of stress fibers. Moreover, the stress fiber arrangements for simple cell-matrix attachments predicted by this model match those observed in experimental studies. Key aspects of the interactions are missing in coarse graining and in approximating discrete macromolecular interactions by a continuum model. Therefore, a multi-scale approach retaining the discrete nature of the macromolecular interactions and keeping material properties in the discrete macromolecular scale is required. Such a model requires taking into account the discrete nature of cytoskeleton as well as the effects of hydrodynamic interactions on the behavior of individual filaments [88,89].

It has been shown that elastic modulus of stress fibers in living endothelial cells is approximately 10–15 KPa which remains constant over large strains up to 12 % [90]. This stiffness is a function of myosin II since disturbing its activity by adding myosin inhibitor blebbistatin causes a 30 % loss of the modulus [91]. The elastic modulus of fibroblast cells also decreases due to the application of this myosin inhibitor [92]. This further confirms the importance of the tension generated by myosin to the overall cellular stiffness. Despite considerable experimental and computational efforts to study basic physical principles of cell contractility due to myosin II motors, our understanding is still limited. It is not yet clear how contractility and pattern formation change with microscopic parameters such as number of myosin motors, number of cross-linkers, and density of actin filaments. It is known that the motor activity inside the cell is a controlled process which results in the formation of the stress fibers in cells on flat substrates and the contractile rings during cytokinesis [93–95]. The contractile rings, which are composed of actin, myosin II, septins, and GTP-binding hetero-oligomers, generate a furrow which partitions one cell into two. There have been some efforts to describe these rings using continuum-level hydrodynamics models whose accuracy have been confirmed by in vitro studies [96, 97]. In these models, the actin network is modeled as an active viscoelastic polar gel which is forced out of its equilibrium state because of the ATP hydrolysis. These studies suggest that the assembly and disassembly of cytoskeletal structures can be tuned via varying the concentration of local myosin II. Furthermore, F-actin cross-linkers such as filamin A and α-actinin should be present for the contraction of F-actin networks induced by myosin II at physiological ATP concentrations. In other words, myosin cannot generate large forces if actin filaments are not cross-linked. The dependence of contractility on the number of cross-linkers and myosin motors per actin filament has been studied by constructing well-controlled model system of purified actin, myosin, and α-actinin. The results of recent theoretical studies show that the force generated by myosin is not sufficiently large for ring contraction during cytokinesis unless the actin filaments are heavily cross-linked.

The study of reconstituted biopolymer networks is an excellent way to model and understand the important mechanical features of living cytoskeleton [79, 97]. These reconstituted networks of filamentous actin combined with myosin motors are a new class of active materials whose mechanical properties can be adjusted and tailored by enzymatic activities [98]. It has been observed that molecular motor proteins can either stiffen or weaken the F-actin solution depending on the density of F-actin cross-linkers. While the presence of actin cross-linker provides sites for mechanical anchorage and accommodates internal tension, active filament sliding occurs in their absence because of myosin activity. Protein myosin permits active control over the mechanical behavior of F-actin network solutions. For instance, the stress relaxation time of actin solutions will shorten upon addition of molecular motor myosin. The interaction of single polymer chains with surrounding polymers induces the viscoelastic behavior. The viscoelasticity of actin-myosin networks can be modulated using the ability of myosin to supersede reptation with sliding motion: myosin II replaces the thermally driven transport of individual polymers (snake-like movement described by reptation model) with active filament sliding motion. In reptation model, a tube represents the topological constraints of neighboring polymers on the movement of a single chain. A modified tube model has been proposed to model the dynamics of polar actin filaments with active, motile centers generating longitudinal motion [99]. Molecular dynamics simulations have also been used to investigate the rheology and the structure of F-actin solution interacting with molecular motors [100].

In summary, this chapter reviewed important features of the cytoskeletal biopolymer network. The physical and mechanical behavior of individual constituents of the cytoskeleton, i.e., F-actin, microtubules, intermediate filaments, binding proteins, and cross-link filaments, was discussed. A summary of the experimental and computational methods that have been commonly used to investigate the rheology and mechanics of the cytoskeleton was also presented. We hope that this chapter would help the readers become more familiar with this field of research and appreciate the fact that despite much have been learned about cytoskeletal mechanics and rheology, still little is known about the underlying molecular mechanisms!

References

1. H. Hatami-Marbini, M.R. Mofrad, *Cellular and Biomolecular Mechanics and Mechanobiology* (Springer, Berlin Heidelberg, 2011), pp. 3–27
2. H. Hatami-Marbini, M.R. Mofrad, *Advances in Cell Mechanics* (Springer, Berlin Heidelberg, 2011), pp. 167–188
3. M.R. Mofrad, Annu. Rev. Fluid Mech. **41**, 433 (2009)
4. M.R. Mofrad, R.D. Kamm, *Cytoskeletal Mechanics: Models and Measurements in Cell Mechanics* (Cambridge University Press, Cambridge, 2006)
5. Y. Tsuda, H. Yasutake, A. Ishijima, T. Yanagida, Proc. Natl. Acad. Sci. **93**(23), 12937 (1996)
6. H. Higuchi, Y.E. Goldman, Biophys. J. **69**(4), 1491 (1995)
7. R. Yasuda, H. Miyata, K. Kinosita Jr, J. Mol. Biol. **263**(2), 227 (1996)
8. F. Gittes, B. Mickey, J. Nettleton, J. Howard, J. Cell Biol. **120**(4), 923 (1993)
9. T. Mitchison, M. Kirschner, Nature **312**(5991), 237 (1984)
10. M.K. Gardner, D.J. Odde, Curr. Opin. Cell Biol. **18**(6), 639 (2006)
11. L. Chang, R.D. Goldman, Nat. Rev. Mol. Cell Biol. **5**(8), 601 (2004)
12. C. Guzman, S. Jeney, L. Kreplak, S. Kasas, A. Kulik, U. Aebi, L. Forro, J. Mol. Biol. **360**(3), 623 (2006)
13. N. Mücke, L. Kreplak, R. Kirmse, T. Wedig, H. Herrmann, U. Aebi, J. Langowski, J. Mol. Biol. **335**(5), 1241 (2004)

14. L. Kreplak, D. Fudge, Bioessays **29**(1), 26 (2007)
15. D.S. Fudge, K.H. Gardner, V.T. Forsyth, C. Riekel, J.M. Gosline, Biophys. J. **85**(3), 2015 (2003)
16. O.P. Hamill, B. Martinac, Physiol. Rev. **81**(2), 685 (2001)
17. A. Hudspeth, Comptes Rendus Biol. **328**(2), 155 (2005)
18. A. Hudspeth, Y. Choe, A. Mehta, P. Martin, Proc. Natl. Acad. Sci. **97**(22), 11765 (2000)
19. G. Chang, R.H. Spencer, A.T. Lee, M.T. Barclay, D.C. Rees, Science **282**(5397), 2220 (1998)
20. J. Gullingsrud, D. Kosztin, K. Schulten, Biophys. J. **80**(5), 2074 (2001)
21. M. Rubinstein, R.H. Colby, *Polymer Physics* (Oxford University Press, Oxford, 2003)
22. J. Howard, *Mechanics of Motor Proteins and the Cytoskeleton* (Sinauer Associates, 2001)
23. J.R. Van der Maarel, *Introduction to Biopolymer Physics* (World Scientific, Singapore, 2008)
24. C.P. Brangwynne, G.H. Koenderink, E. Barry, Z. Dogic, F.C. MacKintosh, D.A. Weitz, Biophys. J. **93**(1), 346 (2007)
25. J. Käs, H. Strey, J. Tang, D. Finger, R. Ezzell, E. Sackmann, P. Janmey, Biophys. J. **70**(2), 609 (1996)
26. D. Head, A. Levine, F. MacKintosh, Phys. Rev. E **68**(6), 061907 (2003)
27. M. Bai, A.R. Missel, A.J. Levine, W.S. Klug, Acta Biomater. **7**(5), 2109 (2011)
28. H. Hatami-Marbini, R. Picu, Phys. Rev. E **77**(6), 062103 (2008)
29. H. Hatami-Marbini, C.R. Picu, *Advances in Soft Matter Mechanics* (Springer, Berlin Heidelberg, 2012), pp. 119–145
30. R. Picu, Soft Matter **7**(15), 6768 (2011)
31. P.K. Purohit, R.I. Litvinov, A.E. Brown, D.E. Discher, J.W. Weisel, Acta Biomater. **7**(6), 2374 (2011)
32. M.C. Wang, E. Guth, J. Chem. Phys. **20**(7), 1144 (2004)
33. M.C. Boyce, E.M. Arruda, Rubber Chem. Tech. **73**(3), 504 (2000)
34. L.R.G. Treloar, *The Physics of Rubber Elasticity* (Oxford University Press, Oxford, 1975)
35. P.A. Janmey, M.E. McCormick, S. Rammensee, J.L. Leight, P.C. Georges, F.C. MacKintosh, Nat. Mater. **6**(1), 48 (2007)
36. M. Gardel, J. Shin, F. MacKintosh, L. Mahadevan, P. Matsudaira, D. Weitz, Science **304**(5675), 1301 (2004)
37. C. Storm, J.J. Pastore, F.C. MacKintosh, T.C. Lubensky, P.A. Janmey, Nature **435**(7039), 191 (2005)
38. H. Hatami-Marbini, R. Picu, Phys. Rev. E **80**(4), 046703 (2009)
39. J.C. Crocker, D.G. Grier, J. Colloid Interface Sci. **179**(1), 298 (1996)
40. A. Pralle, M. Prummer, E.L. Florin, E. Stelzer, J. Hörber, Microsc. Res. Tech. **44**(5), 378 (1999)
41. B. Schnurr, F. Gittes, F. MacKintosh, C. Schmidt, Macromolecules **30**(25), 7781 (1997)
42. J.C. Crocker, M.T. Valentine, E.R. Weeks, T. Gisler, P.D. Kaplan, A.G. Yodh, D.A. Weitz, Phys. Rev. Lett. **85**(4), 888 (2000)
43. A.J. Levine, T. Lubensky, Phys. Rev. Lett. **85**(8), 1774 (2000)
44. A.W. Lau, B.D. Hoffman, A. Davies, J.C. Crocker, T.C. Lubensky, Phys. Rev. Lett. **91**(19), 198101 (2003)
45. P. Janmey, C. Schmidt, *Cytoskeletal Mechanics: Models and Measurements*, ed. by M.R.K. Mofrad, R.D. Kamm (Cambridge University Press, Cambridge, 2006), pp. 18–49
46. M. Gardel, M. Valentine, J.C. Crocker, A. Bausch, D. Weitz, Phys. Rev. Lett. **91**(15), 158302 (2003)
47. P. Valberg, J. Butler, Biophys. J. **52**(4), 537 (1987)
48. A.R. Bausch, F. Ziemann, A.A. Boulbitch, K. Jacobson, E. Sackmann, Biophys. J. **75**(4), 2038 (1998)
49. D. Stamenović, N. Rosenblatt, M. Montoya-Zavala, B.D. Matthews, S. Hu, B. Suki, N. Wang, D.E. Ingber, Biophys. J. **93**(8), L39 (2007)
50. B. Fabry, G.N. Maksym, J.P. Butler, M. Glogauer, D. Navajas, J.J. Fredberg, Phys. Rev. Lett. **87**(14), 148102 (2001)
51. G.T. Charras, M.A. Horton, Biophys. J. **83**(2), 858 (2002)

52. H. Karcher, J. Lammerding, H. Huang, R.T. Lee, R.D. Kamm, M.R. Kaazempur-Mofrad, Biophys. J. **85**(5), 3336 (2003)
53. X. Trepat, M. Grabulosa, L. Buscemi, F. Rico, B. Fabry, J.J. Fredberg, R. Farré, Rev. Sci. Instrum. **74**(9), 4012 (2003)
54. X. Trepat, L. Deng, S.S. An, D. Navajas, D.J. Tschumperlin, W.T. Gerthoffer, J.P. Butler, J.J. Fredberg, Nature **447**(7144), 592 (2007)
55. D. Bray, Dev. Biol. **102**(2), 379 (1984)
56. S.R. Heidemann, S. Kaech, R.E. Buxbaum, A. Matus, J. Cell Biol. **145**(1), 109 (1999)
57. S. Heidemann, P. Lamoureux, R. Buxbaum, (ed.), *The Neuron in Tissue Culture. LW Haynes* (Wiley, London, 1999), pp. 105–119
58. E. Evans, A. Yeung, Biophys. J. **56**(1), 151 (1989)
59. Y. Fung, S. Liu, J. Biomech. Eng. **115**(1), 1 (1993)
60. D.P. Theret, R.M. Nerem, L.T. Wheeler, M. Levesque, M. Sato, J. Biomech. Eng. **110**(3), 190 (1988)
61. D.E. Discher, D.H. Boal, S.K. Boey, Biophys. J. **75**(3), 1584 (1998)
62. R.M. Hochmuth, J. Biomech. **33**(1), 15 (2000)
63. R. Rand, A. Burton, Biophys. J. **4**(2), 115 (1964)
64. S.M. Mijailovich, M. Kojic, M. Zivkovic, B. Fabry, J.J. Fredberg, J. Appl. Physiol. **93**(4), 1429 (2002)
65. H. Hatami-Marbini, *Handbook on Micromechanics and Nanomechanics*, ed. by S. Li, X.-L. Gao Pan (Stanford Publishing Pte Ltd, Singapore, 2013), pp. 273–302
66. F.P. Baaijens, W.R. Trickey, T.A. Laursen, F. Guilak, Ann. Biomed. Eng. **33**(4), 494 (2005)
67. W.R. Trickey, F. Baaijens, T.A. Laursen, L.G. Alexopoulos, F. Guilak, J. Biomech. **39**(1), 78 (2006)
68. W. Lai, J. Hou, V. Mow, J. Biomech. Eng. **113**(3), 245 (1991)
69. D. Boal, D.H. Boal, *Mechanics of the Cell* (Cambridge University Press, Cambridge, 2012)
70. J. Wilhelm, E. Frey, Phys. Rev. Lett. **91**(10), 108103 (2003)
71. F. MacKintosh, J. Käs, P. Janmey, Phys. Rev. Lett. **75**(24), 4425 (1995)
72. H. Hatami-Marbini, A. Shahsavari, R. Picu, Comput. Aided Des. **45**(1), 77 (2013)
73. H. Hatami-Marbini, R. Picu, Acta Mech. **205**(1-4), 77 (2009)
74. T.P. Stossel, J. Condeelis, L. Cooley, J.H. Hartwig, A. Noegel, M. Schleicher, S.S. Shapiro, Nat. Rev. Mol. Cell Biol. **2**(2), 138 (2001)
75. S. Furuike, T. Ito, M. Yamazaki, FEBS Lett. **498**(1), 72 (2001)
76. M. Yamazaki, S. Furuike, T. Ito, *Mechanics of Elastic Biomolecules* (Springer, Netherlands, 2003), pp. 525–534
77. J. Golji, R. Collins, M.R. Mofrad, PLoS Comput. Biol. **5**(5), e1000389 (2009)
78. K.S. Kolahi, M.R. Mofrad, Biophys. J. **94**(3), 1075 (2008)
79. G.H. Koenderink, Z. Dogic, F. Nakamura, P.M. Bendix, F.C. MacKintosh, J.H. Hartwig, T.P. Stossel, D.A. Weitz, Proc. Natl. Acad. Sci. **106**(36), 15192 (2009)
80. K. Burridge, K. Fath, T. Kelly, G. Nuckolls, C. Turner, Annu. Rev. Cell Biol. **4**(1), 487 (1988)
81. K. Burridge, M. Chrzanowska-Wodnicka, Annu. Rev. Cell Dev. Biol. **12**(1), 463 (1996)
82. A. Wechezak, R. Viggers, L. Sauvage, Lab. Invest. **53**(6), 639 (1985)
83. C. Galbraith, R. Skalak, S. Chien, Cell Motil. Cytoskeleton **40**(4), 317 (1998)
84. A.J. Maniotis, C.S. Chen, D.E. Ingber, Proc. Natl. Acad. Sci. **94**(3), 849 (1997)
85. P.L. Chandran, C.B. Wolf, M.R. Mofrad, Cell. Mol. Bioeng. **2**(1), 13 (2009)
86. C.M. Lo, H.B. Wang, M. Dembo, Y.l. Wang, Biophys. J. **79**(1), 144 (2000)
87. A. Saez, M. Ghibaudo, A. Buguin, P. Silberzan, B. Ladoux, Proc. Natl. Acad. Sci. **104**(20), 8281 (2007)
88. P.L. Chandran, M.R. Mofrad, Phys. Rev. E **79**(1), 011906 (2009)
89. P.L. Chandran, M.R. Mofrad, Phys. Rev. E **81**(3), 031920 (2010)
90. L. Lu, S.J. Oswald, H. Ngu, F.C.P. Yin, Biophys. J. **95**(12), 6060 (2008)
91. F. Sbrana, C. Sassoli, E. Meacci, D. Nosi, R. Squecco, F. Paternostro, B. Tiribilli, S. Zecchi-Orlandini, F. Francini, L. Formigli, Am. J. Physiol. Cell Physiol. **295**(1), C160 (2008)
92. J.C. Martens, M. Radmacher, Pflügers Archiv-Euro. J. Physiol. **456**(1), 95 (2008)

93. J.M. Sanger, B. Mittal, M.B. Pochapin, J.W. Sanger, Cell Motil. Cytoskeleton **7**(3), 209 (1987)
94. M. Edlund, M.A. Lotano, C.A. Otey, Cell Motil. Cytoskeleton **48**(3), 190 (2001)
95. A.S. Maddox, L. Lewellyn, A. Desai, K. Oegema, Dev. Cell **12**(5), 827 (2007)
96. K. Kruse, J. Joanny, F. Jülicher, J. Prost, K. Sekimoto, Phys. Rev. Lett. **92**(7), 078101 (2004)
97. F. Backouche, L. Haviv, D. Groswasser, A. Bernheim-Groswasser, Phys. Biol. **3**(4), 264 (2006)
98. D. Humphrey, C. Duggan, D. Saha, D. Smith, J. Käs, Nature **416**(6879), 413 (2002)
99. T. Liverpool, A. Maggs, A. Ajdari, Phys. Rev. Lett. **86**(18), 4171 (2001)
100. F. Ziebert, I.S. Aranson, Phys. Rev. E **77**(1), 011918 (2008)

Chapter 6
Experimental Challenges of Shear Rheology: How to Avoid Bad Data

Randy H. Ewoldt, Michael T. Johnston, and Lucas M. Caretta

Abstract A variety of measurement artifacts can be blamed for misinterpretations of shear thinning, shear thickening, and viscoelastic responses, when the material does not actually have these properties. The softness and activity of biological materials will often magnify the challenges of experimental rheological measurements. The theoretical definitions of rheological material functions are based on stress, strain, and strain-rate components in simple deformation fields. In reality, one typically measures loads and displacements at the boundaries of a sample, and the calculation of true stress and strain may be encumbered by instrument resolution, instrument inertia, sample inertia, boundary effects, and volumetric effects. Here we discuss these common challenges in measuring shear material functions in the context of soft, water-based, and even living biological complex fluids. We discuss techniques for identifying and minimizing experimental errors and for pushing the experimental limits of rotational shear rheometers. Two extreme case studies are used: an ultrasoft aqueous polymer/fiber network (hagfish defense gel) and an actively swimming suspension of microalgae (*Dunaliella primolecta*).

1 Introduction

Rheological properties answer the question, "What happens when I poke it?" A complex material gives a complex answer, e.g., with properties that are functions, not constants.

For a rheologically complex fluid, there is no single value of viscosity or modulus, but instead the dissipative resistance to flow (viscosity) and elastic resistance to

R.H. Ewoldt (✉) • M.T. Johnston
Department of Mechanical Science and Engineering, University of Illinois at Urbana-Champaign, Urbana, IL 61801, USA
e-mail: ewoldt@illinois.edu; mjhnstn2@illinois.edu

L.M. Caretta
Department of Materials Science and Engineering, Massachusetts Institute of Technology, Cambridge, MA 02139, USA
e-mail: caretta@mit.edu

S.E. Spagnolie (ed.), *Complex Fluids in Biological Systems*, Biological and Medical Physics, Biomedical Engineering,
DOI 10.1007/978-1-4939-2065-5_6

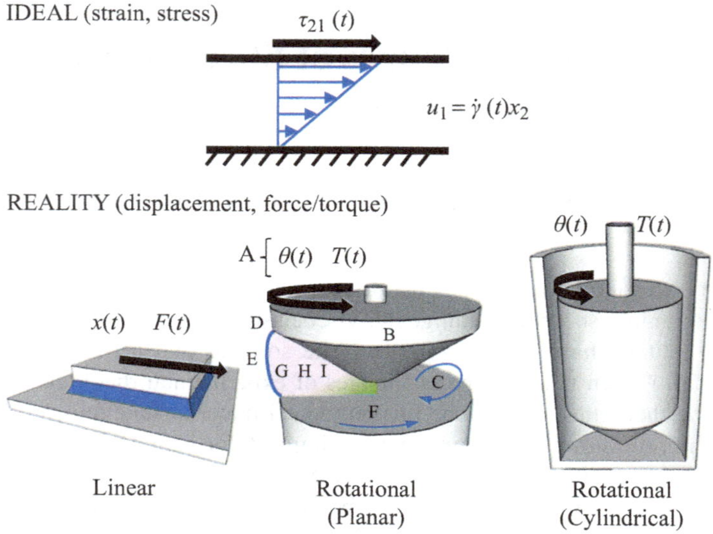

Fig. 6.1 Ideally, shear rheological properties are defined from strain $\gamma(t)$ and stress $\tau_{yx}(t)$ in homogeneous simple shear. In reality, boundary displacements and loads are measured and nonideal experimental artifacts must be considered. The effects labeled as (*A*)–(*I*) are particularly important with biological complex fluids, and include (*A*) resolution/range of measured load and displacement, (*B*) instrument inertia (if load and displacement are measured on same boundary), (*C*) fluid inertia and secondary flows, (*D*) surface tension, (*E*) free surface interfacial rheology, (*F*) slip at boundaries, (*G*) sample underfill or overfill, (*H*) small volume and gap, (*I*) nonhomogeneous sample from settling, migration, or rheotaxis

deformation (modulus) are functions of the loading time scale, loading amplitude, or other environmental factors including temperature, pressure, electromagnetic fields, or the internal activity of living biological systems. These function-valued rheological properties are known as *material functions* [1], and they form the language of descriptive rheology. The descriptive nomenclature is well defined for simple, ideal deformations [1], including simple shear (Fig. 6.1). The input can be either strain or stress, and different time-histories can be used, typically step functions or sinusoidal oscillations [2]. Of course, experimental conditions may be nonideal.

The big idea of this chapter is that properties are defined from ideal deformations (in terms of strain and stress), but experimental techniques measure displacements and loads (such as forces or torques), as outlined in Fig. 6.2. Assumptions are required to convert displacement to strain and, similarly, load to stress, and therein lies the risk that nonideal conditions exist as shown in Fig. 6.1. The effects labeled as (A)–(I) in Fig. 6.1 are particularly important with biological complex fluids and include (A) resolution/range of measured load and displacement, (B) instrument inertia (if load and displacement are measured on same boundary), (C) fluid inertia and secondary flows, (D) surface tension, (E) free surface interfacial rheology, (F) slip at boundaries, (G) sample underfill or overfill, (H) small volume and gap, and (I) non-homogeneous sample from settling, migration, or rheotaxis.

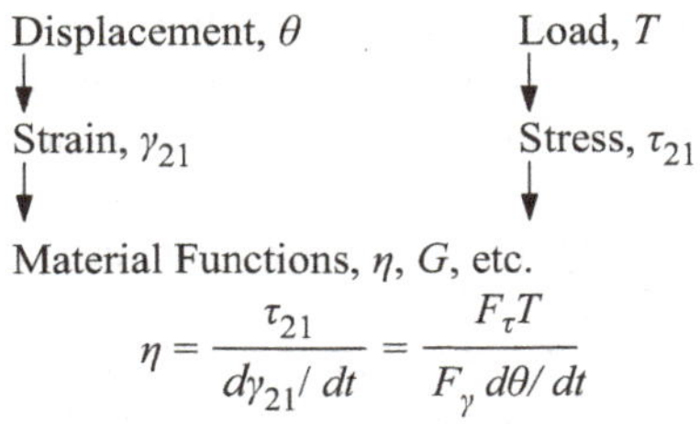

Fig. 6.2 Material properties are related to stress and strain, but these cannot be measured directly. The calculations from measured quantities require assumptions that are commonly violated, especially for biological fluids that tend to be either low viscosity, soft, or slippery. *Arrows* in figure indicate information flow into calculations. The road map is general. Example variables are shown for measuring shear properties with rotational devices

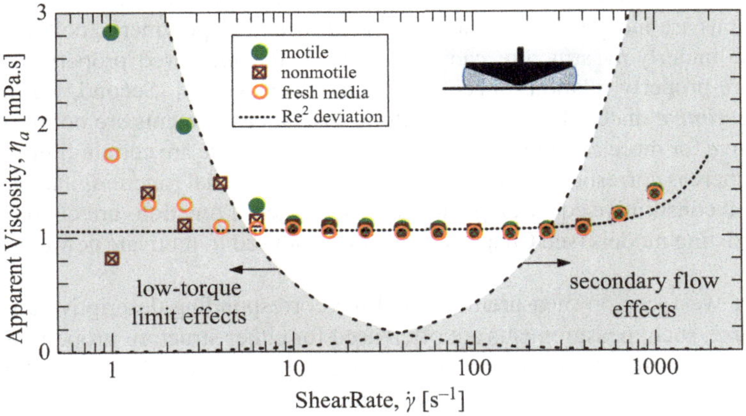

Fig. 6.3 Steady shear flow measurements could be misinterpreted as shear thinning and shear thickening if an experimental window is not identified. Here, shown with dilute suspensions of motile and nonmotile swimming microalgae *Dunaliella primolecta* compared to fresh media (no cells present). The low-torque limit, described in Sect. 3.1, is drawn from Eq. 6.13 using T_{min} = 0.1 μN.m. The secondary flow limit, described in Sect. 3.3, is drawn from Eq. 6.31 using $Re_{max} = 4$. The Re^2 line is from the expected increase in torque due to secondary flow, Eq. 6.29 (Previously unpublished work of authors RHE and LMC)

Nonideal conditions can translate to misinterpretations of results, such as the observation of apparent shear thinning and shear thickening for a fluid that is actually Newtonian within the range of test conditions, as shown in Fig. 6.3.

Figure 6.3 gives examples of rate-dependent shear viscosity measurements, which include data for a living system of microalgae suspended in water, as well as the media without cells. The data show shear thinning at low rates and shear thickening at high rates. But this is not actually the case for the true intensive material property! Note the gray regions in Fig. 6.3. These cover experimental limitations for measuring shear viscosity; in this case a measurement resolution issue at low rates, and at high rates, nonideal flow conditions. Within the experimentally accessible window, i.e., the white central region, the fluids are more confidently interpreted as Newtonian. Such limitations are particularly present when measuring biological complex fluids, since they are often soft, with low viscosity, and may even have activity (such as the swimming microalgae here) or surface-active components that modify the liquid-air interface.

This chapter will outline a checklist and guide for believable experimental measurements, interpretations, and descriptions of complex fluid rheology. The checklist is useful for biological and nonbiological systems alike.

For proper context, two important ideas must be kept in mind. First, rheological material functions are universally applicable to any class of material. They are used to describe polymer liquids, polymer solids, colloidal systems, and any other simple or complex structured fluid of the past, present, or future, so long as the continuum hypothesis is satisfied for the lengthscale of interest. Like other material properties, definitions are independent of the underlying structure (polymeric, colloidal, etc.), yet, the underlying structure can be related to the measured properties through structure-property relations specific to material classes [3]. Second, we note that the *descriptive* material functions resulting from measurements are not necessarily *predictive* for more complex deformations, although there are certain limiting cases where there is correspondence between descriptive material functions and predictive tensorial constitutive equation parameters [4]. Material functions are of course used to fit existing models (see Chap. 1 of this book) or used to motivate new constitutive models.

Here we focus on measurements and the corresponding descriptive quantities. Of course, such measurements are often used for either structure-property relations or model selection/fitting of predictive constitutive models. For those follow-up steps to be successful, the measurements must first be free from errors.

Avoiding bad data is a serious challenge with complex fluids in general and soft biological fluids in particular. Throughout this chapter, three key materials will serve as examples of soft, watery, or active fluids. This includes (i) actively swimming microalgae in a suspension of aqueous media (Fig. 6.3) (see also Chap. 9 of this book on active suspensions), (ii) a biopolymer hagfish defense gel (Fig. 6.4), which involves mucin-like molecules (see Chap. 2 of this book for a discussion of mucins), and (iii) water itself, which is the basis of biological fluids. Material details are outlined in the appendix.

2 Background: Material Functions

The theoretical definitions of material functions are based on stress, strain, and strain-rate components in simple deformation fields. (See Chap. 1 of this book for additional background on stress, and strain-rate tensors.) With real measurements, one typically measures loads and displacements at the boundaries of a sample (Fig. 6.1), and the calculation of true stress and strain may be encumbered by

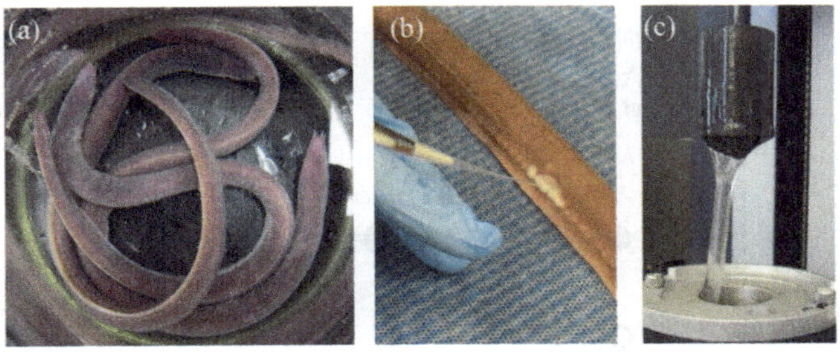

Fig. 6.4 Hagfish defense gel (a.k.a. slime) is one extreme case study used here to outline experimental rheology challenges. Hagfish produce heroic amounts of slime as a predatory defense mechanism, using a very small amount of exudate (0.01 wt% of final gel mass); (**a**) top-down view of three Atlantic hagfish (*Myxine glutinosa*) in a large glass beaker; (**b**) for experiments, exudate can be collected from an anesthetized hagfish with a pipette and then mixed with seawater to form "hagfish slime," an ultra-dilute network of polymeric mucus and fibrous protein-based intermediate filament threads, shown in (**c**) with a rotational rheometer geometry in the raised position after testing (diameter 28 mm). The ultrasoft material pushes experimental limits of low torque, instrument inertia, and sample inertia; measurements also demonstrate interio-elastic ringing (Figure adapted from [5])

the issues labeled A–I in Fig. 6.1. This chapter summarizes key experimental challenges for complex fluids, especially for biological fluids. These experimental challenges may invalidate results and sometimes cause measured properties to incorrectly *appear* nonlinear or non-Newtonian. A useful approach is to identify the experimental windows for proper measurements (Figs. 6.3, 6.5, 6.6, 6.10, 6.11, 6.14, and 6.15). The boundaries of these figures will be described in Sect. 3.

Here we will focus on simple shear deformation, rather than shear-free flows, although many of the experimental challenges will also affect measurements of extensional properties. Ideal simple shear characterization is defined by the velocity field $u = e_1\dot{\gamma}(t)x_2$, giving a homogeneous rate of deformation tensor

$$\dot{\gamma} = \begin{pmatrix} 0 & \dot{\gamma}(t) & 0 \\ \dot{\gamma}(t) & 0 & 0 \\ 0 & 0 & 0 \end{pmatrix}. \tag{6.1}$$

In principle, Eq. 6.1 results in a spatially homogeneous stress tensor

$$\sigma = \begin{pmatrix} -p+\tau_{11}(t) & \tau_{21}(t) & 0 \\ \tau_{21}(t) & -p+\tau_{22}(t) & 0 \\ 0 & 0 & -p+\tau_{33}(t) \end{pmatrix}, \tag{6.2}$$

where symmetry arguments have been used to limit the number of independent stress quantities.

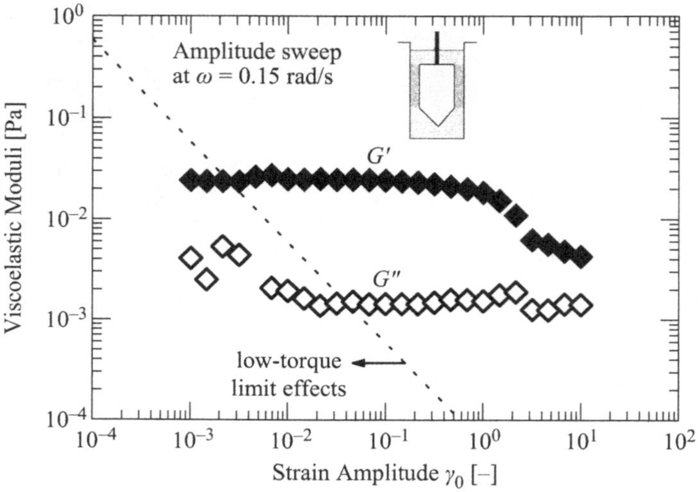

Fig. 6.5 Low-torque limit shown for oscillatory strain-amplitude sweep of hagfish gel. The experimental boundary helps identify the most believable data and explains the increased measurement variability of G'' compared to G' at the same strain amplitude. Low-torque limit drawn from Eq. 6.14 using $T_{min} = 0.003$ μN.m (Data originally published in [5])

Material functions are then defined by reporting an *output* normalized by an *input* amplitude. The controlled input can be either shear deformation (Eq. 6.1) or shear stress (Eq. 6.2), and there are many ways to "poke" a material, even with simple shear, including steps and oscillations.

The simplest rheological characteristic of a fluid is the steady shear flow viscosity, defined from steady shear stress τ_{21} and steady shear rate $\dot{\gamma}$ as

$$\eta(\dot{\gamma}) = \tau_{21}(\dot{\gamma})/\dot{\gamma}. \tag{6.3}$$

A Newtonian response would produce constant $\eta(\dot{\gamma})$, but in general it is a function-valued property. This descriptive material function *happens* to be the same function used in the predictive equation known as the generalized Newtonian fluid (see Chap. 1 of this book). But in general, descriptive material functions need not be identical to constitutive model parameters.

Consider the more general possibility of a transient response. For example, a step input of shear rate $\dot{\gamma}(t) = \dot{\gamma}_0 H(t)$ where $H(t)$ is the Heaviside step function. In this case, the transient shear viscosity is defined as

$$\eta^+(t;\dot{\gamma}_0) = \tau_{21}(t;\dot{\gamma}_0)/\dot{\gamma}_0. \tag{6.4}$$

Transient normal stress differences are also used to define material functions, but we will focus on shear stress. A different way to probe the material is to apply a step stress input $\tau_{21}(t) = \tau_0 H(t)$, from which the creep compliance is defined as

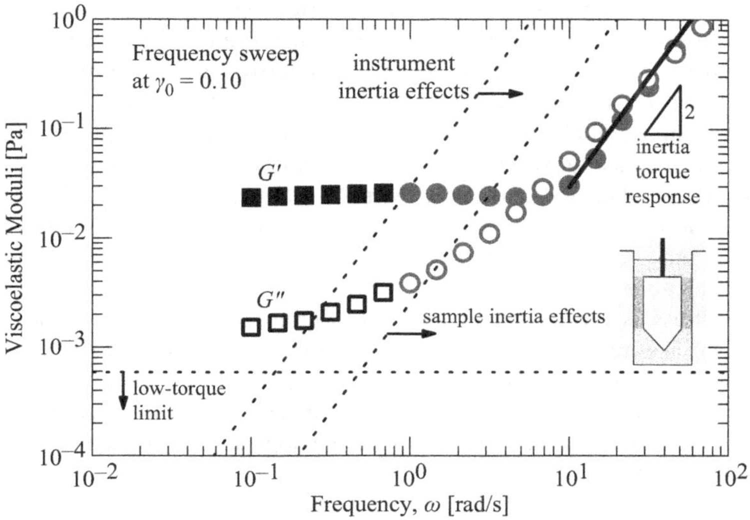

Fig. 6.6 Low-torque and instrument inertia limits shown for oscillatory frequency sweep of hagfish gel. Low-torque limit from Eq. 6.14 with constant γ_0; instrument inertia limit from Eq. 6.18; sample inertia limit from Eq. 6.26. The inertial torque response (*solid line*) is from Eq. 6.20 with $\varepsilon = 0.01$ being the error in the instrument inertia torque correction. *Gray circles* indicate when raw phase angle jumps from $< 15°$ to $> 130°$ which is also an indication that instrument inertia corrections must be made (Data originally published in [5])

$$J(t;\tau_0) = \gamma(t;\tau_0)/\tau_0. \tag{6.5}$$

In this chapter we will also discuss oscillatory inputs [6]. For an input shear rate $\dot{\gamma}(t) = \gamma_0\omega\cos(\omega t)$, the oscillatory shear stress output can be represented by a Fourier series

$$\tau_{21}(t;\omega,\gamma_0) = \sum_n \left\{\tau'_{21,n}(\omega,\gamma_0)\sin n\omega t + \tau''_{21,n}(\omega,\gamma_0)\cos n\omega t\right\}. \tag{6.6}$$

From this, the first harmonic is the most basic feature used to calculate material functions, e.g.,

$$G'_1(\omega,\gamma_0) = \tau'_{21,1}(\omega,\gamma_0)/\gamma_0 \tag{6.7}$$

$$G''_1(\omega,\gamma_0) = \tau''_{21,1}(\omega,\gamma_0)/\gamma_0, \tag{6.8}$$

which are known as the first-harmonic elastic storage and viscous loss moduli, respectively. In the limit of small amplitude oscillatory shear (SAOS), one may expect linear scaling of the stress coefficients with the input amplitude. In this limit of linear viscoelasticity, we need only report the so-called linear viscoelastic storage and loss moduli, $G'(\omega)$ and $G''(\omega)$, respectively.

Shear deformation can be applied with various geometric configurations. We will consider configurations where the stress and strain fields can be calculated without strong assumptions of a particular constitutive model, be it liquid, solid, or nonlinear viscoelastic. For example, indentation tests and embedded particle microrheometry impose nonhomogeneous loading scenarios that change depending on the underlying constitutive model, whereas simplified geometries such as parallel plate, cone and plate and concentric cylinder geometries have well-defined stress and/or strain fields for any material tested (when sample inertia can be neglected; Sect. 3.3), and a constitutive model need not be assumed in order to report the material functions.

Shear stress is the most commonly measured stress component, but normal stresses are also relevant for describing rheological properties in shear. Many of the challenges described here will also apply to normal force measurements, such as experimental windows based on instrument specifications. The reader may find it useful to consult related texts on the topic of rheological measurement for additional details on certain topics [7–9].

3 Challenges

In general, one must keep a system-level perspective to identify bad data and avoid misinterpretations of rheological properties. The following subsections outline particularly problematic sources of error and should serve as a checklist for verifying rheological measurements, especially with soft and slippery biological materials.

3.1 Instrument Specifications

The measurable ranges of load and displacement serve as primary limits to measuring rheological material functions. As shown in Fig. 6.2, the primary measured variables for rotational rheometry include torque T, displacement θ, and rotational velocity Ω. We will use the following notation for conversion factors to calculate stress and deformation

$$\tau_{21} = F_\tau T, \tag{6.9}$$

$$\gamma = F_\gamma \theta, \tag{6.10}$$

$$\dot{\gamma} = F_\gamma \dot{\theta} = F_\gamma \Omega. \tag{6.11}$$

The minimum torque is typically the most important limitation for soft biological systems. Minimum torque is often specified by instrument manufacturers but can often be higher due to other effects (such as surface tension producing torque, Sect. 3.4).

To identify experimental limitations, we will use the approach of drawing boundary lines within the coordinate axes used to report material functions, as done in Fig. 6.3. First, write the reported material functions in terms of the measured quantities and conversion factors. For example, steady shear viscosity from Eq. 6.3 would be

$$\eta(\dot{\gamma}) = \frac{\tau_{21}(\dot{\gamma})}{\dot{\gamma}} = \frac{F_\tau}{F_\gamma}\frac{T(\Omega)}{\Omega}. \tag{6.12}$$

Next, we state the condition for acceptable data that measured torque is above some minimum limit, $T > T_{\min}$. Substituting Eq. 6.12 into the condition $T > T_{\min}$ provides the criteria

$$\eta > \frac{F_\tau T_{\min}}{\dot{\gamma}}, \tag{6.13}$$

for avoiding bad data. This equation was used in Fig. 6.3 for the cone-plate geometry $F_\tau = 3/(2\pi R^3)$ where R is the cone radius and $T_{\min} = 0.1$ μN.m was used.

The appropriate value for $T_{\min}$ can sometimes be larger than instrument specification, e.g., with dilute polymers in aqueous solution [10–12]. Recent results show that surface tension torque may be responsible for torque limits higher than instrument specifications [13], as discussed in Sect. 3.4. The limit of minimum measurable viscosity decreases as the shear rate is increased. This is because the limit corresponds to a minimum measurable shear stress $\tau_{21,\min} = F_\tau T_{\min}$, and viscosity is calculated as shear stress divided by shear rate.

A similar downward-sloping low-torque limit appears for other material functions that are plotted as a function of an amplitude. Consider viscoelastic moduli as a function of strain amplitude (Fig. 6.5), for which the low-torque limit sets the minimum measurable viscoelastic moduli

$$G_{\min} = \frac{F_\tau T_{\min}}{\gamma_0}, \tag{6.14}$$

where $G_{\min}$ refers to either G' or G''. For the concentric cylinder geometry (single gap) used in Figs. 6.5 and 6.6, $F_\tau = 1/(2\pi R^2 L)$ with minimum torque in oscillation $T_{\min} = 0.003$μN.m as specified by the manufacturer (TA Instruments, AR-G2). In Fig. 6.6, the frequency sweep at fixed strain amplitude, the low-torque limit is simply a horizontal line, since $\gamma_0 =$ constant.

The other primary variable measurements of displacement ($\theta_{\min}$, $\theta_{\max}$) and velocity ($\Omega_{\min}$, $\Omega_{\max}$) also provide limits. These can also be plotted as lines within the coordinate axes. In Fig. 6.3 with $\eta(\dot{\gamma})$, the limits ($\Omega_{\min}$, $\Omega_{\max}$) would appear as vertical lines defining the minimum and maximum $\dot{\gamma}$.

Experimental limits also depend on the geometry choice. This is well known among practitioners. As an example, consider a soft material that may not exceed the minimum measurable torque. Writing Eq. 6.14 more generally, the minimum measurable shear modulus could be written as

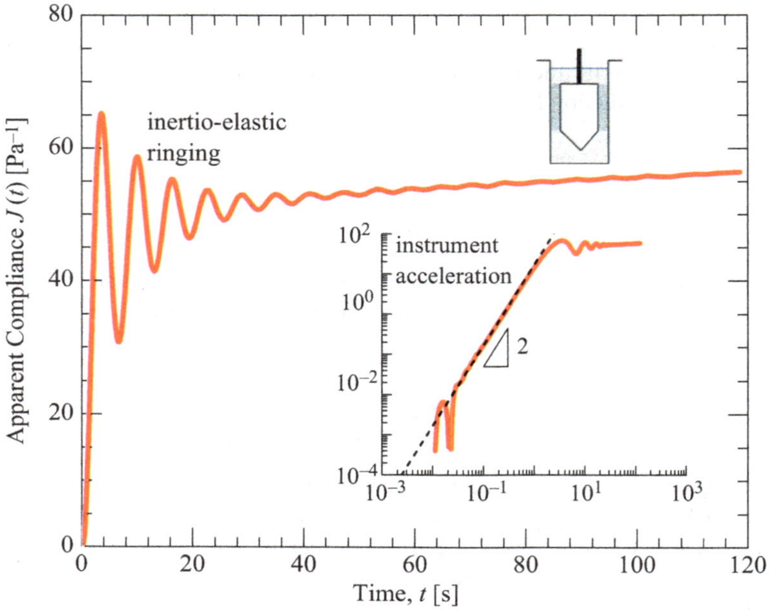

Fig. 6.7 Step torque test with hagfish gel in concentric cylinder geometry. Plotted as apparent compliance $J(t) = F_\gamma \theta(t)/(F_\tau T_0)$, showing instrument inertia effects: acceleration (*inset*) and inertio-elastic ringing. These effects can occur in rheometer designs where load is measured at the moving boundary. The acceleration can be predicted by Eq. 6.23, here using the value $\frac{IF_\tau}{F_\gamma} = 2.9465 \cdot 10^{-2}$ Pa.s^2. The ringing can be used to extract linear and nonlinear viscoelastic information as described in the text. Both effects obscure the true creep compliance of the material (Data originally published in [14])

$$G_{\min} = \frac{F_\tau}{F_\gamma} \frac{T_{\min}}{\theta_{\max}}. \tag{6.15}$$

To measure a very small modulus, one could use a larger displacement amplitude θ_0. However, this may miss the linear viscoelastic regime if the corresponding strain amplitude γ_0 is too large. One can also increase the"gain" in the system that generates the torque from the material resistance. This corresponds to maximizing F_γ/F_τ, e.g., for a cone-plate system $F_\gamma/F_\tau = 2\pi R^3/(3\beta)$, where β is the small cone angle. For a soft material, one may choose a large R to generate sufficient torque to make the measurement or switch to a different geometry with a larger value of F_γ/F_τ, such as concentric cylinders used in Figs. 6.5, 6.6, and 6.7 for the soft hagfish defense gel.

Geometry choices influence other challenges, and there may be trade-offs between different experimental limitations. One issue is inertia of the moving instrument components, if the torque is being measured at the moving boundary. This is outlined in the following section.

3.2 Instrument Inertia

Instrument inertia causes experimental artifacts under *transient conditions*. This includes oscillatory tests (e.g., limiting the high-frequency data in Fig. 6.6) and step tests (e.g., influencing the short-time creep compliance data in Fig. 6.7). This is only a problem if the load is measured at a moving boundary with unsteady motion (as drawn in Fig. 6.1), which is common for many, but not all, commercial rheometers. Unsteady instrument motion involves acceleration, and this requires a load (torque); therefore the measured load is not simply associated with material deformation but also instrument acceleration.

At worst, instrument inertia causes errors that look like real data. Consider yourself warned! For example, the high-frequency data in Fig. 6.6 and short-time data in Fig. 6.7 are not true material properties. In some cases, the effects can be corrected if the instrument inertia is known. At best, the inertial correction is negligible. In still other cases, instrument inertia can cause free oscillations in step load tests. These inertio-elastic free oscillations can actually be used advantageously to characterize material elasticity and dissipation from the ringing frequency and periodic decay, respectively [15–18].

Biomaterials can be exceedingly soft, and instrument inertia artifacts are exaggerated for very soft materials. As examples of softness, the elastic modulus of hagfish gel is $G' \approx 0.2\,\text{Pa}$ (Figs. 6.6 and 6.7), microtubule networks have a plateau modulus $G' \sim 0.4$–$20\,\text{Pa}$ [19], vitreous gel $G' \approx 2\,\text{Pa}$ [20], actin networks as low as $G' \approx 1\,\text{Pa}$ [21, 22], fibrin at low concentration $G' \approx 10\,\text{Pa}$ [23], and collagen–hyaluronic acid interpenetrating polymer network hydrogels $G' \approx 1$–$100\,\text{Pa}$ [24]. In any soft material, instrument inertia must be carefully considered with transient rheological measurements.

To avoid bad data in oscillatory shear, the "material torque" should exceed the "instrument inertia torque." Thus, the criteria for good data is satisfied under the following condition:

$$T_{\text{material}} > T_{\text{inertia}}, \tag{6.16}$$

$$\frac{G\gamma_0}{F_\tau} > I\theta_0\omega^2, \tag{6.17}$$

$$G > \frac{IF_\tau}{F_\gamma}\omega^2, \tag{6.18}$$

where the variable G represents either G' or G'' in oscillation. Equation 6.18 is used to draw the "instrument inertia" boundary in Fig. 6.6 for the onset of instrument-inertia effects. For the experiment in Fig. 6.6, with a concentric cylinder geometry, $\frac{IF_\tau}{F_\gamma} = 2.9465 \cdot 10^{-2}\,\text{Pa.s}^2$. Equation 6.18 corresponds to the jump in raw phase. Instrument inertia corrections can be made beyond this point, and this requires the subtraction of the instrument inertia torque from the single. This is reasonable to a point, but artifacts may eventually appear, e.g., with moduli increasing close to $G' \sim G'' \sim \omega^2$. This signature can be explained by inertia effects.

Some of the most prevalent undiagnosed errors in rheometry involve artifacts in high-frequency oscillatory measurements. One should be very careful when interpreting high-frequency data. For example, without drawing the instrument-inertia boundary line in Fig. 6.6, one might be tempted to interpret a curious power-law scaling of viscoelastic properties as a function of frequency. However, this data at high frequency is completely associated with instrument inertia, and not at all a material property.

An instrument inertia artifact at high frequency is most easily diagnosed by looking at the raw phase difference between the oscillating displacement and torque signals and being critical of data points with raw phase $> 90°$. To see why, consider that a purely elastic material response would have load proportional to displacement, $T \sim \theta$, a purely viscous material gives $T \sim \dot{\theta}$, and a purely inertial effect $T \sim \ddot{\theta}$. For time periodic oscillatory signals $T(t)$ and $\theta(t)$, this corresponds to phase differences of 0°, 90°, and 180°, respectively. Without instrument inertia effects, the phase would be limited to the viscoelastic range $0° < \delta < 90°$. So, when this raw phase is $> 90°$, instrument inertia must be playing a role. Corrections can be made by calibrating the rotational inertia and subtracting the expected inertial torque from the total signal (as done for the data in Fig. 6.6). But, this become exceedingly difficult at large values of raw phase when the inertial torque dominates the total torque signal.

Inertia corrections are not 100 % perfect, and this explains the specific signature at high frequency of $G' \sim G'' \sim \omega^2$. One expects inertial torque $T = I\ddot{\theta}$, and for $\theta(t) = \theta_0 \sin(\omega t)$ this is

$$T_0 = I\theta_0\omega^2. \tag{6.19}$$

This can be subtracted from the measured signal, but if the subtraction is not exact, then some of this inertial torque will remain in the processed signal, say εT_0 where ε is (hopefully) a small number. Translated to material properties, this would produce apparent viscoelastic moduli

$$G = \frac{F_\tau(\varepsilon T_0)}{F_\gamma \theta_0} = \varepsilon \frac{F_\tau I}{F_\gamma}\omega^2. \tag{6.20}$$

With $\varepsilon = 0.01$, Eq. 6.20 explains the high-frequency signature in Fig. 6.6. Therefore, even with inertia corrections, viscoelastic moduli will eventually have frequency-dependent power-law scaling that approaches $G' \sim G'' \sim \omega^2$, since they are calculated from a torque signal that is increasingly dominated by inertia and corrections are only precise to within factor of ε.

Instrument inertia affects high-frequency oscillation data, as well as short-time data in step tests. For creep tests (step load input), the instrument inertia can significantly alter the displacement response (Fig. 6.7, hagfish gel). This includes (i) the time required to accelerate and (ii) free oscillations via "inertio-elastic" ringing, in which the sample elasticity couples with the finite instrument inertia to "ring" at a resonant frequency, just like a mass at the end of a spring [15, 17, 18, 25]. A careful analysis of the inertio-elastic oscillations can reveal both linear and nonlinear viscoelastic properties of the sample [5, 18].

By conservation of momentum, the measured dynamic load must satisfy

$$T(t) = I\ddot{\theta}(t) + \tau_{21}(t)/F_\tau, \tag{6.21}$$

where we have considered a rotational rheometer with instrument inertia I. For a step torque $T(t) = T_0 H(t)$, the initial conditions at $t = 0$ are $\theta = 0$, $\dot{\theta} = 0$, and typically $\tau_{21} = 0$ if starting from rest. Initially, the applied torque is dominated by the acceleration term in Eq. 6.21, since the sample stress term is initially zero and only appears as strain and strain rate increase above zero. The creep response then always has the following form in the limit of short time [17]:

$$\theta(t) = \frac{1}{2}\frac{T_0}{I}t^2 + \cdots . \tag{6.22}$$

Converting this to the apparent material function $J(t)$

$$J(t) = \frac{\gamma(t)}{\tau_0} = \frac{1}{2}\frac{F_\gamma}{F_\tau I}t^2 + \cdots \tag{6.23}$$

which shows the general short-time instrument acceleration artifact, independent of applied torque when plotted as apparent compliance $J(t)$. This is shown in the inset of Fig. 6.7 for the soft hagfish gel.

Inertio-elastic ringing analysis can probe both linear viscoelasticity [15, 17] and nonlinear viscoelasticity [5, 18] in novel ways. Such analysis requires the assumption of an underlying constitutive model for $\tau_{21}(t)$ in Eq. 6.21, e.g., a three-element fluid (Jeffreys), or two-element solid (Kelvin–Voigt). Detailed calculations associated with the inertio-elastic ringing analysis can be found in the references above.

To avoid the instrument inertia effects discussed in this section, one can measure the load (torque) at the *stationary* boundary, e.g., with a force rebalancing transducer, rather than measuring load at the moving boundary, e.g., through a motor. This requires more complex instrumentation to separate the imposed displacement from the measured load, but such separated motor-transducer instruments are commercially available. These setups can eliminate important errors due to instrument inertia including the accurate measurement of stress jumps in response to step displacement inputs [26].

3.3 Fluid Inertia and Secondary Flows

Even if instrument inertia is eliminated, the sample itself will always have finite inertia which can produce artifacts from momentum diffusion, viscoelastic waves, and secondary flows, all of which can violate the assumption of homogeneous simple shear deformation in Eq. 6.1. Purely elastic instability can also produce secondary flows even in the limit of vanishing Reynolds number [27–30]. This section will discuss the symptoms of both wave propagation and secondary flows and how to identify experimental limits due to these artifacts.

3.3.1 Wave Propagation at High Frequencies and Short Timescales

The assumption of homogeneous simple shear strain is violated when there are waves propagating through the material. Propagating waves may come from either viscous momentum diffusion or elastic shear waves or both for viscoelastic materials in general.

The general criteria for approximately homogeneous strain in the velocity gradient direction is that the wavelength l of any propagating wave should be much larger than the geometry gap D

$$l \gg D \tag{6.24}$$

so that, in the gap region, the velocity field is negligibly affected by the propagating wave [31]. Two key questions must be answered: (i) how much smaller must the gap D be for tolerable errors, and (ii) how can the wavelength l be calculated? The wavelength l depends on material properties and the frequency (time scale) of motion. Most importantly, l decreases with high driving frequency, and we therefore expect wave propagation issues at high frequency and short time scales.

Schrag gave a detailed analysis of linear viscoelastic wave propagation [31], showing that linear viscoelastic shear waves between a moving boundary and a fixed reflecting boundary have wavelength

$$l = \frac{1}{\cos(\delta/2)}\left(\frac{|G^*|}{\rho}\right)^{1/2}\frac{2\pi}{\omega}, \tag{6.25}$$

where ω is the driving frequency, $|G^*| = \sqrt{G'^2 + G''^2}$ is the magnitude of the complex modulus, δ is the viscoelastic phase angle, and ρ is the fluid density. The scaling in Eq. 6.25 is $l \sim c\mathcal{T}$ where c is the wavespeed $c \sim (|G^*|/\rho)^{1/2}$ and $\mathcal{T} = 2\pi/\omega$ is the wave period. Using the criteria $l \geq 10D$ to avoid errors of possibly 10 % [31], along with Eq. 6.25, we can identify an approximate edge of the experimental window for plots of viscoelastic moduli,

$$|G^*| > \left(\frac{10}{2\pi}\right)^2 \cos^2(\delta/2)\rho\omega^2 D^2, \tag{6.26}$$

which is used in Fig. 6.6 to identify the "sample inertia limit," using a value $\cos^2(\frac{\delta}{2}) = 1$. Equation 6.26 scales as $|G^*| \sim \rho\omega^2 D^2$, showing the important sensitivity to both driving frequency ω and geometry gap D. Higher frequencies are problematic. Smaller gas are helpful. The numerical front factor has weak dependence on δ, since $\frac{1}{2} < \cos^2(\frac{\delta}{2}) < 1$. The more sensitive number is the factor by which $l > D$. More precise experiments require a larger separation of these lengthscales, as detailed in Schrag [31] (his Table 1). Whatever the front factor, the shape of the experimental limit will still be the same, scaling as $|G^*| \sim \rho\omega^2 D^2$. The sample inertia impacts measurement at high frequency and low modulus, and therefore soft gels and low-viscosity fluids will have greater propensity for sample inertia effects.

Although strain amplitude does not appear explicitly in Eq. 6.26, fluid inertia problems can appear due to large-amplitude oscillations [32], even with constant forcing frequency. For these nonlinear tests, one can conceptually think about $|G^*|$ changing in the nonlinear regime, which would influence the wave propagation speed and therefore the wavelength l. When large-amplitude oscillatory shear strain softens a sample (decreasing $|G^*|$ which is typical for polymer melts), then the sample inertia issue will become more problematic at large strain amplitudes. This is consistent with detailed studies in the literature on flexible polymers [32]. However, if a sample becomes more stiff in the nonlinear regime (increasing $|G^*|$ which is typical of semiflexible biopolymer gels [33]), then one could argue that the inertia artifact may actually be less problematic due to increasing viscoelastic wavelength l. This possibility, however, has not yet been studied in any detail. One challenge for universal analysis of nonlinear viscoelastic measurements with wave propagation is that no universal constitutive equation exists for nonlinear viscoelasticity.

The experimental boundary line defined by Eq. 6.26 should serve as a general guideline to identify possible experimental windows due to shear waves when measuring oscillatory shear material functions. It is useful in linear viscoelastic plots, e.g., $G'(\omega)$,$G''(\omega)$ (as in Fig. 6.6), and may also be useful to estimate the boundary for nonlinear tests, e.g., large-amplitude oscillatory shear (LAOS) tests in terms of $|G_1^*|(\gamma_0)$. For all these cases, the artifact of viscoelastic waves will limit measurement of low modulus and high-frequency data.

3.3.2 Secondary Flows at High Velocity

Sample inertia can also cause nonideal velocity fields during *steady* flow. Even before turbulent flow, high velocities can cause secondary flows superposed on the primary simple shear flow due to finite sample inertia and curved streamlines in unstable configurations. This includes cylindrical geometries with a rotating inner cylinder and planar geometries including cone-plate and parallel disk flow. In each case, secondary flow increases the measured torque and therefore incorrectly increases the apparent viscosity of the fluid. For example, a Newtonian fluid with secondary flow present would incorrectly appear as shear thickening, since the secondary flow effects grow with increasing velocity. This is observed in Fig. 6.3 with the microalgae suspension at high shear rates.

Concentric cylinder measurements have a well-known secondary flow instability that appears when the inner cylinder is rotating at sufficiently large velocity Ω. Known as Taylor–Couette flow after the initial work of Taylor [34], the inertial instability causes axisymmetric vortices. The stability criteria is well established for Newtonian fluids in the limit of small gaps. It is based on a sufficiently small Taylor number Ta [35, 36]:

$$\mathrm{Ta} = \frac{\rho^2 \Omega^2 (R_o - R_i)^3 R_i}{\eta^2} < 1{,}700, \tag{6.27}$$

where R_i is the inner radius moving at angular velocity Ω and R_o is the fixed outer radius. The criteria has been mapped for corotating and counterrotating cylinders as well [88], but the most useful criterion for standard shear rheometry is given in Eq. 6.27. There is some evidence that non-Newtonian polymer solutions increase the critical Taylor number, so that Eq. 6.27 is a conservative estimate for experimental rheological measurements [8]. To draw an experimental boundary line on a plot of viscosity versus shear rate $\eta(\dot{\gamma})$, rearrange Eq. 6.27 and use the definition of shear rate $\dot{\gamma} = \Omega R_i/(R_o - R_i)$. This gives the condition

$$\eta > \frac{(R_o - R_i)^{5/2}}{1{,}700 R_i^{1/2}} \rho \dot{\gamma} \tag{6.28}$$

to avoid Taylor vortices. The criteria emphasizes that low-viscosity fluids are more prone to this secondary flow and that small gaps are very helpful in the geometry design. The scaling $\eta \sim \dot{\gamma}$ defines the shape of the boundary on a plot of $\eta(\dot{\gamma})$ and limits high shear rate measurements. As a quantitative example of an experimental limit for the concentric cylinder geometry, consider properties typical of biological fluids, density $\rho = 10^3$ kg/m^3, and viscosity near water $\eta = 1$ mPa.s. For a nominal concentric cylinder geometry with gap $(R_o - R_i) = 1$ mm and inner radius $R_i = 11.8$ mm (based on the ISO 3219 standard with $R_o/R_i = 1.0847$ [37]), Eq. 6.28 can be rearranged to show the shear rate is limited to $\dot{\gamma} < 5.8 \cdot 10^3\,\mathrm{s}^{-1}$. This is reasonably high, but of course will change depending on the actual viscosity and size of geometry being used.

Cone-plate and parallel disk geometries have a secondary flow that is always present at finite rotational velocity [38] (the critical Taylor number does not apply to these geometries). Here, centrifugal effects create a radial velocity component with outward flow at the rotating boundary. Due to conservation of mass this causes inward flow at the stationary boundary. (Highly elastic liquids can change this scenario as discussed in the following section.) For the Newtonian case, the strength of the flow is based on a Reynolds number. The secondary flow increases the measured torque, and this can be used to set a criteria and draw experimental limits for measurement. For Newtonian fluids with cone-plate or parallel disk geometry, the measured torque T is predicted to depend on the Reynolds number as [39]

$$\frac{T}{T_0} = 1 + \frac{3}{4{,}900}\mathrm{Re}^2, \tag{6.29}$$

where T_0 is the ideal torque due to shear flow alone and Re is the Reynolds number defined as

$$\mathrm{Re} = \frac{\rho \Omega L^2}{\eta_0}, \tag{6.30}$$

where L is the representative gap lengthscale. For cone-plate, $L = \beta R$ where β is the angle between the cone and plate, and for a parallel plate, $L = H$ where H is the gap. For a given error bound on T/T_0, we can identify a critical Reynolds number $\mathrm{Re}_{\mathrm{crit}}$. For example with 1 % error, i.e., $T/T_0 = 1.01$, the critical Reynolds number is $\mathrm{Re}_{\mathrm{crit}} = 4$. This clearly occurs before turbulence could be sustained [38], and therefore sets the experimental boundary for rheological measurements. Using the criteria $\mathrm{Re} < \mathrm{Re}_{\mathrm{crit}}$ and the definition of shear rate $\dot{\gamma} = \Omega R/L$, results in an experimental limit that can be shown on plots of steady shear viscosity $\eta(\dot{\gamma})$,

$$\eta > \frac{L^3/R}{\mathrm{Re}_{\mathrm{crit}}}\rho\dot{\gamma} \tag{6.31}$$

which is used in Fig. 6.3 to draw the "secondary flow limit" line, using $\mathrm{Re}_{\mathrm{crit}} = 4$ and $L = \beta R$ for the cone-plate geometry. Figure 6.3 also shows the expected apparent shear thickening of the shear viscosity, based on Eq. 6.29, and converting to apparent viscosity. Equation 6.31 shows the scaling $\eta \sim \rho\dot{\gamma}$, similar to the shape of the boundary with concentric cylinders and the Taylor–Couette instability, Eq. 6.28.

In all the rotational geometries discussed here, lower viscosity fluids have a smaller experimental window with limitations at high shear rate due to secondary flow.

3.3.3 Purely Elastic Instabilities

Undesirable secondary flows can also be created by purely elastic instabilities, in the limit of vanishingly small Reynolds number. These secondary flows arise from a different physical effect than the previous subsection (i.e., not fluid inertia). For purely elastic instabilities, curved streamlines carrying tension cause instability [27–30]. The effect occurs at high Weissenberg number $\mathrm{Wi} = \lambda_1\dot{\gamma}$ (where λ_1 is the longest relaxation time of the fluid) for elastic liquids in rotational geometries including concentric cylinder, cone-plate, and parallel disk.

The primary symptom of secondary flow is increased torque. Hence, the apparent steady-state viscosity may incorrectly appear to shear thicken (Fig. 6.8). Increased normal force may also occur. The purely elastic secondary flow occurs at a critical Wi which depends on the measurement gap and constitutive behavior of the fluid. Specific predictions are available for certain models including upper-convected

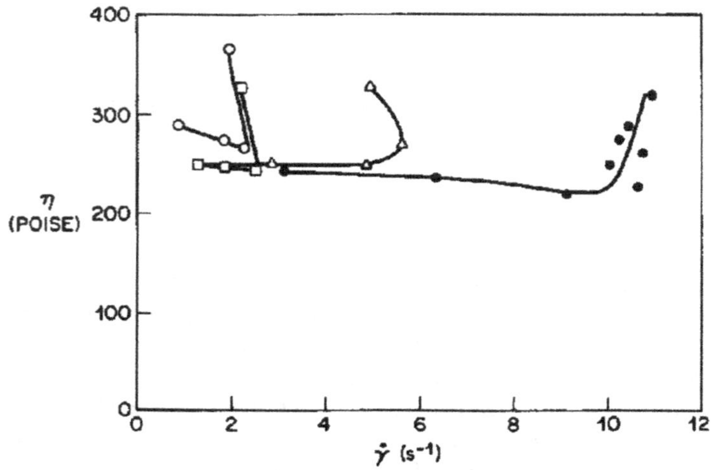

Fig. 6.8 Purely elastic instabilities cause apparent shear thickening of steady shear viscosity. Here with a dilute solution of polyisobutylene with nominal molecular weight $2.7 \cdot 10^6$ from [27]. The thickening appears at a critical shear rate (Weissenberg number); the accessible range of experiments is extended by decreasing the gap, here with a parallel disk geometry at different gaps ranging from 2 mm (*open circles*) to 0.3 mm (*filled circles*). (Reprinted from Magda and Larson [27], Journal of Non-Newtonian Fluid Mechanics, Fig. 10, Copyright (1988), with permission from Elsevier)

Maxwell and Oldroyd-B [28, 29]. Smaller geometry gaps inhibit the instability, pushing the critical shear rate higher as shown in Fig. 6.8. Very large solvent viscosity can also inhibit the instability. The effect is time-dependent, appearing after prolonged shearing [27]. A general instability criteria applicable to complex geometries has been suggested which includes dependence on both Weissenberg and Deborah numbers [29,30]. Purely elastic instabilities in pressure-driven channel flow have also been described [40, 41].

For elastic instabilities to occur before the fluid inertia instabilities (Sect. 3.3.2), the fluid must have a long relaxation time so that the Weissenberg number can be large while Reynolds number or Taylor number is low. For polymeric systems including biological fluids, elastic instabilities are relevant with high-molecular-weight polymers in solution.

For all secondary flows, due to either fluid inertia (Sect. 3.3.2) or fluid elasticity (this subsection), the symptoms are similar: increased viscosity at high shear rates, as seen in Figs. 6.3 and 6.8. These effects limit the high shear rate experimental range for measuring simple shear rheological properties, and tempt misinterpretation of apparent shear thickening at high rates.

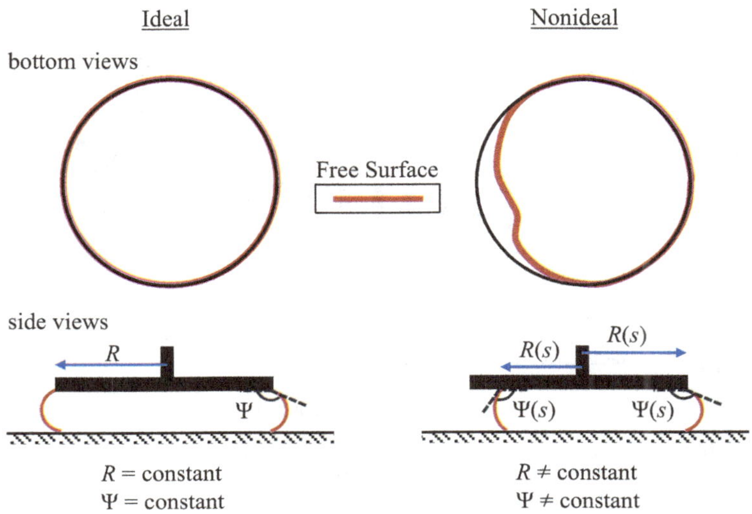

Fig. 6.9 Contact line and interface angle: ideal versus nonideal. Nonideal asymmetries are exaggerated compared to typical loading and can also occur as a result of overfilling. The nonideal condition may create artifacts of apparent shear thinning due to the presence of a constant surface tension torque (Figure adapted from [13])

3.4 Surface Tension Forces

For rotational rheometers, surface tension results in a torque that should not occur in an ideal, rotationally symmetric geometry [13] (Fig. 6.9). It is typically neglected, but the phenomenon may exceed the instrument low-torque limit $T_{\min}$ by orders of magnitude. The effect causes Newtonian fluids, including water, to appear as shear thinning with finite elastic modulus (Figs. 6.10 and 6.11). We discuss the symptoms of the effect, methods for drawing experimental boundaries with a different $T_{\min}$, and techniques to minimize the effect, which is particularly important when measuring aqueous solutions such as biological fluids.

In this section, the focus is rotational geometries where surface tension influences the measurement of shear stress. This is a significant experimental challenge for measuring soft, active, or low-viscosity biological fluids. Related issues not discussed here include (i) normal force from surface tension [42, 43] which is highly dependent on meniscus shape [44–46]; (ii) sliding plate instruments which dilate free surface area and cause surface tension artifacts in shear stress calculations [47–49]; and (iii) surface rheology artifacts from films of surface-active components [50, 51] which will be discussed in Sect. 3.5.

Surface tension torque is caused by traction forces around a material contact line that has broken rotational symmetry, both in terms of geometric location and non-constant contacting angle (or nonconstant surface tension) [13] (Fig. 6.9).

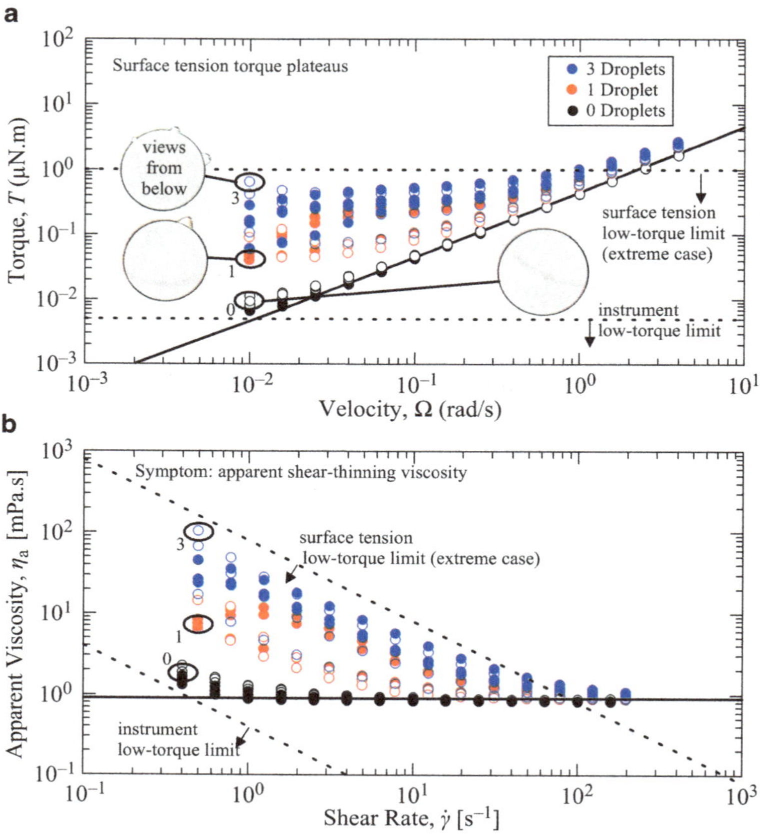

Fig. 6.10 Surface tension can generate torque in *steady* shear (**a**), which could be mistaken as shear thinning (**b**), as shown here with water at room temperature. The effect grows with slight overfilling that increases contact line rotational asymmetry, here with 1 or 3 small extra droplets of water at the boundary, yet the effect can be present even with best practices ("0 droplets") where residual torque plateaus appear above the instrument low-torque limit. Views from below in (**a**) indicate droplet contact lines around the $D = 40$ mm plate, as viewed through a glass bottom plate. Low-torque limits for η_{min} drawn from Eq. 6.13 using $T_{min} = 5$ nN.m (instrument low torque) and $T_{min} = 1\,\mu$N.m (surface tension torque, extreme case) (Figure adapted from [13])

Historically, rotational symmetry has been a primary assumption [2, 8], even when considering effects of surface tension [42, 43]. However, the rotational symmetry assumption can be violated easily. Finite deviations of contact line rotational symmetry, from manufacturing tolerances or sample overfill/underfill, allow surface tension to produce a torque which may dramatically impact measurements of shear rheology, particularly at low shear rates and for low-viscosity fluids.

Symptoms of surface tension torque include apparent viscous shear thinning and elastic shear modulus. In steady shear flow, the effect appears as a superposed constant torque independent of rate (Fig. 6.10a). These torque plateaus would appear

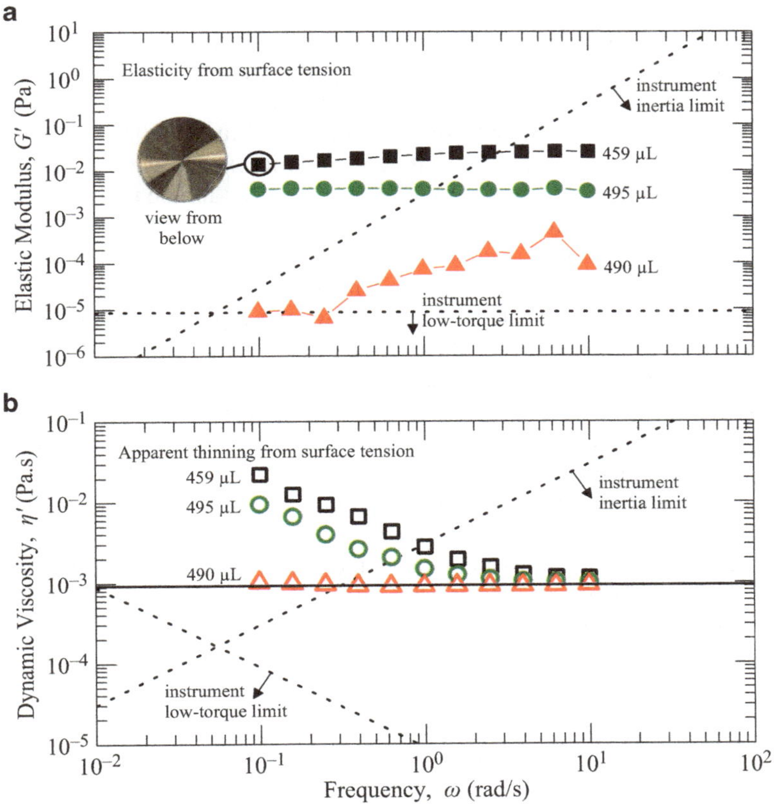

Fig. 6.11 Surface tension can generate torque in *oscillatory* shear, which could be mistaken as shear elasticity (**a**) and nonconstant dynamic viscosity (**b**), as shown here with water at room temperature. Slight underfill or overfill can break contact line rotational symmetry (sample volume 459–495 μL for a $D = 60$ mm steel cone), as shown in the *inset* view from below at 459 μL. Oscillatory strain amplitude $\gamma_0 = 100\,\%$. Instrument low-torque limit for G'_{min} and η'_{min} from Eq. 6.14 with $T_{min} = 0.5$ nN.m in oscillation and instrument inertia limit for G'_{min} and η'_{min} from Eq. 6.18 using $\frac{IF_\tau}{F_\gamma} = 3 \cdot 10^{-3}$ Pa.s^2 (Figure adapted from [13])

inaccurately as apparent shear thinning of water (Fig. 6.10b). This example shows the dramatic impact of slight overfill that breaks the rotational symmetry of the contact line. The impact is to raise the effective T_{min} in the low-torque limit for $\eta(\dot{\gamma})$, based on Eq. 6.13. This may help explain studies showing a practical low-torque limit 20 times larger than that stated by the equipment manufacturer [10–12].

In oscillatory measurements, surface tension artifacts can mistakenly appear as a storage modulus G' plateau. We see this for water in Fig. 6.11a. Figure 6.11b also demonstrates frequency-dependent dynamic viscosity η', which should be constant for water over this range of frequencies. The inset image shows the asymmetric contact line for a slightly underfilled sample. When the geometry is visually properly filled (to within 5 μL, red triangles in Fig. 6.11), the elastic modulus is unmeasurable to within the experimental limits of low torque and instrument inertia, and the proper dynamic viscosity η' is measured down to $\omega = 10^{-1}$ rad/s.

The examples of Figs. 6.10 and 6.11 show that the surface tension torque effect is highly variable due to sample loading, wetting conditions, and contact line asymmetries and cannot be deterministically corrected in experimental measurements. It therefore raises the lower bound of the instrument low-torque limit T_{min}, in some cases by orders of magnitude.

Experimental techniques and careful geometry selection must be used to minimize surface tension effects. The surface tension torque is reduced by maximizing rotational symmetry of the contact line, minimizing evaporation and the migration of the contact line, reducing the radial location of the contact line, and lowering the surface tension. Experimental techniques for controlling the contact line symmetry such as using matched plate geometries are helpful. Reducing evaporation with a solvent trap and precision sample loading with a micropipette can also minimize surface tension torque plateaus.

Identifying and eliminating the surface tension torque is critical for low viscosities, intrinsic viscosities, soft materials, subdominant viscoelastic components, small gaps, and any circumstance where the low-torque limit is experimentally important. This phenomenon should be especially important in aqueous systems, including biological fluids, due to the high surface tension of water.

3.5 Free Surface Films

Biological fluids may contain proteins and other components that are surface active. Such components have an affinity to accumulate at liquid-air interfaces and may create a rigid or semirigid surface film. The film itself can be a fascinating object of study (e.g., see Chap. 4 of this book), but the film formation is a problem when the bulk flow properties of the solution are of interest. This experimental challenge has been known for some time [8, 50]. Care must be taken when measuring rheology using a geometry that involves a free surface, otherwise fluids may appear non-Newtonian when they are not measurably so [50, 51]. This applies to all the rotational geometries discussed in this chapter.

The primary signatures of a free surface film include increased viscosity, enhanced shear thinning, and often the presence of an apparent yield stress. For example, this is shown in Fig. 6.12 for aqueous solutions of the protein bovine serum albumin (BSA) [51]. The figure compares measurements using a cone-plate geometry (with liquid-air interface) and a microchannel pressure-driven flow viscometer (internal flow without a liquid-air interface). The microchannel measurements suggest Newtonian viscosities for this range of protein concentrations and shear rates, and do not match the cone-plate measurements which show higher viscosities and shear thinning at low rates. The increased viscosity and shear thinning are caused by a free surface film of the BSA [51]. A film is undesirable

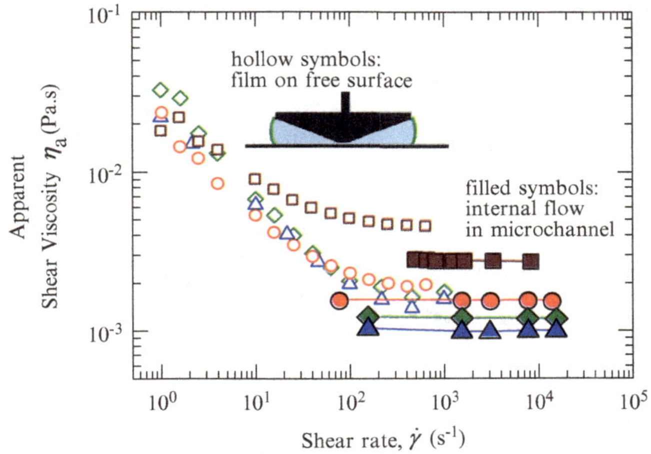

Fig. 6.12 Surface-active components in solution, such as the protein bovine serum albumin (BSA) shown here, may form a film at a free interface. Such films can show apparent yield stress and shear thinning of the apparent bulk viscosity when tested with cone-plate (*hollow symbols*). However, this is an interfacial rheological property and not a true bulk property as shown by the internal channel flow measurements (*filled symbols*). BSA at bulk concentrations of 10–200 mg/mL, *triangles-squares* (Data originally reported by Sharma et al. [51])

when measuring bulk properties. Of course, the presence of a film does provide an opportunity for interfacial surface rheology measurements if this is desired.

Internal flow geometries and guard rings (which eliminate the interface) can avoid the problem (although even in a closed system, there is the possibility of biofilm formation in biological fluids). When these are not available or possible, then one must be mindful of the symptoms of a free surface film. To test for the artifact of surface film rheology, one could make repeated measurements with different geometries and check for reproducibility of apparent material functions. For example, cones with increasing diameters could be used. The larger diameters create a longer film length and larger moment arm to produce torque and would generally be expected to have increased torque effects due to free surface films.

3.6 Slip

In rheological characterization, it is typically assumed that the sample sticks to the contacting boundaries whose motion defines the assumed strain field. In fluids, this is known as the no-slip condition. However, slip can easily occur [52–54], especially with biological gels and tissues. Slip violates the assumptions of standard

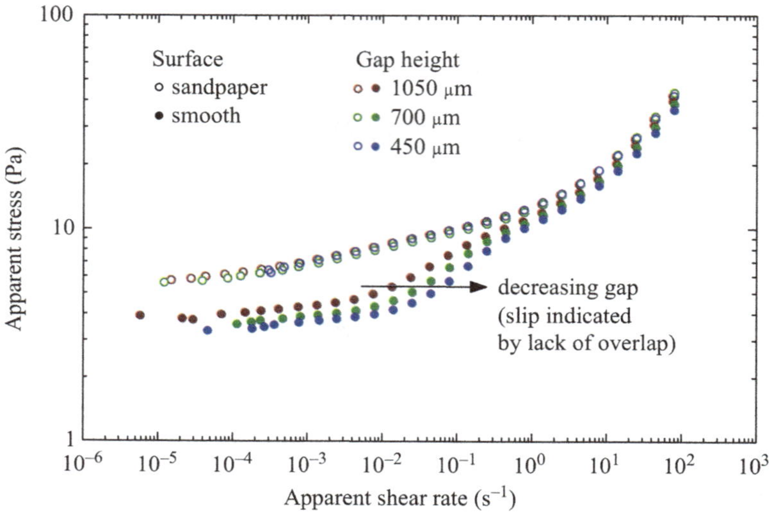

Fig. 6.13 Slip behavior on a smooth geometry surface can decrease flow stress and cause inconsistent gap dependence, as shown here for Nivea Lotion tested with different surfaces and gaps using parallel disks of diameter $D = 40\,\text{mm}$. A sandpaper surface eliminates slip artifacts, showing superposed data for different gap heights (Previously unpublished work of author RHE)

rheological characterization and may cause significant artifacts in the data. This section describes slip artifacts, methods of checking for slip, and techniques for avoiding the problem altogether.

The key signatures of slip include a decreased flow stress and inconsistent apparent stress and strain rate that depend on the geometry gap (gap height with parallel disks or cone angle with cone-plate). Figure 6.13 demonstrates these slip artifacts as seen with a non-Newtonian fluid (Nivea Lotion). The smooth boundary geometry produces artifacts and the true material behavior can be seen with a roughened surface. The steady-state flow sweep is conducted from high to low rates using a combined motor-transducer rotational rheometer (AR-G2, TA Instruments) with a parallel disk of diameter $D = 40\,\text{mm}$ and controlling temperature to $T = 23\,^{\circ}\text{C}$ with a Peltier plate. The sandpaper surfaces are adhesive-back sandpaper, 600 grit (McMaster-Carr Part #47185A51) attached to the standard rheometer geometry on both bounding surfaces. Apparent shear stress is calculated as $\tau_a = F_\tau T$ with $F_\tau = 2/(\pi R^3)$ for the disk, and apparent shear rate from $\dot{\gamma}_a = \Omega/h$ where h is the geometry gap.

Figure 6.13 shows an apparent stress plateau at low rates (an apparent dynamic yield stress) that depends on the geometry being used. The smooth geometry shows a lower apparent yield stress. This is a common experimental artifact that has been discussed in the literature, especially with yield stress fluids [55, 56]. In a recent study with a dense colloidal system, apparent yield stress behavior at low rates was associated with a sub-colloidal lubrication layer at the wall, as confirmed by confocal microscopy [56].

The gap is varied to check for slip in Fig. 6.13. For the rough sandpaper surface, the measurements superpose for all gaps therefore confirming the absence of slip. However, for the smooth plate, the data shifts to higher apparent strain rate as the gap is decreased. This shift is important evidence to indicate slip. To understand why, consider the simple example where an applied stress results in a particular slip velocity at the boundary of a sample. The gap-independent slip velocity contributes a fixed amount to the total velocity Ω. Therefore, as the gap h decreases, the apparent shear rate $\dot{\gamma}_a = \frac{\Omega R}{h}$ will have a numerator that decreases slower than the denominator, therefore increasing $\dot{\gamma}_a$ at small gaps for a fixed stress. This is shown by the arrow in Fig. 6.13 pointing to the right. Varying the geometry checks for the presence of slip but can also be used to correct for slip [57]. With very good control and sensitive instruments, gap-dependent measurements with a linear sliding plate rheometer have been used to characterize the slip itself including slip velocities [58].

Although sandpaper may be sufficient for some biological gels, e.g., as used for biopolymer mucin gels (snail slime) [59], sandpaper roughness is not always sufficient and other techniques must be considered. This includes the addition of grooves [55] or "cleats" [60] in plates, e.g., as used to measure vitreous humor [61]. Vane rotors are also commonly available [62], which are modifications of the concentric cylinder geometry. For more challenging solid materials, such as soft biological tissues, the sample can be squeezed slightly with an applied normal load to prevent slip during shear tests [63]. In extreme cases, gluing the tissue to the plate is required, as shown to be important with porcine kidney tissue, especially in nonlinear tests [64].

3.7 *Small Volume and Small Gap*

Biological fluids may be available only in small quantities. A variety of techniques can be used for measurements on small sample volumes [49]. This section describes a few techniques that have been useful with biological fluids and then focuses on the most widely accessible technique: parallel disks at small gaps. This section then describes the artifacts, corrections, and experimental windows for measurements at small gap.

Several techniques have been used for sample volumes around 10 µL and below. An early example is a capillary rheometer apparatus requiring only about 10 µL in volume [65]. A more recent pressure-driven flow setup (microfluidic cross-slot extensional flow) requires approximately 1 µL volumes [66], as demonstrated with hyaluronic acid and saliva. Boundary-driven flow examples include modification of parallel disks to confine a sample near the outer radius over a small area, approximating sliding plate flow with volumes 1–25 µL [48]. A custom-built linear sliding plate instrument has also been developed for precise, small gap tests (the so-called flexure-based microgap rheometer (FMR) [67, 68]). The FMR has been used to measure microliter quantities of spider silk [69] and sub-microliter quantities of carnivorous plant mucilage [70]. In those studies, samples were also tested with a small-scale extensional instrument based on capillary breakup extensional

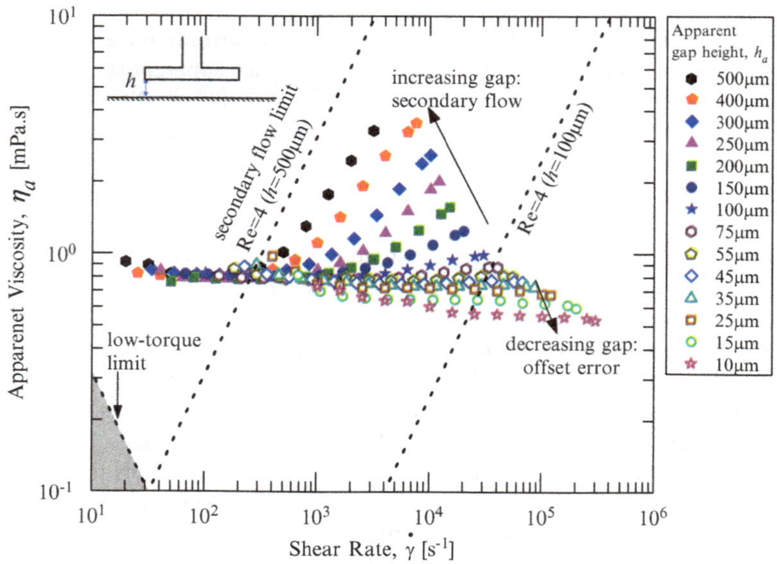

Fig. 6.14 Small gap/volume measurements of water with plate geometry, $D = 20$ mm. Gap offset artifacts are seen for apparent gap heights $h_a < 50$ µm. Inertial secondary flow artifacts appear for sufficiently large Reynolds number, i.e., at larger gaps and high shear rate (Sect. 3.3) (Previously unpublished work of authors RHE and MTJ)

rheometry (CaBER). Embedded probe techniques are also useful. Nanoliter droplets of butterfly saliva have been characterized with an embedded magnetic rod [71]. Of all the demonstrated techniques, standard parallel disks at small gaps may be the most experimentally accessible option for a researcher interested in small volumes of biological fluid.

With parallel disks, the smallest accessible gap will be limited by disk parallelism, precise knowledge of the true gap, and the size of the underlying material structure in the fluid. Confinement effects that violate the continuum hypothesis will not be discussed here (although this is sometimes relevant in biological fluids, such as blood exhibiting a confinement-dependent viscosity [72]). Finite boundary roughness and tribological contact will not be discussed either, since other gap errors are typically encountered first. Gap errors, including parallelism and gap precision, are the primary concern, assuming the continuum hypothesis holds true.

For gap errors, measurement artifacts include a decreasing apparent viscosity at smaller gaps (Fig. 6.14). This occurs under the typical scenario where the true gap h is larger than the apparent gap h_a calibrated by apparent contact of the plates [73]. These symptoms and limitations apply similarly to any boundary-driven drag flow at small gaps, such as the sliding plate FMR [67] and techniques to isolate samples to a small region under a conventional parallel disk geometry [48], although the difference between h and h_a may change depending on the calibration procedure.

Figure 6.14 shows the expected artifacts for small gap measurements, here with water using a disk with diameter $D = 20\,\mathrm{mm}$ down to apparent gap $h_a = 10\ \mu\mathrm{m}$ (down to apparent volume around $3\,\mu\mathrm{L}$). Using small gaps requires less volume and allows for higher shear rates. On average, the viscosity is what we expect for water, $\eta \approx 1$ mPa.s, but there are some issues. For small gaps ($h_a < 50\ \mu\mathrm{m}$), the apparent viscosity decreases as a function of gap. For the larger gaps ($h_a \geq 100\ \mu\mathrm{m}$), the viscosity seems to shear thicken at high shear rates, but at different critical shear rates. These are not true material properties of water, but are artifacts that can be explained.

For larger gaps at high rates, the inertia of the liquid may cause secondary flows (as described in Sect. 3.3). It is common to assume that the liquids will travel in circular stream lines, but centrifugal effects will tend to push fluid outward near a rotating boundary. This secondary flow increases dissipation, resulting in higher measured torque and hence a larger apparent viscosity. The effect increases as a function of Reynolds number, defined as $\mathrm{Re} = \rho \Omega h^2/\eta$, so the effect is evident for higher velocity Ω, larger gaps h, and low-viscosity fluids. Lines for $\mathrm{Re} = 4$ are shown in the figure for two representative gap heights.

For small gaps, the main error is caused by a gap offset ε_h, which is the difference between the apparent calibrated gap h_a and true gap h [73],

$$h = h_a + \varepsilon_h. \tag{6.32}$$

(The term "true" gap means the "effective" or "average" gap since the disks have finite roughness and finite parallelism manufacturing tolerance. Hence the gap is not precisely constant throughout the test geometry.) Since apparent gap h_a is used to calculate apparent viscosity η_a, one expects deviation from the true viscosity to be of the form

$$\eta_a = \eta \frac{h_a}{h} \tag{6.33}$$

which indicates $\eta_a < \eta$ for offset $\varepsilon_h > 0$. The apparent gap h_a is typically calibrated based on contact force at the first point of contact, where h_a is set to zero. Two issues arise to create gap offset error $\varepsilon_h > 0$. (i) A finite force is often observed before solid–solid contact due to viscous resistance of air flow in the squeezing gap. (ii) The parallelism is not perfect, and the average gap will often be larger than the "first point of contact" gap. The nonparallelism contribution generates normal forces [74] and this can be used to identify the relative importance of the two sources of gap offset error. Both of these effects contribute to gap offset error $\varepsilon_h > 0$, so that the actual gap is larger than the apparent value. Typical values for ε_h are on the order of 10–50 μm [75, 76].

Gap offset ε_h can be corrected if Eq. 6.32 holds true [73, 75, 76], although the correction will depend on the uncertainty in calibrating for ε_h. Uncertainly in the calculated viscosity will grow dramatically as the gap approaches the uncertainty of ε_h. Gap offset errors can be minimized by using a smaller radius plate, since

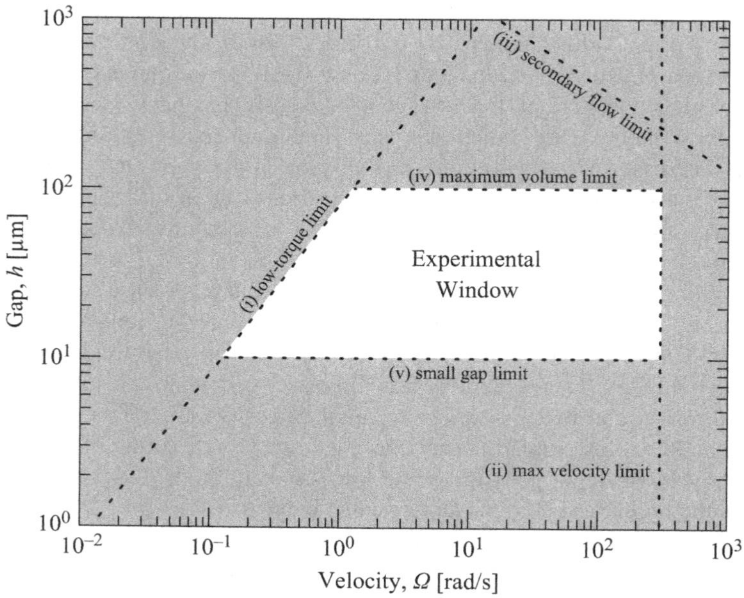

Fig. 6.15 Experimental window for small volume aqueous solutions, using Eqs. 6.34–6.38 to draw the boundaries. Representative values used are $T_{\min} = 5$ nN.m, $\Omega_{\max} = 300$ rad/s, $\mathrm{Re}_{\max} = 4$, $\eta = 1$ mPa.s, $\rho = 1{,}000\ \mathrm{kg/m^3}$, plate diameter $D = 8$ mm, $V_{\max} = 5\ \mu$L, and minimum gap $h_{\min} = 10\ \mu$m

this decreases the viscous squeeze force at apparent contact and also decreases the nonparallelism (angular misalignment) contribution to ε_h due to the smaller radius. However, a smaller radius plate changes other experimental limits such as increasing $\eta_{\min}$ due to the low-torque limit (Eq. 6.13). The experimental window for small gap measurements is therefore bounded by several limitations.

Figure 6.15 is an example experimental window for small volume and small gap measurements, in the operational space of gap h and velocity Ω. Several limitations are considered including (i) minimum torque, (ii) maximum velocity, (iii) secondary flow, (iv) maximum volume available, and (v) small gap limit, e.g., due to gap offset errors. The exact locations of the boundaries will shift depending on the fluid properties and instrument used, but their shapes will not change. Representative values are used in Fig. 6.15 for aqueous fluids. Specific equations for boundary lines come from consideration of each limit. The minimum torque, item (i), is set either by the instrument specification (Sect. 3.1) or surface tension torque which becomes increasingly important at small gaps (Sect. 3.4). Based on the criteria $T > T_{\min}$, where torque from steady viscosity is $T = (\eta\Omega R/h)/F_\tau$, the boundary line is defined by

$$h < \frac{\eta\Omega R}{F_\tau T_{\min}} \qquad (6.34)$$

which shows the scaling $h \sim \Omega$ that appears in Fig. 6.15. The instrument limit of maximum velocity, item (ii), is simply set by the criteria

$$\Omega < \Omega_{\max} \tag{6.35}$$

and therefore appears as a vertical line in the figure. The limit of secondary flow, item (iii), was discussed in Sect. 3.3. There, a maximum Reynolds number $\mathrm{Re}_{\max}$ sets the boundary line and is based on the definition $\mathrm{Re}= \frac{\rho \Omega h^2}{\eta}$ for parallel disks. Then, the criteria $\mathrm{Re} < \mathrm{Re}_{\max}$ can be written

$$h < \left(\frac{\mathrm{Re}_{\max} \eta}{\rho \Omega} \right)^{1/2} \tag{6.36}$$

which shows the scaling $h \sim \Omega^{-1/2}$ seen in the top right of Fig. 6.15. The last two criteria come from this section, considering small volume and small gap limitations. The volume limit is simply $V < V_{\max}$ where $V_{\max}$ is the maximum sample volume available. For parallel disks, $V = \pi R^2 h$, and the boundary line is defined by

$$h < \frac{V_{\max}}{\pi R^2}. \tag{6.37}$$

The final boundary, item (v), is the minimum gap. This boundary line is defined by

$$h > h_{\min}, \tag{6.38}$$

where $h_{\min}$ is set by the gap offset error, or possibly the minimum gap where confinement effects are negligible and the material can still be considered a continuum. Figure 6.15 uses Eqs. 6.34–6.38 with representative parameters given in the caption. The minimum gap $h = 10$ μm is used, assuming uncertainty in gap error much less than 10 μm which would allow for gap offset corrections. Based on the maximum sample volume and minimum gap limits, the experimental window is confined between $h = 10 - 100$ μm for this example. If larger volumes are available, then larger gaps can be used, eventually being limited by the secondary flow (e.g., as seen in Fig. 6.14 at larger gaps). At very large gaps, the experimental window closes. This occurs where the minimum torque and secondary flow boundaries intersect, for gaps $h > 1{,}000\,\mu$ m with this particular geometry.

3.8 Other Issues

Additional challenges, basic and exotic, can also be included on the list of possible ways that rheological measurements can go astray.

One basic but important point is sample volume underfill or overfill. In cone-plate and parallel disk geometries, torque is a very sensitive function of the radial extent of contact $T \sim \frac{1}{R^3}$ [8]. Underfill is more sensitive than overfill, but for both cases one may have the problem of an uncontrolled contact line that loses rotational symmetry which can introduce additional torque due to surface tension forces (Sect. 3.4). Underfill can also develop as a sample evaporates. This is relevant to aqueous biological fluids. Evaporation can be eliminated or reduced by the use of a solvent trap. The use of a micropipette and close attention to fill level can further eliminate the basic issue of sample volume underfill or overfill.

An exotic but relevant issue with some biological fluids is particle settling and migration, in particular with active-swimming microorganisms in suspension. In general, the concentric cylinder geometry is recommended when gravitational particle settling may be an issue, since a depletion layer is not created across the velocity gradient direction as would be the case with cone-plate or parallel disk geometries. But, if the sample volume is not sufficient, parallel disks or the cone-plate geometry must be used. One particularly striking example is with the same microalgae suspension whose flow data is given in Fig. 6.3. Those measurements were made with a cone-plate geometry. The data needed to be collected within the first 2 min of flow due to microalgae rheotaxis (flow-induced swimming), coupled with particle settling and secondary flow. The microalgae migration was visualized through a transparent bottom plate during shear flow, as shown in Fig. 6.16. The migration to a nonhomogeneous state is indicated by the development of a dark green circle toward the center of the geometry. The particles are negatively buoyant and therefore settle to the bottom fixed plate. A finite secondary flow exists which draws fluid radially outward near the moving cone (Sect. 3.3), which is balanced by a radially inward flow near the stationary boundary. This inward flow carries the negatively buoyant microalgae toward the center. The visualization in Fig. 6.16 indicated the time scale of developing a non-homogeneous sample. This set an experimental boundary on the *time* that could be used to collect steady-state shear data, under 2 min for this configuration.

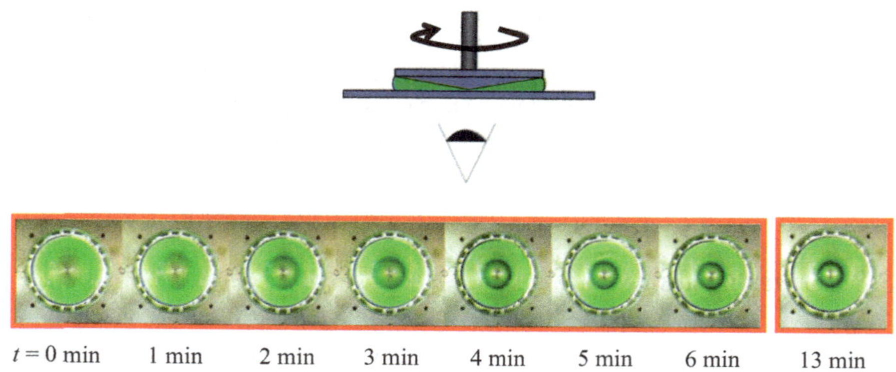

Fig. 6.16 Visualization can be used to check for nonhomogeneous conditions due to rheotaxis, particle settling, and secondary flow. Here, shear flow of a microalgae suspension is viewed from below through a *transparent bottom plate*, showing that a nonhomogeneous distribution develops over time (Previously unpublished work of authors RHE and LMC)

An issue relevant to biological gels, and stiffer biological materials, is edge fracture. This is a well-studied phenomenon, and polymer melts and solutions have tended to receive the most attention [8, 77, 78]. The problem can be identified by visually monitoring the edge of the sample; the experimental artifact is a decreased load (apparent stress) since the true sample contact area is effectively decreased. There are some experimental configurations that minimize edge fracture artifacts, all of which move the free surface away from the transducer surface. For example, a "plate and cup" or "sea of fluid" geometry has been used [79], as well as partitioned plates with the torque transducer plate set inside the larger geometry plate [80–82].

As materials become more viscous and more stiff, additional problems arise. Of all the remaining possible experimental errors [7, 8], viscous heating and instrument compliance are worth mentioning briefly here.

Viscous heating is an issue with higher viscosity fluids. The symptom is a decrease in the apparent viscosity as a function of shear rate, since higher rates further increase temperature and decrease viscosity. The key dimensionless number to check is the Nahme number [2], which can be interpreted as a ratio of viscosity change due to viscous heating compared to the baseline viscosity. Values near zero indicate negligible viscosity change. Smaller gaps help minimize the heating effect, since they decrease the length of the thermal conduction path. This is most important for liquids with high viscosity and low thermal conductivity. For low-viscosity biological fluids, this is less of an issue.

System compliance is an issue with higher stiffness materials. The problem lies in the possibility of finite movement of a "fixed" boundary due to system compliance, or small but finite movement of the load cell, even with modern force rebalancing transducers [83, 84]. Instrument compliance issues have been identified with the dynamic shear measurement of glycerol [85] and polymer melts [86]. Recommendations have been identified for experimental protocol and instrument design to avoid, minimize, and correct for compliance effects [87]. Instrument compliance errors should be considered for stiff, solid materials, or the short-time data from step strain inputs when material stiffness may also be large.

4 Conclusions

The experimental challenges described here will serve as a checklist for troubleshooting and debugging rheological measurements of complex fluids. These challenges are especially evident with biofluids and biological materials. Common artifacts cause a fluid to inaccurately appear as shear thinning, shear thickening, having frequency dependence, time dependence, or having an elastic modulus, when these behaviors are not actually present in the true intensive material response.

We encourage the reader to think critically about experimental rheological measurements, and to ask appropriate and fair questions about the validity of data (including their own and those published in the open literature). This is particularly relevant for biological complex fluids which are soft, have low viscosity, and may contain active components.

For each potential artifact, the system-level perspective can identify an experimental boundary and/or a method of minimizing the effect (Sects. 3.1–3.8). The focus here has been on the most commonly used technique for measuring shear material functions, i.e., drag flow at the boundary ([8] Chap. 5), especially when imposed by rotational geometries such as parallel disks, cone-plate, or concentric cylinders (Fig. 6.1). Other measurement techniques, including those described in this book (Chap. 1, 3–4), will also benefit from a system-level perspective that acknowledges non-ideal conditions for rheological material function measurement (as in Fig. 6.1). The identification of experimental boundaries (similar to Figs. 6.3, 6.5, 6.6, 6.10, 6.11, 6.14, and 6.15) will be especially relevant to biological fluids and materials that are of limited volume, as well as soft, active, and rheologically complex materials more generally.

Acknowledgements This work was supported by the National Science Foundation under Grant No. CBET-1342408. RHE and LMC acknowledge helpful discussions regarding careful rheological measurements with Prof. Christopher Macosko and Dr. David Giles at the University of Minnesota. RHE also thanks Prof. Gareth McKinley at the Massachusetts Institute of Technology for initial discussions on drawing experimental boundaries for rheological measurements. RHE and LMC also acknowledge Prof. Jian Sheng at Texas Tech University (formerly University of Minnesota) for suggesting the study of actively swimming microalgae suspensions, and Mr. Anwar Chengala for preparing those samples.

Appendix: Material Details: Hagfish Gel and Microalgae Suspension

Hagfish gel serves as an example of an ultrasoft biomaterial gel. It is prepared as in [5, 14] and used in Figs. 6.5, 6.6, and 6.7.

The actively swimming microalgae suspension provides an example of a low-viscosity biological solution and is used in Figs. 6.3 and 6.16. The algal species *Dunaliella primolecta* was used. It is a motile, biflagellated, cell-wall-less, unicellular green algae that does not clump. It has slight negative buoyancy, approximate characteristic diameter 11 μm, and natural concentration on the order of $3 \cdot 10^6$ cells/mL. *Dunaliella Primolecta* (UTEX LB 1000) was obtained from UTEX, The Culture Collection of Algae at the University of Texas at Austin. Nonmotile samples were prepared by adding 2 mL of 4 %wt/vol of formaldehyde in phosphate buffered saline (PBS) solution to 25 mL of the bulk sample. The fixed sample was analyzed under light microscope to ensure it was nonmotile.

References

1. J.M. Dealy, J. Rheol. **39**(1), 253 (1995)
2. R.B. Bird, R.C. Armstrong, O. Hassager, *Dynamics of Polymeric Liquids: Volume 1 Fluid Mechanics*, 2nd ed. (Wiley, New York, 1987)
3. R.G. Larson, *The Structure and Rheology of Complex Fluids* (Oxford University Press, New York, 1999)
4. N. Bharadwaj, J.T. Allison, R.H. Ewoldt, in *ASME 2013 International Design Engineering Technical Conferences and Computers and Information in Engineering Conference* (Portland, OR, 2013), paper DETC2013–13,462
5. R.H. Ewoldt, T.M. Winegard, D.S. Fudge, Int. J. Non-Linear Mech. **46**(4), 627 (2011)
6. R.H. Ewoldt, J. Rheol. **57**(1), 177 (2013)
7. K. Walters, *Rheometry* (Wiley, New York, 1975)
8. C.W. Macosko, *Rheology: Principles, Measurements, and Applications* (Wiley-VCH, New York, 1994)
9. R.L. Powell, in *Rheological Measurement*, ed. by A.A. Collyer, D.W. Clegg, 2nd edn., Chap. 9 (Springer, Dordrecht, 1998), pp. 260–298
10. M.S. Oliveira, R. Yeh, G.H. McKinley, J. Non-Newtonian Fluid Mech. **137**(1–3), 137 (2006)
11. L.E. Rodd, T.P. Scott, D.V. Boger, J.J. Cooper-White, G.H. McKinley, J. Non-Newtonian Fluid Mech. **129**(1), 1 (2005)
12. J. Soulages, M. Oliveira, P. Sousa, M. Alves, G. McKinley, J. Non-Newtonian Fluid Mech. **163**(1–3), 9 (2009)
13. M. Johnston, R. Ewoldt, J. Rheol. **57**(6), 1515 (2013)
14. D.S. Fudge, T. Winegard, R.H. Ewoldt, D. Beriault, L. Szewciw, G.H. McKinley, Integr. Comp. Biol. **49**(1), 32 (2009)
15. C. Baravian, D. Quemada, Rheol. Acta **37**(3), 223 (1998)
16. C. Baravian, G. Benbelkacem, F. Caton, Rheol. Acta **46**(5), 577 (2006)
17. R.H. Ewoldt, G.H. McKinley, Rheol. Bull. **76**(1), 4 (2007)
18. N.Y. Yao, R.J. Larsen, D.A. Weitz, J. Rheol. **52**(4), 1013 (2008)
19. Y.C. Lin, G.H. Koenderink, F.C. MacKintosh, D.A. Weitz, Macromolecules **40**(21), 7714 (2007)
20. P. Sharif-Kashani, J.P. Hubschman, D. Sassoon, H.P. Kavehpour, J. Biomech. **44**(3), 419 (2011)
21. C.P. Broedersz, K.E. Kasza, L.M. Jawerth, S. Münster, D.A. Weitz, F.C. MacKintosh, Soft Matter **6**(17), 4120 (2010)
22. H. Lee, J.M. Ferrer, F. Nakamura, M.J. Lang, R.D. Kamm, Acta Biomater. **6**(4), 1207 (2010)
23. K.M. Weigandt, D.C. Pozzo, L. Porcar, Soft Matter **5**(21), 4321 (2009)
24. S. Suri, C.E. Schmidt, Acta Biomater. **5**(7), 2385 (2009)
25. L.C.E. Struik, Rheol. Acta **6**(2), 119 (1967)
26. M.E. Mackay, C.H. Liang, P.J. Halley, Rheol. Acta **31**(5), 481 (1992)
27. J. Magda, R. Larson, J. Non-Newtonian Fluid Mech. **30**(1), 1 (1988)
28. E.S.G. Shaqfeh, Annu. Rev. Fluid Mech. **28**, 129 (1996)
29. G.H. McKinley, P. Pakdel, A. Öztekin, J. Non-Newtonian Fluid Mech. **67**, 19 (1996)
30. P. Pakdel, G. McKinley, Phys. Rev. Lett. **77**(12), 2459 (1996)
31. J.L. Schrag, J. Rheol. **21**(3), 399 (1977)
32. J.A. Yosick, J.A. Giacomin, W.E. Stewart, F. Ding, Rheol. Acta **37**(4), 365 (1998)
33. C. Storm, J.J. Pastore, F.C. MacKintosh, T.C. Lubensky, P.A. Janmey, Nature **435**(7039), 191 (2005)
34. G.I. Taylor, Philos. Trans. R. Soc. Lond. A. Math. Phys. Character **223**(605–615), 289 (1923)
35. S. Chandrasekhar, *Hydrodynamic and Hydromagnetic Stability* (Dover, New York, 1981)
36. R.G. Larson, Rheol. Acta **31**(3), 213 (1992)

37. ISO, 3219, Determination of viscosity using a rotational viscometer with defined shear rate (1993)
38. H.P. Sdougos, S.R. Bussolari, C.F. Dewey, J. Fluid Mech. **138**, 379 (1984)
39. R.M. Turian, Ind. Eng. Chem. Fund. **11**(3), 361 (1972)
40. B. Meulenbroek, C. Storm, A.N. Morozov, W. van Saarloos, J. Non-Newtonian Fluid Mech. **116**(2–3), 235 (2004)
41. L. Pan, A. Morozov, C. Wagner, P.E. Arratia, Phys. Rev. Lett. **110**(17), 174502 (2013)
42. D.F. Griffiths, K. Walters, J. Fluid Mech. **42**(02), 379 (1970)
43. R.W.G. Shipman, M.M. Denn, R. Keunings, Ind. Eng. Chem. Res. **30**(5), 918 (1991)
44. F.M. Orr, L.E. Scriven, A.P. Rivas, J. Fluid Mech. **67**(04), 723 (1975)
45. M.A. Fortes, J. Colloid Interface Sci. **88**(2), 338 (1982)
46. E.J. De Souza, L. Gao, T.J. McCarthy, E. Arzt, A.J. Crosby, Langmuir **24**(4), 1391 (2008)
47. H.M. Laun, J. Meissner, Rheol. Acta **19**(1), 60 (1980)
48. M.E. Mackay, C.A. Cathey, J. Rheol. **35**(2), 237 (1991)
49. M.E. Mackay, in *Rheological Measurement*, ed. by A. Collyer, D. Clegg, 2nd edn., Chap. 20 (Springer, Dordrecht, 1998), pp. 635–665
50. E. Merrill, Physiol. Rev. **49**(4), 863 (1969)
51. V. Sharma, A. Jaishankar, Y.C. Wang, G.H. McKinley, Soft Matter **7**(11), 5150 (2011)
52. R. Buscall, J.I. McGowan, A.J. Morton-Jones, J. Rheol. **37**(4), 621 (1993)
53. H.A. Barnes, J. Non-Newtonian Fluid Mech. **56**(3), 221 (1995)
54. R. Buscall, J. Rheol. **54**(6), 1177 (2010)
55. A. Magnin, J. Piau, J. Non-Newtonian Fluid Mech. **36**, 85 (1990)
56. P. Ballesta, G. Petekidis, L. Isa, W.C.K. Poon, R. Besseling, J. Rheol. **56**(5), 1005 (2012)
57. A. Yoshimura, R.K. Prudhomme, J. Rheol. **32**(1), 53 (1988)
58. C. Clasen, Rheol. Acta **51**(10), 883 (2012)
59. R.H. Ewoldt, C. Clasen, A.E. Hosoi, G.H. McKinley, Soft Matter **3**(5), 634 (2007)
60. C.S. Nickerson, J.A. Kornfield, J. Rheol. **49**(4), 865 (2005)
61. C.S. Nickerson, J. Park, J.A. Kornfield, H. Karageozian, J. Biomech. **41**(9), 1840 (2008)
62. H.A. Barnes, Q.D. Nguyen, J. Non-Newtonian Fluid Mech. **98**(1), 1 (2001)
63. K. Tan, S. Cheng, L. Jugé, L.E. Bilston, J. Biomech. **46**(6), 1060 (2013)
64. S. Nicolle, J.F. Palierne, J. Mech. Behav. Biomed. Mater. **14**(null), 158 (2012)
65. W. Philippoff, C. Han, B. Barnett, M. Dulfano, Biorheology **7**(1), 55 (1970)
66. S.J. Haward, V. Sharma, J.A. Odell, Soft Matter **7**(21), 9908 (2011)
67. C. Clasen, G.H. McKinley, J. Non-Newtonian Fluid Mech. **124**(1–3), 1 (2004)
68. C. Clasen, B.P. Gearing, G.H. McKinley, J. Rheol. **50**(6), 883 (2006)
69. N. Kojić, J. Bico, C. Clasen, G.H. McKinley, J. Exp. Biol. **209**(Pt 21), 4355 (2006)
70. P. Erni, M. Varagnat, C. Clasen, J. Crest, G.H. McKinley, Soft Matter **7**(22), 10889 (2011)
71. A. Tokarev, B. Kaufman, Y. Gu, T. Andrukh, P.H. Adler, K.G. Kornev, Appl. Phys. Lett. **102**(3), 033701 (2013)
72. R. Få hræus, T. Lindqvist, Am. J. Physiol. **96**(3), 562 (1931)
73. R.W. Connelly, J. Greener, J. Rheol. **29**(2), 209 (1985)
74. E. Andablo-Reyes, R. Hidalgo-Álvarez, J. de Vicente, J. Non-Newtonian Fluid Mech. **165**(19–20), 1419 (2010)
75. G.A. Davies, J.R. Stokes, J. Rheol. **49**(4), 919 (2005)
76. C.J. Pipe, T.S. Majmudar, G.H. McKinley, Rheol. Acta **47**(5–6), 621 (2008)
77. R.I. Tanner, M. Keentok, J. Rheol. **27**(1), 47 (1983)
78. M. Keentok, S.C. Xue, Rheol. Acta **38**(4), 321 (1999)
79. J.S. Vrentas, D.C. Venerus, C.M. Vrentas, J. Non-Newtonian Fluid Mech. **40**(1), 1 (1991)
80. T. Schweizer, J. Rheol. **47**(4), 1071 (2003)
81. S.Q. Wang, S. Ravindranath, P.E. Boukany, Macromolecules **44**(2), 183 (2011)
82. F. Snijkers, D. Vlassopoulos, J. Rheol. **55**(6), 1167 (2011)
83. M.E. Mackay, P. Halley, J. Rheol. **35**(8), 1609 (1991)
84. C.Y. Liu, M. Yao, R.G. Garritano, A.J. Franck, C. Bailly, Rheol. Acta **50**(5–6), 537 (2011)

85. K. Schröter, S.A. Hutcheson, X. Shi, A. Mandanici, G.B. McKenna, J. Chem. Phys. **125**(21), 214507 (2006)
86. T. Schweizer, A. Bardow, Rheol. Acta **45**(4), 393 (2006)
87. S.A. Hutcheson, G.B. McKenna, J. Chem. Phys. **129**(7), 074502 (2008)
88. C. Dutcher, S. Muller, Phys. Rev. E. **75**, 047301 (2007)

Part III
Locomotion and Active Matter

Chapter 7
Locomotion Through Complex Fluids: An Experimental View

Josué Sznitman and Paulo E. Arratia

Abstract Recently, there has been renewed interest in the swimming of microorganisms for applications that include artificial swimmers, novel materials, drug delivery, and micro-robotics. Due to small length scales, the fluid mechanics of swimming of microorganisms are governed by low Reynolds number hydrodynamics. In such a regime, linear viscous forces dominate over nonlinear inertial forces. While our current understanding of locomotion at low Reynolds numbers is derived mainly from investigations in simple, Newtonian fluids (e.g., water), many of the fluids in which locomotion occurs contain solids and/or (biological) polymers that are instead not Newtonian. Examples include wet soils, human mucus, and fluids in the cervix and female reproductive track. A major challenge is to understand the propulsion mechanisms in fluids that display complex rheological behavior such as viscoelasticity and shear-thinning viscosity. Here, we will briefly review a few notable swimming experiments in Newtonian fluids and then discuss the latest experimental results on swimming in complex fluids, focusing on viscoelastic fluids.

1 Introduction

Microorganisms are surrounded by fluids. They cope and take advantage of water or wind currents to move, feed, and reproduce. Many, if not most, living organisms live in the realm of low Reynolds numbers [1], usually defined as $\mathrm{Re} = \rho UL/\mu$, where U is a characteristic speed, L a characteristic length (e.g., body size), and ρ and μ are the fluid's density and dynamic viscosity, respectively. For example, the typical Reynolds number for microorganisms such as eukaryotic protozoa

J. Sznitman (✉)
Department of Biomedical Engineering, Technion-Israel Institute of Technology, Haifa, 32000, Israel
e-mail: sznitman@bm.technion.ac.il

P.E. Arratia
Department of Mechanical Engineering and Applied Mechanics, University of Pennsylvania, Philadelphia, PA 19104, USA
e-mail: parratia@seas.upenn.edu

S.E. Spagnolie (ed.), *Complex Fluids in Biological Systems*, Biological and Medical Physics, Biomedical Engineering,
DOI 10.1007/978-1-4939-2065-5_7

(e.g., sperm cells [2–4]), prokaryotes (e.g., bacteria [5]), and even multicellular organisms (e.g., nematodes [5–7]) is much less than unity ($Re \ll 1$) due to the organism's small length scale L. By contrast, humans when swimming in the ocean or water pools can reach Reynolds numbers of approximately 10^4. This means that humans can take advantage of nonlinear inertial forces for propulsion, while microorganisms simply cannot. Small living organisms instead have to overcome the linear viscous forces and drag arising from the fluid in order to achieve any appreciable net motion. The picture that emerges is that moving (and living) at low Reynolds number is drastically different from what we (humans) are accustomed to experience in our everyday lives. For the case of swimming microorganisms in simple fluids such as water, the equations of fluid motion become time-reversible (for a more in-depth discussion on the governing equations, please see Chap. 1). As a result, net locomotion can only be generated from nonreciprocal kinematics in order to break time-reversal symmetry [2,8,9]; this is also known as the "scallop theorem" [1] which states that organisms that rely on reciprocal motion for locomotion cannot achieve net motion in the limit of vanishing Reynolds numbers. A more detailed discussion on the scallop theorem is given in Chap. 8.

Microorganisms have developed diverse strategies to break time-symmetry and create nonreciprocal motion (see Fig. 7.1). Such strategies include body undulations and the presence of moving flagella. For example, the motility of various multicellular organisms including the worm nematode *Caenorhabditis elegans* originates from the propagation of undulatory waves from head to tail as a result of patterns of muscle activation and neuromuscular control [10–12]. Locomotion may also result from flagellar motility where one or several bundled appendages protrude from the cell body of certain prokaryotic and eukaryotic cells. One can typically distinguish between bacterial flagella that are helical filaments (e.g., *Escherichia coli* [13–16]) and eukaryotic flagella that are flexible filaments undergoing "whiplike" motions resulting from the action of molecular motors distributed along the filament length; this latter mode of flagellar actuation is seen for example in many sperm cells [17–19]. Other eukaryotic organisms (e.g., *Paramecium*) have instead their body surface covered with thousands of small hairlike protrusions (*cilia*) that beat in a coordinated manner [17, 20].

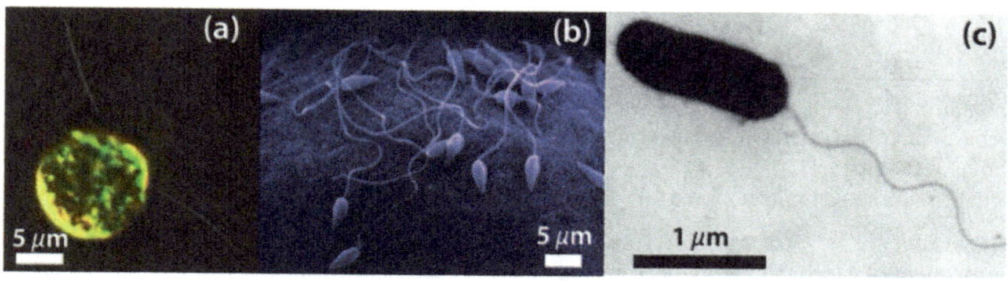

Fig. 7.1 Examples of microorganisms. (**a**) The green algae *Chlamydomonas reinhardtii*, a model eukaryotic organism, (**b**) sperm cells moving next to boundaries, and (**c**) the bacterium *E. coli*, one of the most widely studied prokaryotic model organisms

There are many other cellular environments that are characterized by low Reynolds number locomotion. In mammals, for instance, cells featuring motile cilia include the epithelium of the female Fallopian tubes, where rhythmic beating guarantees translocation of the ovum from the ovary into the uterus [21–23]. Motile cilia are also found in the epithelial lining of the tracheobronchial airways of the lungs [24]. There, the ciliated epithelium prevents mucus accumulation in the airway lumen and serves as an immune barrier against pathogens and foreign particulate matter by the action of the so-called mucociliary escalator [25, 26] characterized by synchronous waves of ciliary beating. These waves effectively transport mucus secretions towards the laryngopharynx for expectoration or swallowing to the stomach.

It is clear from everyday observation and from the few examples cited above that nature has found many fascinating ways to break time-reversibility and achieve net motion at the microscopic scale. And thanks to theoretical, numerical, and experimental investigations, our understanding of swimming of microorganisms at low Reynolds numbers has significantly improved in the past 60 years or so [1,3,8,9,13,27–30]. While much of our efforts have been restricted to swimming in simple, Newtonian fluids (e.g., water), there are many microorganisms that live in (complex) fluids that contain particulates and/or polymers (e.g., human mucus, blood, wet soil, gels, and tissues) and are not Newtonian. Such fluids often possess complex (non-Newtonian) rheological behavior such as shear-thinning viscosity and viscoelasticity. One may expect that swimming in complex fluids such as mucus to be markedly different from swimming in water. But is it? Is swimming in complex fluids at low Reynolds numbers still dominated by purely viscous effects? Does fluid elasticity matter at low Reynolds numbers? Here, we will try to address those questions by examining the recent experimental investigation of swimming of microorganisms in both Newtonian and complex fluids. This chapter is organized as follows: (i) a brief introduction is given on swimming at low Reynolds numbers in simple, Newtonian fluids, (ii) experiments in Newtonian fluids are discussed, (iii) a brief discussion of locomotion in complex fluids is given, (iv) experiments in viscoelastic fluids are discussed, and (v) conclusions and outlook are provided. We begin by introducing some basic principles of swimming at low Reynolds numbers in Newtonian fluids; we note that a more detailed discussion on the basic principles of locomotion in Newtonian and complex fluids can be found in Chaps. 1 and 8.

2 Basic Principles: Fluid Dynamics of Swimming at Low Reynolds Number

Let us begin our discussion of swimming of microorganisms by estimating the Reynolds numbers of the bacterium *E. coli* in water. The shear viscosity μ of water is 1 mPa · s (or 1 cP) and independent of shear rate (i.e., Newtonian). The characteristic size L of *E. coli* is approximately 2 μm, and the bacterium is known to achieve net swimming speeds U of approximately $25\,\mu\text{m/s}$ [31]. Following these parameters, we can estimate the Reynolds number for *E. coli* to be approximately $\text{Re} = \rho UL/\mu = 5 \times 10^{-5} \ll 1$, where ρ is fluid density ($10^3\,\text{kg/m}^3$). Such low value of Re implies that linear viscous forces dominate over inertial forces, and the nonlinear convective term in the Navier–Stokes equation can be safely ignored (see Chap. 1). Additionally, one can compute the frequency-based Reynolds number, typically defined as $\text{Re}_{\text{freq}} = \rho L^2 \omega/\mu$, to assess unsteady flow effects, where ω is the frequency at which the bacterium flagella rotate ($\sim 100\,\text{Hz}$). We find again for *E. coli* that $\text{Re}_{\text{freq}} \ll 0.1$, and thus one can assume the flow to be steady. Under those conditions and assuming that the bacterium is moving in an incompressible ($\nabla \cdot \mathbf{u} = 0$), Newtonian fluid (and ignoring body forces like gravity), the equation of fluid motion reduces to

$$\nabla p = \mu \nabla^2 \mathbf{u}, \tag{7.1}$$

where $\mathbf{u}$ is the velocity vector, p is the pressure, and ∇ is the divergence operator. The above equation is often referred to as the Stokes equation, named after the mathematician Sir George Stokes. Equation (7.1) is the main equation governing the hydrodynamics of swimming microorganism and has some interesting properties. For example, the above equation is instantaneous in the sense that it has no dependence on time other than via boundary conditions. Equation (7.1) is also linear in both velocity and pressure. Furthermore, it is time-reversible in the sense that any time-reversed Stokes flow solves the same equations as the original Stokes flow. This time-reversibility, or kinematic reversibility, forms the hydrodynamic basis of the "scallop theorem" introduced earlier [1]. These properties illustrate that swimming at low Reynolds numbers can seem at first as a highly confined phenomenon, yet microorganisms have found a variety of ways to overcome the constraints of the scallop theorem. In what follows, we briefly review some of the classical theories that have shed light on the hydrodynamic mechanisms leading to net propulsion at low Reynolds numbers. The discussion will be limited but the reader can find a more thorough review in [2, 5, 8, 30] as well as in Chaps. 1 and 8 of this book.

Over half a century ago, Taylor [27, 28] beautifully demonstrated that an infinite waving sheet (see Fig. 7.2a) could swim in an incompressible, Newtonian fluid by generating traveling waves in the absence of inertia or vanishing Reynolds numbers. Note that the hydrodynamics of Taylor's waving sheet is governed by Eq. (7.1). In Taylor's work, the planar sheet oscillates in time *in a prescribed form* according to $y(x,t) = a \sin(kx - \omega t)$, where a is the traveling wave amplitude, ω is the frequency, $\lambda = 2\pi/k$ is the wavelength, $c = \omega/k$ is the traveling wave speed, and k is the wave number. Taylor found that the sheet oscillations induce a forward velocity $U = \omega a^2 k/2 + \mathcal{O}(ka)^4$ [27], where the sheet is propelled in the direction opposite to that of the propagating wave (Fig. 7.2a).

Many important investigations followed Taylor's landmark contribution. Of particular relevance, we highlight the well-known resistive force theory (RFT) introduced by Gray and Hancock in analyzing the locomotion of sperm cells [3]. There, the authors assumed that the hydrodynamic forces experienced by the

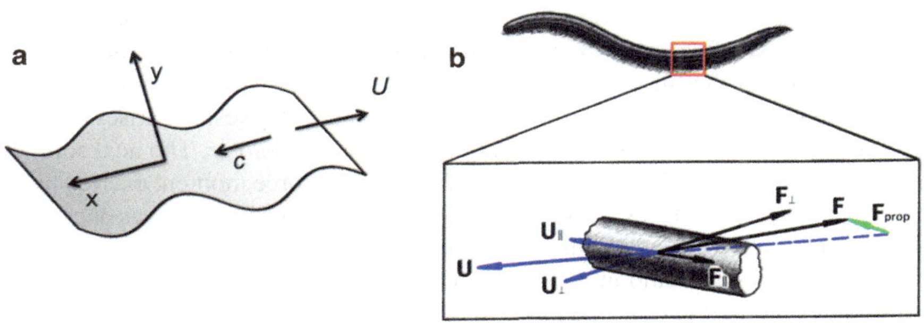

Fig. 7.2 (**a**) Two-dimensional waving sheet in a viscous fluid illustrating the traveling wave of velocity c progressing in the x-direction and the forward swimming speed (U) in opposite direction. (**b**) Resistive force theory (RFT) diagram illustrating the normal and tangential components of the velocity U and force F, and the resulting net propulsive force

organism would be approximately proportional to the local body velocity such that the force exerted by a body or flagellar segment is given by $\mathbf{F} = C_N \mathbf{U_N} + C_T \mathbf{U_T}$, where C corresponds to the local drag coefficient per unit length (dependent on geometry and fluid viscosity) and N and T are the normal and tangential components, respectively (see Fig. 7.2b). Hence, the total thrust can then be obtained by integrating the propulsive force over the entire body or flagellum length. It is namely the anisotropy between the normal and tangential drag coefficients, with $C_N > C_T$, that lies at the origin of the drag-based thrust.

Using RFT, Gray and Hancock obtained (for the case of large-amplitude displacements) a closed-form solution for the swimming speed of an undulating filament given by the expression $U = \pi c(a/\lambda)^2(C_N/C_T - 1)/\ (1 + 2\pi^2(C_N/C_T)(a/\lambda)^2)$. Here, $C_N = 2C_N = 4\pi\mu/\ln(L/a)$ for a straight rod of length L such that the ratio of normal to tangential drag coefficients yields $C_N/C_T = 2$ for a sine wave of wavelength λ where a is the amplitude. For example, more recent experiments using the nematode *C. elegans* estimated this ratio at $C_N/C_T = 1.4$ [7]; such value lies closely with earlier estimates reported by Gray and Lissmann [6] who dropped thin wires into viscous fluids ($C_N/C_T = 1.4$–1.6).

Lighthill [9] later recognized the importance of long-range hydrodynamic interactions and improved RFT by incorporating slender-body approximations. Such improvements led to $C_N/C_T = 1.5$ for the case of an undulating filament swimming in an infinite fluid medium. When incorporating wall effects into the analysis, a significantly larger value of the drag coefficient ratio ($C_N/C_T = 4.1$) was subsequently obtained using the corrections of Katz et al. [32].

But what can we say about the flow fields generated by swimming microorganisms in fluids? A common way to determine flow fields at low Reynolds numbers is to solve the Stokes equation [Eq. (7.1)] with a forcing term replaced by a point force or disturbance [33]. As noted earlier, the flow disturbances driven by the swimming motion of microorganism in a Newtonian fluid depend linearly upon the stresses exerted by the moving body on the fluid [see Eq. (7.1)]. These boundary-driven flows are known to decay very slowly with the distance r away from the body [2, 8, 34]. Most often, such flow disturbances are mathematically cast as linear superpositions of the fundamental solutions of the Stokes equation and decay with inverse powers of r. The first solution, referred to as a "Stokeslet," arises from the net force on the fluid, and has a velocity field that decays as $1/r$. The next solution, also known as a "stresslet" flow, is induced by the first force moment exerted by the body on the fluid and decays more rapidly ($1/r^2$); higher-order solutions decay even more rapidly ($1/r^3$). As a result, linear combinations of the basic solutions of the creeping equations of fluid motion can generate a multitude of complex flow fields, exhibiting contrasting near- and far-field behaviors [34].

3 Experiments in Newtonian Fluids

Experimental studies on low Reynolds number locomotion in Newtonian fluids have undoubtedly complemented early theories on the topic [3, 4, 6, 9, 27, 28]. Many of these works have aimed at addressing the validity of classical theoretical models. In the section below, we briefly review a number of relevant experimental efforts that have helped over time characterize low Reynolds propulsion in Newtonian media.

3.1 From Scale-Up Models to Live Microorganisms

Scale-Up Experiments Experiments with live microorganisms are generally challenging due to difficulties with imaging/optical setups and to the lack of control over the organisms themselves. One attractive experimental approach to circumvent some of these issues relies on leveraging scale-up systems, often designed to mimic the organism's main swimming kinematics (see Fig. 7.3). Scale-up models provide much valuable insight into the main physical mechanisms governing micro-swimming phenomena; they have brought valuable insight in understanding the net motion resulting from traveling waves along elastic tails [35–37], helical flagella [38, 39], and flagellar bundles [40, 41] as well as in uncovering the motility resulting from surface traveling waves along cylindrical shells [42].

Beyond fundamental research into microorganism locomotion, a broad range of scale-up designs has been employed in the context of artificial swimming strategies at low Reynolds numbers [8], including Purcell's seminal "three-link swimmer" which possesses two hinges actuated with both time and phase differences [43] and a flapping body performing reciprocal motions near a deformable free surface [44]. Smaller mechanical systems have also been investigated. For example, the shapes of an oscillating passive actin filament have been experimentally probed [45] and more recently, a three-sphere design has been implemented using colloidal beads and optical tweezers [46]. There is also much interest in artificial micro-swimmers from the robotics and engineering community [47].

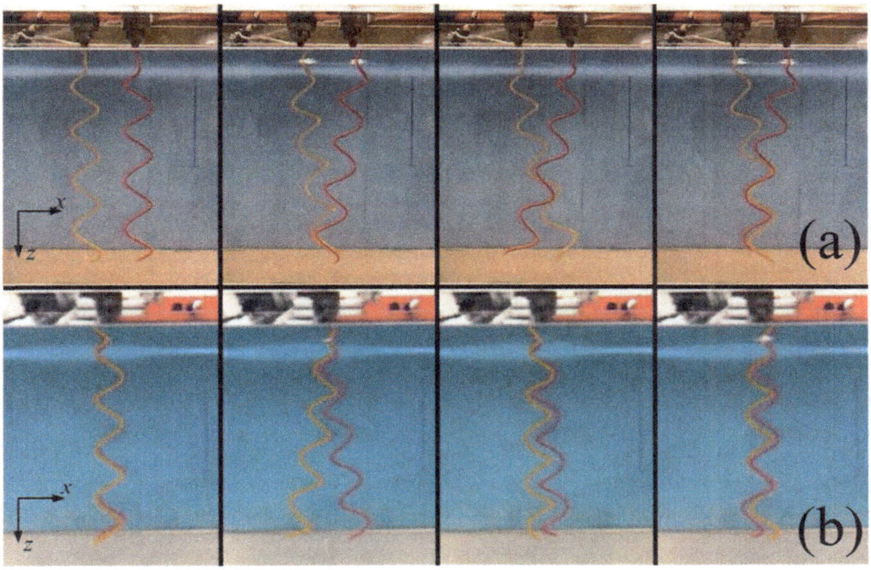

Fig. 7.3 Example of scale-up model of flagellar bundling dynamics [40]. (**a**) Image sequence of semi-coiled helices shown at various instances in time. (**b**) Same helices viewed from the side. The scale bars are 100 mm long; the helices are 310 mm long (from chuck to tip), 4.0 mm in diameter, and turning at 0.1 Hz ($Re_{freq} \approx 3 \times 10^{-5}$). Copyright (2003) National Academy of Sciences, USA

Experiments with Live Microorganisms Despite challenges in working with live microorganisms, microscopy imaging of bacterial flagella has gained tremendous traction following the pioneering work of Berg [13,14,48–50]. Of utmost relevance, flagellar kinematics of individual bacteria have been visualized in real time using fluorescent staining of both cells and flagellar filaments [51]. In more recent years, these initial microscopy techniques have been further developed to obtain time-resolved imaging of flagellar motility using setups with high-speed cameras [52] as well as to track swimming microorganisms three-dimensionally (3D), for instance, in a fluid far from surfaces [53]. These exquisite measurements provide much valuable data for our understanding of swimming and for the development of more realistic models.

Since, however, microorganisms evolve constantly near solid boundaries (e.g., migration of infectious bacteria through tissues), a growing number of experiments have shown that it is important to consider the presence of surfaces; namely, surfaces and wall effects drastically alter the kinematics of swimming microorganisms relative to ideal unbounded swimming conditions. A best-known example is perhaps illustrated for helical flagella (e.g., *E. coli*): swimming trajectories are modified from straight to circular in the vicinity of boundaries, clockwise when the wall is rigid [54] and anticlockwise near a free surface [55]. In particular, solid surfaces not only lead to the reorientation of microorganisms in the direction parallel to the surfaces, they also attract the organism to the closest wall [56].

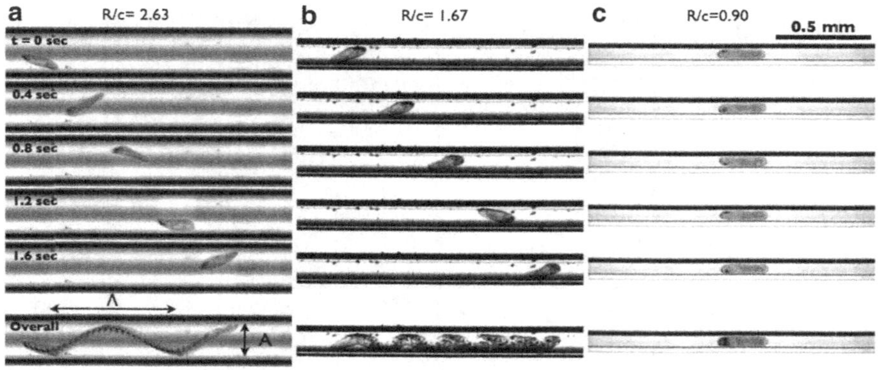

Fig. 7.4 Swimming of *Paramecium* in tubes of different diameters [60]. Here, Λ denotes the wavelength, c denotes the radius of the organism, and A the amplitude of the helical trajectory traced by the organism in tubes of different diameter. (**a**) *Paramecium* swimming in large tube ($R/c = 2.63$) where the trajectory of the motion is helical. (**b**) Small wavelength helices are seen inside tubes of intermediate diameters ($R/c = 1.67$). (**c**) In very small tubes ($R/c = 0.9$), *Paramecium* swims in a *straight line*. Figure reproduced with permission from the American Institute of Physics

Uncovering the fundamental hydrodynamic interactions between surfaces and swimming microorganisms has helped shed light on experimental observations of the accumulation of confined spermatozoa on boundaries [57–59]. It was recently observed [60] that ciliated *Paramecium* swimming in capillary tubes executes helical trajectories that slowly transition to straight lines as the tube diameter decreases (Fig. 7.4). Further experimental studies mimicking bio-locomotion in confined environments (e.g., female reproductive tract) have also revealed that the migration of motile spermatozoa in 3D microchannels is strongly influenced by specific wall shapes, including the turning angles at corners [61]. Beyond single-cell eukaryotic and prokaryotic microorganisms, it has also been noted that multicellular organisms, such as *C. elegans*, may display some finite attraction to the presence of boundaries [62] and certainly exhibit changes in their swimming kinematics under confinement conditions (e.g., setups with parallel-wall cells) [63]. As we can see, the presence of surfaces, solid or fluid-like, can significantly affect the dynamics of swimming microorganisms and cannot be ignored.

3.2 Propulsive Force and Flow Measurements

There has also been much effort in measuring the propulsive forces (i.e., thrust) and energy generated by swimming microorganisms. For instance, optical traps have been used to measure the forces required to tether sperm cells and bacteria [31, 64, 65]. In parallel, atomic force microscopy (AFM) has enabled the measurements of forces exerted by mucus-propelling cilia that lie on the order of

< 1 nN per cilium during the effective stroke [66]. Recently, force measurements using optical tweezers have been obtained on individual bacteria to test the validity of RFT and determine the swimming efficiency of *E. coli* [67]. These latter measurements have revealed for the first time that long-range hydrodynamic interactions are indeed critical in capturing accurately single-cell propulsion; in contrast, relying on RFT assumes a stationary background fluid while ignoring local flows induced from the other moving parts of the cell. Such observations have been most recently corroborated in scale-up models of helical flagella, where the validity of RFT breaks down for increasing pitch angles of the helix [39].

Concurrently, macroscopic (scale-up) experiments with helical bodies of different wavelengths λ, radii R, and lengths L (relevant to bacterial flagella) have highlighted the qualitative and quantitative discrepancies with RFT predictions [68]. This may be ascribed to the fact that RFT only takes into account local effects and neglects any hydrodynamic interactions between different parts of the swimmer. More accurate results can be obtained by using slender-body theories (SBT) that exploit the high aspect ratios seen in swimmer geometries and are based on the use of singularity solutions to the Stokes equations [69–72]. Since nonlocal hydrodynamic interactions between different parts of the swimmer may be incorporated in these theories, they provide much better results. While analytical expressions for the drag forces are generally difficult to obtain (here RFT can be of much value [9]), numerical solutions have been used to choose between different forms of resistance coefficients [73, 74] as well as study swimming in eukaryotic swimmers [75].

Despite the tremendous experimental progress, it is still difficult to measure the flow fields generated by swimming microorganisms due to limited spatial resolution of common velocimetry methods. One of the first attempts of visualizing the flow of swimming organism dates back to the1960s when Gray and Lissmann [6] presented qualitative path lines of freely swimming nematodes (worms) in water seeded with starch grains. With the advent of modern micro-PIV (μPIV) techniques [76] and fast cameras, flow field measurements of swimming microorganisms are within reach. For example, path lines generated by an individual ciliated *Paramecium* have been recently imaged [60], and velocity fields generated by individual unicellular microorganisms [52] and multicellular nematodes [7] have been resolved. In the case of the swimming nematode [7], the authors demonstrated that velocity magnitudes of fluid motion follow closely an exponential decay of the form $\exp(-2\pi r/\lambda)$ as a function of the distance r away from the nematode body (see Fig. 7.5); this analytical solution was originally derived by Lighthill [9] for an undulating sheet of wavelength λ in Stokes flow. Further extension of high-speed imaging techniques using for example tomographic PIV, where multiple cameras image simultaneously the interrogation volume from different angles [77], has enabled measurements of 3D time-resolved flow fields surrounding millimeter-sized copepods (*Calanus finmarchicus*).

There are two recent experiments [78, 79] that deserve much consideration and attention. The experiments by Drescher et al. [78] and Guasto et al. [79] were able

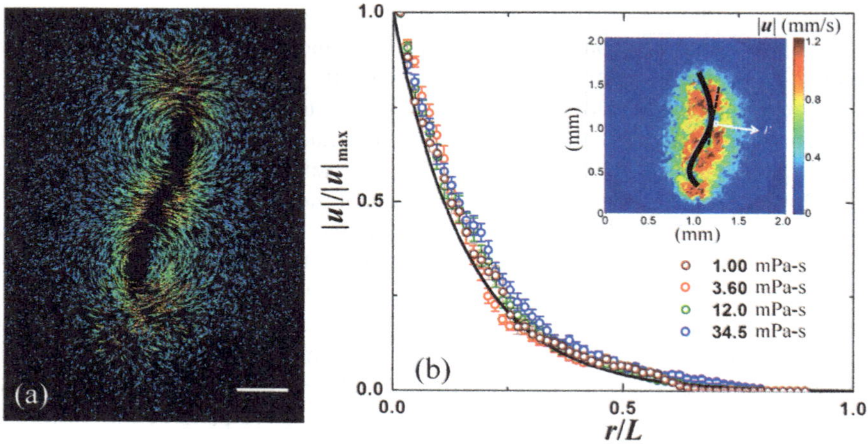

Fig. 7.5 Flow behavior surrounding a swimming *C. elegans*. Adapted from Sznitman et al. [7]. **(a)** Color-coded particle path lines are shown via the trace of dispersed fluorescent particles. Particle trajectories are tracked over ten consecutive frames (~ 0.06 s). Colors are associated with the lengths of trajectories. Two dominant recirculation regions are resolved along the nematode body. Scale bar represents 200 μm. **(b)** Normalized fluid velocity magnitude ($|\mathbf{u}|/|\mathbf{u}|_{max}$) as a function of the dimensionless distance (r/L) away from the nematode body, where L is the nematode body length. Data points of different colors correspond to fluids of different viscosities. The *solid line* corresponds to $\exp(-2\pi r/\lambda)$ [9]. *Inset*: representative velocity magnitude field at a given instant in time surrounding a nematode (marked with a *black line*). Figure reproduced with permission from the American Institute of Physics

to beautifully resolve the flow fields surrounding freely swimming micro-algae and demonstrated that local fluid motions were much more complex than analytical models initially suggested. This observation is particularly true in the near field, where the largest flow velocities occur. For example, Drescher et al. detailed quantitative measurements of time-averaged flows using μPIV for two different types of microalgae: *Volvox carteri*, a ciliated multicellular spherical alga, and *Chlamydomonas reinhardtii*, a unicellular alga featuring two flagella that beat in a breaststroke-like fashion. Guasto et al. resolved rather the oscillatory nature of the flow field driven by *C. reinhardtii* over one period of motion using high-speed imaging, where the swimming microorganisms were confined in thin liquid films. Both studies have emphasized how distinct species are likely to drive qualitatively different disturbance flows, as recently highlighted by Saintillan [34]. This flow feature remains true in the far field as well, where it is commonly assumed that there, flow fields can be described in terms of a stresslet. As a final word, we note that the representation of these driven flows using time-averaged velocity fields falls short of capturing the true nature of the flow, since time fluctuations can be of the same order as the mean.

4 From Newtonian to Complex Fluids

As mentioned before, much of our understanding of low Reynolds number locomotion arises from considerations in simple, Newtonian fluids [2–4, 9]. Many microorganisms, however, evolve and live in complex fluids such as mucus, gels, and wet soil [80–83]. But, what are complex fluids? Here, we define complex fluids as a broad class of materials that are usually homogeneous at the macroscopic scale and disordered at the microscopic scale, but possess structure at an intermediate scale (typically, a few sizes of its particles). Examples include colloidal suspensions, foams and emulsions, polymeric fluids, gels, human mucus, and blood. In colloidal crystals, for example, the intermediate scale is set by the size of the organized crystalline structure; that is, if one considers a cup of a cornstarch suspension, then the microscopic scale is the matrix that includes both the water molecules and the cornstarch grains ($\sim 1\,\mu\text{m}$), and the macroscopic scale is the size of the cup ($\sim 10\,\text{cm}$), while the intermediate scale is set by the structural length scale (if any) of the cornstarch grains. The cornstarch suspension will respond quite differently to an applied stress, depending on the grain size, concentration, and the grain arrangement in the suspending liquid. Thus, the macroscopic flow behavior (or rheology) of complex fluids is a strong function of the fluid microstructure.

Complex fluids are usually not Newtonian, and they often exhibit viscoelasticity and shear-thinning viscosity. Much recent effort has been devoted to the understanding of the effects of non-Newtonian fluid behavior (shear-thinning, viscoelasticity) on the swimming of microorganisms [8, 80, 81, 83, 84]. For example, the nonlinear relationship between shear stress and strain rate that characterizes shear-thinning fluids can have significant consequences to locomotion at low Reynolds numbers including (i) a breakdown of the "scallop" theorem, (ii) kinematic changes in the organism's swimming motion, and (iii) changes in the drag forces experienced by the organism. In fact, it was recently shown by Vélez-Cordero and Lauga [85] for an infinite waving sheet (similar to Taylor's original work) immersed in a shear-thinning fluid that while the sheet's propulsion speed remained the same as in the Newtonian case, the cost of transport was reduced. Simulation studies by Montenegro-Johnson et al. [84] showed that undulatory swimmers with a head or "payload" (similar to a sperm cell) are assisted by shear-thinning viscosity, resulting in increased speed. These recent studies illustrate that even relatively simple non-Newtonian fluid behavior such as shear-thinning may have a significant impact on the swimming behavior of microorganisms.

The effects of fluid elasticity on swimming at low Reynolds numbers have received considerable more attention [83, 86–93] than shear-thinning effects. Viscoelastic fluids possess shear stress that is time dependent and that depends on the history of deformation. Such features give rise to flow behavior in viscoelastic fluids that is markedly different from that of Newtonian fluids even at low Reynolds numbers [94–96], and they can even lead to the breakdown of the scallop theorem [92]. Most of the nonlinear flow behavior observed in the flow of viscoelastic fluids results from the extra elastic stresses due to the presence, of

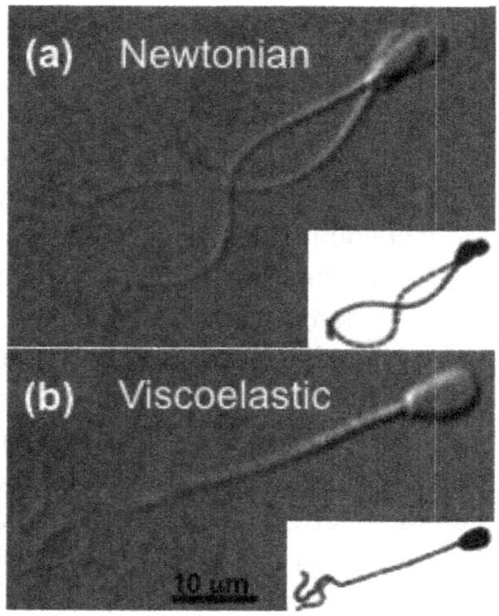

Fig. 7.6 Snapshot of sperm cells moving in (**a**) Newtonian fluid (semen) and viscoelastic fluid (mucus); adapted from Ho and Suarez [97]

usually, polymer molecules in the fluid. Mechanical stresses in viscoelastic fluids are history-dependent and depend namely on a characteristic time λ that in dilute solutions is proportional to the relaxation time of a single polymer molecule. In semi-dilute solutions, λ depends also on molecular interactions. For more details on fluid viscoelasticity and shear-thinning viscosity, please see Chaps. 1, 8, and 10.

The nonlinear response of viscoelastic fluids is expected and has been seen to play a significant role on the swimming behavior of microorganisms. Consider for example the swimming behavior of motile sperm cells [86, 97, 98] that usually swim as a result of (single) flagellar beating. For freely swimming spermatozoa in Newtonian semen (Fig. 7.6a), the flagellum exhibits a regular sinusoidal beating pattern [81]. But once the organism encounters a viscoelastic medium (i.e., cervical mucus), the regular beating pattern is transformed into high-amplitude, asymmetric bending of the flagellum (Fig. 7.6b). This "hyper-activated" sperm is believed to be dramatically influenced by its fluidic environment [81, 97], which in turn can affect human fertility [80, 99]. Other examples of motility in viscoelastic media include the removal of mucus in the human respiratory track by beating cilia [24, 26], the locomotion of bacteria in biofilms [100, 101], and the burrowing of organisms in wet soil [102, 103]. Understanding how microorganisms move in viscoelastic fluids is, therefore, of both scientific and practical importance.

Despite many recent efforts to be discussed below, the effects of bulk fluid elasticity on the motility behavior of live organisms at low Reynolds numbers are still not clear and well understood. In order to provide the reader with some basic insight into this issue, we turn to our favorite dimensionless parameters. The first, of course, is the Reynolds number, which is a measure of the relative importance of the fluid inertia to viscous forces. We already showed that the Reynolds number is approximately 10^{-4} for *E. coli* swimming in water. The effects of fluid elasticity are often estimated using the Deborah number, defined as $\mathrm{De} = \lambda f$ where λ is the fluid relaxation time and f is the organism's beating frequency. Note that $\mathrm{De} = 0$ for

Newtonian fluids and De $\to \infty$ for purely elastic solids. One could imagine that fluid elasticity may begin to play a dominant role for De ≥ 1. If one considers the beating frequency f of sperm cells to range from 20 to 50 Hz and the relaxation time λ of cervical mucus to range from 1 to 10 s (depending on factors like hydration, among others), one can expect fluid elasticity to play a significant role on the motility of spermatozoa since De $\gg 1$. One can also compare the ratio of the (fluid) elastic time scale λ to the (fluid) viscous time scale $\rho L^2/\mu$. This is the so-called elasticity number, defined as El $= \lambda\mu/\rho L^2$. Elastic effects are expected to dominate for El > 1. Because of the nonlinear (squared) dependence of El on the (swimmer) length scale L, one anticipates the effects of fluid elasticity to become increasingly important for swimming microorganisms.

4.1 *Swimming in Viscoelastic Fluids: Expectations*

In 1979, Chaudhury [87] attempted to incorporate the effects of fluid elasticity on swimming using a second-order fluid and a series of expansions similar to Taylor's analysis. It was then predicted that fluid elasticity could either increase or decrease the propulsion speed of the waving sheet (Fig. 7.2), depending on the value of Re. Later, inspired by experimental observations of spermatozoa swimming in mucus [81, 97], the effects of elasticity on beating flagellar structures were considered in Stokes flow using the Maxwell model [104]. It was shown that self-propulsion was not affected by viscoelasticity even at large Deborah numbers (De), where De $= \lambda f$ and λ is the fluid relaxation time and f is the beating frequency. However, the total work decreased with increasing De. It was then suggested that a microorganism could swim faster in a viscoelastic fluid with the same expenditure of energy compared with a Newtonian fluid.

More recently, Lauga [83] showed that, for a 2D waving sheet (Fig. 7.2), elastic stresses could significantly alter the organism speed and the work required to achieve net motion. Using nonlinear viscoelastic fluid models such as the Oldroyd-B and the FENE-P models (see Chap. 1), Lauga [83] showed that the sheet's forward speed U in a purely elastic fluids is given by

$$\frac{U}{U_N} = \frac{1 + \mathrm{De}^2(\eta_s/\eta)}{1 + \mathrm{De}^2}, \tag{7.2}$$

where U_N is the swimming speed of the sheet in a viscous Newtonian fluid (i.e., Taylor's original result) and η_s is the solvent viscosity; note that the solution viscosity η is assumed to be the sum of the solvent viscosity and the polymer viscosity such that $\eta = \eta_s + \eta_p$. Hence, for a given (i.e., prescribed) swimming gait $U_N \geq U$, that is, elastic stresses reduce the overall speed of the waving sheet. Equation (7.2) is an elegant and interesting result and has spurred much of the recent interest in swimming in viscoelastic fluids. See Chap. 8 for a more detailed discussion on the derivation of Eq. (7.2).

A similar result to Eq. (7.2) was also obtained for a waving cylinder by Fu et al. [88]. These important results imply that fluid elasticity can reduce the swimming speed of microorganisms when compared to simple, Newtonian fluids. An important caveat, of course, is that organisms may compensate the reduction in velocity by increasing their beating frequency and/or concurrently decreasing their body wavelength. In other words, microorganisms can alter their swimming kinematics to adjust or adapt to varying fluidic environments. It is worth noting that the analysis by Lauga [83] and Fu et al. [88] reveals the net magnitude of locomotion scales quadratically with the amplitude of the local oscillatory motion, and therefore nonlinear terms in viscoelastic constitutive relationships cannot be neglected (as in the case of [104]).

Numerical simulations have also been used to address the role of fluid elasticity on the swimming behavior of microorganisms. In particular, Teran and co-workers [89] considered two-dimensional swimming "free" sheets (i.e., with free head and tail) of *finite* length L in viscoelastic fluids. The simulations were performed by solving Stokes equation using the Oldroyd-B model as the constitutive equation using an immersed boundary method. The simulations show that, for accentuated tail motions, the sheet swims faster at De ≈ 1 than in a Newtonian fluid. This regime corresponds to where "swimmer" stroke frequency matches the fluid relaxation time. This is a fascinating result and is unlike Eq. (7.2) which predicts that the swimmer speed in viscoelastic fluids is always slower than in Newtonian fluids. The simulations do show that for De > 1, the swimming speed decreases as De increases (Fig. 7.7).

As briefly discussed above, fluid elasticity can strongly affect both the swimming dynamics and kinematics of microorganisms even at low Reynolds numbers. Recent analytical works predict that fluid elasticity hinders swimming speed while numerical simulations show that it is possible to obtain an enhancement in self-propulsion

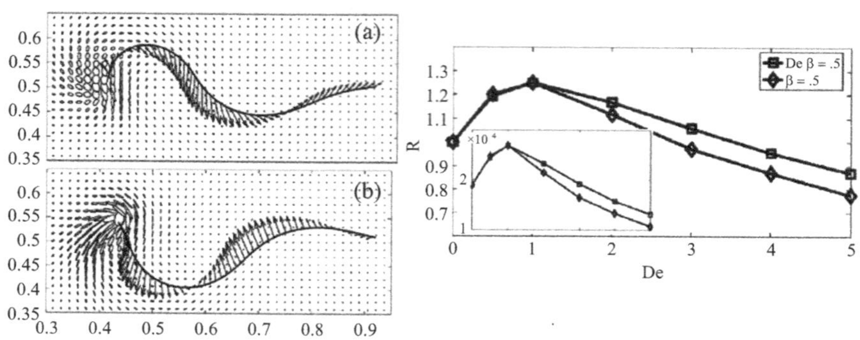

Fig. 7.7 (**a,b**) Polymer stress tensor fields for a finite length, two-dimensional undulating sheet. The *ellipses* in the figures represent the directions and degree of distension of the polymer field. The *arrows* represent the fluid velocity on the immersed filament. (**c**) Normalized average swimming speed of the sheet as a function of the Deborah number. Note the enhancement in propulsion at De ≈ 1. Figures adapted from [89]

in a regime where the fluid relaxation time matches the swimmer stroke frequency, that is De $\approx$ 1. Despite such important advances, it is still not clear whether elastic stresses enhance or hamper self-propulsion since theoretical and numerical results are model dependent. So the question still stands: does fluid elasticity enhance or hinder self-propulsion at low Re? Perhaps experiments will shed more light into this important question.

5 Experiments in Viscoelastic Fluids

5.1 Scale-Up Experiments

Swimming experiments in complex fluids are hard to come by and *systematic* investigations are scarce in the literature. Part of the problem is undoubtedly the difficulty in identifying a model organism or swimmer that is both able to move in different types of media and relatively easy to image and track. To circumvent some of these difficulties, many investigators choose to build instead macroscopic-scaled versions of the microorganisms' propulsion mechanism [35, 41, 90, 105]. In this section, we discuss an interesting experimental setup proposed by Liu and coworkers [90] in which a scale-up model of bacterial filaments is investigated.

The experimental setup is shown in Fig. 7.8 and consists of a large cylindrical tank filled with either a viscous Newtonian fluid or a viscoelastic fluid. The Newtonian fluid is silicon oil and the polymeric solution is a mixture of polyisobutylene

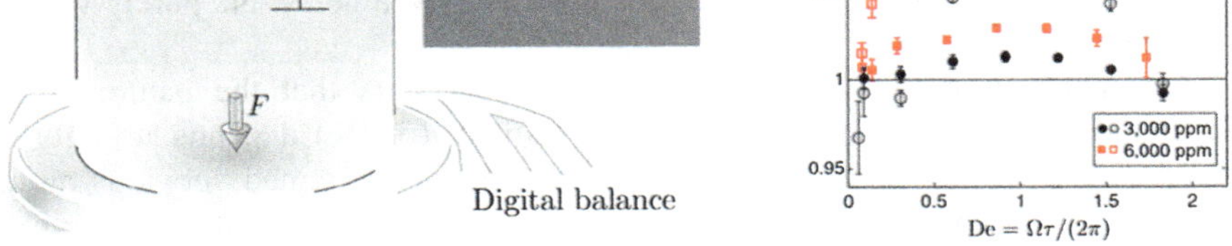

Fig. 7.8 (**a**) Scale-up mechanical apparatus used for measuring the motility of a rotating helix [90]. The helical structure, shown in (**b**), is slowly immersed into a Newtonian or viscoelastic fluid and rotates about the vertical direction. The net hydrodynamic force on the helix is determined by a laboratory balance beneath the tank as shown in (**a**). (**c**) Normalized propulsion speed as a function of the Deborah number for two polymeric solutions and helices with different pitch angles. Copyright (2011) National Academy of Sciences, USA

(PIB) in polybutene solvent. More details on the polymeric solutions are given in the next paragraph and in [90]. A rigid helix rotating at speed ω is slowly immersed or plunged at a constant speed into fluid-filled tank. Helices of varying pitch angles are used to mimic the geometry of bacterial flagellar filaments (e.g., *E. coli*). The hydrodynamic force exerted on the helices by the fluid is measured by placing the tank on top of a sensitive digital balance. In these experiments, zero-force swimming is achieved by adjusting the translation speed until the measured axial force is zero. Because the helix is inserted from above, a positive vertical force on the helix represents a drag, and a negative vertical force on the helix is a thrust. The force-free swimming speed is measured as a function of helix rotation rate, helix geometry (i.e., pitch angle), and fluid properties (i.e., Newtonian vs. viscoelastic).

Because the fluids are very viscous, the Reynolds number is well below 0.01 and inertial effects are negligible. In order to decouple the effects of fluid elasticity from those of rate-dependent viscosity (e.g., shear-thinning) common in polymeric solutions, a nearly constant-viscosity, elastic fluid was prepared. Such fluids are often called "Boger fluids" in reference to David Boger who first proposed the use of such model fluids [106]. Boger fluids are constructed by adding a small amount (usually in part per millions) of high molecular weight (MW) flexible polymer to a very viscous solvent. Because the polymer contribution to the overall solution viscosity is small, the solution viscosity is overwhelmed by the viscosity of the Newtonian solvent. But the addition of high MW flexible polymers is able to add elasticity to the fluids. As a result, Boger fluids have nearly constant viscosity while still possessing elasticity. In the work of Liu et al. [90], Boger fluids are prepared by dissolving either 3,000 ppm or 6,000 ppm of PIB in polybutene (solvent). The average relaxation time λ for both polymeric solutions is approximately 0.6 s.

The main result of this very clever experiment is shown in Fig. 7.8 (rightmost panel). The investigators find an enhancement of the measured swimming speed of a rotating helix in a viscoelastic fluid near De $= 1$, where De $= \omega\lambda/2\pi$. This result is similar to the enhancement observed in numerical simulations of 2D flexible filaments in Oldroyd-B fluids by Teran and coworkers [89] and of helical filaments [93], but is in contrast with the decrease observed in analytical calculations [83, 107] and experiments with live organisms [91]. As the rotating speed (and De) increases, the helix propulsion speed decreases even below the purely viscous Newtonian speed.

An important take-away message is that it appears that the nature of the dependence of propulsion speed on fluid elasticity (or De) depends strongly on the geometry of the waveform used for swimming. This is made obvious by the sensitivity of the peak enhancement of swimming speed on the pitch angle of the helix, as shown in Fig. 7.8 (rightmost panel). This is an important point which we will further discuss later in this chapter.

While scale-up experiments can provide much useful information, they cannot fully capture the complexity of live organisms. Therefore, it is important to perform systematic studies using live organisms. In the next section, we will discuss experiments using a well-known biological model system, namely, the nematode *C. elegans*.

5.2 *Experiments with Live Organisms*

Despite recent advances, the effects of fluid elasticity on the swimming behavior of microorganisms are still not clear [108], in part due to the lack of systematic experiments with live organisms. In this section, we will discuss swimming experiments in viscoelastic fluids with a biological model system, namely the nematode *C. elegans*. Model organisms are nonhuman species which are extensively studied to understand particular biological phenomena. Examples include the zebra fish, *E. coli*, fruit fly (*Drosophila melanogaster*), and mice, among many others. The idea is that discoveries made in model organisms will provide insight into the workings of other non-model organisms.

5.2.1 *C. elegans*: An Attractive Model Organism for Swimming Studies

An interesting model system that has received much attention in the biological community is nematode *C. elegans*, which is a small, multicellular, free-living roundworm found in soil environments. Much is known about the nematode's genetics and physiology; its genome has been completely sequenced [109] and a complete cell lineage has been established [110]. These nematodes are equipped with 95 muscle cells that are highly similar in both anatomy and molecular makeup to vertebrate skeletal muscle [11]. Their neuromuscular system controls their body undulations which allows *C. elegans* to swim, dig, and crawl through diverse environments. The wealth of biological knowledge accumulated to date makes *C. elegans* ideal candidates for investigations that combine aspects of biology, biomechanics, and the fluid mechanics of propulsion.

Figure 7.9 shows an image of an adult, wild-type *C. elegans* swimming in a water-like buffer (M9) solution. The nematode is characterized by a relatively long and quasi-cylindrical body shape (Fig. 7.9). Its length can vary from 50 μm (embryonic stage) to 1 mm (adult stage) while its radius is approximately 80 μm. The nematode length scale is an important feature they are large enough that flow fields can be accurately obtained.

5.2.2 Swimming Experiments with *C. elegans*: Dilute Polymeric Solutions

We now discuss swimming experiments using the nematode *C. elegans* in some detail. Experiments with the two main types of fluids, namely Newtonian and viscoelastic, are discussed. Experiments are performed in small channels that are made of acrylic and are 1.5 mm wide and 500 μm deep; they are sealed with a thin (0.13 mm) cover glass. In order to minimize three-dimensional motion, the channels are relatively shallow, yet the nematode is able to freely move in all directions. The swimming motion of *C. elegans* is imaged using standard bright-field microscopy and a fast CMOS camera. The image acquisition rate is kept constant at 125

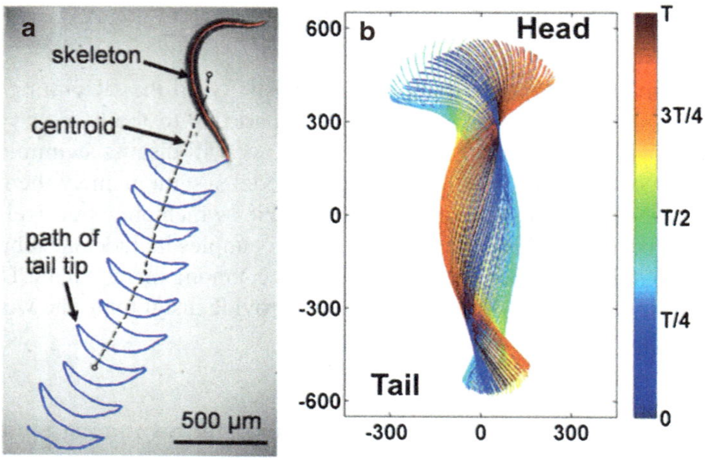

Fig. 7.9 (**a**) Snapshot of the nematode *C. elegans* moving in a water-like buffer solution. The *red line* along the nematode corresponds to the body centerline or "skeleton," obtained using image analysis. The centroid (*dashed*) and the trajectory of the tail (*blue*) are tracked at a sampling rate of 125 frames per second. (b) *Body-shape lines* as a function of time color-coded by time over one beating period

frames per second to guarantee small linear displacements along the nematode's body between consecutive frames. All data presented here pertain to nematodes swimming at the center plane of the fluidic channel. Out-of-plane recordings are discarded.

An important consideration in swimming experiments with live organisms is the fluid medium. Fluids must be developed such that they possess the desirable rheological property (elasticity, shear-thinning, etc.) but without being toxic to the organism. Here, Newtonian fluids of different shear viscosities are prepared by mixing two low molecular weight oils (halocarbon oil, Sigma-Aldrich). Viscoelastic fluids are prepared by adding small amounts of carboxymethyl cellulose (CMC, 7×10^5 MW) into deionized water. CMC is a long, flexible polymer with an overlap concentration (c^*) of approximately 10^4 ppm. In order to rule out the effects of shear-rate-dependent viscosity, an aqueous solution of the stiff polymer xanthan gum (XG) that is shear-thinning but possesses negligible elasticity is also used in experiments.

Fluid Rheology: Viscosity Data An important step in these experiments is fluid rheological characterization. How viscous or elastic are the fluids? (A very useful discussion on fluid characterization and rheology is given in Chap. 6.) A strain-controlled rheometer RSF III (TA Instruments) with a cone-and-plate geometry is used to characterize the rheological properties of the CMC and XG solutions. Figure 7.10 shows the viscosity curves of both CMC and XG solutions. The viscosity data are fitted with the power-law fluid model of the type $\eta = m(\dot{\gamma})^{n-1}$,

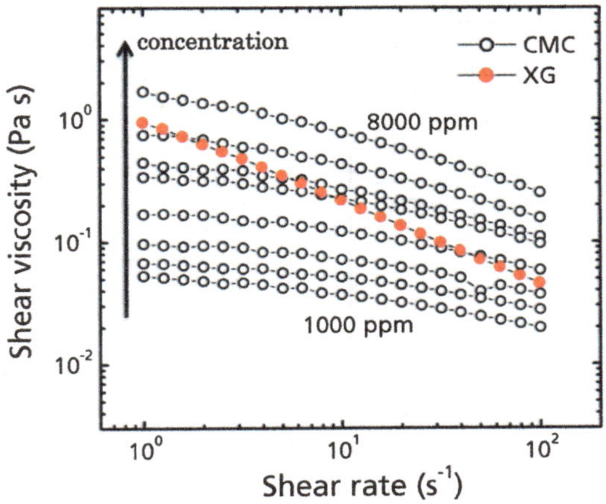

c (ppm)	n
1500	0.83
2000	0.81
3000	0.80
4500	0.76
5000	0.74
6000	0.70
8000	0.65

Fig. 7.10 (*Left*) Fluid shear viscosity curves for the flexible carboxymethyl cellulose (CMC) and semi-rigid xanthan gum (XG) solutions. The concentrations of CMC solution ranges from 1,000 ppm to 8,000 ppm by weight, from *bottom* to *top* in the plot. *Solid circles* represent the 3,000 ppm XG aqueous solution. The values of the power law index n are 0.65 and 0.35 for the 8,000 ppm CMC and the 3,000 ppm XG solutions, respectively. Table: the power law indexes of the CMC aqueous solutions

where m is a flow consistency factor and n is the power law index. Results from the fits are shown in the table in Fig. 7.10. The CMC solutions show a relatively weak shear-thinning behavior, particularly in the shear rate range of 1–20 s^{-1}. This is the range of shear rates produced by the swimming *C. elegans* in fluids. As the CMC concentration in solution increases, shear-thinning effects also increases. In the most concentrated CMC solution, i.e., at 8,000 ppm, the power law index n is approximately 0.65. As a comparison, the xanthan gum solution at 3,000 ppm shows much stronger shear-thinning behavior ($n = 0.35$). Note that the mixture of low molecular weight halocarbon oils shows constant shear viscosity and is not shown.

Fluid Rheology: Relaxation Times Shear viscosity or flow curves are not sufficient to describe the material properties of viscoelastic fluids. An important quantity used to characterize viscoelastic fluids is the fluid relaxation time λ. Measuring λ is not a trivial task for several reasons including the fact that most real viscoelastic fluids have not one but a spectrum of relaxation times; λ can also be shear rate dependent. What is usually reported in the literature (and used in many models) is the longest, most dominant value of λ. There are several ways to obtain λ including (i) measurements of the first normal stress difference N_1 from steady rheology combined with an appropriate constitutive model, (ii) oscillatory or frequency-dependent measurements in which both the viscous G'' and elastic G' moduli are measured for small strains, and (iii) stress relaxation experiments. For more detailed information on rheological measurements and applications, please see Chap. 6.

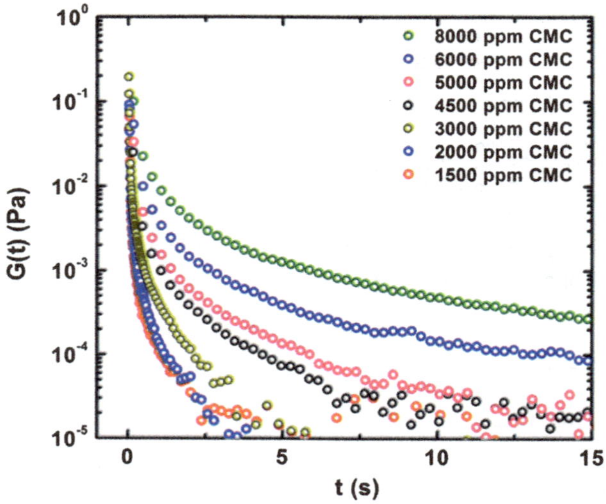

c (ppm)	λ (s)
1500	0.4
2000	0.6
3000	0.8
4500	2.1
5000	2.3
6000	3.4
8000	5.6

Fig. 7.11 (*Left*) Stress relaxation data and (table) fluid relaxation time λ for all polymeric (CMC) solutions

The values of λ for all the viscoelastic CMC solutions are obtained using a stress relaxation technique after a sudden applied strain or shearing displacement. This technique is sometimes referred to as step-strain stress relaxation. In the experiment, the time decay of the shear stress is described by a relaxation modulus $G(t)$ (Fig. 7.11), which is fitted with the generalized linear viscoelastic model of a single relaxation time of the type $G(t) = G_o e^{-t/\lambda}$. By varying the polymer concentration in solution, the values of λ can range by as much as an order of magnitude from 0.4 s for the most dilute concentration (1,500 ppm) to about 5.6 s for the most concentrated solution (8,000 ppm). The values of λ for all CMC solutions are shown in the table in Fig. 7.11.

Swimming Kinematics Now that the fluids have been characterized, we can begin to discuss the swimming experiments using *C. elegans*. Because it is important to establish a baseline, results obtained with the viscoelastic fluids (CMC solutions) are compared to swimming in Newtonian fluids (halocarbon oils). An important quantity that is used to characterize the swimming behavior of undulatory swimmers such as the nematode *C. elegans* is the bending curvature, defined as $\kappa(s,t) = d\phi/ds$=. Here, ϕ is the angle made by the tangent to the x-axis in the laboratory frame at each point along the body centerline, and s is the arc length coordinate spanning the head of the nematode ($s = 0$) to its tail ($s = L$). Note that the y-axis corresponds to the dimensionless position s/L along the nematode's body where $s = 0$ is the head and $s = L$ the tail and the x-axis cuts across the nematode as shown in Fig. 7.12a. The spatiotemporal evolution of the nematode's body curvature $\kappa(s,t)$ for 3T, or 3 swimming cycles is shown in Fig. 7.12a. The contour plots show the existence of periodic, well-defined diagonally oriented lines characteristic of

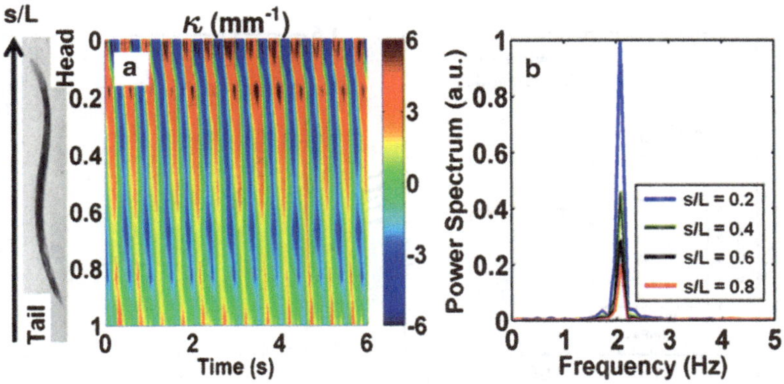

Fig. 7.12 The kinematics of swimming *C. elegans* at low Re number in viscous Newtonian fluids (Re ≈ 0.1). (**a**) Contour plot of the measured curvature (κ) along the nematode's "skeleton" or body centerlines as a function of time. The y-axis corresponds to the dimensionless position s/L along the nematode's body where $s = 0$ is the head and $s = L$ the tail; the x-axis cuts across the nematode. (**b**) Frequency spectra of κ at different selected positions s/L. The nematode's beating frequency peaks at a single value (2.0 Hz), irrespective of the location s/L

bending waves, which propagate in time along the nematode body length. By taking the Fourier transform of the contour plots (along the axis of time), a single peak at 2 Hz is found, indicating that the nematode beating is periodic in time (Fig. 7.12b). Other kinematic metrics such as wavelength (1 mm) and wave speed (5 mm/s) can also be extracted from the contour plots.

Propulsion Speed—Newtonian vs. Viscoelastic Now, it is possible to address the question of whether fluid elasticity hinders or enhances the propulsion speed of live organisms using *C. elegans*. The average nematode forward velocity U is calculated by differentiating the nematode's centroid position with respect to time (Fig. 7.9). For nematodes swimming in a Newtonian fluid of shear viscosity μ of 5 mPa · s (or 5× the viscosity of water), the value of U is approximately 0.4 mm/s and the Reynolds number (Re $= \rho UL/\mu$) is approximately 0.05. Hence, the model organism *C. elegans* can be considered a low Re swimmer.

The nematode's swimming speed as a function of fluid viscosity for both Newtonian and viscoelastic (CMC) solutions is shown in Fig. 7.13a. For relatively low viscosity values, the swimming speed is independent of fluid viscosity μ and the values of U are nearly identical for both cases. For $\mu > 30$ mPa · s, the swimming speed decreases with increasing μ even for Newtonian fluids. This decrease in U is most likely due to the nematode's finite power. Note that, for a nematode swimming with constant power at low Re, $P \sim \mu U^2$ where P is power. Results show that, over the admittedly limited range of μ, the nematode's propulsion speed shows a decay that is slower than $\mu^{-1/2}$, which strongly suggests that the nematode does not swim with constant power. The maximum power generated by the organism is approximately 200 pW, calculated for $\mu = 30$ mPa · s.

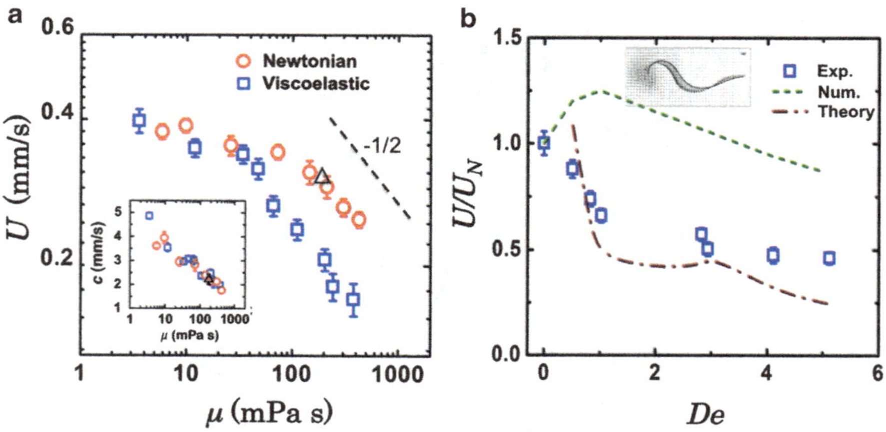

Fig. 7.13 (**a**) Nematode's swimming speed U as a function of shear viscosity μ for Newtonian (*red circle*) and viscoelastic (*blue square*) fluids. *Triangle symbol* represents the nonelastic xanthan gum solution. The data shows that fluid elasticity decreases the nematode's swimming speed when compared to a Newtonian fluid of same viscosity. For $\mu > 30\,\text{mPa}\cdot\text{s}$, the nematode's swimming speed decreases, indicating a limit in power for this type of organism. Inset in (**a**) shows the nematode's wave speed as a function of viscosity for all fluids. There is no apparent difference between the different fluids at a given viscosity value. (**b**) Normalized swimming speed as a function of the Deborah number. Experimental data is plot, together with numerical [89] and theoretical predictions [83]. Note that swimming speed decreases as De is increased

Importantly, the values of U for viscoelastic fluids are found to be 35 % lower than the Newtonian fluid of same shear viscosity (Fig. 7.13a). For example, the nematode's swimming speeds for the viscoelastic and Newtonian cases are 0.18 mm/s and 0.25 mm/s, respectively, even though the shear viscosity for both fluids is 300 mPa · s (Fig. 7.13a). The decrease in swimming speed in CMC (polymeric) solutions does not seem to be due to shear-thinning effects since nematode swimming in the non-viscoelastic, shear-thinning fluid (XG) showed no apparent decrease in propulsion speed (Fig. 7.13a, triangle symbol) compared to the Newtonian case. This result with XG is in agreement with the recent theoretical analysis of a waving sheet in a shear-thinning fluids [85].

So far, experiments using an undulatory, low Re swimmer (i.e., *C. elegans*) show that for similar viscosities fluid elastic stresses seem to hinder the organisms' propulsion speed. An important question is whether the organism is responding or adapting to the extra elastic stresses present in the fluid or are the polymer molecules toxic to the organism? In other words, is the observed decrease in swimming speed due to hydrodynamics or biology? This is a difficult question to answer with certainty, but one can address it at least in part by comparing the swimming phenotypic behavior (i.e., kinematics) between Newtonian and viscoelastic fluids. The wave speed c produced by the nematode is of particular interest since it has been shown that the values of c seem to change significantly once the nematode adopts a different swimming gait, i.e., swimming versus crawling [111]. The wave speed can be easily measured from the curvature contour plots. The inset in Fig. 7.13a shows the nematode's bending wave speed c as a function of fluid viscosity. Results indicate that viscoelasticity (and polymer molecules) has negligible effect on the nematode's swimming kinematics. That is, the changes in kinematics including the decrease in beating frequency and wave speed are due to viscous effects only. In addition, there is no evidence of change in motility gait (e.g., swimming to crawling) as μ increases since the beating amplitudes remain constant ($A = 0.26$ mm) even for the most viscous fluid ($\mu = 400$ mPa · s).

The effects of fluid elasticity on the nematode's swimming behavior are best illustrated by plotting the normalized swimming speed U/U_N as a function of the Deborah number ($\mathrm{De} = \lambda f$), where U_N is the Newtonian speed. Figure 7.13b shows that the normalized swimming speed decreases monotonically with De and reaches an asymptotic value of 0.4 as De is further increased. In other words, as the elastic stresses increase in magnitude in the fluid, it introduces a larger resistance to propulsion, therefore decreasing the nematode's swimming speed.

Comparing Experiments to Calculations We can now compare the experimental results to the numerical and theoretical predictions discussed earlier. Of course, such comparisons are not quite fair because there are significant differences between the experiments and the calculations. For example, most calculations are two-dimensional (2D) and for small-amplitude displacements while the nematode is allowed to swim in 3D and can bend quite a bit. Most importantly, while the calculations assume prescribed kinematics or waveform, the nematode is free to choose its own. Nevertheless, qualitative assessments can be made.

We begin by noting that for all the experiments presented in this section, the ratio of the solvent viscosity to the total solution viscosity is below 0.05, which is similar to the calculations [83, 107]. As shown in Eq. (7.2), the swimming speed U of an undulating sheet is predicted to decrease with increasing De. While the experimental data supports the predicted trend, at least qualitatively, there are quantitative discrepancies between the experimental and theoretical results as shown in Fig. 7.13b. Some of the possible reasons for the observed discrepancies may be the finite length of the swimmer and the assumption of small beating amplitude in the theoretical works. That is, only small deflections are considered for both the waving sheet and cylinder while the nematode shows significant bending. Nevertheless, the theoretical models are able capture the main trends in the experimental data and perform surprisingly well.

How do the experimental results compare to the numerical simulations of Teran et al. [89]? Remember that the simulations predict an interesting enhancement of the sheet swimming speed at $\mathrm{De} = 1$. The experimental results do not reveal such swimming speed enhancement, although a scale-up mechanical experiment did find such enhancement [90]. Nevertheless, for $\mathrm{De} > 1$, the simulation predicts a gradual decrease in U as elasticity is increased. The discrepancies between the experiment and the simulations are most likely due to the difference in the swimming beating patterns. While simulations used a left-moving traveling wave with an amplitude that increased from head to tail, the experiments with *C. elegans* reveal a traveling wave with an exponential decay from head to tail. Therefore, it is a bit unfair to directly compare the numerical simulations with the *C. elegans* work since the swimming kinematics are *very* different. As a side note, a recent three-dimensional numerical simulation of helical bodies in viscoelastic fluids by Spagnolie et al. [93] found that an enhancement in propulsion speed is indeed possible at $\mathrm{De} = 1$, in good agreement with scale-up experiments [90]. Clearly, swimming kinematics matter, and Chap. 10 provides an interesting discussion on the effects of swimming beating patterns on propulsion in viscoelastic fluids.

A Possible Mechanism: The Role of Extensional Viscosity So what could explain the decrease in swimming speed for nematodes moving in viscoelastic fluids? One possible explanation may be related to the *extensional viscosity* of polymeric fluids. The reader may be very familiar with the concept of shear viscosity μ, which is the fluid resistance to a shear deformation. The concept of extensional viscosity may be less familiar because it is not usually taught in standard fluid dynamics textbooks. Simply put, extensional viscosity η_e is the fluid resistance to an extensional deformation. Pure extensional flows are devoid of shear; such flows are often referred to as shear-free flows. Examples of shear-free flows also include biaxial stretching and elongation flows.

For Newtonian fluids, the extensional viscosity is equal to three times the shear viscosity such that $\eta_e = 3\mu$. This result was first reported by Trouton over a century ago in 1906 [112]. The quantity η_e/μ is often referred as the Trouton ratio; for Newtonian liquids, the Trouton ratio is constant ($\mathrm{Tr} = \eta_e/\mu = 3$). Viscoelastic fluids, however, can exhibit an enhancement in η_e compared to Newtonian fluids due to the extra elastic or polymeric stresses. Many experiments have shown that the extensional viscosity of liquids containing flexible polymers can be orders of magnitude larger than the extensional viscosity of Newtonian fluids [113, 114]. This is true even for viscoelastic and Newtonian fluids that have similar values of μ. In addition, while Newtonian fluids exhibit η_e values that are independent of strain, viscoelastic fluids show strain-hardening behavior. It should be evident that viscoelastic and Newtonian fluids will behave quite differently in flows with a strong extensional component [115].

But, how can extensional viscosity explain the reduced swimming speed of *C. elegans* in viscoelastic fluids? A clue may be in the velocity fields produced by the swimming nematodes. Figure 7.14 shows typical streamlines computed from experimentally measured velocity fields using particle-tracking methods for both the (a) Newtonian and (b) viscoelastic cases. Overall, the streamlines display large recirculation flow structures, or vortices, that are attached to the nematode's body. While the large-scale patterns are similar for both cases, detailed inspections shows the appearance of a distinct *hyperbolic point* near the nematode for the viscoelastic case. A recent computational effort by Guy and Thomases also found the existence of such hyperbolic points in the velocity fields of swimming nematodes (Fig. 10.9 in Chap. 10 of this book). It is important to note that a flow near such hyperbolic points is purely extensional. The hypothesis is that the decrease in U (in the nematode case, at least) is mostly likely due to the sudden increase of elastic stresses near the

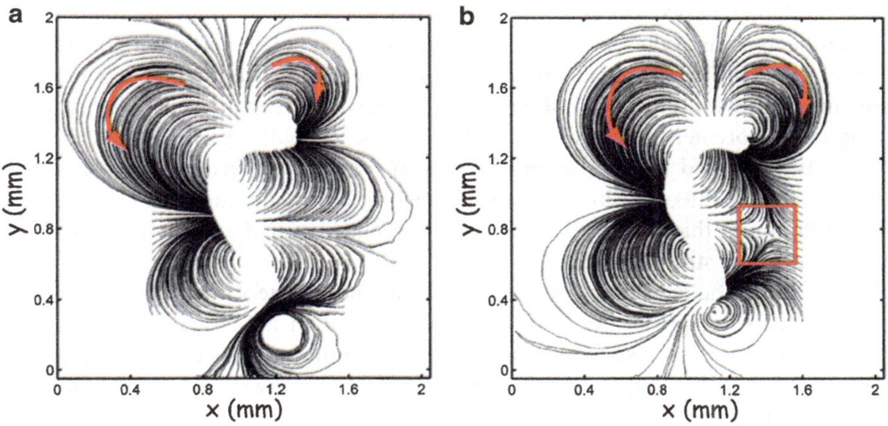

Fig. 7.14 (**a**) Streamlines computed from instantaneous velocity fields of Newtonian ($Re < 10^{-3}$) and (**b**) polymeric ($Re < 10^{-3}; De = 3.0$) fluids. *Arrows* in (**a, b**) indicate the flow direction and the box in (**b**) shows a hyperbolic point in the flow

regions of high velocity gradients such as hyperbolic points. Near such regions, the extensional viscosity of a solution of flexible polymers can be orders of magnitude larger than a Newtonian fluid, resulting in an additional resistance to fluid transport and swimming.

In summary, experiments with the nematode *C. elegans* show that fluid elasticity can hinder its swimming speed. Further, it appears that the nematode's swimming speed decreases with increasing fluid elasticity, that is, U decreases as the Deborah number is increased. This trend is predicted by both numerical simulations [89, 93] and theory [83, 107], but the agreement is only qualitative. Hence, there is plenty of room for refining the experiments, theory, and simulations. It is clear that knowledge of the flow fields is important in determining how fluid elasticity (and other rheological properties) affects the organism's swimming kinematics and dynamics. In the case of complex fluids in particular, one is interested in the interactions between the fluid microstructure (e.g., polymer molecules, networks) and the flow fields produced by swimming microorganisms. (Of course different swimming kinematics usually result in different flow fields.) In the next section, we will discuss the undulatory swimming of *C. elegans* in non-dilute (i.e., semi-dilute and concentrated) solutions, where polymer networks rather than single molecules are of interest.

5.2.3 Swimming Experiments with *C. elegans*: Beyond the Dilute Regime

In this section, we will briefly discuss the swimming behavior of *C. elegans* in semi-dilute and concentrated polymeric solutions. Such solutions are characterized by the formation of polymer networks. The interplay between the fluid's internal structure (e.g., polymer networks) and self-propulsion is critical to many biological processes such as reproduction [116], bacterial infection [117], and biodegradation in soil [118]. Early experimental observations have revealed that polymer networks can enhance the swimming speed of flagellated bacteria moving in solutions containing long-chain polymer molecules [119, 120]. For these small organisms ($L < 10\,\mu$m), it has been argued that the main mechanism for this propulsion enhancement is due to the benefits of pushing against a quasi-rigid polymer network [119, 121]. It is worth noting that the exact mechanism responsible for the observed propulsion enhancement is still not clear.

The role of the mechanical properties of fluid internal networks on an organism's swimming behavior has recently been investigated in numerical [121–124] and theoretical [125] studies. Numerical studies of swimming in structured fluids have postulated that the shapes and dynamics of internal networks are accounted for by two effective anisotropic viscosities [121, 124], which qualitatively explain some of the observed propulsion enhancement in microorganisms [119, 120]. Such anisotropic viscosities, however, are difficult to measure and apply to quantitative analysis. In heterogeneous, gel-like environments, modeled by embedding stationary objects in an incompressible viscous fluid, the swimming speed of a microorganism can be enhanced by the underlying structures in the fluid [125]. For internal networks made of small molecules, such as a binary blend of two intermixed fluids, a two-fluid model predicts an enhancement in swimming speed for stiff and compressible networks [122] and a reduction in swimming speed when local distributions of volume fractions of the two phases scale differently for thrust and drag [123]. Overall, the observed propulsion speed variations in these studies underscore the important role of the fluid internal structures on the swimming behavior of microorganisms.

But let's see how the fluid internal structures, in this case polymer networks, affect the swimming behavior of *C. elegans*; more details can be found elsewhere [126]. Polymer networks are formed by controlling the concentration of the biocompatible rodlike polymer xanthan gum (XG) in water. Polymer concentration ranges from 300 ppm to 5,000 ppm by weight. These XG solutions transition from the semi-dilute to the concentrated regime at a concentration of approximately 3,000 ppm [126]. This is made clear by plotting the solutions' zero-shear viscosity μ_0 as a function of polymer concentration, as shown in Fig. 7.15a. Note that the values of μ_0 increase as the polymer concentration c is increased, as expected. But we find a change in slope as the solution transitions from the semi-dilute to the concentrated regime at a concentration of approximately 2,800 ppm. This transition is commonly interpreted as a structural transition [127, 128]. In concentrated solution, the shape and dynamic properties of polymer networks dominate flow behaviors; in semi-dilute solution, the hydrodynamic interactions among individual polymers dominate flow behaviors [128].

Effects of Polymer Networks on Swimming Speed Fig. 7.15b shows that U remains relatively constant for polymer concentrations below 3,000 ppm. Surprisingly, however, the data shows sudden increase in U for concentrations above

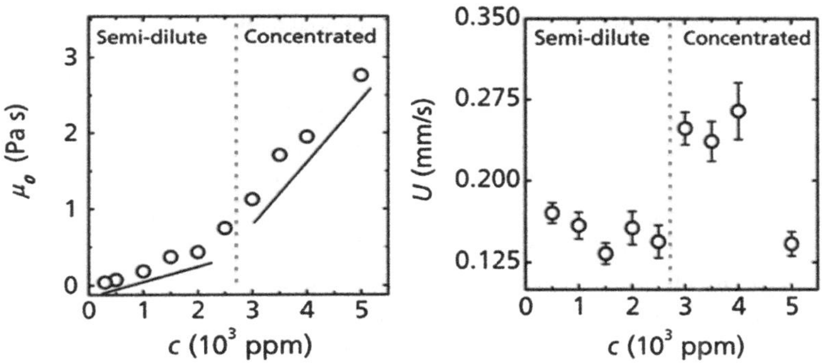

Fig. 7.15 (*Left*) Xanthan gum's (XG) zero-shear viscosity μ_0 as a function of polymer concentration c. The values of μ_0 increase with polymer concentration, as expected. The change in slope at $c \approx 2,800$ ppm is often associated with a structural transition. (*Right*) Nematode swimming speed U as a function of concentration. Swimming speed exhibits a rapid increase as the solution enters the concentrated regime at approximately 2,800 ppm

3,000 ppm. The values of U are maintained around 0.15 mm/s in semi-dilute solutions ($c < 2,800$ ppm), but they quickly rise by 65 % to about 0.25 mm/s in concentrated solutions ($c > 2,800$ ppm) *despite a significant increase in solution viscosity*. As expected, the swimming speed ultimately decreases as the concentration is further increased due to the nematode's finite power output [91]. A recent theoretical work suggests that such increase may be due to the presence of polymer networks in the media and that microorganisms may be able to push against such quasi-static networks and move more efficiently [121]. However, because of the large difference in length scales between the nematode ($\approx$ 1 mm) and the polymer networks ($\approx$ 10 μm), this notion does not adequately explain the observed propulsion enhancement.

A possible explanation for this enhancement in U in concentrated solutions is given in [126], in which the authors argue that the phenomenon is most likely due to shear-induced fluid anisotropy. That is, the increase in U observed in Fig. 7.15b is probably due to the anisotropic response of the fluid microstructure to applied stress due to the nematode's swimming motion. In short, the undulatory swimming motion of *C. elegans* induces a structural anisotropy which leads to an increase in the effective drag coefficient ratio C_n/C_t (see RFT equation in Sect. 1) and consequently an enhancement in U.

The above results show that the nematode *C. elegans* can swim faster in concentrated solutions than in semi-dilute solutions. This is an unexpected result since the fluid viscosity increases as polymeric solution transitions from the semi-dilute to the concentrated regime. This sudden increase in U is thought to be connected to the anisotropic response of the fluid microstructure to applied shear stresses due the nematode's motion [126]. While intriguing, the proposed mechanism is speculative due to the difficulty to measure or visualize the polymer microstructure during swimming. In these cases, numerical simulations and theoretical calculations can provide much needed clarity and understanding to the problem.

5.2.4 Swimming Experiments with *C. elegans*: Final Remarks

In this section, we discussed the swimming behavior of the nematode *C. elegans* in Newtonian and viscoelastic (i.e., polymeric) fluids. For dilute polymeric solutions, experiments show that fluid elasticity hinders the swimming speed of nematodes by 40 % compared to Newtonian fluids. The swimming speed is also shown to decrease as elasticity (i.e., Deborah number) increases. On the other hand, for concentrated polymeric solutions, the presence of polymer networks seem to enhance swimming speed by as much as 65 % when compared to semi-dilute and dilute polymeric solutions. These results underscore the importance of the fluid microstructure and its interactions with the applied stresses generated by the swimmer. Perhaps the main message so far is that it is difficult to quantitatively describe a priori the motility of microorganisms in complex fluids without knowledge of the interactions between the fluid microstructure and the applied stresses. It becomes clear that these non-trivial interactions need to be accounted for in theoretical calculations and numerical simulations.

5.3 Fluid-Assisted Locomotion in Complex Fluids: Artificial Swimmers

As shown above, fluid elasticity can significantly affect the swimming behavior of live organisms. In this section, we will explore a different, and perhaps simpler, question: *Can fluid elasticity enable propulsion?*

To answer the above question, one needs to think back to the "scallop theorem," which tells us that only nonreciprocal deformations of the swimmer can break time-reversal symmetry and result in net motion [1]. The main assumptions of the theorem is that the swimmer is moving at low Reynolds numbers ($Re < 0.1$) and that the fluid is purely viscous or Newtonian. So, in order to break the "scallop theorem" or kinematic reversibility one needs to increase the amount of inertia in the system (i.e., increase Re) or alternatively use fluids that possess nonlinear rheological behavior. In this section, we will review recent experiments [92] in which artificial particles with reciprocal swimming strategies are able to break the scallop theorem once immersed in complex fluids. The experiments focus mainly on the role of viscoelasticity, and two main types of fluids will be used: (i) dilute polymeric solutions [92] and (ii) wormlike micellar (WLM) solutions [129].

Before we begin, it is worth noting that the possibility that fluid elasticity can *enable* rather than modify propulsion circumventing the scallop theorem is still largely unexplored. Propulsion enabled by fluid elasticity has been predicted for the three special cases of reciprocal motion: a flapping surface extending from a plane [130, 131]; a sphere which generates small-amplitude sinusoidal motion of fluid along its surface [8]; and a "wriggling" cylinder with reciprocal forward and backward strokes at different rates [88]. However, there remains little experimental demonstration, and such propulsion of free, finite-amplitude swimmers has been seldom studied.

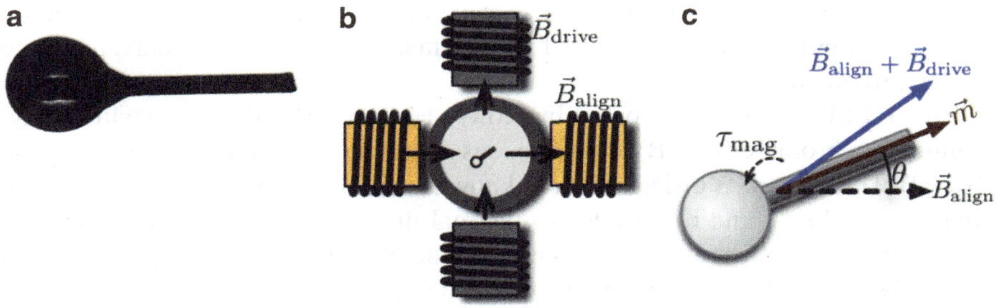

Fig. 7.16 (**a**) Snapshot of the polar dimer. An epoxy bead is attached to a steel wire to form polar (asymmetric) dimers. (**b**) *Top view* of the experiment. Two aligning electromagnets at constant current are orthogonal to two driving magnets, controlled by a computer. (**c**) The dimer with magnetization **m** experiences torque τ_{mag} to align with the magnetic field. Dimer orientation $\hat{a}$ oscillates around $\langle \hat{a} \rangle$, which is parallel to $\mathbf{B}_{\text{align}}$

5.3.1 Experiments with Reciprocal Swimmers: Can Fluid Elasticity Enable Propulsion?

Here, we briefly discuss recent experiments [92, 129] in which a single rigid object, in this case a dumbbell particle or dimer, is actuated in a reciprocal manner in very viscous fluids. In the experiments, the dimer such as the one shown in Fig. 7.16 is immersed in a fluid and repeatedly reoriented by a magnetic field. The effects of inertia are absent due to the high fluid viscosity ($\sim 10\,\text{Pa}\cdot\text{s}$), resulting in $\text{Re} \ll 0.1$ comparable to that of a swimming microorganism. By applying only magnetic torques, the apparatus reciprocally actuates just one degree of freedom in the system, the dimer's orientation $\hat{a}$. For a purely viscous Newtonian fluid at low Reynolds numbers, the authors found no net motion because $\hat{a}(t)$ is cyclic; this is as expected.

Experimental Setup Before diving into the discussion, let us briefly describe the experimental setup. More details can be found elsewhere [92, 129]. The artificial swimmer is a polar (asymmetric) dimer (Fig. 7.16a); symmetric dimers are also used for control but no net motion is expected. The polar dimer consists of a piece of carbon steel wire of length $2R_{\text{dimer}} = 2.5$–$3\,\text{mm}$ and diameter $230\,\mu\text{m}$, with an epoxy bead of diameter $2R_{\text{dimer}} \sim 500\,\mu\text{m}$ at one end. The dimer is then immersed in a fluid bath that is surrounded by four electromagnets; a schematic of the apparatus is shown in Fig. 7.16b. The dimer has orientation $\hat{a}$ and is magnetized with moment $\mathbf{m} = \hat{a}m$, so that a uniform magnetic field $\mathbf{B}$ reorients it with torque $\text{ø}_{\text{mag}} = \mathbf{m} \times \mathbf{B}$, as depicted in Fig. 7.16c.

Working Fluids The dimer is immersed in a container (50 mm tall, 30 mm in diameter) of either Newtonian or viscoelastic fluid (Fig. 7.16b). The Newtonian fluid is a 96 %-corn syrup aqueous solution (by mass) with a kinematic viscosity μ/ρ of approximately 4×10^4 cSt. Two viscoelastic solutions are prepared: a dilute polymeric solution and a WLM solutions. The polymeric solution is made by adding 0.17 % (by mass) of high-molecular-weight polyacrylamide (PAA, $M_W = 1 \times 10^6$)

to a viscous Newtonian solvent (93 %-corn syrup aqueous solution). The solution has nearly constant shear viscosity of approximately 50Pa · s and a relaxation time λ of approximately 2 s.

The WLM solution is prepared by slowly adding 130 mM hexadecyltrimethylammonium bromide (CTAB) to an aqueous solution of 130 mM sodium salicylate (NaSal). This type of WLM solutions is known to form long wormlike micelles that continuously break and reform due to thermal fluctuations leading to viscoelastic stress relaxation. The relaxation time λ of the WLM solution is approximately 1.5 s [129]. Here, the Deborah number is defined as De$f_{\text{drive}}\lambda$.

Swimming with Reciprocal Motion Now that the methods are in place, we can ask an important question: Can the nonlinear rheological properties of a given fluid enable propulsion at low Re? Before we discuss the results, it is worth making sure that artifacts that could lead to net motion are not being introduced in the experiment. To that end, the investigators [129] calibrated their results against a polar and symmetrical dimer immersed in a viscous Newtonian fluid with similar shear viscosity as the polymeric and WLM solutions. As shown in Fig. 7.17a, they found negligible net displacement, comparable to the effects of magnetic drift and sedimentation when driving is turned off altogether.

Evidence of purely elastic propulsion is shown in Fig. 7.17b–d for the polar dimer immersed in dilute polymeric (b) and WLM solutions (c, d). Overall, the data show a striking contrast between performing reciprocal motion in Newtonian and in complex, viscoelastic fluids. For example, Fig. 7.17b shows that in dilute polymeric solutions, far from any boundaries, the polar dimer at $\text{De} = 5.7(f_{\text{drive}} = 2.8\,\text{Hz}; \text{Re} = 1.2 \times 10^{-4})$ is able to achieve net motion at constant speed even under reciprocal forcing. The same is true for the WLM solution (Fig. 7.17b, d). It is interesting to note that, unlike the polymeric solution case, the swimming direction in the WLM solutions seems to depend on De. For low elasticity values ($\text{De} < 1$), the particle moves towards the bead while the particle moves preferentially towards the rod for cases in which actuating frequency is approximately similar to the fluid relaxation time ($\text{De} \sim 1$). The dimer directional dependence on De in WLM solutions is still not understood.

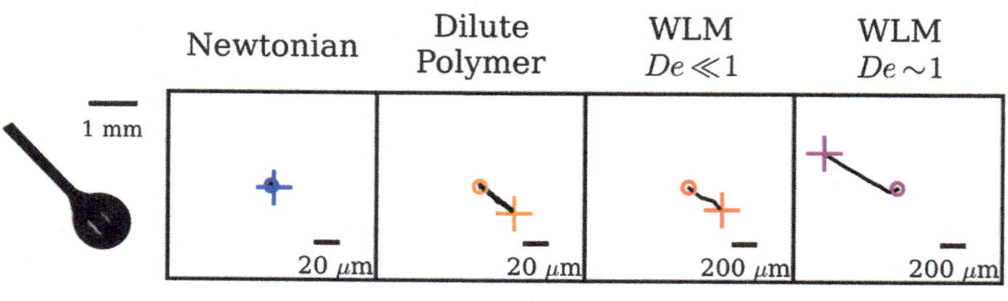

Fig. 7.17 Centroid trajectories from ∘ to + reciprocally actuated polar dimer (*leftmost panel*) in Newtonian, dilute polymeric, and wormlike micellar (WLM) solutions at low and high Deborah numbers. No appreciable net motion is found in the Newtonian case. Appreciable net motion is found once small amounts of polymers or surfactants (WLM) are added to a Newtonian solvent

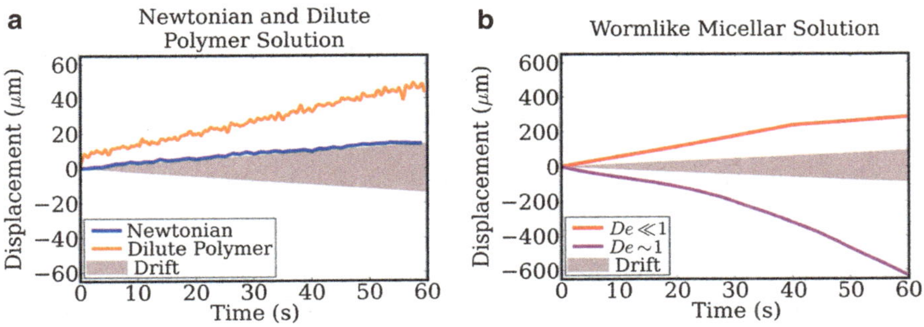

Fig. 7.18 Dimer net displacement as a function of time. The reciprocally actuated dimer is immersed in (**a**) polymeric and (**b**) wormlike micellar solutions. *Shaded areas* correspond to the dimer drift due to small but finite magnetic gradients

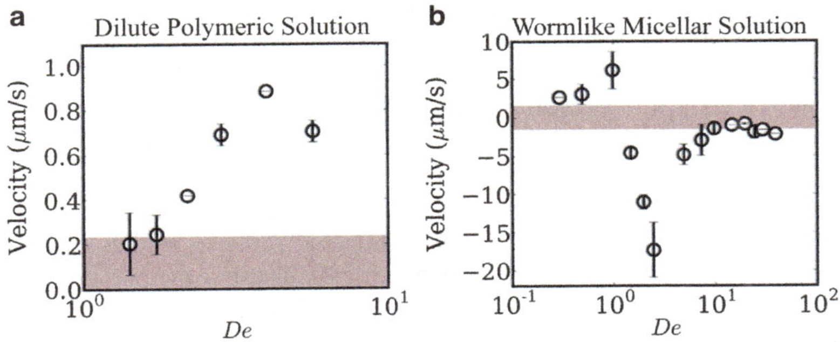

Fig. 7.19 Dependence of mean propulsion velocity on Deborah (De) for a reciprocally actuated dimer immersed in (**a**) dilute polymeric and (**b**) wormlike micellar (WLM) solutions

Figure 7.18a, b shows the net displacement for the dilute polymeric and WLM solutions, respectively. Both cases show dimer displacements well above the noise or drift level (shaded area). Clearly, a much larger displacement can be achieved with the WLM solutions, although the mechanisms are still unknown. The data also shows that the dimer can move either towards the bead (positive velocity) or towards the rod (negative velocity) depending on the De as discussed above.

In order to gain further understanding on the effects of fluid elasticity on the dimer motion, it is helpful to plot the displacement data as a function of De. Figure 7.19a, b shows the dimer velocity as a function of De for the dilute polymeric and WLM solutions, respectively. Recall that the positive velocity means that the polar dimer is moving towards its bead and negative velocity means that the dimer is moving towards its rod. For the polymeric solution case, the velocity increases monotonically as a function of De. In fact, the velocity seems to obey a De^2 scaling.

The effects of elasticity are less obvious for the WLM solutions. While it is clear that the polar dimer speed increases as De increases, the velocity trend is more complex. It seems that for low elasticity values ($0 < De < 1$, the trend and dimer swimming direction resemble the polymeric solution data. For $\mathrm{De} \geq 1$, the dimer swimming direction reverses, and for $\mathrm{De} \gg 1$ the dimer comes to a complete stop. While the exact mechanisms leading to such behaviors in WLM solutions are not well understood, the data underscore the rich nonlinear behavior encountered in complex fluids.

In summary, we find that polar particles (dimers) undergoing reciprocal motion can indeed achieve locomotion at low Reynolds numbers in complex, viscoelastic fluids. This is fascinating because it opens the door for a new way to achieve locomotion, one that relies on the fluid itself. For the case of dilute polymeric solutions, the net motion achieved by the dimers results from elastic stresses due to flow-induced changes in polymer conformation. These elastic stresses are history-dependent and do not entirely cancel out over one forcing period, but instead have a small rectified component that accumulates. In fact, Keim and Arratia [92] claim that the combination of the fluid first normal stress difference N_1 and the curved streamlines produced by the actuated dimer leads to a volume force (or "hoop stress") that is able to displace the dimer. The proposed mechanism is reasonable but it is yet to be validated. The picture is less clear for WLM solutions where shear-bands are known to occur in addition to elasticity.

Finally, this type of work is a proof-of-concept for an artificial "swimmer" that moves through complex fluids with only reciprocal actuation, a simple body shape, and no moving parts—a less complicated design than for other propulsive strategies [47, 132]. There may thus be a practical route to studies of collective phenomena among large numbers of such particles. These principles could also be applied to microfluidic pumps [130, 131] or to exploiting other types of nonlinear fluid rheology. Further understanding of this effect and similar ones could greatly simplify fabrication of microswimmers for many artificial environments or for biological settings where viscoelasticity is ubiquitous.

6 Conclusions and Outlook

In this chapter, we discussed swimming in complex fluids. The discussion was given from an experimental point of view and was centered on a few fundamental experiments both in living and in artificial systems that we hope illustrate the rich dynamics encountered by organisms moving in complex fluids, particularly in viscoelastic fluids. Much of the investigations in viscoelastic fluids was driven by a simple question: does elasticity hinder or enhance swimming speed?

Theoretical analysis on idealized swimming models, following the seminal work of Taylor [27], showed that in general fluid elasticity hinders propulsion [83, 107]. Numerical simulations [89, 93], however, found that fluid elasticity could in fact enhance propulsion under certain conditions ($\mathrm{De} \sim 1$). Experiments [90, 91] seem to confirm such predictions, at least qualitatively. For example, the swimming speed of *C. elegans* effectively decreases once the organism is immersed in dilute viscoelastic fluids, while propulsion speed increased for De ~ 1 in scale-up mechanical experiments with helical bodies. In more concentrated polymeric solutions, the nematode *C. elegans* showed a remarkable increase in speed for solutions possessing polymer networks. (There are of course many other impressive experimental investigations of swimming in anisotropic media such as liquid crystals [133–135] that could not be covered here.) Overall we find that it is very hard to know a priori whether fluid elasticity enhances or hinders the propulsion speed of microorganisms. In fact, it may be an unfair question altogether as the answer depends on the type of swimming kinematics (i.e., waveform) and the interactions of the swimmer with the fluid microstructure (i.e., polymer or particles). That is, the local details may matter quite a bit. Nevertheless, it is clear from both experiments and simulations that fluid elasticity can significantly affect the motility behavior of microorganisms.

Parallel to these considerations, we also discussed an exciting avenue of new developments, that is, the possibility of fluid-assisted propulsion. This type of propulsion is driven by the nonlinear rheological properties of the fluid such as viscoelasticity and formation of shear-bands. We showed that even for reciprocally actuated polar particles, the extra elastic stresses of polymeric fluids can lead to net motion at low Reynolds numbers. It is important to note that under the same conditions, the same particle in a Newtonian fluid achieves no net motion due to kinematic reversibility. These findings are exciting because such discovery opens a new mechanism for the study of "active" particles and collective behavior in complex fluids.

So where do we go from here? Of course there is still much to be done and understood. The question of how microorganisms move in complex fluids is still largely unanswered. We still do not know how microorganisms move in rheologically complex materials that possess rate-dependent viscosity, yield stress, and thixotropy. But a picture is starting to emerge, and it seems that the interaction of the swimmer with the fluid microstructure is very important. That is not surprising since microorganisms with different swimming kinematics will likely produce different velocity fields in a given fluid, and the bulk response will be determined by the way the fluid microstructure interacts with the velocity field. For example, it is well known that extensional flows are more efficient in stretching and aligning polymer molecules than shear flows, and swimmers that produce flows with a large extensional component are dominated by the fluid extensional viscosity. On the other hand, swimmers that produce large amounts of curvature may experience a viscoelastic instability due to hoop stresses. So knowledge of the velocity field is quite important, but equally important is how the molecules or particles respond to the imposed velocity field. In a way, the swimmer may be thought of as a local probe for fluid rheology. An exciting and important direction that would allow us to gain a more complete understanding of swimming in complex fluids is to determine how single polymer molecules, particles, and networks interact with microorganisms. This direction will ultimately take us from the continuum to a statistical mechanics description of the problem. The challenge lies in connecting the statistical representation with the continuum (bulk) description.

Finally, one can also ask the question of how multiple swimmers interact in complex fluids. One would anticipate that the fluid microstructure will significantly alter the way multiple microorganisms interact. The type of work is related to the field of active matter or fluid in which "active" or live particles are present in the fluid medium. Active fluids differ from their passive counterpart in that the active particles have the ability to absorb and dissipate energy and to generate motion and mechanical stresses in the fluid medium (see Chap. 9). Importantly, these active particles can drive the system out of equilibrium even in the absence of external forcing. Active fluids exhibit novel properties not seen in conventional (passive) complex fluids such as large-scale flows and collective motion on length scales much greater than the particle dimensions [136], anomalous shear viscosity [137], giant density fluctuations [138], and enhanced fluid mixing [139]. While much recent progress has been made, the dynamics and flow behavior (rheology) of such active complex fluids are still poorly understood. In this case, one can take advantage of a vast body of knowledge developed to understand passive complex fluids.

Acknowledgment The authors would like to thank David Gagnon, Nathan Keim, Arvind Gopinath, Alexander Leshansky, and Xiaoning Shen for help with the text and in drafting illustrations. This work was supported by the US-Israel Binational Science Foundation (BSF grant nr. 2011323) and J. Sznitman was supported in part by the European Commission (FP7 Program) through a Career Integration Grant (PCIG09-GA-2011-293604). P.E. Arratia was supported in part by the Army Research Office through award W911NF-11-1-0488 and by NSF-CBET-Career Award.

References

1. E.M. Purcell, Am. J. Phys. **45**, 3 (1977)
2. C. Brennen, H. Winet, Annu. Rev. Fluid Mech. **9**(1), 339 (1977)
3. J. Gray, J. Exp. Biol. **32**(4), 775 (1955)
4. J. Gray, G. Hancock, J. Exp. Biol. **32**(4), 802 (1955)
5. N. Cohen, J.H. Boyle, Contemp. Phys. **51**(2), 103 (2010)
6. J. Gray, H.W. Lissmann, J. Exp. Biol. **41**(1), 135 (1964)
7. J. Sznitman, X. Shen, R. Sznitman, P.E. Arratia, Phys. Fluids (1994–present) **22**, 121901 (2010)
8. E. Lauga, T.R. Powers, Rep. Prog. Phys. **72**(9), 096601 (2009)
9. J. Lighthill, SIAM Rev. **18**(2), 161 (1976)
10. N.A. Croll, *The Behaviour of Nematodes: Their Activity, Senses and Responses* (Edward Arnold, London, 1970)
11. J.G. White, E. Southgate, J.N. Thomson, S. Brenner, Philos. Trans. R. Soc. Lond. B Biol. Sci. **314**, 1 (1986)
12. W.B. Wood, *The Nematode Caenorhabditis elegans* (Cold Spring Harbour Laboratory, New York, 1987)
13. H.C. Berg, *E. coli in Motion* (Springer, New York, 2004)
14. H.C. Berg, Phys. Today **53**(1), 24 (2000)
15. H.C. Berg, Biochemistry **72**(1), 19 (2003)
16. H.C. Berg, Curr. Biol. **18**(16), R689 (2008)
17. D. Bray, *Cell Movements: From Molecules to Motility* (Garland Science, New York, 2001)

18. D.R. Mitchell, J. Phycol. **36**, 261 (2000)
19. M. Werner, L.W. Simmons, Biol. Rev. **83**, 191 (2008)
20. I. Gibbons, J. Cell Biol. **91**(3), 107s (1981)
21. R. Lyons, O. Djahanbakhch, T. Mahmood, E. Saridogan, S. Sattar, M. Sheaff, A. Naftalin, R. Chenoy, Hum. Reprod. **17**, 584 (2002)
22. R. Lyons, E. Saridogan, O. Djahanbakhch, Hum. Reprod. Update **12**, 363 (2006)
23. T. Nakahari, A. Nishimura, C. Shimamoto, A. Sakai, H. Kuwabara, T. Nakano, S. Tanaka, Y. Kohda, H. Matsumura, H. Mori, Biomed. Res. **32**, 321 (2011)
24. P. Satir, M.A. Sleigh, Annu. Rev. Physiol. **52**, 137 (1990)
25. A. Korngreen, Z. Priel, Biophys. J. **67**(1), 377 (1994)
26. A.L. Oldenburg, R.K. Chhetri, D.B. Hill, B. Button, Biomed. Opt. Express **3**, 1978 (2012)
27. G. Taylor, Proc. R. Soc. Lond. Ser. A Math. Phys. Sci. **209**, 447 (1951)
28. G. Taylor, Proc. R. Soc. Lond. Ser. A Math. Phys. Sci. **211**, 225 (1952)
29. I. Aranson, Physics **6**, 61 (2013)
30. J.S. Guasto, R. Rusconi, R. Stocker, Annu. Rev. Fluid Mech. **44**, 373 (2012)
31. S. Chattopadhyay, R. Moldovan, C. Yeung, X. Wu, Proc. Natl. Acad. Sci. **103**(37), 13712 (2006)
32. D. Katz, J. Blake, S. Paveri-Fontana, J. Fluid Mech. **72**(03), 529 (1975)
33. J. Happel, H. Brenner, *Low Reynolds Number Hydrodynamics: With Special Applications to Particulate Media*, vol. 1 (Springer, Berlin, 1983)
34. D. Saintillan, Physics **3**, 84 (2010)
35. S.A. Koehler, T.R. Powers, Phys. Rev. Lett. **85**(22), 4827 (2000)
36. B. Qian, T.R. Powers, K.S. Breuer, Phys. Rev. Lett. **100**, 078101 (2008)
37. S.Y. Tony, E. Lauga, A. Hosoi, Phys. Fluids (1994–present) **18**, 091701 (2006)
38. M.S. Sakar, C. Lee, P.E. Arratia, Phys. Fluids **21**, 91107 (2009)
39. S. Zhong, K.W. Moored, V. Pinedo, J. Garcia-Gonzalez, A.J. Smits, Exp. Therm. Fluid Sci. **46**, 1 (2013)
40. M. Kim, J.C. Bird, A.J. Van Parys, K.S. Breuer, T.R. Powers, Proc. Natl. Acad. Sci. **100**(26), 15481 (2003)
41. M.J. Kim, M.J. Kim, J.C. Bird, J. Park, T.R. Powers, K.S. Breuer, Exp. Fluids **37**(6), 782 (2004)
42. E. Setter, I. Bucher, S. Haber, Phys. Rev. E **85**, 066304 (2012)
43. P. Weiss, Sci. News **169**, 107 (2006)
44. R. Trouilloud, S.Y. Tony, A. Hosoi, E. Lauga, Phys. Rev. Lett. **101**, 048102 (2008)
45. C.H. Wiggins, D. Riveline, A. Ott, R.E. Goldstein, Biophys. J. **74**, 1043 (1998)
46. M. Leoni, J. Kotar, B. Bassetti, P. Cicuta, M.C. Lagomarsino, Soft Matter **5**(2), 472 (2009)
47. J.J. Abbott, K.E. Peyer, M.C. Lagomarsino, L. Zhang, L. Dong, I.K. Kaliakatsos, B.J. Nelson, Int. J. Rob. Res. **28**(11), 1434 (2009)
48. H.C. Berg, Rev. Sci. Instrum. **42**(6), 868 (1971)
49. H.C. Berg, Adv. Opt. Elect. Microsc. **7**, 1 (1978)
50. H.C. Berg, D.A. Brown, Nature **239**(5374), 500 (1972)
51. L. Turner, W.S. Ryu, H.C. Berg, J. Bacteriol. **182**, 2793 (2000)
52. S.C. Lenaghan, C.A. Davis, W.R. Henson, Z. Zhang, M. Zhang, Proc. Natl. Acad. Sci. **108**(34), E550 (2011)
53. K. Drescher, K.C. Leptos, R.E. Goldstein, Rev. Sci. Instrum. **80**(1), 014301 (2009)
54. E. Lauga, W.R. DiLuzio, G.M. Whitesides, H.A. Stone, Biophys. J. **90**(2), 400 (2006)
55. R. Di Leonardo, D. Dell'Arciprete, L. Angelani, V. Iebba, Phys. Rev. Lett. **106**(3), 038101 (2011)
56. A.P. Berke, L. Turner, H.C. Berg, E. Lauga, Phys. Rev. Lett. **101**(3), 038102 (2008)
57. J. Cosson, P. Huitorel, C. Gagnon, Cell Motil. Cytoskeleton **54**(1), 56 (2003)
58. L. Rothchild, Nature **198**, 1221 (1963)
59. D. Woolley, Reproduction **126**, 259 (2003)
60. S. Jana, S.H. Um, S. Jung, Phys. Fluids (1994–present) **24**(4), 041901 (2012)

61. P. Denissenko, V. Kantsler, D.J. Smith, J. Kirkman-Brown, Proc. Natl. Acad. Sci. **109**(21), 8007 (2012)
62. R. Ghosh, J. Sznitman, J. Vis. **15**, 1–3 (2012)
63. F. Lebois, P. Sauvage, C. Py, O. Cardoso, B. Ladoux, P. Hersen, J.M. Di Meglio, Biophys. J. **102**(12), 2791 (2012)
64. K. Konig, L. Svaasand, Y. Liu, G. Sonek, P. Patrizio, Y. Tadir, M. Berns, B. Tromberg, Cell. Mol. Biol. (Noisy-le-Grand, France) **42**(4), 501 (1996)
65. Y. Tadir, W. Wright, O. Vafa, T. Ord, R. Asch, M. Berns, Fertil. Steril. **53**, 944 (1990)
66. Z. Teff, Z. Priel, L.A. Gheber, Biophys. J. **92**, 1813 (2007)
67. S. Chattopadhyay, X.L. Wu, Biophys. J. **96**(5), 2023 (2009)
68. B. Rodenborn, C.H. Chen, H.L. Swinney, B. Liu, H. Zhang, Proc. Natl. Acad. Sci. **110**, E338 (2013)
69. G. Batchelor, J. Fluid Mech. **44**(03), 419 (1970)
70. A.T. Chwang, T. Wu, J. Fluid Mech. **67**(04), 787 (1975)
71. J.B. Keller, S.I. Rubinow, J. Fluid Mech. **75**(04), 705 (1976)
72. R.E. Johnson, J. Fluid Mech. **99**(02), 411 (1980)
73. J. Higdon, J. Fluid Mech. **90**(04), 685 (1979)
74. J. Higdon, J. Fluid Mech. **94**(02), 331 (1979)
75. H. Kurtuldu, D. Tam, A. Hosoi, K. Johnson, J. Gollub, Phys. Rev. E **88**(1), 013015 (2013)
76. S.T. Wereley, C.D. Meinhart, Annu. Rev. Fluid Mech. **42**, 557 (2010)
77. D. Murphy, D. Webster, J. Yen, Limnol. Oceanogr. Methods **10**, 1096 (2012)
78. K. Drescher, R.E. Goldstein, N. Michel, M. Polin, I. Tuval, Phys. Rev. Lett. **105**(16), 168101 (2010)
79. J.S. Guasto, K.A. Johnson, J.P. Gollub, Phys. Rev. Lett. **105**(16), 168102 (2010)
80. L.J. Fauci, R. Dillon, Annu. Rev. Fluid Mech. **38**, 371 (2006)
81. S. Suarez, H. Ho, Reprod. Domest. Anim. **38**, 119 (2003)
82. E.R. Trueman, *Locomotion of Soft-Bodied Animals* (Edward Arnold, London, 1975)
83. E. Lauga, Phys. Fluids **19**(8), 083104 (2007)
84. T.D. Montenegro-Johnson, D.J. Smith, D. Loghin, Phys. Fluids (1994–present) **25**, 081903 (2013)
85. J.R. Vélez-Cordero, E. Lauga, J. Non-Newtonian Fluid Mech. **199**, 37 (2013)
86. D. Katz, R. Mills, T. Pritchett, J. Reprod. Fertil. **53**(2), 259 (1978)
87. T. Chaudhury, J. Fluid Mech. **95**(01), 189 (1979)
88. H.C. Fu, C.W. Wolgemuth, T.R. Powers, Phys. Fluids **21**(3), 033102 (2009)
89. J. Teran, L. Fauci, M. Shelley, Phys. Rev. Lett. **104**, 038101 (2010)
90. B. Liu, T.R. Powers, K.S. Breuer, Proc. Natl. Acad. Sci. **108**, 19516 (2011)
91. X. Shen, P.E. Arratia, Phys. Rev. Lett. **106**, 208101 (2011)
92. N.C. Keim, M. Garcia, P.E. Arratia, Phys. Fluids (1994–present) **24**(8), 081703 (2012)
93. S.E. Spagnolie, B. Liu, T.R. Powers, Phys. Rev. Lett. **111**, 068101 (2013)
94. R.B. Bird, R.C. Armstrong, O. Hassager, *Dynamics of Polymeric Liquids. Vol. 1: Fluid Mechanics* (Wiley, New York, 1987)
95. R.G. Larson, *The Structure and Rheology of Complex Fluids*, vol. 4 (Oxford University Press New York, 1999)
96. E.S. Shaqfeh, Annu. Rev. Fluid Mech. **28**, 129 (1996)
97. H.C. Ho, S.S. Suarez, Reproduction **122**(4), 519 (2001)
98. D. Katz, T. Bloom, R. Bondurant, Biol. Reprod. **25**(5), 931 (1981)
99. L.A. McPartlin, S.S. Suarez, C.A. Czaya, K. Hinrichs, S. Bedford-Guaus, Biol. Reprod. **81**, 199 (2009)
100. A. Houry, M. Gohar, J. Deschamps, E. Tischenko, S. Aymerich, A. Gruss, R. Briandet, Proc. Natl. Acad. Sci. **109**(32), 13088 (2012)
101. S. Yazdi, A.M. Ardekani, Biomicrofluidics **6** (2012)
102. S. Jung, A.G. Winter, A. Hosoi, Int. J. Non-Linear Mech. **46**(4), 602 (2011)
103. A.G. Winter, R.L. Deits, A.E. Hosoi, J. Exp. Biol. **215**, 2072 (2012)
104. G.R. Fulford, D.F. Katz, R.L. Powell, Biorheology **35**(4), 295 (1998)

105. M.J. Kim, K.S. Breuer, Small **4**(1), 111 (2008)
106. D. Boger, J. Non-Newtonian Fluid Mech. **3**(1), 87 (1977)
107. H.C. Fu, T.R. Powers, C.W. Wolgemuth, Phys. Rev. Lett. **99**(25), 258101 (2007)
108. S.K. Lai, Y.Y. Wang, D. Wirtz, J. Hanes, Adv. Drug Deliv. Rev. **61**(2), 86 (2009)
109. S. Brenner, Genetics **77**(1), 71 (1974)
110. L. Byerly, R. Cassada, R. Russell, Dev. Biol. **51**(1), 23 (1976)
111. X. Shen, J. Sznitman, P. Krajacic, T. Lamitina, P. Arratia, Biophys. J. **102**, 2772 (2012)
112. F.T. Trouton, Proc. R. Soc. London Ser. A **77**, 426 (1906)
113. S.L. Anna, G.H. McKinley, J. Rheol. (1978–present) **45**(1), 115 (2001)
114. G.H. McKinley, T. Sridhar, Annu. Rev. Fluid Mech. **34**, 375 (2002)
115. P.E. Arratia, C. Thomas, J. Diorio, J.P. Gollub, Phys. Rev. Lett. **96**(14), 144502 (2006)
116. D.S. Guzick, J.W. Overstreet, P. Factor-Litvak, C.K. Brazil, S.T. Nakajima, C. Coutifaris, S.A. Carson, P. Cisneros, M.P. Steinkampf, J.A. Hill, New England J. Med. **345**(19), 1388 (2001)
117. C. Josenhans, S. Suerbaum, Int. J. Med. Microbiol. **291**(8), 605 (2002)
118. M. Alexander, *Introduction to Soil Microbiology* (Wiley, New York, 1991)
119. H.C. Berg, L. Turner, Nature **278**, 349 (1979)
120. W.R. Schneider, R. Doetsch, J. Bacteriol. **117**, 696 (1974)
121. Y. Magariyama, S. Kudo, Biophys. J. **83**, 733 (2002)
122. H.C. Fu, V.B. Shenoy, T.R. Powers, EPL (Europhys. Lett.) **91**(2), 24002 (2010)
123. J. Du, J.P. Keener, R.D. Guy, A.L. Fogelson, Phys. Rev. E **85**(3), 036304 (2012)
124. S. Nakamura, Y. Adachi, T. Goto, Y. Magariyama, Biophys. J. **90**, 3019 (2006)
125. A. Leshansky, Phys. Rev. E **80**(5), 051911 (2009)
126. D. Gagnon, X. Shen, P. Arratia, EPL (Europhys. Lett.) **104**(1), 14004 (2013)
127. A. Rodd, D. Dunstan, D. Boger, Carbohydr. Polym. **42**, 159 (2000)
128. M. Doi, *The Theory of Polymer Dynamics* (Oxford university press, Oxford, 1988)
129. D.A. Gagnon, N.C. Keim, X. Shen, P.E. Arratia, Fluid-induced propulsion of rigid particles in wormlike micellar solutions. Phys. Fluids **26**, 103101 (2014)
130. T. Normand, E. Lauga, Phys. Rev. E **78**, 061907 (2008)
131. O.S. Pak, T. Normand, E. Lauga, Phys. Rev. E **81**, 036312 (2010)
132. R. Dreyfus, J. Baudry, M.L. Roper, M. Fermigier, H.A. Stone, J. Bibette, Nature **437**(7060), 862 (2005)
133. S. Zhou, A. Sokolov, O.D. Lavrentovich, I.S. Aranson, Proc. Natl. Acad. Sci. **111**, 1265 (2014)
134. P.C. Mushenheim, R.R. Trivedi, H.H. Tuson, D.B. Weibel, N.L. Abbott, Soft Matter **10**, 88 (2014)
135. A. Kumar, T. Galstian, S.K. Pattanayek, S. Rainville, Mol. Cryst. Liq. Cryst. **574**(1), 33 (2013)
136. L.H. Cisneros, R. Cortez, C. Dombrowski, R.E. Goldstein, J.O. Kessler, Exp. Fluids **43**(5), 737 (2007)
137. A. Sokolov, I.S. Aranson, Phys. Rev. Lett. **103**, 148101 (2009)
138. V. Narayan, S. Ramaswamy, N. Menon, Science **317**, 105 (2007)
139. H. Kurtuldu, J.S. Guasto, K.A. Johnson, J.P. Gollub, Proc. Natl. Acad. Sci. **108**(26), 10391 (2011)

Chapter 8
Theory of Locomotion Through Complex Fluids

Gwynn J. Elfring and Eric Lauga

Abstract Microorganisms such as bacteria often swim in fluid environments that cannot be classified as Newtonian. Many biological fluids contain polymers or other heterogeneities which may yield complex rheology. For a given set of boundary conditions on a moving organism, flows can be substantially different in complex fluids, while non-Newtonian stresses can alter the gait of the microorganisms themselves. Heterogeneities in the fluid may also be characterized by length scales on the order of the organism itself leading to additional dynamic complexity. In this chapter we present a theoretical overview of small-scale locomotion in complex fluids with a focus on recent efforts quantifying the impact of non-Newtonian rheology on swimming microorganisms.

1 Introduction

Many microorganisms swim through fluids that display non-Newtonian characteristics. For example, as spermatozoa make their journey through the female reproductive tract they encounter several complex fluids including glycoprotein-based cervical mucus in the cervix [1], mucosal epithelium inside the fallopian tubes, and an actin-based viscoelastic gel outside the ovum [2, 3]. These complex fluids often have dramatic effects on the locomotion of microorganisms. Human sperm flagella beat with higher frequency but smaller amplitude and wavelength in cervical mucus compared to semen. This results in roughly the same swimming speed but along straighter paths due to a reduction of wobbling and end effects [1]. Sperm hyperactivation (larger amplitude, asymmetric beating patterns) increases

G.J. Elfring (✉)
Department of Mechanical Engineering, The University of British Columbia, Vancouver, BC V6T 1Z4, Canada
e-mail: gelfring@mech.ubc.ca

E. Lauga
Department of Applied Mathematics and Theoretical Physics, University of Cambridge, Trinity Ln, Cambridge CB2 1TN, UK
e-mail: e.lauga@damtp.cam.ac.uk

S.E. Spagnolie (ed.), *Complex Fluids in Biological Systems*, Biological and Medical Physics, Biomedical Engineering,
DOI 10.1007/978-1-4939-2065-5_8

the ability to penetrate viscoelastic fluids [4, 5]. Experimental evidence suggests therefore that spermatozoa both passively and actively modulate their swimming kinematics due to the presence of non-Newtonian stresses. In contrast, the bacterium *Helicobacter pylori* actively modulates the viscoelastic properties of its environment in order to move [6]. *H. pylori* lives in the human stomach and produces urease which leads to a drastic reduction of viscoelastic moduli, allowing the bacterium to swim freely.

In order to understand these, and related, effects, one must develop a theory for locomotion in complex fluids. The equations of motion governing the flow of most non-Newtonian fluids are nonlinear and hence classical Stokes flow methods involving the superposition of fundamental solutions are invalid. As a result, useful properties which constrain locomotion in a Newtonian fluid, such as the kinematic reversibility of the field equations, break down in viscoelastic fluids. The presence of time-dependent stresses, normal stress differences, and shear-dependent material functions in complex fluids are able to fundamentally alter the physics of locomotion [7, 8]. In this chapter we present a very general overview of the theoretical framework used to describe the effect of complex fluids on the locomotion of microorganisms. In Sect. 2 we elucidate the mathematical framework used to study locomotion in fluids and review well-established principles governing swimming in Newtonian fluids. In Sect. 3 complex constitutive relations are introduced and considered in this framework. Section 4 presents analytical results obtained for geometrically simple model swimmers in complex fluids, and comparisons with numerical simulation and theory are made. Finally we close this chapter by offering our perspective on the direction of research in this area in Sect. 5.

2 Locomotion in Fluids

Experience may furnish the reader with intuition on swimming in fluids but as illustrated in the previous chapter of this book (Chap. 7), locomotion in fluids is quite different for humans than it is for microorganisms. In this section we present a mathematical definition of locomotion in fluids, elucidate what it means to swim in a fluid if one is very small, and demonstrate consequences if the fluid is Newtonian.

2.1 Boundary Motion

In order to swim, a body undergoes (periodic) changes in its surface $S(t)$ (see Fig. 8.1). When in a fluid, this surface deformation leads to stresses exerted from the fluid on the body and, in general, motion. Periodic deformations may be described as deviations from a reference surface S_0. The position, $\mathbf{x}^S$, of a point on the surface of a swimmer, $S(t)$, may be decomposed as

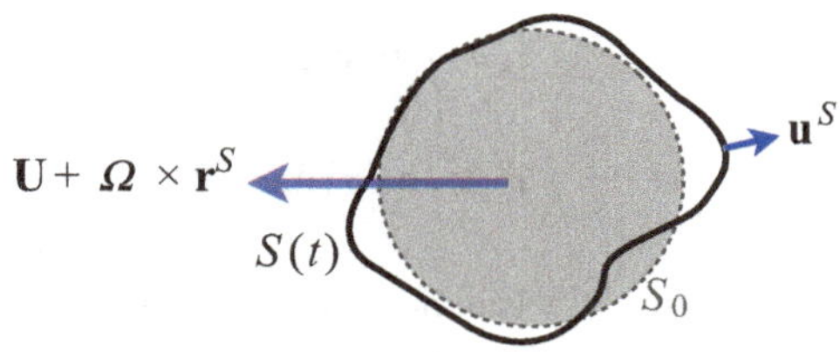

Fig. 8.1 Schematic representation of a general swimmer, adapted from Ref. [9]. A swimmer is defined as body with whose surface deforms in time thereby effecting an instantaneous rigid-body translation, **U**, and rotation, Ω

$$\mathbf{x}^S(t) = \mathbf{x}_0(t) + \mathbf{r}^S(t), \tag{8.1}$$

where $\mathbf{x}_0$ is a body-fixed position (the center of mass). The swimming gait of the body is defined using a body-fixed frame as follows

$$\mathbf{r}^S(t) = \mathbf{R}(t) \cdot \mathbf{r}(t), \tag{8.2}$$

where the rotation operator, **R**, orients the reference frame as

$$\frac{\mathrm{d}R_{ik}}{\mathrm{d}t} R_{jk} = -\varepsilon_{ijk}\Omega_k, \tag{8.3}$$

in which Ω is the angular velocity. Upon differentiation of the position of a point on the body we obtain the velocity

$$\frac{\mathrm{d}\mathbf{x}^S}{\mathrm{d}t} = \frac{\mathrm{d}\mathbf{x}_0}{\mathrm{d}t} + \frac{\mathrm{d}\mathbf{R}}{\mathrm{d}t} \cdot \mathbf{R}^\top \cdot \mathbf{R} \cdot \mathbf{r} + \mathbf{R} \cdot \frac{\mathrm{d}\mathbf{r}}{\mathrm{d}t} = \mathbf{U} + \Omega \times \mathbf{r}^S + \mathbf{u}^S. \tag{8.4}$$

Stresses imparted on the body by the fluid may lead to an instantaneous rigid-body translation, **U**, and/or rotation, Ω, due to Newton's second law. Mathematically, one can freely move between the lab frame and the body-fixed frame, the sole difference being that the rigid-body motion of the body is either reflected on the body or at infinity. In non-inertial frames, additional inertial forces have to also be considered in general.

If a typical time scale of the periodic deformation is ω^{-1}, for a body of size L, and density ρ_b, in a fluid of dynamic viscosity η and density ρ, the Stokes number, $\rho_b \omega L^2/\eta$, determines the magnitude of inertial forces versus viscous forces on the body. Similarly the ratio of gravitational body forces to viscous fluid forces is given by the Archimedes number, $\mathrm{Ar} = gL(\rho_b - \rho)/\eta\omega$. For microorganisms of sufficiently small length scales and which are close to density matched with the surrounding fluid, we can ignore the inertial and body force terms and thus the body is instantaneously force- and torque-free,

$$\mathbf{F} = \int_S \mathbf{n} \cdot \boldsymbol{\sigma} \, \mathrm{d}S = \mathbf{0}, \tag{8.5}$$

$$\mathbf{L} = \int_S \mathbf{r}^S \times (\mathbf{n} \cdot \boldsymbol{\sigma}) \, \mathrm{d}S = \mathbf{0}, \tag{8.6}$$

where the surface S is implicitly a function of time and the normal to the surface $\mathbf{n}$ points into the fluid. The stress tensor $\boldsymbol{\sigma}$ can be decomposed into an isotropic and deviatoric part $\boldsymbol{\sigma} = -p\mathbf{I} + \boldsymbol{\tau}$ (see Chap. 2 in this book for more details). We assume here the fluid is incompressible, $\nabla \cdot \mathbf{u} = 0$, and that the Reynolds number, $\mathrm{Re} = \rho\omega L^2/\eta$, is small and so any fluid parcel is in instantaneous mechanical equilibrium

$$\nabla \cdot \boldsymbol{\sigma} = \mathbf{0}. \tag{8.7}$$

In a Newtonian fluid the deviatoric stress is linearly proportional to the strain-rate tensor

$$\boldsymbol{\tau} = \eta \dot{\boldsymbol{\gamma}} = \eta (\nabla \mathbf{u} + \nabla \mathbf{u}^\top), \tag{8.8}$$

where $^\top$ is the transpose. Upon substitution of this constitutive equation into Eq. 8.7 we have, together with incompressibility, the Stokes equations

$$\eta \nabla^2 \mathbf{u} = \nabla p \tag{8.9}$$

$$\nabla \cdot \mathbf{u} = 0. \tag{8.10}$$

The linearity of the Stokes equations allows the superposition of solutions and so we can think of the problem of swimming as the sum of two, conceptually simpler problems. In the first, a body is held fixed and so the boundary motion is due only to the swimming gait $\mathbf{u}(\mathbf{x}^S) = \mathbf{u}^S$. Because $\mathbf{U} = \mathbf{0}$ and $\boldsymbol{\Omega} = \mathbf{0}$, there may arise a non-zero hydrodynamic force, $\mathbf{F}_1$, and torque, $\mathbf{L}_1$, on the body (indicated with a 1 subscript). In the second problem, conversely, an undeforming swimmer with the same instantaneous shape is subjected to rigid-body motion and so $\mathbf{u}(\mathbf{x}^S) = \mathbf{U} + \boldsymbol{\Omega} \times \mathbf{r}^S$. This rigid-body motion leads to drag forces (indicated by a 2 subscript). The force and the torque on the body may be written as a linear equation

$$\begin{pmatrix} \mathbf{F}_2 \\ \mathbf{L}_2 \end{pmatrix} = -\begin{pmatrix} \mathbf{R}_{FU} & \mathbf{R}_{F\Omega} \\ \mathbf{R}_{LU} & \mathbf{R}_{L\Omega} \end{pmatrix} \cdot \begin{pmatrix} \mathbf{U} \\ \boldsymbol{\Omega} \end{pmatrix}, \tag{8.11}$$

where the resistance tensors, $\mathbf{R}_{FU}$, $\mathbf{R}_{F\Omega}$, $\mathbf{R}_{LU}$, and $\mathbf{R}_{L\Omega}$ connect the kinematics to the force and torque. We write this relationship more compactly in terms of six-dimensional tensors as $\mathsf{F}_2 = -\mathsf{R} \cdot \mathsf{U}$. Naturally, since swimming is force- and torque-free, we must have $\mathsf{F}_1 + \mathsf{F}_2 = \mathsf{0}$, (essentially thrust balances drag) and so the rigid-body-motion arising during free swimming is simply

$$\boldsymbol{U} = \boldsymbol{R}^{-1} \cdot \boldsymbol{F}_1. \tag{8.12}$$

This simple form belies the fact that we still must solve for both the hydrodynamic forces $\boldsymbol{F}_1$ and the rigid-body resistance tensor $\boldsymbol{R}$ at each instant for a deforming body, a considerable task even in a Newtonian fluid.

The rate of work (power) expended by a swimming organism by deforming its surface in time in the fluid is instantaneously equal to the total energy dissipation rate in the fluid exterior to S,

$$P = \int_S -\mathbf{n} \cdot \boldsymbol{\sigma} \cdot \mathbf{u} \, \mathrm{d}S = \int_V \boldsymbol{\tau} : \nabla \mathbf{u} \, \mathrm{d}V. \tag{8.13}$$

The contraction $\mathbf{A} : \mathbf{B} \equiv A_{ij} B_{ji}$ in this chapter. If the fluid is Newtonian we may write

$$P = \frac{\eta}{2} \int_V \dot{\boldsymbol{\gamma}} : \dot{\boldsymbol{\gamma}} \, \mathrm{d}V. \tag{8.14}$$

2.2 The Lorentz Reciprocal Theorem

Solving for the motion of the swimmer is complicated by the fact that the boundary conditions are not fully prescribed but must satisfy the integral constraints of force- and torque-free motion for all times. Stone and Samuel showed in Ref. [10] that determining the rigid-body motion of the free swimmer, $\mathbf{U}$ and $\boldsymbol{\Omega}$, may be greatly simplified by appealing to the Lorentz reciprocal theorem [11].

We denote $\mathbf{u}$ and $\boldsymbol{\sigma}$ as the velocity field and its associated stress tensor for a force- and torque-free swimmer of surface S, while $\hat{\mathbf{u}}$ and $\hat{\boldsymbol{\sigma}}$ the velocity field and its associated stress tensor for a body of the same instantaneous shape subject to rigid-body translation and rotation with speeds $\hat{\mathbf{U}}$ and $\hat{\boldsymbol{\Omega}}$. Each fluid is in mechanical equilibrium $\nabla \cdot \boldsymbol{\sigma} = \nabla \cdot \hat{\boldsymbol{\sigma}} = \mathbf{0}$ so by the equality of virtual powers

$$\int_V \nabla \cdot \boldsymbol{\sigma} \cdot \hat{\mathbf{u}} \, \mathrm{d}V = \int_V \nabla \cdot \hat{\boldsymbol{\sigma}} \cdot \mathbf{u} \, \mathrm{d}V = 0, \tag{8.15}$$

where the volume of fluid V is external to the surface S with normal $\mathbf{n}$ into the fluid. Invoking the divergence theorem we obtain

$$\int_S \mathbf{n} \cdot \boldsymbol{\sigma} \cdot \hat{\mathbf{u}} \, \mathrm{d}S - \int_V \boldsymbol{\sigma} : \nabla \hat{\mathbf{u}} \, \mathrm{d}V = \int_S \mathbf{n} \cdot \hat{\boldsymbol{\sigma}} \cdot \mathbf{u} \, \mathrm{d}S - \int_V \hat{\boldsymbol{\sigma}} : \nabla \mathbf{u} \, \mathrm{d}V. \tag{8.16}$$

Because the swimmer is force- and torque-free, the first term on the left-hand side of Eq. (8.16) is zero and hence so is the second term on the left-hand side by construction,

$$\int_V \boldsymbol{\sigma} : \nabla \hat{\mathbf{u}} \, \mathrm{d}V = 0. \tag{8.17}$$

The first term on the right-hand side of Eq. (8.16) may be expanded by using the boundary motion on S from Eq. (8.4), meanwhile the stress tensor $\boldsymbol{\sigma} = -p\mathbf{I} + \boldsymbol{\tau}$, and so assuming the fluids to be incompressible

$$\hat{\mathbf{F}} \cdot \mathbf{U} + \hat{\mathbf{L}} \cdot \boldsymbol{\Omega} = -\int_S \mathbf{n} \cdot \hat{\boldsymbol{\sigma}} \cdot \mathbf{u}^S \, \mathrm{d}S + \int_V \hat{\boldsymbol{\tau}} : \nabla \mathbf{u} \, \mathrm{d}V. \tag{8.18}$$

Here $\hat{\mathbf{F}}$ and $\hat{\mathbf{L}}$ represent the force and torque resulting from the rigid-body motion of S. As we shall show, if both fluids are Newtonian, then the last term vanishes and one is left simply with the swimming kinematics $\mathbf{U}$ and $\boldsymbol{\Omega}$ as a function of known quantities $\mathbf{u}^S$ and the auxiliary rigid-body problem. For the purposes of this work we will always assume that the fluid in the rigid-body problem is Newtonian, a dramatic simplification that is without penalty. Taking $\hat{\boldsymbol{\tau}} = \hat{\eta}\hat{\dot{\boldsymbol{\gamma}}}$ we can write

$$\hat{\mathbf{F}} \cdot \mathbf{U} + \hat{\mathbf{L}} \cdot \boldsymbol{\Omega} = -\int_S \mathbf{n} \cdot \hat{\boldsymbol{\sigma}} \cdot \mathbf{u}^S \, \mathrm{d}S + \hat{\eta} \int_V \dot{\boldsymbol{\gamma}} : \nabla \hat{\mathbf{u}} \, \mathrm{d}V, \tag{8.19}$$

where we have used the identity $\hat{\dot{\boldsymbol{\gamma}}} : \nabla \mathbf{u} = \dot{\boldsymbol{\gamma}} : \nabla \hat{\mathbf{u}}$. Due to the linearity of the Stokes equations we may write $\hat{\mathbf{u}} = \hat{\mathsf{G}} \cdot \hat{\mathsf{U}}$, $\hat{\boldsymbol{\sigma}} = \hat{\mathsf{T}} \cdot \hat{\mathsf{U}}$ while $\hat{\mathsf{F}} = -\hat{\mathsf{R}} \cdot \hat{\mathsf{U}}$ (using six-dimensional tensors for compactness, and so $\hat{\mathsf{T}}$ is $[3 \times 3] \times 6$). The resistance tensor takes the form

$$\hat{\mathsf{R}} = -\int_S \begin{bmatrix} \mathbf{n} \cdot \hat{\mathsf{T}} \\ \mathbf{r} \times \left(\mathbf{n} \cdot \hat{\mathsf{T}}\right) \end{bmatrix} \mathrm{d}S. \tag{8.20}$$

Substituting into the reciprocal theorem yields

$$-\mathsf{U} \cdot \hat{\mathsf{R}} \cdot \hat{\mathsf{U}} = -\int_S \mathbf{u}^S \mathbf{n} : \hat{\mathsf{T}} \cdot \hat{\mathsf{U}} \, \mathrm{d}S + \hat{\eta} \int_V \dot{\boldsymbol{\gamma}} : \nabla \hat{\mathsf{G}} \cdot \hat{\mathsf{U}} \, \mathrm{d}V. \tag{8.21}$$

where $\mathbf{u}^S \mathbf{n} : \hat{\mathsf{T}} \cdot \hat{\mathsf{U}} = u_j^S n_i \mathsf{T}_{ijk} \mathsf{U}_k$. Discarding the arbitrary vector $\hat{\mathsf{U}}$ and using the symmetry of the resistance tensor $\hat{\mathsf{R}}$ [12], we finally arrive at a general integral theorem for swimming

$$\mathsf{U} = \hat{\mathsf{R}}^{-1} \cdot \left[\int_S \mathbf{u}^S \mathbf{n} : \hat{\mathsf{T}} \, \mathrm{d}S - \hat{\eta} \int_V \dot{\boldsymbol{\gamma}} : \nabla \hat{\mathsf{G}} \, \mathrm{d}V \right]. \tag{8.22}$$

The volume integral in Eq. (8.22) only contributes if the fluid in the swimming problem is *not* Newtonian and hence is a measure of the modification of the swimming dynamics due to the presence of non-Newtonian stresses.

There remain two difficulties however: the first is that the surface $S(t)$ is changing and so the auxiliary rigid-body problem must be solved for all possible shapes; the second is the unknown integrand in the volume integral. We show below that both difficulties may be tackled by taking a perturbative approach.

Note that the reciprocal theorem above may be extended to N swimmers by taking $S = \bigcup_\alpha S^\alpha$ which gives

$$\boldsymbol{U} = \hat{\boldsymbol{R}}^{-1} \cdot \left[\left(\sum_\alpha \int_{S^\alpha} \mathbf{u}^S \mathbf{n} : \hat{\boldsymbol{T}} \, \mathrm{d}S \right) - \hat{\eta} \int_V \dot{\boldsymbol{\gamma}} : \nabla \hat{\boldsymbol{G}} \, \mathrm{d}V \right], \tag{8.23}$$

except now the tensors are $6N$ in size where N is the number of bodies.

2.3 *Swimming in Newtonian Fluids*

When swimming in a Newtonian fluid, Eq. (8.17) dictates that

$$\int_V \dot{\boldsymbol{\gamma}} : \nabla \hat{\mathbf{u}} \, \mathrm{d}V = 0, \tag{8.24}$$

and so the velocity of the swimmer, as given by Eq. (8.22), reduces simply to

$$\boldsymbol{U} = \hat{\boldsymbol{R}}^{-1} \cdot \left[\int_S \mathbf{u}^S \mathbf{n} : \hat{\boldsymbol{T}} \, \mathrm{d}S \right]. \tag{8.25}$$

This is a marked simplification because now the swimming motion, $\boldsymbol{U} = [\mathbf{U} \ \boldsymbol{\Omega}]^\top$, can be resolved without the knowledge of the entire flow field $\mathbf{u}$, rather one must simply ascertain the solution of the auxiliary field $\hat{\mathbf{u}}$ and its associated stress tensor. Swimming in a Newtonian fluid is therefore reduced to solving, pointwise, the stress on a body of the same shape undergoing instantaneous rigid-body translation. We observe that the term in the brackets is simply the hydrodynamic force and torque exerted by the fluid on the swimmer if it was held instantaneously in place (see Eq. (8.12) for comparison). In practice one must solve the auxiliary problem of rigid-body translation and rotation in a Newtonian fluid for each shape of $S(t)$ for all time t at a resolution desired for integrating the instantaneous velocity. Note that because of the relationship between $\hat{\boldsymbol{R}}$ and $\hat{\boldsymbol{T}}$, Eq. (8.25) implies that the locomotion of the body is independent of the viscosity of the fluid. As an example, if the body is a sphere of radius a, then the rigid-body problem is well known, the resistance matrix is diagonal and hence easily invertible, $\hat{\mathbf{R}}_{FU} = 6\pi\eta a\mathbf{I}$, $\hat{\mathbf{R}}_{L\Omega} = 8\pi\eta a^3\mathbf{I}$, $\hat{\mathbf{R}}_{F\Omega} = \mathbf{0}$, and $\hat{\mathbf{R}}_{LU} = \mathbf{0}$, while $\mathbf{n} \cdot \hat{\boldsymbol{T}} = -\frac{3\eta}{2a}[\mathbf{I} \ 2\Theta]$ where $\Theta_{ij} = \varepsilon_{ijk} r_k$. With these the swimming speed for a sphere with only tangential deformations is given by

$$\boldsymbol{U} = -\frac{1}{4\pi a^2} \int_S \begin{bmatrix} \mathbf{I} \\ \frac{3}{2a^2}\Theta^\top \end{bmatrix} \cdot \mathbf{u}^S \, \mathrm{d}S, \tag{8.26}$$

as shown by Stone and Samuel [10].

Swimming gaits are typically periodic in time (with period T), and often we are only interested in the steady swimming speed given by a time-average of Eq. (8.25)

$$\langle \boldsymbol{U} \rangle = \left\langle \hat{\boldsymbol{R}}^{-1} \cdot \left[\int_S \mathbf{u}^S \mathbf{n} : \hat{\boldsymbol{T}} \, \mathrm{d}S \right] \right\rangle, \tag{8.27}$$

where $\langle \boldsymbol{U} \rangle \equiv T^{-1} \int_T \boldsymbol{U} \, \mathrm{d}t$.

2.3.1 The Scallop Theorem

Because the Stokes equations are linear and independent of time, time only enters the problem as a parameter in the boundary conditions and the locomotion is instantaneously linear in $\mathbf{u}^S$ (Eq. 8.25). This has two profound implications for locomotion in Newtonian fluids. The first is that the rate of actuation of the boundaries is irrelevant to the distance traveled over a period—in other words it does not matter if the actuation rate is fast or slow over a period [7]. This is because re-parameterizing time $t' = f(t)$ changes the velocity $\mathbf{u}^S \to \dot{f}\mathbf{u}'^S$ but also the interval $dt \to \dot{f}^{-1} dt'$ to no net effect. This also implies that if a swimmer reverses its sequence of shapes after a period it goes back to where it started, regardless of the rate of motion. Finally if a swimmer goes through the same sequence of shapes whether forwards and backwards in time (time-reversible) then it will have no net motion. More precisely stated, if the deformation of the swimmer shape over a period, $t_2 - t_0$, is such that there exists a $t_1 \in (t_0, t_2)$ where the sequence of shapes after t_1 is exactly reversed, then the net motion is zero [8]. We can show this by invoking Eq. (8.27) as follows:

$$\begin{aligned} \langle \boldsymbol{U} \rangle &= \frac{1}{t_2 - t_0} \int_{t_0}^{t_2} \int_{S(t)} \mathbf{u}^S \mathbf{n} : \hat{\boldsymbol{T}} \cdot \hat{\boldsymbol{R}}^{-1} \, \mathrm{d}S \, \mathrm{d}t, \qquad (8.28) \\ &= \frac{1}{t_2 - t_0} \left[\int_{t_0}^{t_1} \int_{S(t)} \mathbf{u}^S \mathbf{n} : \hat{\boldsymbol{T}} \cdot \hat{\boldsymbol{R}}^{-1} \, \mathrm{d}S \, \mathrm{d}t + \int_{t_1}^{t_2} \int_{S(t')} \mathbf{u}'^S \mathbf{n} : \hat{\boldsymbol{T}} \cdot \hat{\boldsymbol{R}}^{-1} \, \mathrm{d}S \, \mathrm{d}t' \right], \\ &= \frac{1}{t_2 - t_0} \left[\int_{t_0}^{t_1} \int_{S(t)} \mathbf{u}^S \mathbf{n} : \hat{\boldsymbol{T}} \cdot \hat{\boldsymbol{R}}^{-1} \, \mathrm{d}S \, \mathrm{d}t + \int_{t_1}^{t_0} \int_{S(t)} \mathbf{u}^S \mathbf{n} : \hat{\boldsymbol{T}} \cdot \hat{\boldsymbol{R}}^{-1} \dot{f}^{-1} \, \mathrm{d}S \dot{f} \, \mathrm{d}t \right], \\ &= \mathbf{0}. \end{aligned}$$

This statement is often called the *scallop theorem* because the physical actuation of a scallop was used by Purcell to first elucidate this principle [7] (for a detailed mathematical treatment see [13]). The scallop is an example of a swimmer with only one degree of freedom, namely its hinge. In general, if a body has only one degree of freedom it can only execute time-reversible motion if it is to perform a cyclical gait and thus no such body can swim in a Newtonian fluid in the absence of inertia [9].

This impediment means often that the simplest designs of swimmers in the inertial realm, for example, objects with a rigid flapping tail, cannot locomote at small scales in Newtonian fluids. We will see below that in non-Newtonian fluids, the mathematical details which lead to the scallop theorem, namely linearity and an independence of time, are no longer present and time-reversible swimmers can generally self-propel.

2.4 *Small-Amplitude Motion*

When a swimming motion results from small deformations of a body from a reference surface S_0, as is the case for ciliated organisms [14], one can describe this motion theoretically by writing

$$\mathbf{r}^S = \mathbf{R} \cdot \left[\mathbf{r}_0 + \sum_m \varepsilon^m \mathbf{r}_m(\mathbf{r}_0, t) \right], \tag{8.29}$$

where $\varepsilon \ll 1$ is a dimensionless measure of the gait amplitude. The position of a point on the reference surface $\mathbf{x}^{S_0} = \mathbf{x}_0 + \mathbf{R} \cdot \mathbf{r}_0$ and $\mathbf{x}^S - \mathbf{x}^{S_0} = \varepsilon \mathbf{R} \cdot \mathbf{r}_1 + \mathcal{O}(\varepsilon^2)$. Such a gait presents several mathematical advantages for calculating the swimming kinematics. In general, in order to use the reciprocal formulation, one must know the solution to the auxiliary problem $\hat{\boldsymbol{\sigma}}$ for all $S(t)$, which is typically impractical for a nontrivial gait. When $S(t)$ deviates only slightly from S_0 we can, through Taylor series expansions, recast the problem onto S_0 [15,16]. As the shape of S_0 is invariant in time we then need only the resolution of a single auxiliary problem. Furthermore, as discussed below, by posing the problem as a perturbation expansion in ε we are able to tackle the constitutive relationships in a systematic fashion.

2.4.1 Recasting the Problem onto S_0

The swimming gait, $\mathbf{u}^S(\mathbf{x}^S)$, represents motion of the material points, $\mathbf{x}^S$, on the swimmer surface. To represent the boundary condition on a surface, S_0, which is not material, we Taylor expand the flow field $\mathbf{u}$,

$$\mathbf{u}(\mathbf{x}^S) = \mathbf{u}(\mathbf{x}^{S_0}) + (\mathbf{x}^S - \mathbf{x}^{S_0}) \cdot \nabla \mathbf{u}|_{\mathbf{x}_0^S} + \mathcal{O}(\varepsilon^2), \tag{8.30}$$

$$\Rightarrow \mathbf{u}(\mathbf{x}^{S_0}) = \mathbf{U} + \boldsymbol{\Omega} \times \mathbf{R} \cdot (\mathbf{r}_0 + \varepsilon \mathbf{r}_1) + \mathbf{u}^S - \varepsilon (\mathbf{R} \cdot \mathbf{r}^S) \cdot \nabla \mathbf{u}|_{\mathbf{x}^{S_0}} + \mathcal{O}(\varepsilon^2), \tag{8.31}$$

where we have used $\mathbf{u}(\mathbf{x}^S) = \mathbf{U} + \boldsymbol{\Omega} \times \mathbf{r}^S + \mathbf{u}^S$ and where $|_{\mathbf{x}^{S_0}}$ means that derivatives are evaluated at $\mathbf{x}^{S_0}$. The first two terms in Eq. (8.31) represent rigid-body rotation of the undeforming surface S_0, and the remaining terms are the boundary condition that one must impose on S_0 to obtain the appropriate solution on S. We refer to these terms as $\mathbf{u}^{S_0}$, the swimming gait defined on S_0, and so write

$$\mathbf{u}(\mathbf{x}^{S_0}, t) = \mathbf{U} + \boldsymbol{\Omega} \times \mathbf{r}^{S_0} + \mathbf{u}^{S_0}, \tag{8.32}$$

where $\mathbf{r}^{S_0} = \mathbf{R} \cdot \mathbf{r}_0$ and

$$\mathbf{u}^{S_0} = \varepsilon \left\{ \mathbf{R} \cdot \frac{\mathrm{d}\mathbf{r}_1}{\mathrm{d}t} + \Omega \times (\mathbf{R} \cdot \mathbf{r}_1) - (\mathbf{R} \cdot \mathbf{r}_1) \cdot \nabla \mathbf{u}\big|_{\mathbf{x}^{S_0}} \right\} + \mathscr{O}(\varepsilon^2). \tag{8.33}$$

This formulation leads to a swimming problem defined entirely on S_0 which satisfies the correct boundary conditions on S.

In order to solve for the swimming kinematics, we then apply the reciprocal theorem on S_0 (namely assuming a rigid-body translation and rotation of S_0 for the auxiliary field). The motion of a point on S_0 in the swimming problem is given by $\mathbf{u}(\mathbf{x}^{S_0}) = \mathbf{U} + \Omega \times \mathbf{r}^{S_0} + \mathbf{u}^{S_0}$ while for rigid-body motion $\hat{\mathbf{u}}(\mathbf{x}^{S_0}) = \hat{\mathbf{U}} + \hat{\Omega} \times \mathbf{r}^{S_0}$. In the swimming problem the force and torque on S are both zero and because the fluid stress in the volume between S_0 and S is divergence free, $\nabla \cdot \boldsymbol{\sigma} = \mathbf{0}$, the total force and torque on S_0 must also be zero. Applying the integral theorem for swimming, Eq. (8.22), on the surface S_0 we get

$$\mathsf{U} = \hat{\mathsf{R}}^{-1} \cdot \left[\int_{S_0} \mathbf{u}^{S_0}\mathbf{n} : \hat{\mathsf{T}}\, \mathrm{d}S - \hat{\eta} \int_{V_0} \dot{\boldsymbol{\gamma}} : \nabla \hat{\mathsf{G}}\, \mathrm{d}V \right]. \tag{8.34}$$

The great benefit of this approach is that one need only resolve the rigid-body translation for a single shape, that of the undeforming surface S_0. Note that we do not get something for nothing. In particular the swimming gait on S_0, $\mathbf{u}^{S_0}$, depends on gradients of the (unknown) flow field $\mathbf{u}$ and the rotation rate of the swimmer Ω. However, because these terms are $\mathscr{O}(\varepsilon)$ upon a perturbation expansion they will vanish to leading order.

3 Locomotion in Non-Newtonian Fluids

The motion of a swimming body in a viscous fluid is characterized, in general, by the integral relationship given by Eq. (8.22). In a non-Newtonian fluid, the expression for the deviatoric stress, $\boldsymbol{\tau}$, is not linear in the strain-rate tensor, $\dot{\boldsymbol{\gamma}}$, and hence the term in brackets on the right-hand side of Eq. (8.22) will not vanish even as we continue to take the auxiliary fluid to be Newtonian $\hat{\boldsymbol{\tau}} = \eta \hat{\dot{\boldsymbol{\gamma}}}$. Let us assume for the moment, for illustrative purposes, that the stress tensor may be decomposed into a Newtonian contribution and an additional non-Newtonian part, $\boldsymbol{\tau} = \eta \dot{\boldsymbol{\gamma}} + \mathbf{A}(\mathbf{x}, t)$. With this form of constitutive equation, Eq. (8.17) yields

$$\int_V \dot{\boldsymbol{\gamma}} : \nabla \hat{\mathbf{u}}\, \mathrm{d}V = -\frac{1}{\eta} \int_V \mathbf{A} : \nabla \hat{\mathbf{u}}\, \mathrm{d}V \tag{8.35}$$

and so substitution into Eq. (8.22) leads to

$$U = \hat{R}^{-1} \cdot \left[\int_S \mathbf{u}^S \mathbf{n} : \hat{T} \, \mathrm{d}S + \frac{\hat{\eta}}{\eta} \int_V \mathbf{A} : \nabla \hat{G} \, \mathrm{d}V \right], \tag{8.36}$$

or substitution into (8.34), for the problem recast onto S_0, yields

$$U = \hat{R}^{-1} \cdot \left[\int_{S_0} \mathbf{u}^{S_0} \mathbf{n} : \hat{T} \, \mathrm{d}S + \frac{\hat{\eta}}{\eta} \int_{V_0} \mathbf{A} : \nabla \hat{G} \, \mathrm{d}V \right]. \tag{8.37}$$

Swimming in a non-Newtonian fluid involves the solution of a second integral which is, in general, a function of the complex flow field and may depend on the history of the deformations $S(t)$. For example, if the sequence of shapes is time-reversible, then there may still exist net locomotion due to that non-Newtonian integral, a breakdown of the scallop theorem [17].

The dissipation due to swimming would likewise be modified by nonlinearities as

$$P = \frac{\eta}{2} \int_V \dot{\boldsymbol{\gamma}} : \dot{\boldsymbol{\gamma}} \, \mathrm{d}V + \int_V \mathbf{A} : \nabla \mathbf{u} \, \mathrm{d}V. \tag{8.38}$$

In order to derive a more precise statement about swimming in non-Newtonian fluids we have to specify a constitutive relationship that gives rise to non-Newtonian behavior. For modeling purposes we assume here that the deviatoric stress tensor can be decomposed into a sum of relaxation modes $j, \boldsymbol{\tau} = \sum \boldsymbol{\tau}^{(j)}$. We write the relationship between each stress mode and the velocity field very generally as

$$\mathscr{A}_j \boldsymbol{\tau}^{(j)} = \eta_j \mathscr{B}_j \dot{\boldsymbol{\gamma}} + \mathbf{N}_j(\mathbf{u}, \boldsymbol{\tau}^{(j)}), \tag{8.39}$$

where η_j is the zero-shear-rate viscosity for the j-th mode, $\mathscr{A}_j$ and $\mathscr{B}_j$ are linear operators in time, and $\mathbf{N}_j$ is a symmetric tensor which depends nonlinearly on the velocity and stress and represents the transport and stretching of the polymeric microstructure by the flow. This general constitutive relationship includes, in particular, all classical models of polymeric fluids [9].

From this point forward it can be very difficult to make analytical progress in large part because of the presence of the nonlinear operators $\mathbf{N}_j$. One way to make progress in light of this difficulty is to consider the constitutive relationship perturbatively. In this manner at each order, the nonlinear terms are functions of the previous order solutions only. We detail this approach below.

3.1 Small-Amplitude Perturbations

As shown in Sect. 2.4, when the deformation of the body is small, $\varepsilon \ll 1$, one can recast the problem onto a body whose shape is not deforming, S_0 (recall that ε is a dimensionless measure of the amplitude deformation of the surface $S(t)$).

By employing perturbation expansions, the nonlinear non-Newtonian constitutive equations are linearized order-by-order facilitating analytical solution.

Expanding all fields formally in a regular perturbation series, e.g., $\mathbf{u} = \sum_m \varepsilon^m \mathbf{u}_m$, the boundary condition, Eq. (8.32), becomes at each order m

$$\mathbf{u}_m(\mathbf{x}_0^S) = \mathbf{U}_m + \boldsymbol{\Omega}_m \times \mathbf{r}^{S_0} + \mathbf{u}_m^{S_0}, \tag{8.40}$$

where

$$\mathbf{u}_1^{S_0} = \mathbf{u}_1^S, \tag{8.41}$$

$$\mathbf{u}_2^{S_0} = \mathbf{u}_2^S + \boldsymbol{\Omega}_1 \times (\mathbf{R} \cdot \mathbf{r}_1) - (\mathbf{R} \cdot \mathbf{r}_1) \cdot \nabla \mathbf{u}_1 |_{\mathbf{x}^{S_0}} . \tag{8.42}$$

The constitutive relation at each order m becomes

$$\mathscr{A}_j \boldsymbol{\tau}_m^{(j)} = \eta_j \mathscr{B}_j \dot{\boldsymbol{\gamma}}_m + \mathbf{N}_m^{(j)}[\mathbf{u}_1, \ldots, \mathbf{u}_{m-1}], \tag{8.43}$$

where $\mathbf{N}$ is a functional of previous order solutions, for example, $\mathbf{N}_1^{(j)} = \mathbf{0}$ and $\mathbf{N}_2^{(j)} \equiv \mathbf{N}_2^{(j)}[\mathbf{u}_1]$.

3.1.1 Fourier Series

Given a swimmer with a time-periodic swimming gait, it is a reasonable assumption to write the flow and stress fields as time periodic and expand them in a Fourier series in time as

$$\mathbf{u} = \sum_n \mathbf{u}^{(n)} e^{in\omega t}. \tag{8.44}$$

As we shall describe below, this assumption means we neglect the influence of a particular initial stress state in the fluid, but is suitable for determining the steady swimming speed of a microorganism, and all harmonic oscillations around it. The constitutive relationship for each Fourier mode, n, is

$$\begin{aligned} \boldsymbol{\tau}^{(j,n)} &= \eta_j \frac{B_j(n)}{A_j(n)} \dot{\boldsymbol{\gamma}}^{(n)} + \frac{1}{1+A_j(n)} \mathbf{N}^{(j,n)}, \\ &= \eta_j^*(n) \dot{\boldsymbol{\gamma}}^{(n)} + \mathbf{A}^{(j,n)}, \end{aligned} \tag{8.45}$$

where $A_j(n)$ and $B_j(n)$ are the characteristic polynomials of the differential operators (i.e., $e^{-in\omega t}\mathscr{A}_j[e^{in\omega t}]$) while $A_j(0) = B_j(0) = 1$. Summing over all relaxation modes, j, we then have

$$\boldsymbol{\tau}^{(n)} = \eta^*(n) \dot{\boldsymbol{\gamma}}^{(n)} + \mathbf{A}^{(n)}. \tag{8.46}$$

For each Fourier mode we thus have a linear response with complex viscosity, $\eta^*(n)$, and a nonlinear term which depends on the solutions at previous orders.

We may now decompose the boundary conditions into Fourier modes and solve for the flow field order-by-order in ε using the aforementioned small-amplitude expansion about the static surface S_0. Upon substitution of (8.46) into the reciprocal relationship for swimming, Eq. (8.37), we obtain, order-by-order,

$$\boldsymbol{U}_m^{(n)} = \hat{\boldsymbol{R}}^{-1} \cdot \left[\int_{S_0} \mathbf{u}_m^{S_0,(n)} \mathbf{n} : \hat{\boldsymbol{T}} \, \mathrm{d}S + \frac{\hat{\eta}}{\eta^*(n)} \int_{V_0} \mathbf{A}_m^{(n)} : \nabla \hat{\boldsymbol{G}} \, \mathrm{d}V \right], \tag{8.47}$$

while the mean swimming speed is given by the zeroth Fourier mode, $\langle \mathbf{U} \rangle = \mathbf{U}^{(0)}$, and thus satisfies

$$\langle \boldsymbol{U}_m \rangle = \hat{\boldsymbol{R}}^{-1} \cdot \left[\int_{S_0} \langle \mathbf{u}_m^{S_0} \rangle \, \mathbf{n} : \hat{\boldsymbol{T}} \, \mathrm{d}S + \frac{\hat{\eta}}{\eta^*(n)} \int_{V_0} \langle \mathbf{A}_m \rangle : \nabla \hat{\boldsymbol{G}} \, \mathrm{d}V \right], \tag{8.48}$$

where we have used the notation that $\eta_0 = \eta^*(0)$ is the zero-shear-rate viscosity. The values of both $\hat{\eta}$ and η_0 do not affect $\boldsymbol{U}$ and so we may set $\hat{\eta} = \eta_0$ for the sake of convenience.

At leading order, $m = 1$, $\mathbf{A}_1 = \mathbf{0}$ and so the swimming speed, as shown by Eq. (8.47), is the Newtonian one for all times. This is expected because we simply have a set of Stokes equations for each Fourier mode

$$\nabla p_1^{(n)} = \eta^*(n) \nabla^2 \mathbf{u}_1^{(n)}. \tag{8.49}$$

The velocity field is independent of the viscosity and identical to the Newtonian solution. Furthermore, because the kinematics of the swimmer are periodic with zero mean, by Eq. (8.48) the swimmer will have zero mean translation or rotation to leading order. Using the integral equation to determine the kinematics of the body, $\mathbf{U}_1^{(n)}$ and $\Omega_1^{(n)}$, we may then proceed to solve the Stokes equations and obtain the entire flow field, $\mathbf{u}_1$.

At second order, $m = 2$, one need not solve for the full velocity field, $\mathbf{u}_2$, in order to find the mean swimming velocity $\langle \mathbf{U}_2 \rangle$. One must simply compute the mean of the nonlinear tensor $\mathbf{A}_2^{(0)}[\mathbf{u}_1]$ and the mean of the gait at second order,

$$\left\langle \mathbf{u}_2^{S_0} \right\rangle = \left\langle \mathbf{u}_2^S + \Omega_1 \times \mathbf{x}_1^S(\mathbf{x}_0^S, t) - \mathbf{x}_1^S(\mathbf{x}_0^S, t) \cdot \nabla \, \mathbf{u}_1 |_{\mathbf{x}^{S_0}} \right\rangle, \tag{8.50}$$

both of which depend only on the first order solution, $\mathbf{u}_1$.

Now consider a swimmer with a time-reversible gait. Such a swimmer will have zero net motion in a Newtonian fluid as demonstrated in Sect. 2.3.1, and hence its velocity in a complex fluid will be entirely determined by non-Newtonian stresses. Consider for example a sphere with tangential surface motion only, a model known as a squirmer. The shape is not deforming, and if the gait is time-reversible then $\left\langle \mathbf{u}^{S_0} \right\rangle = \mathbf{0}$. The leading-order swimming speed is given by the integral

$$\langle U \rangle = \varepsilon^2 \frac{\hat{\eta}}{\eta_0} \int_{V_0} \langle \mathbf{A}_2 \rangle : \nabla \hat{G} \cdot \hat{R}^{-1} \, \mathrm{d}V + \mathscr{O}(\varepsilon^4), \tag{8.51}$$

where for a sphere

$$\hat{G} \cdot \hat{R}^{-1} = \frac{1}{8\pi\eta} \left[\left(1 + \tfrac{a^2}{6} \nabla^2 \right) \mathbf{G} \;\; \tfrac{1}{|\mathbf{x}|^3} \Theta \right], \tag{8.52}$$

and $\mathbf{G} = \frac{1}{|\mathbf{x}|} \left(\mathbf{I} + \frac{\mathbf{xx}}{|\mathbf{x}|^2} \right)$ is the Oseen tensor. We see that there is an $\mathscr{O}(\varepsilon^2)$, strictly non-Newtonian swimming speed which arises from the nonlinear terms in a given constitutive relationship, $\mathbf{A}_2$. Because we chose a gait which does not locomote in a Newtonian fluid, any net motion is then a measure of the non-Newtonian rheology of the fluid. For a generic gait which achieves locomotion in a Newtonian fluid, there is instead a non-Newtonian correction to the swimming speed at quadratic order in amplitude.

3.1.2 Linear Viscoelasticity

Here we address an important point on locomotion in linearly viscoelastic fluids. As we have seen, the swimming speed of a microorganism is Newtonian at linear order in amplitude while non-Newtonian effects do not appear until quadratic order. If the fluid is linearly viscoelastic then $\mathbf{A}_m = \mathbf{0}$ at all orders and hence the fluid yields no change in swimming speed from that of a Newtonian fluid for prescribed kinematics, as found in Ref. [18].

It is also typical for organisms executing time-periodic gaits to exhibit $\varepsilon \to -\varepsilon$ symmetry. In such a case the Newtonian swimming speed itself is at least quadratic in amplitude and hence it is desirable to keep nonlinearities in the constitutive relation which would emerge at the same order as the swimming speed itself [19].

3.1.3 Transients

Specifying periodicity in time, as done above, neglects any transient regime in which stresses and velocities evolve from an initial condition, $\{\boldsymbol{\tau}(0), \mathbf{u}(0)\}$, to a periodic steady state. This simplification is desirable when we are concerned with determining the time-averaged steady-state swimming speed of an organism. Swimming organisms are however intrinsically unsteady organisms in the sense that while the swimming gait may be periodic over a short period, the swimmers may stop and start or change direction, as exemplified by bacteria executing *run-and-tumble* motion [20]. Here we examine briefly the effects on locomotion of an arbitrary initial stress state. We note that we will still assume that the swimmer is instantaneously force-free, in other words that the inertial time scale is still much smaller than the relevant relaxation time scale.

Taking the Laplace transform of the constitutive relation, Eq. (8.39), we obtain

$$\widetilde{\mathscr{A}_j \boldsymbol{\tau}_m^{(j)}} = \eta_j \widetilde{\mathscr{B}_j \dot{\boldsymbol{\gamma}}_m} + \widetilde{\mathbf{N}_m^{(j)}}, \tag{8.53}$$

where the tilde indicates a unilateral Laplace transform $\tilde{f}(s) = \int_0^\infty f(t)e^{-st}\,\mathrm{d}t$. Rearranging and summing over all relaxation modes, j, we may write

$$\tilde{\boldsymbol{\tau}}_m = \eta^*(s)\widetilde{\dot{\boldsymbol{\gamma}}}_m + \tilde{\mathbf{A}}_m + \tilde{\mathbf{B}}_m, \tag{8.54}$$

where

$$\eta^*(s) = \sum_j \eta_j \frac{B_j(s)}{A_j(s)}, \tag{8.55}$$

$$\tilde{\mathbf{A}}_m = \sum_j \frac{1}{A_j(s)} \widetilde{\mathbf{N}_m^{(j)}}. \tag{8.56}$$

The tensor $\tilde{\mathbf{B}}$ represents the effect of the initial condition on the stress. For example, if our fluid of interest is a single-mode Boger fluid [21], then for the Oldroyd-B equations have $\mathscr{A} = 1 + \lambda_1 \partial_t$ while $\mathscr{B} = 1 + \lambda_2 \partial_t$, where λ_1 is the relaxation time and λ_2 is the retardation time, and hence

$$\eta^*(s) = \eta_0 \frac{1+\lambda_2 s}{1+\lambda_1 s}, \tag{8.57}$$

$$\tilde{\mathbf{A}}_m = \frac{1}{1+\lambda_1 s}\widetilde{\mathbf{N}_m}, \tag{8.58}$$

$$\tilde{\mathbf{B}}_m = \frac{\lambda_1}{1+\lambda_1 s}\left[\boldsymbol{\tau}_m(t=0) - \eta_0(\lambda_2/\lambda_1)\dot{\boldsymbol{\gamma}}_m(t=0)\right]. \tag{8.59}$$

In order to understand the effect the constitutive equation (8.54) has on locomotion we appeal to the reciprocal theorem for swimming,

$$\boldsymbol{U}_m = \hat{\boldsymbol{R}}^{-1} \cdot \left[\int_{S_0} \mathbf{u}_m^{S_0}\mathbf{n} : \hat{\boldsymbol{T}}\,\mathrm{d}S + \hat{\eta}\int_{V_0} \mathscr{L}^{-1}\left(\frac{\tilde{\mathbf{A}}_m + \tilde{\mathbf{B}}_m}{\eta^*(s)}\right) : \nabla\hat{\boldsymbol{G}}\,\mathrm{d}V\right], \tag{8.60}$$

where $\mathscr{L}^{-1}$ indicates an inverse Laplace transform. Again we see that we need not know the solution for the flow field at order m to determine the swimming kinematics at that order. The nonlinear term, $\mathbf{A}_m$, is a function of the flow field at previous orders while the tensor $\mathbf{B}_m$ consists entirely of the initial conditions.

At leading order the nonlinear contribution vanishes, $\mathbf{A}_1 = \mathbf{0}$. In an Oldroyd-B fluid the contribution at leading order may thus be written as

$$U_1 = \hat{R}^{-1} \cdot \left[\int_{S_0} \mathbf{u}_1^{S_0} \mathbf{n} : \hat{\boldsymbol{T}} \, \mathrm{d}S + \frac{\hat{\eta}}{\eta_0} e^{-t/\lambda_2} \int_{V_0} \left(\frac{\lambda_1}{\lambda_2} \boldsymbol{\tau}_1(0) - \eta_0 \dot{\boldsymbol{\gamma}}_1(0) \right) : \nabla \hat{\boldsymbol{G}} \, \mathrm{d}V \right]. \tag{8.61}$$

We see that the influence of the initial condition decays exponentially on the retardation time scale of the fluid. That time scale needs then to be compared with the other relevant time scales in each specific swimming problem.

3.2 *Slowly Varying Flows*

We have demonstrated the utility of small-amplitude deformations as a method to probe the nonlinear effects of a particular non-Newtonian fluid in the context of small-scale locomotion. A particular benefit of such an approach is that the actuation time scales of the microorganisms can be arbitrary in comparison to the relaxation time scales of the non-Newtonian medium in which they are moving [9].

Conversely, one may be interested in gaits which do not display small-amplitude motions but instead lead to slowly varying flows. In such cases, one can resort to the use of the second-order fluid model which describes the first non-Newtonian behavior in an expansion of stress in strain rate. In the weakly nonlinear regime the deviatoric stress for almost all complex fluids can be represented by

$$\boldsymbol{\tau} = \eta \dot{\boldsymbol{\gamma}} - \frac{1}{2} \Psi_1 \overset{\triangledown}{\dot{\boldsymbol{\gamma}}} + \Psi_2 \dot{\boldsymbol{\gamma}} \cdot \dot{\boldsymbol{\gamma}}, \tag{8.62}$$

where Ψ_1 and Ψ_2 are the first and second normal stress coefficients [22, 23], and so

$$\mathbf{A} = -\frac{1}{2} \Psi_1 \overset{\triangledown}{\dot{\boldsymbol{\gamma}}} + \Psi_2 \dot{\boldsymbol{\gamma}} \cdot \dot{\boldsymbol{\gamma}}. \tag{8.63}$$

If we scale strain rates as $\dot{\boldsymbol{\gamma}} = \omega \dot{\boldsymbol{\gamma}}'$ and stresses as $\boldsymbol{\tau} = \eta \omega \tau'$, where ω is the characteristic actuation frequency of the body, then we have in dimensionless form (primes indicate dimensionless quantities)

$$\boldsymbol{\tau}' = \dot{\boldsymbol{\gamma}}' - \mathrm{De} \left(\overset{\triangledown}{\dot{\boldsymbol{\gamma}}'} + B \dot{\boldsymbol{\gamma}}' \cdot \dot{\boldsymbol{\gamma}}' \right). \tag{8.64}$$

The Deborah number, $\mathrm{De} = \omega \Psi_1 / 2\eta$, is the ratio of the relaxation time scale of the fluid compared to the time scale of actuation and $B = -2\Psi_2/\Psi_1 \geq 0$.

If we now assume a regular perturbation expansion of the flow field in Deborah number, $\mathbf{u}' = \mathbf{u}_0' + \mathrm{De}\,\mathbf{u}_1' + \ldots$, then at leading order we have simply a Newtonian fluid

$$\boldsymbol{\tau}_0' = \dot{\boldsymbol{\gamma}}_0{}', \tag{8.65}$$

while at first order we have

$$\boldsymbol{\tau}'_1 = \dot{\boldsymbol{\gamma}}'_1 - \left(\overset{\nabla}{\dot{\boldsymbol{\gamma}}'_0} + B\dot{\boldsymbol{\gamma}}'_0 \cdot \dot{\boldsymbol{\gamma}}'_0\right) \equiv \dot{\boldsymbol{\gamma}}'_1 + \mathbf{A}'_1[\mathbf{u}'_0]. \tag{8.66}$$

3.2.1 Locomotion

Using the above constitutive relation in the integral theorem for swimming (8.22) we obtain

$$\boldsymbol{U}' = \hat{\boldsymbol{R}}'^{-1} \cdot \left[\int_{S'} \mathbf{u}'^S \mathbf{n} : \hat{\boldsymbol{T}}' \, \mathrm{d}S' + \mathrm{De} \int_{V'} \mathbf{A}'_1 : \nabla' \hat{\boldsymbol{G}}' \, \mathrm{d}V'\right] + \mathscr{O}(\mathrm{De}^2). \tag{8.67}$$

Hence in order to compute the $\mathscr{O}(\mathrm{De})$ correction to the swimming speed we need only solve for the Newtonian flow field, $\mathbf{u}_0(\mathbf{x})$.

Consider the counterrotation of two connected but different axisymmetric bodies (e.g., two unequal spheres). Such a swimmer will not locomote in Stokes flow due to the kinematic reversibility of the field equations but may have net motion in a second-order fluid due to the $\mathscr{O}(\mathrm{De})$ term. Because we explicitly chose a gait which would not swim in a Newtonian fluid, the rigid-body motion of the swimmer is given by the non-Newtonian contribution only,

$$\boldsymbol{U}' = \mathrm{De} \int_{V'} \mathbf{A}'_1 : \nabla' \hat{\boldsymbol{G}}' \cdot \hat{\boldsymbol{R}}'^{-1} \, \mathrm{d}V' + \mathscr{O}(\mathrm{De}^2). \tag{8.68}$$

The swimming speed is at best linear in Deborah number. The dimensional swimming speed scales thus as $\mathbf{U} \sim \omega L \mathrm{De}$, where L is the typical length scale. Note this is invariant under a reversal of actuation $\omega \to -\omega$ (because $\mathrm{De} \propto \omega$), as expected due to the axisymmetry of the body. The Newtonian problem may not be available analytically, but one could obtain the solution numerically using the boundary integral method, for instance.

Now due to the symmetry of this body, we expect the only rigid-body motion to be translation along the axis connecting the spheres. For this reason it is unnecessary to compute the full resistance tensor. In order to find the swimming speed one needs only to solve the auxiliary problem of rigid-body translation of the same body in the direction of swimming and the integration reduces to

$$U' = \mathrm{De} R'^{-1}_{FU} \int_{V'} \mathbf{A}'_1 : \nabla' \hat{\mathbf{u}}' \, \mathrm{d}V' + \mathscr{O}(\mathrm{De}^2), \tag{8.69}$$

where $\hat{\mathbf{u}}' = \hat{\mathbf{u}}/\hat{U}$ is the dimensionless flow field due to rigid-body translation along the axis of rotation.

3.2.2 Rheology

In the previous section we were able to construct locomotion which depended on the presence of non-Newtonian stresses. Rather than calculating the swimming speed for a known constitutive equation, we may use this technique instead to infer unknown rheological properties, namely the normal stress coefficients Ψ_1 and Ψ_2, of a fluid through experimental measurement of the swimming kinematics of deforming bodies [24] based on theory proposed by Khair and Squires [23]. This idea is summarized below.

One can recast Eq. (8.69) as a linear equation in the normal stress coefficients

$$U = \frac{\omega^2 L}{\eta}\left[C_1\Psi_1 + C_2\Psi_2\right], \tag{8.70}$$

where

$$C_1 = -\frac{1}{2}\hat{R}_{FU}^{-1}\int_{V'}\overset{\nabla}{\dot{\gamma}}{}'_0 : \nabla'\hat{\mathbf{u}}'\,\mathrm{d}V', \tag{8.71}$$

$$C_2 = \hat{R}_{FU}^{-1}\int_{V'}(\dot{\gamma}'_0\cdot\dot{\gamma}'_0) : \nabla'\hat{\mathbf{u}}'\,\mathrm{d}V', \tag{8.72}$$

are so-called coupling coefficients, functions of the zeroth order solution, independent of the non-Newtonian properties of the flow and hence are functions of the geometry alone.

Since typically $\Psi_1 \gg |\Psi_2|$, a reasonable approximation of the first normal stress coefficient would be

$$\Psi_1 \approx \frac{\eta U}{C_1\omega^2 L}. \tag{8.73}$$

By measuring swimming kinematics in a non-Newtonian fluid (one which does not display net motion in a Newtonian fluid) we may thus obtain a good approximation of the first normal stress coefficient.

In order to unambiguously determine both the first and second normal stress coefficients, two independent measurements must be made. One may, for example, devise another independent swimmer or one could measure kinematics other than a swimming speed. An example of the latter would be to measure the relative displacement of two equal counterrotating spheres. Experimentally this could be accomplished by an axial coupling that transmits torque but allows free translation. In such a case it is helpful to decompose the body as $S = S_1 \cup S_2$ where S_1 and S_2 are the two sphere surfaces. By symmetry we expect only a relative displacement of the spheres $\mathbf{U}^R = \mathbf{U}^1 - \mathbf{U}^2 = 2\mathbf{U}^1$, equal and opposite, along the axis of rotation (with resistance to this motion, defined by $\hat{R}_{FU}^R$) and so this is the only rigid-body motion that needs resolution. By kinematic reversibility we know the rotation of the spheres produces no net displacement in a Newtonian fluid so the reciprocal theorem yields directly

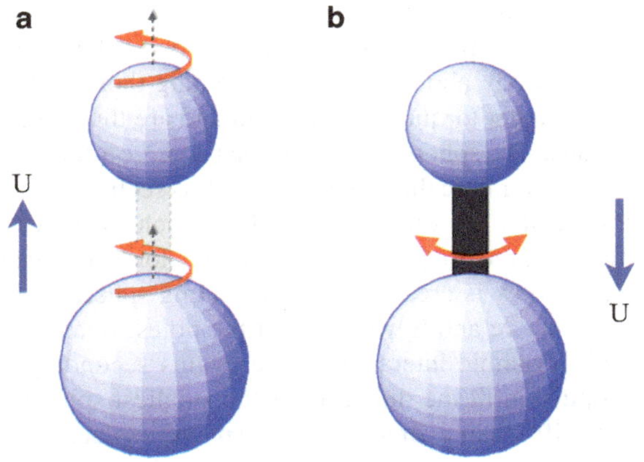

Fig. 8.2 (**a**) Two unequal spheres rotating together as a rigid body will move in the direction of the smaller sphere due to an imbalance of hydrodynamic interactions driven by viscoelastic stresses [24]. (**b**) Experiments on two rigidly connected spheres rotating about an out-of-plane axis leads to net viscoelastic locomotion in the direction of the larger sphere [25]

$$U'^R = \mathrm{De}\frac{1}{\hat{R}'^R_{FU}}\int_{V'} \mathbf{A}'_1 : \nabla'\hat{\mathbf{u}}'\,\mathrm{d}V' + \mathcal{O}(\mathrm{De}^2), \tag{8.74}$$

where $\hat{\mathbf{u}}' = \hat{\mathbf{u}}/(\hat{U}^R/2)$ is the dimensionless flow field due to rigid-body translation of two equal spheres away from one another with relative speed $\hat{U}^R$. With this second kinematic measurement one has two equations for the unknown normal stress coefficients of the fluid. We can write this simply in matrix form

$$\begin{pmatrix} U \\ U^R \end{pmatrix} = \frac{L\omega^2}{\eta}\begin{pmatrix} C_1 & C_2 \\ C_1^R & C_2^R \end{pmatrix}\begin{pmatrix} \Psi_1 \\ \Psi_2 \end{pmatrix}. \tag{8.75}$$

The normal stress coefficients are obtained by inverting the matrix of coupling coefficients, **C**. Practically, one may manipulate the geometries in the experiment so as to minimize the condition number of **C** and hence the potential error in the values of the normal stress coefficients obtained through the measurements of U and U^R which are subject to experimental error [23] (Fig. 8.2).

Although we have emphasized here experimental observation of the kinematics of force- and torque-free bodies as means to decipher the normal stress coefficients, one could alternatively directly measure the forces on bodies which move with prescribed kinematics. In the method first proposed by Khair and Squires [23], one would pull two equal spheres in the direction joining their two centers, $\mathbf{U}_{\|}$, in a first experiment and in a direction orthogonal to the line joining their centers, $\mathbf{U}_{\perp}$, in a second experiment, each time measuring the force on each sphere. The difference between the forces on each sphere is then a measure of the non-Newtonian stresses, because in each case these forces would be identical in a Newtonian fluid. The suggested experimental method would be to trap two particles in a dual optical trap, then translate the bulk fluid uniformly and measure the difference in the trapping force on each probe [23].

Alternatively Pak and Lauga proposed a setup involving force-free but not torque-free spheres. In their model, two spheres are externally rotated together as a rigid body about the axis connecting their centers. In that case the spheres will impart a net torque onto the fluid and are thus not strictly swimmers. Nevertheless, because the spheres impart no net force on the fluid, the equations for translation precisely mimic those presented in this section for a force- and torque-free swimmer. Physically, the direction in which such an object moves can be understood by means of the hoop stresses generated along curved streamlines. A secondary, purely elastic flow is created by each rotating sphere, contracting in along the equator of each sphere and flowing out of the poles. Because the spheres are unequal in size, hydrodynamic interactions due to this secondary flow are unbalanced leading to propulsion in the direction of the smallest sphere. The same mechanism causes a net drift when a single rotating sphere is placed near a wall in a viscoelastic fluid. The rotating sphere drives fluid radially out in the direction of its poles and if a no-slip wall is placed opposite of one pole, the sphere is driven away from the wall.

A similar object, two unequal spheres connected by a rigid rod, was used in experiments by Keim et al. [25]. In that case the spheres oscillate together as a rigid body about an axis orthogonal to the line connecting their centers. This motion is time-reversible and hence does not yield time-averaged locomotion in a Newtonian fluid, yet in a viscoelastic fluid there is net migration. These experiments also demonstrated that a wall acts as a symmetry-breaking mechanism for the propulsion of a dimer with equal spheres. Analytical studies have also indicated that time-reversible motions of anchored bodies (flapping a rigid rod for instance) may pump fluid [26,27] in a viscoelastic fluid when such net flow is not possible in a Newtonian fluid due to the scallop theorem.

4 Infinite Models

In the previous section we looked at a formulation for the swimming speed based on a modification of the reciprocal theorem for locomotion in a viscoelastic fluid. Historically the simplest possible models for understanding swimmers in Newtonian fluids have been infinite models with reduced dimensionality, in particular the canonical Taylor swimming sheet [28]. Here again, the reciprocal theorem may be used to solve for the swimming speed of a two-dimensional sheet and yield insight into the effects of complex fluids interacting with nontrivial boundary actuation.

The analysis of the correction to the swimming speed of the swimming sheet due to a non-Newtonian fluid was first performed by Chaudhury using a second-order fluid (or Rivlin-Erickson fluid of grade 2) [29] and later by Sturges for a second-order fluid of grade N [30]. In both cases no change in swimming speed was observed in the zero Reynolds number limit. The analysis was then performed for a variety of nonlinear non-Newtonian constitutive equations by Lauga [31]. There, it was shown that, for fixed swimming kinematics, the swimming speed for the swimming sheet systematically decreases compared to the Newtonian case for all

Oldroyd-type fluids to leading order in small-amplitude waveforms. Subsequently the same result was obtained for planar undulating filaments [32] and helical waves [33] in Oldroyd-B fluids.

In the following we give an overview of the use of the reciprocal theorem to obtain, in a direct fashion, the influence of complex flows on the swimming sheet in a general waveform. We also derive viscoelastic corrections to the swimming speed of a general sheet near walls.

4.1 Taylor Swimming Sheet

Consider a two-dimensional sheet which propagates traveling waves. In a frame moving with the traveling waves the sheet has the static shape $y_1 = ag(\xi)$ where g is written very generally as

$$g(\xi) = \sum_n c_n e^{in\xi}, \tag{8.76}$$

with a being the dimensional wave amplitude and $\xi = kx - \omega t$ with wavenumber k and frequency ω. In Fig. 8.3 we illustrate a sinusoidal swimming sheet which has only one mode $c_1 = -i/2$. Ostensibly the sheet generates vorticity which is oppositely signed from peak to trough, and thus for a waveform traveling to the right the fluid forces act to drive the sheet to the left [8].

The boundary conditions for an inextensible sinusoidal sheet were first described by Taylor [28] and catalogued in detail in Ref. [34] for general waveforms. Following Taylor, we take the approach that the amplitude of transverse oscillations of the sheet is small and expand all fields a regular perturbation series in $ak = \varepsilon \ll 1$. The boundary conditions are redefined onto the surface $y = 0$ (S_0) through Taylor expansion. Finally we introduce a no-slip wall at $y = h$, and by taking the limit $h \to \infty$ we will obtain swimming in an unbounded fluid. We take the auxiliary flow to be simple shear flow between $y = 0$ and $y = h$, for which the reciprocal theorem simplifies dramatically to give

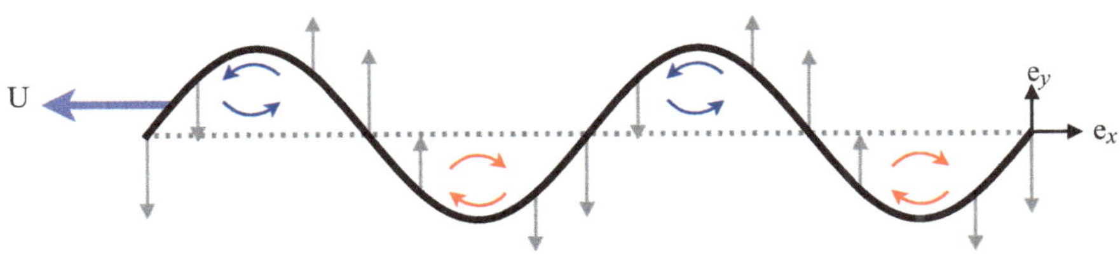

Fig. 8.3 A sinusoidal swimming sheet propagates waves of transverse oscillation in a Newtonian fluid (adapted from Ref. [8]). The *grey arrows* indicate the transverse oscillations of the wave which travels to the right at speed ωk^{-1}. The vorticity generated drives the sheet to the left at speed U

$$U = -u_x^{S_0,(0)} + \frac{1}{\eta_0}\int_0^h A_{xy}^{(0)}\,\mathrm{d}y, \tag{8.77}$$

where $u_x^{S_0,(0)} = \langle \mathbf{u}^{S_0} \rangle \cdot \mathbf{e}_x$ while $A_{xy}^{(0)} = \langle \mathbf{A} \rangle : \mathbf{e}_y\mathbf{e}_x$ (with basis vectors as shown in Fig. 8.3). This framework can also be used for a sheet unequally spaced between two walls at $y = h_1$ and $y = -h_2$, In that case, due to the lack of symmetry both the upper and lower surfaces of the sheet and the volume of fluid above and below each surface must be accounted for and the reciprocal theorem then yields

$$U = \frac{1}{h_1+h_2}\left[-h_2 u_x^{S_1,(0)} - h_1 u_x^{S_2,(0)} + \frac{1}{\eta_0}\left(h_2\int_0^{h_1} A_{xy}^{(0)}\,\mathrm{d}y + h_1\int_{-h_2}^{0} A_{xy}^{(0)}\,\mathrm{d}y\right)\right]. \tag{8.78}$$

Coming back to the one-wall case, we consider as an example an Oldroyd-B fluid [22]. The complex viscosity is given by

$$\eta^* = \eta_0 \frac{1+in\omega\lambda_2}{1+in\omega\lambda_1}. \tag{8.79}$$

At leading order in ε, $\mathbf{A}_1 = \mathbf{0}$ and $\boldsymbol{\tau}_1^{(n)} = \eta^*(n)\dot{\boldsymbol{\gamma}}_1^{(n)}$, while at second order

$$\mathbf{A}_2^{(0)} = \sum_q \left[\frac{\eta_0}{iq\omega}\left(1-\frac{\eta^*(q)}{\eta_0}\right)\left(\nabla\mathbf{u}_1^{(\text{-q})\mathrm{T}}\cdot\dot{\boldsymbol{\gamma}}_1^{(q)} + \dot{\boldsymbol{\gamma}}_1^{(q)}\cdot\nabla\mathbf{u}_1^{(-q)} - \mathbf{u}_1^{(-q)}\cdot\nabla\dot{\boldsymbol{\gamma}}_1^{(q)}\right)\right]. \tag{8.80}$$

Following the framework described in Sect. 3, we seek a perturbative solution to the swimming speed as $U(\varepsilon) = \sum \varepsilon^m U_m$ and similarly expand all fields in powers of ε. Referring to Eq. (8.77) one can find immediately that $U_1 = 0$.

In order to obtain the swimming speed at quadratic order, U_2, we must find the nonlinear contribution at second order, $\mathbf{A}_2$, the boundary conditions, $\mathbf{u}_2^{S_0}$, and the full leading-order flow field, $\mathbf{u}_1$. The mean of the boundary condition on S_0 is given by

$$\mathbf{u}_2^{S_0,(0)} = \omega k^{-1}\sum_n n^2 \Upsilon(nh) c_n c_n^\dagger \mathbf{e}_x, \tag{8.81}$$

where $\Upsilon(x) = (\sinh^2(x)+x^2)/(\sinh^2(x)-x^2)$. In a Newtonian fluid the swimming speed is given directly as

$$\mathbf{U}_2 = -\left\langle \mathbf{u}_2^{S_0} \right\rangle, \tag{8.82}$$

and in particular when $h \to \infty$ we have $\Upsilon \to 1$. For Taylor's swimming sheet ($c_1 = -i/2$) we recover the classical result,

$$\mathbf{U} = -\frac{1}{2}\omega k^{-1}\varepsilon^2\mathbf{e}_x + \mathscr{O}(\varepsilon^4). \tag{8.83}$$

Resolving $\mathbf{A}_2$ in the Oldroyd-B case we have

$$\frac{1}{\eta_0}\int_0^h A_{2,xy}^{(0)}\,\mathrm{d}y = \omega k^{-1}\sum_n \left[1-\frac{\eta^*(n)}{\eta_0}\right] n^2 c_n c_n^\dagger \Upsilon(nh). \tag{8.84}$$

Here we see that the nonlinear forcing term, for each mode, is proportional to the boundary condition on S_0 at second order. With this result the swimming speed is expressed as

$$U_2 = -\omega k^{-1}\sum_n n^2 c_n c_n^\dagger \Upsilon(nh)\frac{\eta^*(n)}{\eta_0}. \tag{8.85}$$

Each component of the Newtonian swimming speed is therefore simply scaled by the dimensionless viscous modulus. Now if the sheet is unevenly spaced between two walls, at distances h_1 and h_2, application of Eq. (8.78) yields

$$U_2 = -\omega k^{-1}\sum_n n^2 c_n c_n^\dagger \frac{h_2\Upsilon(nh_1)+h_1\Upsilon(nh_2)}{h_1+h_2}\frac{\eta^*(n)}{\eta_0}. \tag{8.86}$$

In both cases we see that the factor which differentiates the result from the Newtonian swimming speed, namely that which arises strictly from non-Newtonian stresses, is simply the dimensionless loss modulus for each mode

$$\mathscr{R}\left[\frac{\eta^*(n)}{\eta_0}\right] = \frac{1+n^2\beta\mathrm{De}^2}{1+n^2\mathrm{De}^2}, \tag{8.87}$$

where $\mathrm{De}=\omega\lambda_1$ and $\beta=\lambda_2/\lambda_1$, $\beta<1$ as it indicates the ratio of the suspending solvent viscosity to total viscosity in the fluid. When $\mathrm{De}\ll 1$, the fluid is probed on time scales much longer than it takes to relax, and the Newtonian behavior is recovered $\mathscr{R}[\eta^*/\eta_0]\to 1$. When the Deborah number is nonzero the swimming speed is always less than the Newtonian one. In particular in the $\mathrm{De}\gg 1$ limit, we obtain $\mathscr{R}[\eta^*/\eta_0]\approx\beta$ and the swimming speed is simply a factor of the Newtonian swimming speed, $U=\beta U_N$.

The results obtained above indicate that in a Oldroyd-B fluid, each Fourier mode in the swimming speed is rescaled by the respective (dimensionless) viscosity for that mode. It may appear obvious that the linear response of the fluid should affect the swimming speed but it is not. If the fluid were simply linearly viscoelastic there would be no change in the swimming speed from that of a Newtonian fluid. It is thus the nonlinear response that leads to change in the locomotion speed. As the flow field scales linearly with the sheet amplitude, the first nonlinear correction (the tensor $\mathbf{A}_2$) is at the origin of the modification in swimming. The tensor $\mathbf{A}_2$, oddly enough, conspires to give the correction of rescaling by the linear viscoelastic modulus.

For the sinusoidal swimming sheet originally considered by Taylor, the only nonzero coefficient is $c_1=-i/2$ leading to a swimming speed of $-\omega/2k$ in a

Newtonian fluid [28]. The viscoelastic correction to this result, $\mathscr{R}[\eta^*(1)/\eta_0]$, was first derived in Ref. [31]. The leading-order result is, written in a dimensional form,

$$U = -\frac{1}{2}a^2 k\omega \frac{1+\beta\omega^2\lambda_1^2}{1+\omega^2\lambda_1^2} + \mathscr{O}(\varepsilon^4). \tag{8.88}$$

The swimming speed is no longer linear in the actuation frequency ω, rather it is strictly decreasing with Deborah number from the Newtonian speed, $U = U_N$ as $\mathrm{De} \to 0$, to $U = \beta U_N$ as $\mathrm{De} \to \infty$, as illustrated in Fig. 8.4.

The rate at which a certain gait is performed now has a direct impact on the distance, ΔD, traveled over a period,

$$\Delta D = -\pi a^2 k \frac{1+\beta\omega^2\lambda_1^2}{1+\omega^2\lambda_1^2}, \tag{8.89}$$

clear evidence of the breakdown of Purcell's scallop theorem due to the presence of viscoelastic stresses [33].

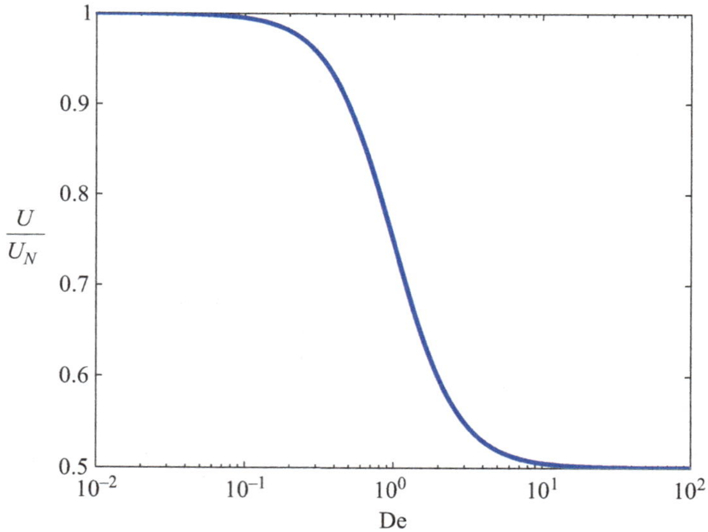

Fig. 8.4 The leading-order viscoelastic swimming speed scaled by the Newtonian solution, U/U_N, as a function of Deborah number, $\mathrm{De} = \omega\lambda_1$, for a Taylor swimming sheet in an Oldroyd-B fluid with $\beta = 0.5$. The swimming speed is Newtonian, $U = U_N$, in the small Deborah number limit, $\mathrm{De} \to 0$, while $U = \beta U_N$ as $\mathrm{De} \to \infty$

Beyond kinematics, one might need to compute the power expended by the swimming motion. The energy dissipated in the fluid due to the actuation of the swimmer is given by an integration of $\boldsymbol{\tau}_1 : \dot{\boldsymbol{\gamma}}_1$ over the whole fluid domain. Because the swimmer propagates periodic traveling waves, only the mean value is important and it is given by

$$\langle \boldsymbol{\tau}_1 : \dot{\boldsymbol{\gamma}}_1 \rangle = \sum_n \eta^*(n) \dot{\boldsymbol{\gamma}}_1^{(n)} : \dot{\boldsymbol{\gamma}}_1^{(-n)}. \tag{8.90}$$

We see again that the only difference between the viscoelastic dissipation here and that in a Newtonian fluid is the complex viscosity, $\eta^*(n)$, for each Fourier mode in comparison to η_N in the Newtonian fluid. Note that it matters, in a comparison of the dissipation, what we choose the Newtonian dissipation to be. As shown above as De $\gg 1$ (or $n \gg 1$) $\mathscr{R}[\eta^*(n)/\eta_0] \approx \beta$ and a comparison to the energy dissipated in a Newtonian fluid with viscosity η_N yields a factor of $\eta_0\beta/\eta_N$ which is less than one if $\eta_N = \eta_0$ (i.e., if we compare the dissipation in the complex fluid to that of a Newtonian fluid of the same viscosity). If the Newtonian fluid has the viscosity of the solvent in the Oldroyd-B fluid then $\eta_N = \beta\eta_0$, and we see by comparison that the viscoelastic energy dissipated is always higher, meaning the addition of elasticity to a solvent always increases energy dissipation. This effect is diminished for higher Fourier modes and at higher actuation rates and eventually, for De $\gg 1$, only the solvent is dissipating energy as the network does not have time to flow [31].

4.2 Large-Amplitude Deformations

One of the interesting debates in the literature in the field concerns the question of whether some swimming gaits could see a swimming speed increase in viscoelastic fluids. It is obviously possible to construct a swimmer which has no net locomotion in a Newtonian fluid yet displays net motion in a non-Newtonian fluid and we discussed this strategy as a measurement technique to recover rheological properties of a fluid. Less obvious is the effect on swimming speed for a gait that already leads to nonzero net motion in a Newtonian fluid. We showed above that a swimming sheet of arbitrary geometry will always swim slower in a viscoelastic fluid. However, that result and the theory presented above rely on small gait amplitudes in order to take advantage of perturbation expansions. Clearly, small-amplitude theory need not hold for gaits that are significantly straining the fluid.

Recent experiments by Liu et al. [35] showed a modest increase in swimming speed of a force-free (but not torque-free) rotating helix in a Boger fluid near De $= 1$ and showed that this enhancement is independent of end effects. In contrast, prior asymptotic analysis by Fu et al. [33] showed that, like the swimming sheet, the leading-order swimming speed (in a small-amplitude perturbation series) of a body propagating helical waves in an Oldroyd-B fluid is always slower than in a non-Newtonian fluid. This discrepancy between the small-amplitude asymptotics and large-amplitude experiments was resolved in numerical work by Spagnolie et al. for a helix in an Oldroyd-B fluid [36]. They showed that there is a smooth transition between small-amplitude hindered swimming and large-amplitude enhanced locomotion near $\mathrm{De} = 1$. The authors argued that a reason the speed enhancement occurs at an actuation rate that is on the same order as the relaxation rate of the fluid, $\omega^{-1} \approx \lambda_1$, may be because this is the time scale in which the flagellum revisits the viscoelastic wake it creates upon rotation.

A question that one might ask is whether this mechanism translates to other unsteady swimming gaits. Numerical simulations by Teran et al. [37] showed that a finite two-dimensional swimmer, propagating deformation waves of increasing amplitude head to tail, sees an increase in swimming speed near $\mathrm{De} = 1$ in an Oldroyd-B fluid with $\beta = 1/2$. The authors rationalize that the increase in swimming speed results as a consequence of the highly strained fluid localized at the swimmers tail. Alternatively, in a set of experiments with the nematode *C. elegans* swimming in a Boger fluid, Shen and Arratia [38] found that non-Newtonian stresses strictly decrease the swimming speed. This nematode swims by propagating traveling waves with amplitudes which decay from head to tail. The functional dependence of the swimming speed on the Deborah number in their experiments [38] is reminiscent of the systematic decay found for a small-amplitude swimming sheet in an unbounded fluid [31]. Indeed numerical simulations documented later in this book (Chap. 10) show that if the gait of the nematode is reversed (reflection of the wavevector), yielding an increasing amplitude head to tail, then the nematode would experience a speed enhancement similar to the results presented by Teran et al. Additionally, numerical simulations of potential squirmers [39] and pushers or pullers [40] showed a speed decrease in a Giesekus fluid versus a Newtonian one.

In an effort to understand these large-amplitude results, we extend the small-amplitude result for Taylor's swimming sheet to higher order by deriving the next two orders of the perturbation series for the swimming velocity, $U = \varepsilon^2 U_2 + \varepsilon^4 U_4 + \varepsilon^6 U_6 + \mathscr{O}(\varepsilon^8)$ (the perturbation series contains only even terms because of the $\varepsilon \to -\varepsilon$ symmetry). In a Newtonian fluid the first two terms were found by Taylor [28], while the third was later derived by Drummond [41]. The series was recently resolved to arbitrarily high order by Sauzade et al. [42] who showed that the series converges only for small ε and then only slowly, but methods to accelerate convergence prove very effective enabling accurate prediction up of the swimming speed for order-one amplitudes.

The leading-order steady swimming speed in an Oldroyd-B fluid, U_2, was computed in Eq. (8.88). The next two nonzero orders in the asymptotic series, U_4 and U_6, can be found with a straightforward, but laborious, application of the formalism presented in this chapter. At fourth order we obtain analytically

$$U_4 = \omega k^{-1} \frac{\left(1+\mathrm{De}^2\beta\right)}{128\left(1+\mathrm{De}^2\right)^3\left(1+\mathrm{De}^2\beta^2\right)}$$
$$\times \left[76+50\mathrm{De}^2+47\mathrm{De}^4+\mathrm{De}^2\left(102+29\mathrm{De}^2\right)\beta\right.$$
$$\left.+\mathrm{De}^2\left(76+45\mathrm{De}^2+42\mathrm{De}^4\right)\beta^2+\mathrm{De}^4\left(107+34\mathrm{De}^2\right)\beta^3\right], \tag{8.91}$$

while at sixth order we find a much lengthier but still entirely analytical formula (not shown).

The swimming speed found using a conventional sum of the first three terms of the perturbation series for a Newtonian fluid is inaccurate past $\varepsilon \approx 0.5$, but by using the method of Padé approximants [43], with just three terms in the series, the $P_1^1(\varepsilon)$ approximate is accurate up to $\varepsilon \approx 1$ (within 1 % of the computational result determined using the boundary integral method) [42]. We follow the same tactic for the coefficients of the viscoelastic swimming speed based on the assumption that the Padé $P_1^1(\varepsilon)$ approximate is accurate for a larger range of amplitudes. In Fig. 8.5 we plot the swimming speed scaled by the wavespeed, $U' = U/\omega k^{-1}$, as

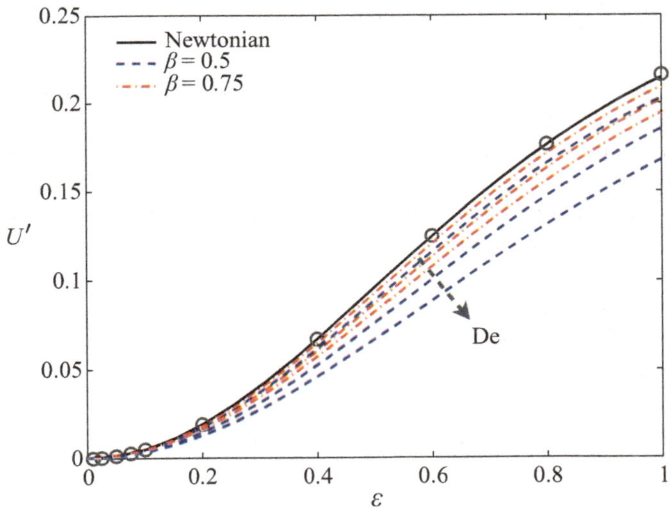

Fig. 8.5 Dimensionless swimming speed, $U' = U/\omega k^{-1}$, as a function of dimensionless amplitude, $\varepsilon = ak$, for a Taylor swimming sheet in an Oldroyd-B fluid. Plotted is the $P_1^1(\varepsilon)$ Padé approximate of a 3-term perturbation series valid up to $\mathcal{O}(\varepsilon^6)$. The *solid line* shows the Newtonian swimming speed, while the *dashed line* indicates the non-Newtonian result with $\beta = 0.5$ and *dashed-dot* is $\beta = 0.75$, with Deborah numbers $\mathrm{De} = 0.5, 1, 1.5$. The *circles* are numerical results from boundary integral computation for $\mathrm{De} = 0$. Viscoelasticity strictly decreases swimming speeds in this range of amplitudes

a function of dimensionless amplitude $\varepsilon = ak$. Under our modeling approach, we see that viscoelasticity serves to strictly hinder the swimming speed of the sheet even for dimensionless amplitudes on the order of $\varepsilon = 1$. This result is in agreement with the experiments for the undulatory nematode but in contrast to numerical and experimental work on the propulsion of a rotating helix.

Recent computational work confirms that the Taylor sheet is indeed hindered at all amplitudes by the presence of viscoelastic stresses [44]. This was interpreted as due to stagnation points in the stress near the peak and trough of the sheet, which act to retard the motion of the sheet. Exploiting this insight, sheets asymmetric about the horizontal axis, reminiscent of hyperactivated sperm flagella, may be constructed which see a speed increase in an Oldroyd-B fluid [44].

Recent experiments using a cylindrical variant of a Taylor sheet do show a speed increase in a Boger fluid (but not for a shear-thinning one) [45]. In these experiments a Couette-like device has a flexible inner cylinder which passes angular traveling waves of radial deformation while the outer cylinder rotates freely supported by low-friction bearings and hence at steady state yields an approximation of a swimming sheet near a wall. A noticeable speed increase is observed if a Boger fluid is used instead of a Newtonian fluid for all wavespeeds but with an increasing difference for larger wavespeeds—in stark contrast to the results for a planar sheet.

4.3 Shear-Dependent Viscosity

Many biological fluids through which microorganisms might swim, such as mucus, are not only viscoelastic but also have shear-dependent viscosity. Studies on the effects of a variable viscosity on the swimming speed of microorganisms, like those of elasticity, show mixed results. Shen and Arratia's experiments with *C. elegans* show that shear-thinning fluids have no noticeable effect on swimming speeds [38]. In contrast, experiments on the cylindrical Taylor sheet show a marked decrease in the swimming speed [45].

A recent theoretical study on the Taylor swimming sheet in a Carreau fluid was undertaken by Vélez-Cordero and Lauga [46]. A Carreau fluid has a shear-rate-dependent deviatoric stress given by

$$\boldsymbol{\tau} = \eta_0 \left[1 + \lambda_t^2 |\dot{\boldsymbol{\gamma}}|^2\right]^N \dot{\boldsymbol{\gamma}}, \tag{8.92}$$

with $|\dot{\boldsymbol{\gamma}}|^2 = \dot{\boldsymbol{\gamma}} : \dot{\boldsymbol{\gamma}}/2$ and $N = (n-1)/2$ where n is the so-called power-law index for the fluid and λ_t is the relaxation time of fluid. Physically, we see from Eq. (8.92) that at high shear rates, the typical shear stress scales as $\tau \propto \dot{\gamma}^n$ and thus the number n is the power-law dependence in the stress/shear rate relationship at high shear rates. If the typical shear rate in the flow scales as ω then the Carreau number $\mathrm{Cu} = \omega\lambda_t$ is a dimensionless measure of the extent to which the fluid viscosity is altered; if $\mathrm{Cu} \ll 1$ the fluid behaves as Newtonian with (zero-shear-rate) viscosity η_0.

In the small-amplitude study of Ref. [46] it was found that a shear-thinning fluid has actually no effect on the swimming speed of the sheet if it deforms inextensibly. If instead the motion of the material is extensible, then there is an additional higher order non-Newtonian contribution, $U - U_N \sim \pm\varepsilon^4 N\mathrm{Cu}^2$, with a sign which depends on the details of the waving kinematics. Recent numerical work at high amplitude and for finite swimmers by Montenegro-Johnson et al. confirms that the (often weak) effects of a shear-thinning fluid depend on the gait of the swimmer with examples of both faster and slower swimming given for a variety of model swimmers [47,48].

In addition, we note that the swimming sheet model has also been used to address swimming near a wall in shear-rate-dependent and yield-stress fluids [49–51]. In these studies, it was assumed that the separation between the sheet and the wall was much smaller than the wavelength of the sheet, $hk \ll 1$, thus taking advantage of the long-wavelength lubrication approximation.

4.4 Prescribed Forcing

Instead of imposing the wave kinematics, an alternative modeling approach to the problem of locomotion consists in prescribing the internal forcing which, through a dynamic balance, leads to the deformation of the body. Fu and Powers investigated theoretically the effects of viscoelasticity on the shape of a beating eukaryotic flagellum [32]. Flagellar beat patterns are determined by an interplay between the mechanical properties of the flagellum, the internal action from the dynein motor proteins which produce active bending moment, and the hydrodynamic forces [52]. Using a simplified sliding filament model for a sperm flagellum, the investigation in Ref. [32] showed that the introduction of viscoelasticity into the fluid can dramatically affect the shape of the waving flagellum.

One approach to model the impact of this change in kinematics on locomotion is to consider the body of swimmer to be composed of a repeated series of simple shapes, such as spheres. Najafi and Golestanian proposed the simplest such swimmer which can locomote in Newtonian fluid [53], consisting of three equal co-linear spheres with prescribed, periodically varying, relative displacements (as if connected by hydrodynamically invisible rods of time-varying lengths). A system of two such spheres cannot self-propel in a Newtonian fluid as it has but a single degree of freedom, but with three spheres, one can produce time-irreversible motion and swimming [53].

Curtis and Gaffney recently endeavored to determine the motion of the three-sphere swimmer in an Oldroyd-B fluid [54]. In general the superposition of far-field singularity solutions, which is the standard method to solve the N-sphere Newtonian problem, is not possible in a viscoelastic fluid due to the nonlinearities in the governing equations. However, if the flow field is resolved as a perturbation expansion in small-amplitude disturbances then the constitutive relation can be linearized order-by-order. After expanding to quadratic order in small-amplitude motion (relative to the sphere radii) and for prescribed kinematics, the net displacement over a period is found to be identical to that of a Newtonian fluid [54]. This surprising result, at odds with the swimming sheet result, might be a consequence of the far-field approximation. However, the simplicity of this model allows for a straightforward implementation of prescribed-forcing case, rather than prescribed kinematics. In this case, the relative forces exerted by each sphere on its neighbor are prescribed and the gait and swimming kinematics are then determined. For prescribed forcing the time-averaged swimming speed U differs from the Newtonian swimming speed U_N by a factor

$$\frac{U}{U_N} = \frac{1+\mathrm{De}^2}{1+\beta^2\mathrm{De}^2}, \tag{8.93}$$

which is always greater than one since $\beta \leq 1$. Prescribing the internal forcing of a swimmer rather than its kinematics leads to a qualitatively different response and a speed increase ensues. It will be interesting to extend this approach in the future to more realistic models of swimming cells.

We close by also noting that numerical simulations performed using the immersed boundary method by Chrispell et al. [55] involve specifying preferred waving kinematics with the true kinematics then resolved as a balance between elastic and fluid forces (in a sense, a hybrid between specifying kinematics and specifying the internal forces).

4.5 Two-Fluid Models

Microorganisms may swim in environments where heterogeneities in the fluid exist on length scales similar to the swimmers themselves. Berg and Turner found propulsion enhancement in gel-like environments (methylcellulose solutions) where the solute forms a loose, quasi-rigid network [56]. They postulated the microorganisms could then push directly on this network and that a helical flagellum would move as would a corkscrew through a cork, with less circumferential slip than in a Newtonian fluid. In order to properly capture this behavior the material should possess a nonzero zero-frequency elastic shear modulus, and the swimmer must be able to slip past the solid medium.

Fu et al. used a two-fluid model to explore the effects of a cross-linked network on the swimming speed of Taylor's waving sheet [57] (illustrated in Fig. 8.6). In this model a drag force density, proportional to the relative local velocity,

$$\mathbf{f}_\mathrm{d} = \Gamma\left(\frac{\mathrm{d}}{\mathrm{d}t}\mathbf{X} - \mathbf{u}\right), \tag{8.94}$$

couples the elastic network (displacement field $\mathbf{X}$) to the Newtonian solvent (velocity field $\mathbf{u}$),

$$\nabla\cdot\boldsymbol{\sigma}^n = \mathbf{f}_\mathrm{d}, \quad \nabla\cdot\boldsymbol{\sigma}^s = -\mathbf{f}_\mathrm{d}, \tag{8.95}$$

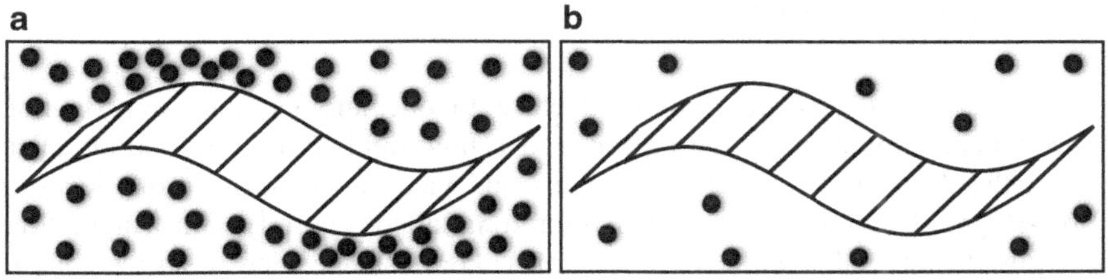

Fig. 8.6 Schematic representation of a Taylor swimming sheet in a two-fluid model fluid [57]. (**a**) in a dense network the sheet interacts directly with the elastic network; (**b**) if the network is dilute the interactions between the sheet and the network are mediated through the fluid

where n indicates the network and s indicates the solvent. The friction coefficient Γ introduces an intrinsic length scale into the problem typically known as the screening length, $l_s = \sqrt{\eta/\Gamma}$. The ratio of the screening length to the physical length scale of the sheet, kl_s, is a dimensionless measure of the interaction between the fluid flow and the network and as expected if $kl_s \gg 1$ then the sheet effectively sees only a Newtonian fluid.

If the network is dense as illustrated in Fig. 8.6a, the sheet comes into direct contact with the network which is then deformed while permitting tangential slip via a Navier friction law

$$\mathbf{t} \cdot \boldsymbol{\sigma}^n(y = y_1) \cdot \mathbf{n} = \Xi \mathbf{t} \cdot \left(\frac{\mathrm{d}}{\mathrm{d}t} \mathbf{X} - \mathbf{u} \right) \bigg|_{y=y_1}, \tag{8.96}$$

with slip coefficient Ξ. If the network is dilute as illustrated in Fig. 8.6b, then no contact is made and hence no traction is applied to the sheet, $\boldsymbol{\sigma}^n(y = y_1) \cdot \mathbf{n} = \mathbf{0}$.

The rigidity of the network has a dramatic effect on the swimming speed, and stiff networks enhance swimming speed compared to a Newtonian fluid while compliant networks retard swimming speed, regardless of whether the network contact is direct or solvent mediated [57]. If the network is considered immobile then the model above reduces to a Brinkman fluid [58]. In this limit the swimming speed is seen to be systematically enhanced by a factor

$$\frac{U}{U_N} = \sqrt{1 + \Gamma/(\eta k^2)}, \tag{8.97}$$

as shown by Leshansky [59]. As expected if we take $kl_s \gg 1$ then the swimming speed reduces to that in a Newtonian fluid.

A two-fluid model was also used by Du et al. to study two intermixed Newtonian fluids of different viscosities [60]. In this case it was shown that swimming in a mixture of two Newtonian fluids is always slower and less efficient than swimming in a single viscous fluid.

4.6 Collective Effects

When microorganisms are swimming close to one another, they may interact hydrodynamically. Namely the presence of a flow field created by one organism affects the dynamics of nearby swimmers and vice versa. Because the decay of a low-Reynolds number flow field is often long tailed, hydrodynamic interactions may significantly alter collective dynamics. For instance, dense suspensions of microswimmers display transient ordered flow structures with much larger length and velocity scales than the organisms themselves [61]. Hydrodynamic interactions also enable synchronous flagellar beating for the alga *Chlamydomonas*, synchrony that then facilitates directed locomotion [62].

Non-Newtonian fluids affect not only how microorganisms self-propel individually but also impact hydrodynamic interactions between microorganisms in close proximity. A striking example of hydrodynamic synchronization occurs in the phase locking of two (or several) spermatozoa flagella when they are near one another [63]. This synchronization leads to a speed increase of the group of cells and thereby a competitive advantage.

Taylor first attempted to model synchronization in the Newtonian case by solving for the hydrodynamic interaction between two sinusoidal sheets. He found that energy dissipation in the fluid is minimized if the two sheets are oscillating with no phase difference [28]. However, in this symmetric sinusoidal setup, there can be no evolution of the phase from an arbitrary initial condition, due to the kinematic reversibility of the Stokes flow-field equations [64, 65]. If two such sheets are swimming in a viscoelastic fluid, such as those present along the path through the female reproductive system, then kinematic reversibility no longer constrains the dynamics. Two Taylor sheets are found to systematically synchronize to an inphase conformation in an Oldroyd-B fluid [66].

Here again, the reciprocal theorem may be used to study the interactions between two general sheets in a complex fluid. Taking into account the force-free motion of both sheets we find that in an Oldroyd-B fluid the phase, ϕ, between two sheets evolves in time as

$$\frac{\mathrm{d}\phi}{\mathrm{d}t} = -\varepsilon^2\omega\sum_n 2c_n c_n^\dagger n\sin(n\phi)C(nh)\frac{G^*(n)}{\eta_0\omega}, \tag{8.98}$$

where $C(x) = (x\sinh x + x^2\cosh x)/(\sinh^2 x - x^2)$. The rate of sheet synchronization is thus dependent on the elastic modulus of the of the fluid, $\mathscr{R}[G^*/\eta_0\omega] = n^2(1-\beta)\mathrm{De}/(1+n^2\mathrm{De}^2)$ where $G^*(n) = in\omega\eta^*(n)$, rather than the viscous modulus which affects the collective locomotion speed. Clearly, in a Newtonian fluid ($\mathrm{De} = 0$) there is no evolution of phase in time while the addition of viscoelastic forces ($\mathrm{De} \neq 0$ and $\beta \neq 1$) leads to the evolution of all initial configurations to an in-phase state, $\phi = 0$ [34]. Similar results were recently demonstrated numerically [55], a study that also addressed the transient evolution of stress from an initial condition.

A variety of numerical and analytical studies have examined the evolution of suspensions of swimmers in Newtonian fluids, as reviewed for example by Saintillan and Shelley in the following chapter of this book (Chap. 9). Recent studies have adapted the mean-field theories for suspensions in Newtonian fluids [67–70] to quantify the impact of the addition of viscoelasticity [71, 72]. Specifically a polymeric stress is incorporated into the mean-field description of the flow which is forced by a configurational average of the force dipoles exerted by the swimmers on the fluid. Under these conditions the stability of an isotropic suspension is qualitatively unchanged for up De $\approx \mathcal{O}(1)$ while showing significant variation for larger Deborah number.

We finally note that the addition of swimmers (more generally, a suspension of active particles) modifies the apparent rheology of the fluid when viewed from a continuum perspective [73]. That swimmers should modify the bulk rheology of a system is already clear by simply looking at the passive analogue. As discussed earlier in this book (Chap. 3), a dilute suspension of rigid spheres leads to the famous Einstein correction to viscosity [74], while the addition of Brownian motion and weakly anisotropic particles leads to non-Newtonian rheological features [75]. Passive rod-shaped Brownian particles tend to align with imposed shear on average and hence a suspension of rods, while possessing a larger zero-shear rate, displays shear thinning at higher shear rates. The preferred alignment of the swimmers with the shear direction then leads, in the active case, to additional stresslets imposed on the fluid and further impact the rheological properties.

5 Perspective

Despite the fact that the first studies on the locomotion of microorganisms in non-Newtonian fluids are over thirty years old, the bulk of the field has developed very recently. In spite of recent progress in developing a theory to describe the effects which non-Newtonian fluids have on swimming kinematics, many subtleties remain to be parsed out. What we do know is that there is no simple answer as to the impact of viscoelasticity on locomotion. Changes in the gait of a swimmer seem to lead to drastic changes in the non-Newtonian effects. For example, when propagating helical waves, a swimmer is retarded by a viscoelastic (Boger) fluid if the helical amplitude is small but enhanced if the helical amplitude is large. In contrast, for swimmers propagating sinusoidal undulatory waves, elastic stresses only serve to slow it down regardless of the amplitude of the motion, yet if the amplitude is increasing tip to tail on finite swimmers, a speed increase may yet be obtained.

In general complex fluids are both shear-dependent and viscoelastic, and biological swimming gaits show a great deal of diversity. Developing an understanding of the differences which arise amongst various gaits and the non-Newtonian response of a particular fluid will lead to insight into the swimming strategies observed in different natural environments. In particular, future work should help shed light on how organisms passively or actively modulate their behavior to cope with complex stresses. This understanding may also lead to more effective designs of artificial microswimmers for use in biological environments such as in therapeutic delivery devices. As is often the case in physics, progress in this field so far has been achieved through careful analysis of the locomotion of simple model swimmers which then raises questions to be addressed by numerical simulation and experiment and we hope that in the future, work will be also fueled by novel biological experimental insights.

Acknowledgment GE gratefully acknowledges funding from the Natural Science and Engineering Research Council of Canada while EL thanks the European Union (through a CIG Grant) for partial support.

References

1. D.F. Katz, R.N. Mills, T.R. Pritchett, J. Reprod. Fertil. **53**, 259 (1978)
2. P.F. Dunn, B.F. Picologlou, Biorheol. **13**, 379 (1976)
3. S.S. Suarez, A.A. Pacey, Human Reprod. Update **12**, 23 (2006)
4. S.S. Suarez, D.F. Katz, D.H. Owen, J.B. Andrew, R.L. Powell, Biol. Reprod. **44**, 375 (1991)
5. S.S. Suarez, X.B. Dai, Biol. Reprod. **46**, 686 (1992)
6. J.P. Celli, B.S. Turner, N.H. Afdhal, S. Keates, I. Ghiran, C.P. Kelly, R.H. Ewoldt, G.H. McKinley, P. So, S. Erramilli, R. Bansil, Proc. Natl. Acad. Sci. USA **106**, 14321 (2009)
7. E.M. Purcell, Am. J. Phys. **45**, 11 (1977)
8. E. Lauga, T.R. Powers, Rep. Prog. Phys. **72**, 096601 (2009)
9. E. Lauga, Europhys. Lett. **86**, 64001 (2009)
10. H.A. Stone, A.D.T. Samuel, Phys. Rev. Lett. **77**, 4102 (1996)
11. J. Happel, H. Brenner, *Low Reynolds Number Hydrodynamics* (Prentice-Hall, Inc., New Jersey 1965)
12. H. Brenner, Chem. Eng. Sci. **18**, 1 (1963)
13. K. Ishimoto, M. Yamada, SIAM J. Appl. Math. **72**, 1686 (2012)
14. J.R. Blake, J. Fluid Mech. **46**, 199 (1971)
15. B. Felderhof, R. Jones, Phys. A **202**, 94 (1994)
16. B. Felderhof, R. Jones, Phys. A **202**, 119 (1994)
17. E. Lauga, Soft Matter **7**, 3060 (2011)
18. G.R. Fulford, D.F. Katz, R.L. Powell, Biorheol. **35**, 295 (1998)
19. D.J. Smith, J.R. Blake, E.A. Gaffney, Proc. R. Soc. A. **465**, 2417 (2009)
20. H.C. Berg, D.A. Brown, Nature **239**, 500 (1972)
21. D.F. James, Annu. Rev. Fluid Mech. **41**, 129 (2009)
22. R.B. Bird, R.C. Armstrong, O. Hassager, *Dynamics of Polymeric Liquids. Fluid Mechanics*, 2nd edn. vol. 1 (Wiley-Interscience, New York, NY, 1987)
23. A.S. Khair, T.M. Squires, Phys. Rev. Lett. **105**, 156001 (2010)
24. O.S. Pak, L. Zhu, L. Brandt, E. Lauga, Phys. Fluids **24**, 103102 (2012)
25. N.C. Keim, M. Garcia, P.E. Arratia, Phys. Fluids **24**, 081703 (2012)
26. T. Normand, E. Lauga, Phys. Rev E **78**, 061907 (2008)
27. O.S. Pak, T. Normand, E. Lauga, Phys. Rev. E **81**, 036312 (2010)
28. G.I. Taylor, Proc. R. Soc. Lond. A **209**, 447 (1951)
29. T.K. Chaudhury, J. Fluid Mech. **95**, 189 (1979).
30. L.D. Sturges, J. Non-Newtonian Fluid Mech. **8**, 357 (1981)
31. E. Lauga, Phys. Fluids **19**, 083104 (2007)
32. H.C. Fu, T.R. Powers, C.W. Wolgemuth, Phys. Rev. Lett. **99**, 258101 (2007)

33. H.C. Fu, C.W. Wolgemuth, T.R. Powers, Phys. Fluids **21**, 033102 (2009)
34. G.J. Elfring, E. Lauga, Phys. Fluids **23**, 011902 (2011)
35. B. Liu, T.R. Powers, K.S. Breuer, Proc. Natl. Acad. Sci. USA **108**, 19516 (2011)
36. S.E. Spagnolie, B. Liu, T.R. Powers, Phys. Rev. Lett. **111**, 068101 (2013)
37. J. Teran, L. Fauci, M. Shelley, Phys. Rev. Lett. **104**, 038101 (2010). DOI 10.1103/PhysRevLett.104.038101
38. X.N. Shen, P.E. Arratia, Phys. Rev. Lett. **106**, 208101 (2011)
39. L. Zhu, M. Do-Quang, E. Lauga, L. Brandt, Phys. Rev. E **83**, 011901 (2011)
40. L. Zhu, E. Lauga, L. Brandt, Phys. Fluids **24**, 051902 (2012)
41. J.E. Drummond, J. Fluid Mech. **25**, 787 (1966)
42. M. Sauzade, G.J. Elfring, E. Lauga, Phys. D **240**, 1567 (2011).
43. C. Bender, S. Orszag, *Advanced Mathematical Methods for Scientists and Engineers* (McGraw-Hill, Nweyork, 1978)
44. A. Morozov, Swimming of Taylor wavy sheets in viscoelastic fluids. Bull. Am. Phys. Soc. **59**, (2014)
45. M. Dasgupta, B. Liu, H.C. Fu, M. Berhanu, K.S. Breuer, T.R. Powers, A. Kudrolli, Phys. Rev. E **87**, 013015 (2013)
46. J.R. Vélez-Cordero, E. Lauga, J. Non-Newton. Fluid Mech. **199**, 37 (2013).
47. T.D. Montenegro-Johnson, A.A. Smith, D.J. Smith, D. Loghin, J.R. Blake, Euro. Phys. J. E **35**, 1 (2012)
48. T.D. Montenegro-Johnson, D.J. Smith, D. Loghin, Phys. Fluids **25**, 081903 (2013)
49. B. Chan, N.J. Balmforth, A.E. Hosoi, Phys. Fluids **17**, 113101 (2005).
50. E. Lauga, A.E. Hosoi, Phys. Fluids **18**, 113102 (2006).
51. N.J. Balmforth, D. Coombs, S. Pachmann, Q.J. Mechanics Appl. Math. **63**, 267 (2010)
52. I.H. RiedelKruse, A. Hilfinger, J. Howard, F. Jülicher, HFSP J. **1**, 192 (2007)
53. A. Najafi, R. Golestanian, Phys. Rev. E **69**, 062901 (2004)
54. M.P. Curtis, E.A. Gaffney, Phys. Rev. E **87**, 043006 (2013)
55. J.C. Chrispell, L.J. Fauci, M. Shelley, Phys. Fluids **25** (2013)
56. H.C. Berg, L. Turner, Nature **278**, 349 (1979)
57. H.C. Fu, V.B. Shenoy, T.R. Powers, Europhys. Lett. **91**, 24002 (2010)
58. H.C. Brinkman, Appl. Sci. Res. **1**, 27 (1949)
59. A.M. Leshansky, Phys. Rev. E **80**, 051911 (2009)
60. J. Du, J.P. Keener, R.D. Guy, A.L. Fogelson, Phys. Rev. E **85**, 036304 (2012)
61. C. Dombrowski, L. Cisneros, S. Chatkaew, R.E. Goldstein, J.O. Kessler, Phys. Rev. Lett. **93**, 098103 (2004)
62. M. Polin, I. Tuval, K. Drescher, J.P. Gollub, R.E. Goldstein, Science **325**, 487 (2009)
63. D.M. Woolley, R.F. Crockett, W.D.I. Groom, S.G. Revell, J. Exp. Biol. **212**, 2215 (2009)
64. G.J. Elfring, E. Lauga, Phys. Rev. Lett. **103**, 088101 (2009)
65. G.J. Elfring, E. Lauga, J. Fluid Mech. **674**, 163–173 (2011)
66. G.J. Elfring, O.S. Pak, E. Lauga, J. Fluid Mech. **646**, 505 (2010)
67. D. Saintillan, M.J. Shelley, Phys. Rev. Lett. **100**, 178103 (2008)
68. D. Saintillan, M.J. Shelley, Phys. Fluids **20** (2008)
69. C. Hohenegger, M.J. Shelley, Phys. Rev. E **81**, 046311 (2010)
70. D. Saintillan, M.J. Shelley, C. R. Physique **14**, 497 (2013)
71. Y. Bozorgi, P.T. Underhill, Phys. Rev. E **84**, 061901 (2011)
72. Y. Bozorgi, P.T. Underhill, J. Rheol. **57**, 511 (2013)
73. Y. Hatwalne, S. Ramaswamy, M. Rao, R.A. Simha, Phys. Rev. Lett. **92**, 118101 (2004)
74. A. Einstein, *Investigations on the Theory of the Brownian Movement* (Courier Dover Publications, Newyork, 1956)
75. L.G. Leal, E.J. Hinch, J. Fluid Mech. **55**, 745 (1972)

Chapter 9
Theory of Active Suspensions

David Saintillan and Michael J. Shelley

Abstract Active suspensions, of which a bath of swimming microorganisms is a paradigmatic example, denote large collections of individual particles or macromolecules capable of converting fuel into mechanical work and microstructural stresses. Such systems, which have excited much research in the last decade, exhibit complex dynamical behaviors such as large-scale correlated motions and pattern formation due to hydrodynamic interactions. In this chapter, we summarize efforts to model these systems using particle simulations and continuum kinetic theories. After reviewing results from experiments and simulations, we present a general kinetic model for a suspension of self-propelled rodlike particles and discuss its stability and nonlinear dynamics. We then address extensions of this model that capture the effect of steric interactions in concentrated systems, the impact of confinement and interactions with boundaries, and the effect of the suspending medium rheology. Finally, we discuss new active systems such as those that involve the interactions of biopolymers with immersed motor proteins and surface-bound suspensions of chemically powered particles.

1 Background

The emerging field of *soft active matter* has excited much research in the last decade in areas as diverse as biophysics, colloidal science, fluid mechanics, and statistical physics [7–9]. Broadly speaking, an active matter system consists of a large collection of individual agents, such as particles or macromolecules, that convert some form of energy (typically chemical) into mechanical work. This work, in turn,

D. Saintillan (✉)
Department of Mechanical and Aerospace Engineering, University of California, San Diego, La Jolla, CA 92093, USA
e-mail: dstn@ucsd.edu

M.J. Shelley
Courant Institute of Mathematical Sciences, New York University, New York, NY 10012, USA
e-mail: shelley@cims.nyu.edu

S.E. Spagnolie (ed.), *Complex Fluids in Biological Systems*, Biological and Medical Physics, Biomedical Engineering,
DOI 10.1007/978-1-4939-2065-5_9

leads to microstructural changes in the system, either via direct contact interactions or through long-ranged nonlocal interactions mediated by a suspending medium. Dramatic manifestations of these interactions include spontaneous unsteady flows on mesoscopic length scales, the formation of complex spatiotemporal patterns, and the emergence of directed collective motion. A wide variety of biological and physical systems fall into this broad definition, including (see Fig. 9.1): suspensions of self-propelled microorganisms such as motile bacteria and microscopic algae [10–12], the cell cytoskeleton and cytoplasm [13–15], solutions of motor proteins and biological filaments such as actin [4] and microtubules [3, 16, 17], reactive and driven colloidal suspensions [18–22], reactive emulsions [5], and shaken granular materials [23, 24]. A central question in all of these systems is the relation between the mechanics and interactions on the scale of individual particles and the ensuing self-organization and collective dynamics on the system scale [25].

Of particular interest to us here are so-called *wet* active systems, or *active suspensions*, in which the active particles are suspended in a viscous fluid and long-ranged hydrodynamic interactions are important. Numerous experiments have focused on the dynamics in suspensions of swimming bacteria. Some of the observations that have been made on this system include: the emergence of

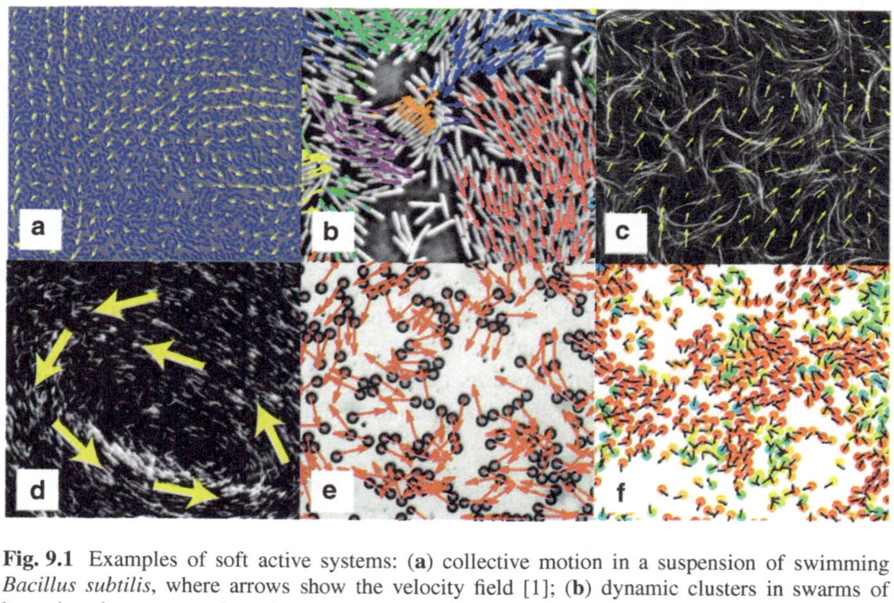

Fig. 9.1 Examples of soft active systems: (**a**) collective motion in a suspension of swimming *Bacillus subtilis*, where arrows show the velocity field [1]; (**b**) dynamic clusters in swarms of bacteria, where arrows show the direction of motion of the particles [2]; (**c**) spontaneous motion in a suspension of microtubules and kinesin motors confined at a two-dimensional interface [3]; (**d**) large-scale swirling motion in a suspension of actin filaments transported by wall-tethered myosin molecular motors [4]; (**e**) swarming of self-propelling liquid droplets in a Hele-Shaw cell [5]; (**f**) long-range order of vibrated polar disks on a two-dimensional substrate [6]. (Reproduced with permission)

complex chaotic flows on length scales much greater than the particle dimensions and characterized by unsteady whirls and jets [10–12, 26, 27], enhanced particle velocities [10], a transition to collective motion when the bacterial density exceeds a certain threshold [11], local polar ordering [11], complex patterns and density fluctuations [28], enhanced swimmer and passive tracer diffusion [29–32], efficient fluid mixing [28, 33, 34], and bizarre rheologies created by particle activity [35–38].

The key ingredient to understanding how hydrodynamic interactions can yield such phenomena is the fluid flow set up by an isolated swimming particle. Because of their small sizes and the highly viscous environments in which they live, biological swimmers such as bacteria and microphytes move in the realm of low Reynolds numbers, where inertial forces are negligible and viscous stresses dominate [39]. In this regime, typical macroscopic mechanisms for locomotion are inefficient (or even inoperative) and novel strategies have evolved that are based on so-called non-reciprocal shape deformations [40]. Common locomotion mechanisms are based on the beating or rotation of flagellar appendages or the propagation of metachronal waves on the surface of ciliated cells [39,41]. In the case of flagellar propulsion, which is the typical mode of locomotion in motile bacteria such as *Bacillus subtilis* and *Escherichia coli* as well as certain types of microalgae including *Chlamydomonas reinhardtii*, the cyclic nonreciprocal deformation of the flagella imparts a net propulsive thrust $\mathbf{F}_p$ on the surrounding fluid in the direction opposite the net swimming motion, which we henceforth characterize in terms of a unit vector $\mathbf{p}$. As microorganisms are typically neutrally buoyant, or nearly so, the net force on a particle must vanish in the limit of zero Reynolds number unless an external field is applied, and therefore an equal and opposite viscous drag force $\mathbf{F}_d = -\mathbf{F}_p$ is also exerted on the fluid by the other parts of the organism (typically the cell body). This simple description of the forces on a microorganism suggests that their net effect on the suspending fluid is a force dipole, which drives a long-ranged flow with slow $1/r^2$ spatial decay in three dimensions, where r is the distance from the particle center. In the far field, the fluid velocity at relative position $\mathbf{r}$ from the particle can be expressed as

$$\mathbf{u}^d(\mathbf{r}|\mathbf{p}) = \mathbf{S}(\mathbf{p}) : \nabla \mathbf{J}(\mathbf{r}) \tag{9.1}$$

in terms of the fundamental solution $\mathbf{J}(\mathbf{r}) = (1/8\pi\eta)(\mathbf{I} + \hat{\mathbf{r}}\hat{\mathbf{r}})/r$ of the Stokes equations or response to a localized point force [44]. In Eq. (9.1), η denotes the viscosity of the fluid, and the second-order tensor $\mathbf{S}$, called the *stresslet*, is the symmetric first moment of the stresses exerted by the particle on the fluid and can be obtained as $\mathbf{S}(\mathbf{p}) = \sigma_0 \mathbf{pp}$. Its magnitude is given by $\sigma_0 = \pm|\mathbf{F}_p|\ell$, where ℓ is the distance between the points of application of the thrust and drag forces and scales with the particle length. The sign of σ_0 depends on the position of the thrust and drag forces relative to the swimming direction: $\sigma_0 < 0$ for a *pusher* particle that exerts a thrust with its tail (such as *B. subtilis* and *E. coli*), whereas $\sigma_0 > 0$ for a head-actuated *puller* particle (such as *C. reinhardtii*). A simple force balance on the cell body and based on Stokes drag also shows that the stresslet is linearly related to the swimming speed V_s of the particle as $\sigma_0 \propto V_s \eta \ell^2$. Schematic diagrams of pusher and

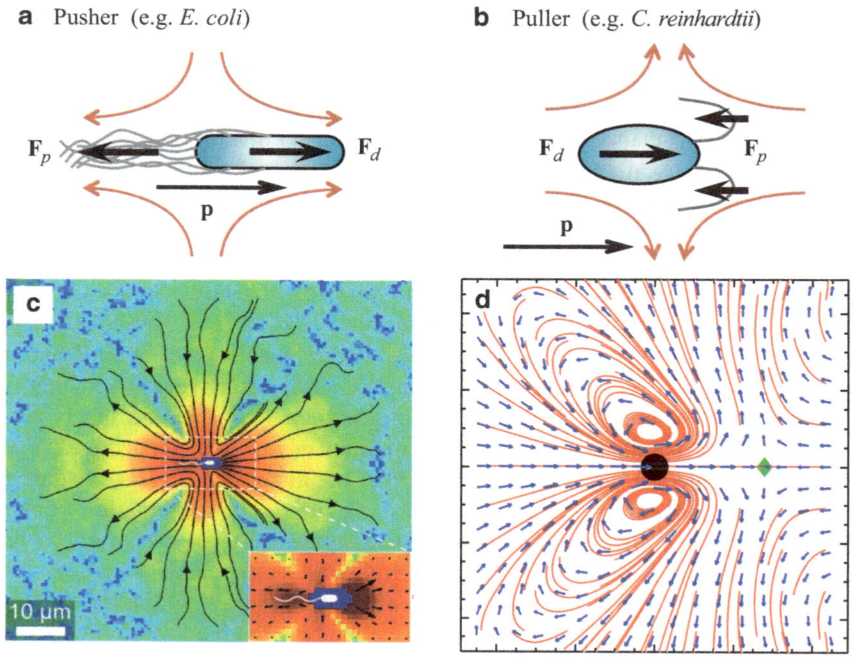

Fig. 9.2 (**a**)–(**b**) Schematic diagrams of pusher and puller particles, of which *E. coli* and *C. reinhardtii* are paradigmatic examples. The swimming direction is indicated by **p**, and the *red arrows* show the direction of the induced fluid flow. (**c**) Experimental measurement of the flow field near an individual *E. coli*, showing good agreement with an extensile dipole flow [42]. (**d**) Experimental measurement of the time-averaged flow field near an individual *C. reinhardtii*, showing a complex flow in the near field and good agreement with a contractile dipole flow in the far field [43]. (Parts (**c**) and (**d**) reproduced with permission)

puller particles and their flows are shown in Figs. 9.2a–b. It is interesting to note that the notion of pushers and pullers is not limited to self-propelled particles. One can indeed envisage a particle that exerts a force dipole on the fluid ($\sigma_0 \neq 0$) but does not swim ($V_s = 0$), and so-called shakers, which can be either pushers or pullers, have been proposed as a very basic model for the dynamics of suspensions of microtubule bundles that extend in length due to motor-protein activity [3]. Immotile force dipoles could also be produced by elaborations of the synthesis process that produces the motile chemically-powered motors discussed in Sect. 4.2.

The elementary description of a swimming particle in terms of a force dipole has been tested experimentally with relative success. As illustrated in Fig. 9.2c, Drescher et al. [42] used particle-image velocimetry to measure the flow field around isolated *E. coli* cells and found good agreement with the flow field predicted by Eq. (9.1) with a negative stresslet, although strong noisy fluctuations were reported in the far field where the velocity field is the weakest. Similarly, Guasto et al. [43] and Drescher et al. [45] observed the flow field driven by *C. reinhardtii*: they uncovered a complex near-field flow structure that is best captured by a set of three off-centered point forces (corresponding to the cell body and two anterior flagella), but confirmed that the far-field flow can again be modeled as a dipole flow with a net positive stresslet. Guasto et al. [43] also noted that the flow around *C. reinhardtii* is time-periodic, with period equal to the duration of a swimming stroke, and in fact even reverses direction over the course of one stroke: we will not consider such time-dependence in the following discussion and only focus on the effect of the net time-averaged flow. Recent theoretical models, however, have suggested that such unsteady dynamics can lead to synchronization and novel instabilities as a result of hydrodynamic interactions [46, 47].

Knowledge of the velocity field driven by an isolated particle forms the basis for the modeling and study of hydrodynamic interactions between swimmers. While this description has been used to consider pair interactions [48–50], it can also be deployed to model large-scale suspensions. Hernández-Ortiz et al. [51] developed a minimal swimmer model in which a self-propelled microorganism is represented as a rigid bead-rod dumbbell. Propulsion arises as a result of a "phantom flagellum" exerting a force on the fluid at an off-centered point along the swimmer axis, causing the translation of the dumbbell at a velocity $V_s = |\mathbf{F}_p|/2\zeta$, where $\zeta = 6\pi\eta a$ is the drag coefficient of each bead of radius a, and where hydrodynamic interactions between the two beads have been neglected. Because the propulsive force exerted by the flagellum is exactly balanced by the total drag on the dumbbell, the leading effect on the fluid is again that of a force dipole. This model was applied to simulate confined suspensions of many swimmers [51, 52], where it was shown to capture many qualitative features observed in experiments on bacterial suspensions, including enhanced diffusivities and large-scale correlated flows. More elaborate simulation models have also been developed over the years, though at the cost of increased computational complexity. This includes Pedley and coworkers' Stokesian dynamics simulations of spherical *squirmers* [53], which propel as a result of a prescribed surface slip velocity and are an appropriate model for ciliated microorganisms. These simulations also showed enhanced motile particle and tracer diffusion [54, 55], as well as the development of large-scale coherent structures [56–58]. In recent work, we also developed detailed simulations of active suspensions based on a slender-body model for hydrodynamically interacting rodlike particles [59, 60]. In this model, the particles propel themselves by exerting a prescribed tangential stress on some part of their surfaces, and both pusher and puller particles can be modeled by an appropriate choice of the stress distribution. In semi-dilute suspensions of pushers, large-scale chaotic flows taking place near the system size were observed (Fig. 9.3), together with increased swimming speeds and strong particle diffusion; no such dynamics were found in suspensions of pullers, which always remained uniform and isotropic.

Such particle simulations are useful for testing models and for detailed comparison with experiments, but are often too costly to simulate systems of realistic sizes and only yield limited analytical insight into the physical mechanisms involved. Another approach, which circumvents these limitations, consists in modeling the suspension as a continuum. Several continuum models have been developed for

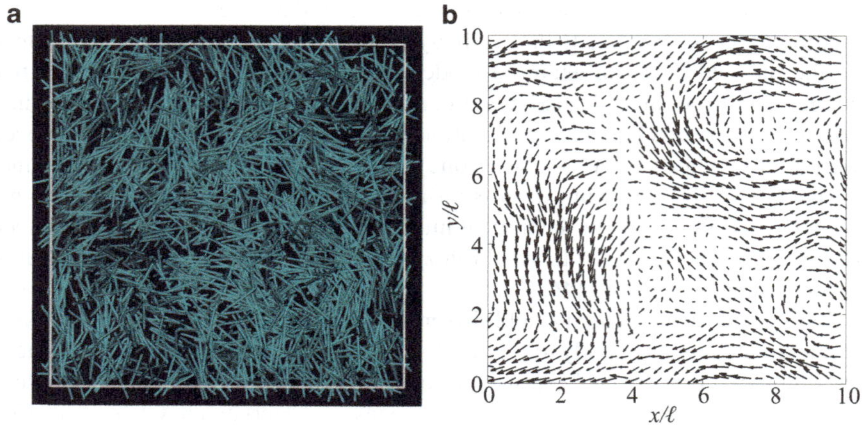

Fig. 9.3 Numerical simulation of a semi-dilute suspension of self-propelled slender-rods above the onset of collective motion [59]: (**a**) snapshot of the particle distribution, showing coherent dynamic clusters with local orientational order; (**b**) hydrodynamic velocity field in a plane. (Adapted with permission)

active suspensions, using a variety of approaches. In a seminal paper, Simha and Ramaswamy [61] extended phenomenological models for passive polar liquid crystals to account for activity. They wrote down an evolution equation for the polarization field $\mathbf{n}(\mathbf{r},t)$, which will be defined more precisely in Sect. 2.1, in which terms accounting for self-propulsion, diffusion, and rotation by the mean-field flow were included. This evolution equation was coupled to the Navier-Stokes equations for the fluid motion, forced by an active stress term capturing the effect of the force dipoles on the fluid. Based on this model, they predicted in the Stokesian limit a long-wave instability of globally aligned suspensions. Other phenomenological models have been proposed to account for additional effects such as steric interactions, which are included via *ad hoc* terms constructed based on symmetries [8, 12, 62–65].

In another related approach, which is the focus of this chapter, kinetic equations are self-consistently derived from a first-principles mean-field description of interactions between particles using coarse-graining. Such a model was introduced in our previous work [66, 67] and is based on a Smoluchowski equation for the conservation of the particle probability distribution function, in which the fluxes describe the linear and angular motions of the particles in the mean-field hydrodynamic flow driven by self-propulsion. This flow is obtained by solution of the Stokes equations forced by a coarse-grained active stress tensor similar to that used in the model of Simha and Ramaswamy [61]. This coupled system of partial differential equations can then be analyzed theoretically in the vicinity of theoretically relevant base states, or integrated numerically to investigate dynamics in the nonlinear regime. Extensions to include more complex effects such as an external flow [68], chemotaxis in a chemical field [69, 70], or steric interactions at high concentrations [71] have also been described.

In this chapter, we review our recent theoretical work on the continuum modeling of active suspensions. We begin in Sect. 2 by deriving a basic kinetic model for a suspension of slender swimmers interacting via force-dipole hydrodynamic interactions, where we show that the dynamics can be captured by a Smoluchowski equation for the particle distribution function, coupled to the Stokes equations for the fluid velocity in which the effect of the force dipoles on the flow is shown to amount to an effective active stress. After discussing theoretical and computational results on this model, a number of extensions are presented in Sects. 3 and 4, and we conclude in Sect. 5.

2 A Simple Kinetic Model

2.1 Smoluchowski Equation

In this section, we present the basic kinetic model introduced in our previous work [66, 67], which shares similarities with classic models for suspensions of passive rodlike particles [72, 73]. A very similar theory for active suspensions was independently proposed by Subramanian and Koch [74].

In the present model, we describe the configuration of the suspension at time t in terms of the probability distribution function $\Psi(\mathbf{r},\mathbf{p},t)$ of finding a particle with center-of-mass position $\mathbf{r}$ and orientation $\mathbf{p}$ (with $|\mathbf{p}|^2 = 1$). It is normalized as

$$\frac{1}{V}\int_V \int_\Omega \Psi(\mathbf{r},\mathbf{p},t)\,d\mathbf{p}\,d\mathbf{r} = n, \tag{9.2}$$

where V is the volume of the system, Ω is the unit sphere of orientations, and $n = N/V$ is the mean number density in a suspension of N particles. Conservation of particle number requires that $\Psi(\mathbf{r},\mathbf{p},t)$ satisfy the Smoluchowski equation [72]

$$\partial_t \Psi + \nabla_r \cdot (\dot{\mathbf{r}}\Psi) + \nabla_p \cdot (\dot{\mathbf{p}}\Psi) = 0, \tag{9.3}$$

where $\nabla_p = (\mathbf{I} - \mathbf{pp}) \cdot (\partial/\partial\mathbf{p})$ denotes the gradient operator on the unit sphere. The flux velocities $\dot{\mathbf{r}}$ and $\dot{\mathbf{p}}$ describe the linear and angular motions of the particles in the suspension. The linear velocity of a particle is expressed as the sum of the single-particle swimming velocity $V_s\mathbf{p}$ (assumed to be unaffected by interactions) and the local background fluid velocity $\mathbf{u}(\mathbf{r},t)$ and also includes a contribution from translational diffusion with diffusivity D (assumed to be isotropic):

$$\dot{\mathbf{r}} = V_s\mathbf{p} + \mathbf{u} - D\nabla_r \ln \Psi. \tag{9.4}$$

The rotational velocity of a swimmer is modeled as

$$\dot{\mathbf{p}} = (\mathbf{I} - \mathbf{pp}) \cdot (\beta \mathbf{E} + \mathbf{W}) \cdot \mathbf{p} - d\nabla_p \ln \Psi. \tag{9.5}$$

The first term on the right-hand side captures rotation of an anisotropic particle in the local flow according to Jeffery's equation [75], where $\mathbf{E} = (\nabla\mathbf{u} + \nabla\mathbf{u}^T)/2$ and $\mathbf{W} = (\nabla\mathbf{u} - \nabla\mathbf{u}^T)/2$ denote the rate-of-strain and vorticity tensors, respectively. The parameter β characterizes the shape of the particle, with $\beta = (a^2 - 1)/(a^2 + 1)$ for a spheroid of aspect ratio a and $\beta \approx 1$ for a slender particle [76]. Rotational diffusion is also included with diffusivity d.

Equations (9.4)–(9.5), and in particular the contributions from the mean-field flow $\mathbf{u}(\mathbf{r},t)$, are strictly valid for a linear flow field and provide an accurate estimate of the velocities if the characteristic length scale of the flow is much greater than the particle size, a good approximation in a sufficiently dilute suspension. If velocity variations on the scale of a particle are significant, these can be captured using the more accurate Faxén laws for a slender body [77] or a spheroidal particle [44].

The physical origin of the diffusive terms in Eqs. (9.4)–(9.5) deserves some discussion. While Brownian diffusion due to thermal fluctuations can be significant in colloidal systems [19, 21], it is generally negligible in suspensions of biological swimmers. As demonstrated in experiments [42,78], diffusion still occurs in biological systems owing to shape imperfections or noise in the swimming actuation. In a dilute suspension, these effects can be described in terms of constant diffusion coefficients D_0 and d_0. We note, however, that rotational diffusion alone leads to a random walk in space owing to its coupling with the swimming motion, resulting in enhanced spatial diffusion at long times by a mechanism similar to generalized Taylor dispersion [79, 80], with a net translational diffusivity given by $D = D_0 + V_s^2/6d$ in three dimensions [60, 81]. In addition to diffusion due to noise, fluid-mediated hydrodynamic interactions between particles can also result in hydrodynamic diffusion in semi-dilute and concentrated systems. At fairly low concentrations ($n\ell^3 \lesssim 1$), a simple argument based on pair interactions suggests that $d \propto n\ell^3$, from which $D \propto (n\ell^3)^{-1}$ [74, 82], and such scalings have indeed been verified in particle simulations [60].

While the distribution function $\Psi(\mathbf{r},\mathbf{p},t)$ fully characterizes the configuration of the particles in the suspension, it is often useful to consider its orientational moments, which have easy physical interpretations. Of particular interest are the zeroth, first, and second moments, which correspond respectively to the concentration field $c(\mathbf{r},t)$, polar order parameter $\mathbf{n}(\mathbf{r},t)$, and nematic order parameter $\mathbf{Q}(\mathbf{r},t)$. These are defined as

$$c(\mathbf{r},t) = \langle 1 \rangle, \quad \mathbf{n}(\mathbf{r},t) = \frac{\langle \mathbf{p} \rangle}{c(\mathbf{r},t)}, \quad \mathbf{Q}(\mathbf{r},t) = \frac{\langle \mathbf{pp} - \mathbf{I}/3 \rangle}{c(\mathbf{r},t)}, \tag{9.6}$$

where $\langle \cdot \rangle$ denotes the orientational average:

$$\langle h(\mathbf{p}) \rangle = \int_\Omega h(\mathbf{p}) \Psi(\mathbf{r},\mathbf{p},t)\, d\mathbf{p}. \tag{9.7}$$

Evolution equations for c, $\mathbf{n}$, and $\mathbf{Q}$ can be obtained by taking moments of the Smoluchowski equation (9.3):

$$\mathscr{D}_t c = -V_s \nabla \cdot (c\mathbf{n}) + D\nabla^2 c, \tag{9.8}$$

$$\begin{aligned}\mathscr{D}_t(c\mathbf{n}) = &-V_s [\nabla \cdot (c\mathbf{Q}) + (1/3)\nabla c] + D\nabla^2 (c\mathbf{n}) \\ &+ (c\mathbf{I}\mathbf{n} - \langle \mathbf{ppp} \rangle) : (\beta \mathbf{E} + \mathbf{W}) - 2dc\mathbf{n},\end{aligned} \tag{9.9}$$

$$\begin{aligned}\mathscr{D}_t(c\mathbf{Q}) = &-V_s[\nabla \cdot \langle \mathbf{ppp} \rangle - (\mathbf{I}/3)\nabla \cdot (c\mathbf{n})] + D\nabla^2 (c\mathbf{Q}) \\ &+ \beta c[\mathbf{E} \cdot (\mathbf{Q} + \mathbf{I}/3) + (\mathbf{Q} + \mathbf{I}/3) \cdot \mathbf{E}] \\ &+ c[\mathbf{W} \cdot \mathbf{Q} - \mathbf{Q} \cdot \mathbf{W}] - 2\beta \langle \mathbf{pppp} \rangle : \mathbf{E} - 6dc\mathbf{Q},\end{aligned} \tag{9.10}$$

where $\mathscr{D}_t \equiv \partial_t + \mathbf{u} \cdot \nabla$ is the material derivative. Unsurprisingly perhaps, these equations involve the third and fourth moments $\langle \mathbf{ppp} \rangle$ and $\langle \mathbf{pppp} \rangle$ of the distribution function and can therefore only be used together with a closure model, unlike the more general and self-contained description in terms of $\Psi(\mathbf{r}, \mathbf{p}, t)$. Several closure models have been proposed in the past, usually based on various approximations such as weak or strong flow or near isotropy (see references in Saintillan and Shelley [9]) or by interpolating between such states [83].

2.2 *Mean-Field Flow and Active Stress Tensor*

Evolution of the Smoluchowski equation requires knowledge of the mean-field hydrodynamic velocity in the suspension. While this velocity could include a contribution from an external flow [68], we are primarily interested in the flow driven by the suspended particles themselves as they propel through the fluid. In a dilute suspension, the velocity $\mathbf{u}(\mathbf{r}, t)$ can then be obtained as the superposition of all the point dipole flows induced by individual particles. For a given distribution $\Psi(\mathbf{r}, \mathbf{p}, t)$, the velocity at point $\mathbf{r}$ is therefore expressed as a convolution:

$$\mathbf{u}(\mathbf{r}, t) = \int_V \int_\Omega \mathbf{u}^d(\mathbf{r} - \mathbf{r}_0 | \mathbf{p}) \Psi(\mathbf{r}_0, \mathbf{p}, t)\, d\mathbf{p}\, d\mathbf{r}_0, \tag{9.11}$$

where $\mathbf{u}^d(\mathbf{r}|\mathbf{p})$ is the single-particle dipolar flow given in Eq. (9.1). This single-particle flow can be shown to satisfy the Stokes equations forced by a dipole singularity as

$$-\eta \nabla^2 \mathbf{u}^d(\mathbf{r}) + \nabla q^d(\mathbf{r}) = \sigma_0 \mathbf{pp} \cdot \nabla \delta(\mathbf{r}), \quad \nabla \cdot \mathbf{u}^d(\mathbf{r}) = 0, \tag{9.12}$$

where $\delta(\mathbf{r})$ is the three-dimensional Dirac delta function and q^d denotes the pressure. By combining Eqs. (9.11) and (9.12), it is straightforward to show that the mean-field velocity $\mathbf{u}(\mathbf{r}, t)$ and its associated pressure field $q(\mathbf{r}, t)$ satisfy

$$-\eta\nabla^2\mathbf{u}(\mathbf{r},t)+\nabla q(\mathbf{r},t)=\int_V\int_\Omega \sigma_0\mathbf{pp}\cdot\nabla\delta(\mathbf{r}-\mathbf{r}_0)\Psi(\mathbf{r}_0,\mathbf{p},t)\,d\mathbf{p}\,d\mathbf{r}_0, \qquad (9.13)$$

together with the incompressibility condition $\nabla\cdot\mathbf{u}(\mathbf{r},t)=0$. After manipulations, this can be rewritten

$$-\eta\nabla^2\mathbf{u}(\mathbf{r},t)+\nabla q(\mathbf{r},t)=\nabla\cdot\langle\sigma_0\mathbf{pp}\rangle. \qquad (9.14)$$

The second-order tensor inside the divergence on the right-hand side is the local configurational average of the particle stresslet: $\langle\sigma_0\mathbf{pp}\rangle=\langle\mathbf{S}(\mathbf{p})\rangle$. Following classic theories for the stress in particle suspensions [84, 85], it can be interpreted as an extra stress induced by the particles, which we term *active stress* and define more precisely as

$$\Sigma^a(\mathbf{r},t)=\langle\sigma_0(\mathbf{pp}-\mathbf{I}/3)\rangle. \qquad (9.15)$$

We have made the tensor traceless by removing an isotropic tensor that only modifies the pressure but has no effect on the flow. It can be seen that the active stress is related to the nematic order parameter as: $\Sigma^a(\mathbf{r},t)=\sigma_0 c(\mathbf{r},t)\mathbf{Q}(\mathbf{r},t)$, implying that active stresses vanish in the isotropic state and are caused by the nematic alignment of the swimmers. We also note that the active stress has the same tensorial form as the Brownian stress $\Sigma^b(\mathbf{r},t)=\langle 3kT(\mathbf{pp}-\mathbf{I}/3)\rangle$ in suspensions of passive rodlike polymers [72]. One significant difference, however, lies in the sign of the stresslet strength σ_0, which is negative for pusher particles. Note also that the expression for the active stress tensor (9.15), which we derived here for a distribution of point dipoles, is in fact more general and can be used for a suspension of finite-sized axisymmetric particles such as swimming rods [9], though these more detailed derivations yield additional contributions of higher order in volume concentration [71, 86].

In the following, we find it useful to nondimensionalize lengths by the characteristic scale $l_c=\ell/\nu$, where $\nu=N\ell^3/V=n\ell^3$ is an effective volume fraction, and time by $t_c=V_s/l_c$. The distribution function Ψ is also scaled by the mean number density n. Upon these scalings, the dipole strength σ_0 appearing in the active stress tensor is replaced by a dimensionless signed coefficient $\alpha=\sigma_0/V_s\eta\ell^2$. After nondimensionalization, the basic kinetic system is given by

$$\partial_t\Psi+\nabla_r\cdot(\dot{\mathbf{r}}\Psi)+\nabla_p\cdot(\dot{\mathbf{p}}\Psi)=0, \qquad (9.16)$$

$$\dot{\mathbf{r}}=\mathbf{p}+\mathbf{u}-D\nabla_r\ln\Psi, \qquad (9.17)$$

$$\dot{\mathbf{p}}=(\mathbf{I}-\mathbf{pp})\cdot(\beta\mathbf{E}+\mathbf{W})\cdot\mathbf{p}-d\nabla_p\ln\Psi, \qquad (9.18)$$

$$-\nabla^2\mathbf{u}+\nabla q=\nabla\cdot\langle\alpha\mathbf{pp}\rangle \text{ and } \nabla\cdot\mathbf{u}=0, \qquad (9.19)$$

whose nondimensional coefficients, aside from the shape factor β, are the signed $O(1)$ parameter α and the rescaled diffusion coefficients D and d.

This model is very similar structurally to those developed by Doi and coworkers to describe the dynamics of passive rod suspensions [72, 87]. The primary differences are the additional contribution (i.e., $\mathbf{p}$) to Eq. (9.17) for $\dot{\mathbf{r}}$ coming from locomotion and that $\alpha > 0$ for the dipolar extra stress in passive rod suspensions. That said, the origins of the dipolar stress are very different in the two cases. For active suspensions it arises from the swimming of pushers or pullers, while in the passive case it arises from rotational thermodynamic fluctuations which we have neglected here. For both passive and active systems, the existence of global "entropy solutions" has recently been proved by Chen and Liu [88].

2.3 The Conformational Entropy

Much insight can be gained into the differences between pusher and puller suspensions by consideration of the system's conformational entropy [67], which we define in terms of the distribution function as

$$\mathscr{S}(t) = \int_V \int_\Omega \frac{\Psi}{\Psi_0} \ln\left(\frac{\Psi}{\Psi_0}\right) d\mathbf{p}\, d\mathbf{r}, \tag{9.20}$$

where $\Psi_0 = 1/4\pi$ denotes the constant value of Ψ for a uniform isotropic suspension. It is straightforward to show that the entropy is a positive quantity and that it reaches its minimum of zero only for $\Psi \equiv \Psi_0$. The entropy $\mathscr{S}(t)$ therefore provides a global measure of the level of fluctuations in the system, both orientational and spatial. When linearized about the uniform isotropic state Ψ_0, it reduces to the squared $\mathscr{L}_2$ norm in $\mathbf{r}$ and $\mathbf{p}$. Using the kinetic equations above, one can derive an expression for its rate of change:

$$4\pi \frac{d}{dt}\mathscr{S}(t) = -\frac{6}{\alpha}\int_V \mathbf{E}:\mathbf{E}\, d\mathbf{r} - \int_V \int_\Omega \left[D|\nabla_r \ln\Psi|^2 + d|\nabla_p \ln\Psi|^2\right] \Psi\, d\mathbf{p}\, d\mathbf{r}. \tag{9.21}$$

The last term in Eq. (9.21), which is always negative, arises due to diffusive processes which tend to homogenize the suspension and decrease the entropy. However, the first term on the right-hand side, which arises from active stresses in the fluid, can be either positive or negative depending on the sign of α. In a suspension of pullers ($\alpha > 0$), this term is negative definite and drives the system towards equilibrium. In the case of pushers ($\alpha < 0$), however, the active stress term becomes positive and can increase fluctuations in the system by driving $\mathscr{S}$ away from zero. This suggests that pusher suspensions may be subject to the spontaneous growth of fluctuations whereas pullers are not, and this fundamental difference between the role of active stresses in pusher and puller suspensions is further examined by a more detailed stability analysis as described next.

2.4 Stability of the Uniform Isotropic State

The uniform isotropic state $\Psi \equiv \Psi_0 = 1/4\pi$ is an exact steady solution of the above continuum model, whose stability can be investigated. Perturbing Ψ_0 by a plane wave of the form $\tilde{\Psi}(\mathbf{p},\mathbf{k})\exp(i\mathbf{k}\cdot\mathbf{r}+\lambda t)$ and linearizing the governing equations yields an eigenvalue problem for the growth rate λ and eigenmode $\tilde{\Psi}$ that can be solved numerically [66, 67]. In agreement with the analysis on the configurational entropy in Sect. 2.3, solutions of the eigenvalue problem reveal fundamentally different dynamics in suspensions of rear- and front-actuated swimmers. In puller suspensions, the real growth rate $\mathrm{Re}(\lambda)$ is found to be negative at all wavenumbers, indicating that the uniform isotropic state is stable to infinitesimal perturbations. This is indeed borne out by particle simulations in the dilute and semi-dilute regimes [59], which never show the emergence of collective motion. On the other hand, the solution of the eigenvalue problem for a suspension of pushers, which is shown in Fig. 9.4a, shows a positive growth rate at low wavenumbers, suggesting that long-wavelength fluctuations can amplify as a result of hydrodynamic interactions. Moreover, Fig. 9.4a shows that the fastest growing linear modes occur near $k = 0$, implying that the linear analysis does not yield a dominant length scale independent of system size.

Consideration of the eigenmodes demonstrates that this linear instability is not associated with the growth of concentration fluctuations ($\tilde{c} = 0$), but rather with the local nematic alignment of the particles. More precisely, the nematic order tensor parameter for the unstable eigenfunctions can be shown to be of the form

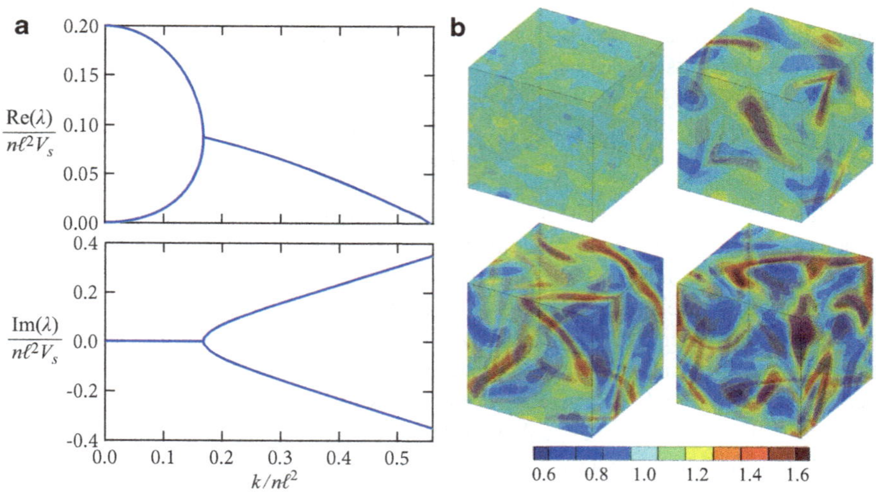

Fig. 9.4 (**a**) Real and imaginary parts of the complex growth rate λ for a plane-wave perturbation with respect to the uniform isotropic state as function of wavenumber k in the absence of diffusion. (**b**) Evolution of the concentration field $c(\mathbf{r},t)$ in a three-dimensional periodic simulation of a suspension of pushers, starting near the state of uniform isotropy

$$\tilde{\mathbf{Q}}(\mathbf{k}) = \hat{\mathbf{k}}\hat{\mathbf{k}}_\perp + \hat{\mathbf{k}}_\perp\hat{\mathbf{k}}, \tag{9.22}$$

where $\hat{\mathbf{k}}$ denotes the wave direction and $\hat{\mathbf{k}}_\perp$ is any direction orthogonal to $\hat{\mathbf{k}}$. In the language of liquid crystals, such spatial fluctuations of the nematic order parameter correspond to "bend" modes, as also predicted in other studies based on moment equations [89], and such bend modes are indeed visible in particle simulations such as that of Fig. 9.3a.

An important result also shown in Fig. 9.4a is the decay of the growth rate with increasing wavenumber, which even results in stabilization at high k. As discussed by Hohenegger and Shelley [90], this indicates that there exists a critical system size above which pusher suspensions become unstable. In dimensional variables, this criterion is written more specifically in terms of the linear system size $L = V^{1/3}$ as

$$\frac{L\nu}{\ell} \geq \frac{2\pi}{k_c}, \tag{9.23}$$

where k_c is the dimensionless wavenumber for which $\lambda(k_c) = 0$ and is a function of α and of the diffusion coefficients. The condition (9.23) states that instability occurs either in dense systems (large ν) or in large systems (large L/ℓ), and this criterion was systematically tested and confirmed in our previous particle simulations [59], where good agreement was found for the value of k_c.

Another interesting interpretation for this instability involves the active power input generated by the swimming particles in the fluid. The global power input $\mathscr{P}(t)$ was introduced in our previous work [67], where we also used an energy balance on the momentum equation (9.19) to show that it equates the rate of viscous dissipation in the fluid:

$$\mathscr{P}(t) = -\alpha \int_V \int_\Omega [\mathbf{E}(\mathbf{r},t) : \mathbf{pp}]\Psi(\mathbf{r},\mathbf{p},t)\, d\mathbf{p}\, d\mathbf{r} = \int_V 2\mathbf{E}(\mathbf{r},t) : \mathbf{E}(\mathbf{r},t)\, d\mathbf{r}. \tag{9.24}$$

Assuming a cubic periodic domain of unit length L, a simple application of Parseval's identity allows one to rewrite $\mathscr{P}(t)$ in terms of the Fourier coefficients $\tilde{\mathbf{E}}(\mathbf{k},t)$ of the rate-of-strain tensor, which themselves can be related to the Fourier coefficients $\tilde{\mathbf{Q}}(\mathbf{k},t)$ of the nematic order tensor parameter:

$$\mathscr{P}(t) = L^3 \sum_{\mathbf{k}} |\tilde{\mathbf{E}}(\mathbf{k},t)|^2 \approx \frac{L^3\alpha^2}{2} \sum_{\mathbf{k}} |(\mathbf{I} - \hat{\mathbf{k}}\hat{\mathbf{k}}) \cdot \tilde{\mathbf{Q}}(\mathbf{k},t) \cdot \hat{\mathbf{k}}|^2, \tag{9.25}$$

where the last term was obtained assuming that $c(\mathbf{r},t) \approx 1$, a valid approximation in the linear regime. From the form of the right-hand side, it is clear that only Fourier modes of the form of (9.22) will contribute to the power input, which can be interpreted as the total energy of the unstable bend modes in the system. The growth of $\mathscr{P}(t)$ therefore provides a direct measure of instability. Its evolution in particle simulations was considered in our previous work [59] and is shown in Fig. 9.5 for suspensions of pushers and pullers at various concentrations. In agreement

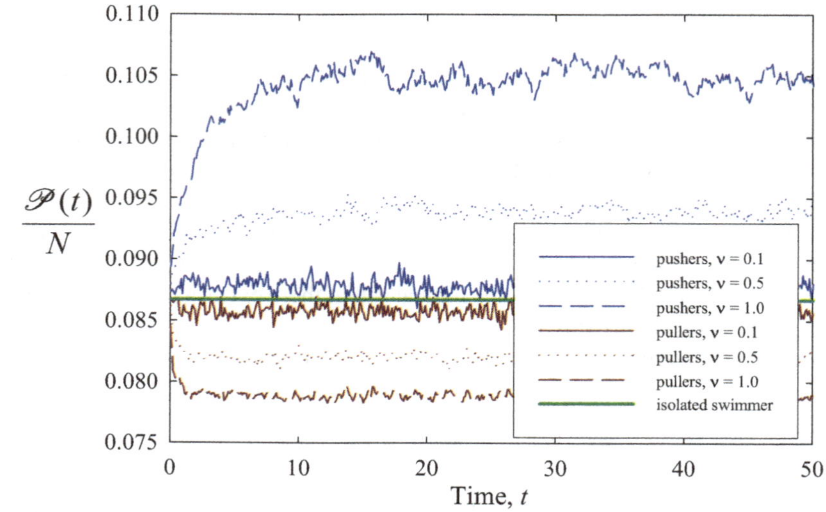

Fig. 9.5 Time evolution of the active power input per particle in direct numerical simulations of pushers and pullers in a cubic periodic box of size $L = 10\ell$ at various volume fractions ν. (Adapted with permission from Saintillan and Shelley [59])

with the theoretical prediction, the power only grows in sufficiently concentrated suspensions of pushers. In suspensions of pullers, it decreases below the dilute value corresponding to isolated swimmers, suggesting that particles in fact reorganize in a subtle way so as to suppress bend modes in the system.

After the initial transient growth, a statistical steady state is reached as a result of diffusive processes, which counteract the instability as expected from Eq. (9.21). No steady solution is observed in simulations, which instead show unsteady chaotic dynamics with the formation of dense and nematically aligned particle clusters that quasi-periodically form and break up over time as shown in Fig. 9.4b. As we argued above, the growth of concentration fluctuations is not predicted by the linear analysis, but it can be explained as a result of nonlinearities. Equation (9.8) for the concentration field, which is written in dimensionless variables as

$$\partial_t c + \mathbf{u} \cdot \nabla c - D\nabla^2 c = -\nabla \cdot (c\mathbf{n}), \tag{9.26}$$

shows that concentration fluctuations can only grow through the source term on the right-hand side, which arises from self-propulsion. The mechanism can be understood as follows [67]: (i) the linear instability for the nematic order parameter first causes local alignment of the particles, which is primarily nematic but also generally involves some weak polarity due to random fluctuations in the initial condition; (ii) this net polarity then leads to concentration of particles as a result of self-propulsion in regions where $\nabla \cdot (c\mathbf{n}) < 0$. Interestingly, this also suggests that no concentration fluctuations would arise in a suspension of either apolar or non-self-propelled active particles (so-called shakers), for which the source term on the right-hand side of Eq. (9.26) is strictly zero.

As a side note, the case of nonmotile shakers is particularly revealing as to the linear structure of an active suspension. As discussed in Sect. 1, a shaker suspension is simply an ensemble of immotile force dipoles that are moved by whatever velocity field they produce by their collective flows (i.e., set $V_s = 0$ in Eq. (9.4) while retaining the active stress in the momentum equation). This simple model again satisfies the entropy equality (9.21). Betterton et al. [91] further showed that in this case the linearized model simplifies remarkably by introducing a vector stream function $\boldsymbol{\Phi}$ and the vorticity ω. Then we have

$$\nabla^2 \boldsymbol{\Phi} = \omega \quad \text{and} \quad \nabla^2 \omega = \mathbf{h}, \tag{9.27}$$

where Betterton et al. showed that $\mathbf{h}$ satisfies the simple dynamics

$$\partial_t \mathbf{h} = -\left(\frac{\alpha}{5} + 6d\right)\mathbf{h} + D\nabla^2 \mathbf{h}. \tag{9.28}$$

For plane-wave perturbations, this equation has the simple growth-rate relation $\lambda(k) = -(\alpha/5 + 6d) - Dk^2$ and hence can show instability only if particle extensile flows are sufficiently strong to overcome rotational diffusion, that is, when $\alpha < -30d$. More to the point though, Eq. (9.28) shows that an active suspension has a very elementary underlying linear structure of a simple exponential growth, driven by activity and damped by rotational diffusion and regularized by spatial diffusion. Note that unlike the motile case [90], there is no loss of solutions, at a finite k, to the plane-wave eigenvalue problem when $d = 0$.

3 Extensions and Applications

3.1 Concentrated Suspensions

The kinetic model described above is based on a dilute assumption and only includes mean-field hydrodynamic interactions between particles. While this approximation is valid at sufficiently low volume fractions [59], it is likely to break down in concentrated systems in which particle-particle contact interactions become significant. Including such interactions is important to accurately capture dynamics in bacterial suspensions, as the onset of collective motion in experiments is typically observed at high densities [11]; in fact, it has sometimes even been suggested that contacts may be the dominant effect leading to collective dynamics. While accounting for contacts in particle simulations is feasible [60, 92], albeit at a high computational cost, it is not as straightforward within the context of our kinetic theory as such interactions are discrete and pairwise. Aranson et al. [62] proposed a continuum model to account for steric interactions based on a collision operator having the effect of aligning contacting particles, in qualitative agreement with experimental observations. To correctly account for collisions, however, their model requires knowledge of the pair distribution function in the suspension, which was approximated as the product of two singlet distributions. Similarly, Baskaran and Marchetti [93] developed a kinetic theory for self-propelled hard rods in two dimensions accounting for pairwise collisions. They were able to show that the leading effect of collisions is to modify the orientational flux by addition of an aligning torque of the same form as the classic Onsager potential for excluded-volume interactions in passive rodlike polymer suspensions [94].

Based on this observation, Ezhilan et al. [71] adapted the kinetic model discussed in Sect. 2 to account for contact interactions in a mean-field framework similar to that used in classic theories for passive rods. Specifically, following the work of [87], we account for contacts by including an effective steric torque derived from a potential U:

$$U(\mathbf{r},\mathbf{p},t) = \int_{\Omega} \Psi(\mathbf{r},\mathbf{p}',t)K(\mathbf{p},\mathbf{p}')\,d\mathbf{p}', \tag{9.29}$$

where the interaction kernel is taken to be the phenomenological Maier-Saupe kernel: $K(\mathbf{p},\mathbf{p}') = -U_0(\mathbf{p}\cdot\mathbf{p}')^2$ with strength constant U_0 [95]. Inserting the expression for K into Eq. (9.29) and taking the orientational gradient yield a new expression for the rotational velocity:

$$\dot{\mathbf{p}} = (\mathbf{I}-\mathbf{pp})\cdot(\beta\mathbf{E}+\mathbf{W}+2U_0c\mathbf{Q})\cdot\mathbf{p} - d\nabla_p \ln\Psi, \tag{9.30}$$

where it can be seen that the new term causes alignment of $\mathbf{p}$ along the principal axes of the nematic order tensor parameter $\mathbf{Q}$ associated with positive eigenvalues, i.e., along the local preferred directions of nematic alignment.

The first effect of this additional torque is to allow for non-isotropic nematic base states as volume concentration increases. As shown by Ezhilan et al. [71], the transition from isotropy to nematic alignment, which is the same as that occurring in liquid crystalline systems, is governed by the dimensionless group $\xi = 2U_0\nu/d$ representing the ratio of the steric alignment torque to rotational diffusion. All spatially uniform base states can be shown to be axisymmetric and of the Boltzmann form

$$\Psi(\mathbf{r},\mathbf{p},t) = \Psi_0(\theta) = \frac{\exp(\delta\cos 2\theta)}{2\pi\int_0^{\pi}\exp(\delta\cos 2\theta')\,d\theta'}, \tag{9.31}$$

where θ denotes the angle between $\mathbf{p}$ and the direction of nematic alignment, which must be specified. Here, the parameter δ governs the shape of the orientation distribution and is a zero of the function

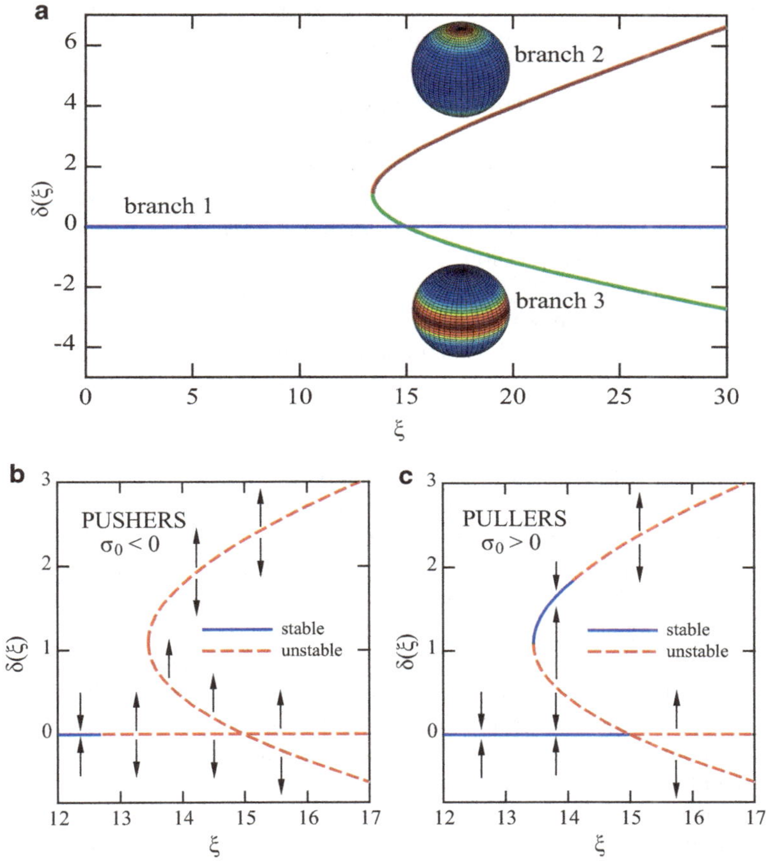

Fig. 9.6 (**a**) Isotropic-to-nematic transition in a concentrated suspension as the parameter $\xi = 2U_0\nu/d$ is increased. Branch 1 is isotropic, whereas branches 2 and 3 are nematic. (**b**)–(**c**) Stability diagrams for pusher and puller suspensions, showing the stability of the three branches in each case. (Adapted with permission from Ezhilan et al. [71])

$$g(\delta) = \delta - \frac{\int_0^\pi \sin\theta(3\cos^2\theta - 1)\exp(\delta\cos 2\theta)\,d\theta}{\int_0^\pi \sin\theta\exp(\delta\cos 2\theta)\,d\theta}. \tag{9.32}$$

It is easy to see that $\delta = 0$ is a solution, which corresponds to the isotropic base state. However, when $\xi \geq \xi_c \approx 13.46$, there exist two other zeroes corresponding to nematic orientation distributions, as illustrated in Fig. 9.6a showing the three branches of the function $\delta(\xi)$. Positive values of δ are achieved on branch 2, which corresponds to the strongest nematic alignment; negative values are also possible along branch 3 and indicate a preferential alignment in the plane normal to the axis of symmetry. A simple consideration of the total steric interaction energy on each branch suggests that above ξ_c the energy is minimized on branch 2, which is consistent with the concept of an isotropic-to-nematic transition as concentration increases.

This energy argument, however, does not imply that such base states are hydrodynamically stable. To investigate the stability of these branches, the kinetic model of Sect. 2 must also be modified to account for additional stresses that arise at finite concentrations: first, passive viscous stresses due to the interactions of the particles with the local flow [83] have to be included and lead to an effective increase in the viscosity of the suspension, unlike active stresses that tend to decrease it in suspensions of pushers [37]. Second, steric interactions also lead to an additional stress contribution which was previously calculated for slender particles [71, 72]. Using this model, Ezhilan et al. [71] numerically studied the stability of the various base states, and results are summarized in Figs. 9.6b–c for both pusher and puller particles. In the case of pushers, the isotropic base state (branch 1) becomes unstable with increasing concentration before the isotropic-to-nematic transition occurs: this instability is of hydrodynamic origin and simply corresponds to the basic instability described in Sect. 2.4. The case of pullers, however, is more interesting. It is found that the isotropic base state loses stability at $\xi = 15$ when branch 1 intersects branch 3 as a result of steric interactions only. Branch 3, however, is always unstable. Branch 2, which has the lowest steric interaction energy, is stable at first but eventually also loses stability when ξ increases as result of hydrodynamic modes. This high-concentration instability of puller suspensions is quite surprising and is corroborated by numerical simulations. To our knowledge it has never been observed in experiments, perhaps because biological pullers are scarce and the few species that exist, including *Chlamydomonas*, have nearly spherical bodies and are therefore unlikely to undergo the isotropic-to-nematic transition.

While the model described above provides a qualitative understanding of the effect of collisions on the dynamics, it is based on a number of strong approximations and on a phenomenological mean-field description of steric interactions in terms of the Maier-Saupe potential. First, the validity of this description can be questioned and should ideally be tested using particle simulations. These are quite expensive in the concentrated regime even for passive rods [96] and have yet to be developed for self-propelled particles in three dimensions. Second, the description of the stresses typically used in the kinetic models such as those discussed herein is based on a dilute assumption, resulting in stresses that depend linearly or quadratically on density; a more accurate description of these stresses should account for multiple reflections between particles as well as multi-body interactions, though models of this type have been limited to passive rod suspensions [97]. Finally, the kinetic theory outlined above used a single velocity field $\mathbf{u}$ to describe the transport of the fluid and particle phases. This is a good approximation in the dilute limit but is likely to break down at high volume concentrations, where a two-fluid approach would be more appropriate [63, 98].

3.2 Confinement

Experimental evidence suggests that interactions with rigid boundaries in confined environments can be highly complex and play an important role in the dynamics and transport properties. Some examples of the complex effects that have been reported under confinement include: accumulation of particles at boundaries [38,99, 100], upstream swimming in channel flows [101], unexpected scattering dynamics [102, 103], modified diffusivities [104], circular swimming trajectories [105], and spontaneous flow transitions [106, 107]. Modeling efforts on the role of boundaries and the effects of confinement, however, have been relatively scarce. Analytical models and numerical simulations indeed predict concentration at boundaries [51, 52, 108, 109], both as a result of hydrodynamic interactions [110] and because of particle self-propulsion, though models for collision rules are often chosen in an *ad hoc* manner.

The modeling of wall interactions in continuum theories has also been relatively limited. In phenomenological theories for active liquid crystals, boundary conditions are often formulated in terms of anchoring conditions for the nematic order parameter [111–113], which are borrowed from classic liquid crystal theories but do not account for the unique nature of wall interactions due to self-propulsion. Within the context of the kinetic model of Sect. 2, a natural boundary condition to enforce at impenetrable walls consists in prescribing zero net translational flux in the wall-normal direction (with unit normal $\mathbf{N}$): $\mathbf{N}\cdot\dot{\mathbf{r}} = 0$. Inserting Eq. (9.17) for the flux velocity, this translates into a Robin boundary condition

$$V_s(\mathbf{p}\cdot\mathbf{N})\Psi = D\mathbf{N}\cdot\nabla_r\Psi, \tag{9.33}$$

which expresses the balance between the swimming flux towards the wall and the diffusive flux away from it, and this simple boundary condition has been shown to capture many features observed in experiments. Note that Eq. (9.33) neglects the finite size of the particles, which forbids configurations near the boundaries leading to overlap of the particles with the wall and is expected to result in a thin depletion layer as seen in experiments [114]; more complex boundary conditions that account for excluded-volume interactions have also been formulated, both in the context of passive suspensions [115] and also recently for active particles [116].

As a simple example of application of Eq. (9.33), Ezhilan and Saintillan [116] analyzed the case of a suspension confined between two parallel flat plates separated by a gap $2H$ in the limit of infinite dilution where hydrodynamic interactions can be entirely neglected. Assuming the Taylor dispersion relation $D = V_s^2/6d$ for the translational diffusivity, the dimensionless Smoluchowski equation (9.16) reduces at steady state to a simple partial differential equation expressing the balance of self-propulsion and translational and rotational diffusion:

$$\mathrm{Pe}_s \cos\theta\,\partial_z\Psi - \tfrac{1}{3}\mathrm{Pe}_s^2\,\partial_{zz}\Psi = \tfrac{1}{2}\nabla_p^2\Psi. \tag{9.34}$$

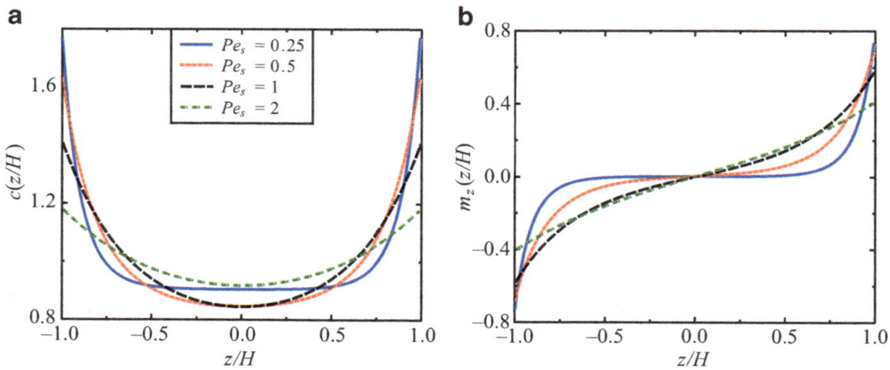

Fig. 9.7 (**a**) Concentration profile and (**b**) wall-normal polarization at steady state in a dilute suspension of particles confined between two parallel flat plates separated by H. Profiles were obtained by numerical solution of Eqs. (9.34)–(9.35) for various values of $\mathrm{Pe}_s = V_s/2Hd$

Here, $z \in [-1,1]$ is the wall-normal coordinate, $\theta = \cos^{-1}(\mathbf{p}\cdot\mathbf{N})$ is the polar angle measured with respect to the wall-normal direction, and we have introduced a swimming Péclet number comparing the relative magnitude of self-propulsion to rotational diffusion: $\mathrm{Pe}_s = V_s/2Hd$. Equation (9.34) should be solved subject to the boundary condition (9.33), which simplifies to

$$\Psi \cos\theta - \tfrac{1}{3}\mathrm{Pe}_s\, \partial_z \Psi = 0 \quad \text{at} \quad z = \pm 1. \tag{9.35}$$

A numerical solution of Eqs. (9.34)–(9.35) was obtained by Ezhilan and Saintillan [116] and is shown in Fig. 9.7, where both the concentration $c(z)$ and wall-normal polarization $m_z(z) = c(z)n_z(z)$ are plotted. A net accumulation of particles is observed near both boundaries, in agreement with experiments [99] and simulations [51, 109]. This accumulation is accompanied by a net polarization towards the boundaries, and both effects are found to become stronger as Pe_s decreases, which corresponds to an effective decrease in translational diffusivity owing to the Taylor dispersion scaling. This accumulation and corresponding polarization are easily understood physically: any particle inside the channel will tend to swim to the wall towards which it points and accumulate there until rotational diffusion causes it to reverse polarity. Note that this accumulation is not a result of hydrodynamic interactions with the boundaries, though it has been suggested that hydrodynamic interactions can reinforce migration in pusher suspensions due to the reflection of the dipolar flow driven by the swimmers [99]. One interesting consequence of this net polarization occurs when a pressure-driven flow is applied between the two plates: particles near the walls rotate under the flow in such a way that they preferentially point upstream, causing them to swim against the flow. This curious prediction is consistent with experimental observations using bacteria in microfluidic devices [101, 117, 118] and has also been observed in simulations [109, 119].

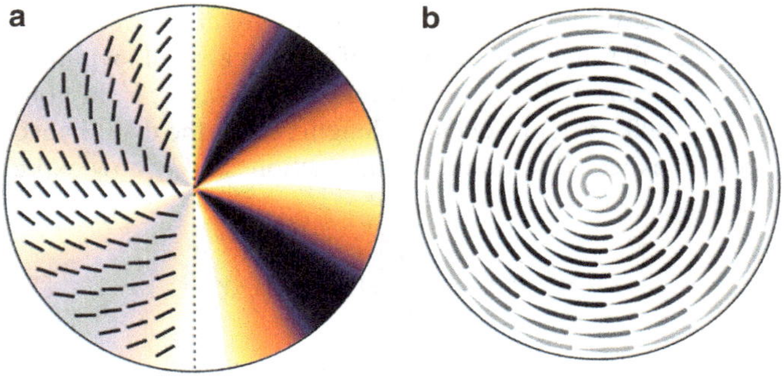

Fig. 9.8 Numerical simulation by Woodhouse and Goldstein [106] of the spontaneous flow in a two-dimensional shaker suspension confined in a circular domain and modeled using a closure approximation: (**a**) shows Schlieren patterns of the nematic order director, whereas (**b**) shows flow streamlines, with *darker streamlines* indicating faster flow. (Adapted with permission)

The effect of confinement has also been studied theoretically in closed domains. Woodhouse and Goldstein [106] posited that a suspension of shakers could serve as a basic model for the dynamics of biopolymers moved by immersed motor proteins. To simplify the model, they assumed a uniform particle concentration (which is an allowable state of the system) and employed a classical moment closure scheme of Hinch and Leal [83] to find an approximate dynamical equation for the nematic order parameter $\mathbf{Q}$. Evolving this equation in a circular two-dimensional domain, they assumed a no-slip boundary condition on the background velocity and the boundary condition $\partial\mathbf{Q}/\partial\mathbf{N} = \mathbf{0}$ for the nematic tensor, which is consistent with Eq. (9.33) after setting $V_s = 0$. Using numerical simulations, they identified the existence of a bifurcation, with increasing active stress strength α, from an isotropic state with no flow to a unipolar vortical flow driven by suspension activity, as shown in Fig. 9.8. Higher levels of active stress can lead to successive bifurcations with yet more complex vortical flows, sometimes oscillatory, and with nematic orientation singularities. Such vortical flows are indeed observed in experiments on confined drops of cytoskeletal extracts [15], as well as in drops of confined suspensions [107]. Jhang and Shelley [120] found consistent results using the full unapproximated suspension model of Sect. 2, again using the no-slip condition and the no-flux boundary condition $\partial\Psi/\partial\mathbf{N} = 0$ on the boundary of a circular domain.

Not only does confinement affect particle distributions in the dilute regime, but it also modifies the way particles interact hydrodynamically. This is particularly apparent in the case of strong confinement where the size of the particles is on the order of the geometric dimensions of the domain, say the gap width in a Hele-Shaw geometry. This situation was analyzed theoretically by Brotto et al. [121] using a similar continuum kinetic theory as in Sect. 2 in the case of a two-dimensional monolayer of particles confined between two flat plates. As explained in their study, the leading effects of confinement are twofold. Firstly, as is well known from studies on passive suspensions [122, 123], momentum screening by the rigid boundaries leads to the rapid decay as $1/r^3$ of the flow driven by the force dipole due to self-propulsion. On the other hand, the motion of the finite-sized swimmers relative to the fluid results in a mass dipole which now decays as $1/r^2$ in confinement, as opposed to $1/r^3$ in bulk systems. On large length scales, the disturbance flow due to this mass dipole is expected to dominate interactions and is now expressed in two dimensions as

$$\mathbf{u}^m(\mathbf{r}|\mathbf{p}) = \frac{1}{2\pi r^2}(2\hat{\mathbf{r}}\hat{\mathbf{r}} - \mathbf{I}) \cdot \chi. \tag{9.36}$$

The dipole strength is proportional to the relative velocity between the swimmer and the suspending fluid: $\chi = \chi_0[\dot{\mathbf{r}} - \mathbf{u}(\mathbf{r})]$, where the prefactor χ_0 scales as the particle surface area in the plane of the flow and is independent of the propulsion mechanism. Secondly, Brotto *et al* also argued that confinement can modify the way particles respond to a given flow field: in particular, fore-aft asymmetric particles (such as much flagellated swimmers) are expected to align not only with the velocity gradient as in Jeffery's equation (9.18) but also with the velocity itself as a result of the lubricated friction with the neighboring walls. To capture this effect, they derived a modified equation for the rotational flux velocity to read

$$\dot{\mathbf{p}} = \beta(\mathbf{I} - \mathbf{pp}) \cdot \nabla_r \mathbf{u} \cdot \mathbf{p} + \beta'(\mathbf{I} - \mathbf{pp}) \cdot \mathbf{u} - d\nabla_p \ln \Psi. \tag{9.37}$$

Here, the parameter β' depends on particle shape: it is zero for a fore-aft symmetric particle, positive for a "large-tail" swimmer that aligns with the flow, and negative for a "large-head" swimmer that aligns against the flow. Based on these effects, they derived a kinetic model similar to that of Sect. 2 and analyzed the stability of the uniform isotropic base state. They uncovered a long-wave instability in suspensions of large-head swimmers (for which $\beta' < 0$), which pertains to splay components of the nematic tensor above a certain level of activity. Their analysis was subsequently refined by Lefauve and Saintillan [124], who also performed direct numerical simulations of point particles in two-dimensional geometries. Above the threshold of instability, complex dynamics illustrated in Fig. 9.9 were observed that differed significantly from those observed in unconfined suspensions: when $\beta' < 0$, particles were found to arrange in longitudinal polarized waves with a net curvature indicative of splay, whereas for $\beta' > 0$ they converged into active lanes circulating around large-scale vortices. Recent numerical work on the same system by Tsang and Kanso [125] also predicted the formation of stable clusters when $\beta' < 0$ and proposed an interpretation of β' in terms of flagellar activity.

3.3 Chemotaxis

The ability of swimming microorganisms to detect and respond to external stimuli such as chemical fields is critical to biological functions such as nutrient and oxygen uptake, toxin avoidance, colony growth, and cell-cell communication for gene

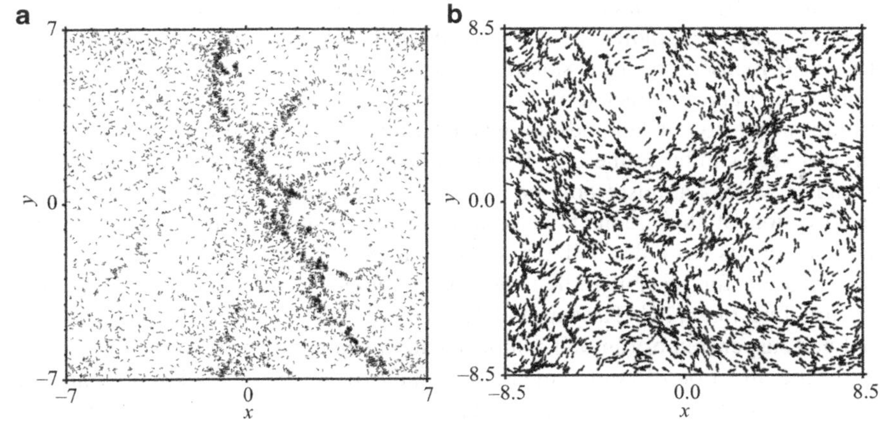

Fig. 9.9 Particle simulations of two-dimensional swimmer suspensions in a Hele-Shaw geometry, based on the model of Brotto et al. [121]: (**a**) large-head swimmers ($\beta' < 0$) form polarized density waves with splay, whereas (**b**) large-tail swimmers ($\beta' > 0$) organize into active lanes circulating around large-scale vortices. (Adapted with permission from Lefauve and Saintillan [124])

regulation or aggregation. The method used by many bacteria to perform chemotaxis (or directed migration along a chemical gradient) is a modulated run-and-tumble dynamics [126]. Here, "runs" of directed bacterial swimming are interspersed with random reorientations, or "tumbles," resulting from the unbundling and rebundling of their flagella. These arbitrary changes in swimming direction lead to a random walk in space [81], which can be biased towards a particular direction by modulating the frequency λ of tumbles. More specifically, a bacterium that tumbles less frequently when it swims in the direction of increasing chemical concentration will on average drift towards regions of higher concentration. Run-and-tumble dynamics can be easily incorporated into the kinetic framework discussed here. In particular, the Smoluchowski equation is modified to

$$\partial_t \Psi + \nabla_r \cdot (\dot{\mathbf{r}}\Psi) + \nabla_p \cdot (\dot{\mathbf{p}}\Psi) = -\lambda(\tilde{\mathscr{D}}_t C)\Psi + \frac{1}{4\pi}\int_\Omega \lambda(\tilde{\mathscr{D}}_t C) K(\mathbf{p},\mathbf{p}')\Psi(\mathbf{r},\mathbf{p}',t)\,d\mathbf{p}'. \tag{9.38}$$

Here λ is the tumbling frequency away from orientation $\mathbf{p}$ and depends upon $\tilde{\mathscr{D}}_t C = \partial_t C + (\mathbf{u} + V_s\mathbf{p}) \cdot \nabla C$, which is the rate of change of the chemical concentration C along the swimming path. The function $K(\mathbf{p},\mathbf{p}')$ is called the "turning kernel" and captures correlations between pre- and post-tumbling orientations. One expects K to be independent of frame orientation and so depend only upon $\mathbf{p}\cdot\mathbf{p}'$. Subramanian et al. [127] proposed the form $K(\mathbf{p},\mathbf{p}') = B\exp(B\mathbf{p}\cdot\mathbf{p}')/4\pi\sinh B$, which yields small changes in orientation for large values of B, and an uncorrelated, uniform post-tumble orientation as $B \to 0$. The fluxes and stresses are taken to be unchanged from Sect. 2.

Motivated by experiments [28], recent studies have considered chemotaxis in an externally imposed gradient of a chemoattractant, say oxygen. Koch and coworkers [127–129] focused on linear stability analyses in the case where the oxygen field is prescribed and unaffected by the flow. Subramanian et al. [127] considered an infinite suspension of run-and-tumble bacteria in a uniform oxygen gradient, using a model very similar to the one described here. They showed that run-and-tumble dynamics yields an anisotropic orientation distribution, with a net polarization in the direction of the gradient. Performing a linear stability analysis, they found that chemotaxis has a destabilizing effect and tends to reduce the critical bacterial concentration for instability that they had previously derived in the absence of the tumbling bias [74]. Kasyap and Koch [128, 129] analyzed the stability of a confined suspension of run-and-tumble bacteria when the chemoattractant gradient lies in the direction of confinement, as in the experiments of Sokolov et al. [28]. In this geometry they showed that the base state is one of inhomogeneous bacterial concentration and stress, both increasing exponentially across the channel. These inhomogeneous base states allow new couplings within the linearized dynamics. Kasyap and Koch [128] performed a long-wavelength analysis and showed a quadratic increase of the perturbation growth rate with wavenumber, and that active stresses drive flows that tend to reinforce density fluctuations in the plane of the film. Kasyap and Koch [129] later presented a more complete analysis and showed the existence of a linear mode of maximal growth, providing quite good quantitative agreement with the transition to instability observed in the experiments of Sokolov et al. [28].

The effects of oxygen transport were also modeled in numerical simulations by Ezhilan et al. [69]. They considered the dynamics of aerotactic bacteria confined to liquid films suspended in an oxygen-rich environment and swimming towards sources of oxygen while simultaneously consuming it. The dynamics they observed were quite similar to the experiments of Sokolov et al. [28] and are illustrated in Fig. 9.10(a): first, bacteria and oxygen concentration approached steady profiles in thin films, but above a critical film thickness, three-dimensional chaotic dynamics were observed with dense plumes of bacteria penetrating the bulk. In very thick films, a dense bacterial layer was also observed near the film centerline and was explained by the nearly uniform oxygen concentration in that region, where chemotaxis ceases. The onset of instability in these nonlinear simulations compared favorably to the prediction of the linear stability analysis of Kasyap and Koch [128].

A very different situation arises when the chemoattractant is secreted by the swimming bacteria themselves, as was recently modeled by Lushi et al. [70] who were inspired by studies showing bacterial self-concentration as a result of chemotactic focusing [130], as well as communication processes in bacterial colonies via quorum sensing [131, 132]. This situation has been studied using the celebrated Keller–Segel model [133] and its many variants, though all neglected the effect of the fluid flow generated by the swimmers. Lushi et al. [70] coupled the run-and-tumble chemotaxis model to a transport equation for the chemoattractant concentration that modeled chemoattractant production, depletion, and diffusion. One steady state for this system is uniform isotropy for the swimmers, with a balance

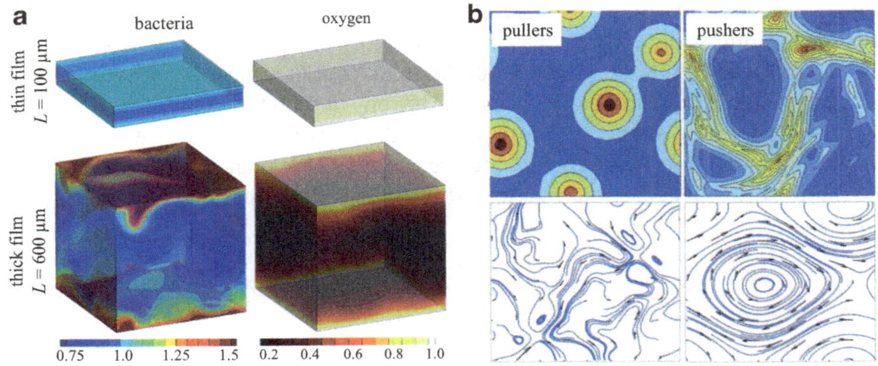

Fig. 9.10 (**a**) Dynamics in films of aerotactic bacteria in the continuum simulations of Ezhilan et al. [69]: in thin films (*top row*), both bacterial and oxygen concentrations reach steady profiles; as the film thickness is increased (*bottom row*), a transition to unsteady dynamics is observed, with the formation of bacterial plumes and enhanced oxygen transport. (**b**) Structure and dynamics of swimmer concentration in autochemotactic suspensions in simulations by Lushi et al. [70]. Lower figures show the corresponding flow streamlines and polar order parameter. (Adapted with permission)

of production and degradation in the chemoattractant concentration. Linearizing around this state for a simplified version of the model, Lushi et al. [70] found two uncoupled stability problems for chemotactically driven aggregation and alignment-driven large-scale flows. As a function of tumbling frequency, they identified different regimes where instabilities to aggregation, or alignment, or both, are dominating. Their nonlinear simulations showed, for pushers, that chemotactically driven aggregation was halted by the eruption of local hydrodynamic instabilities, while for pullers the competition of aggregation and active stresses yielded steady-state spots of finite size.

3.4 Fluid Viscoelasticity

The effect of non-Newtonian fluid response upon microorganism locomotion has been studied intensely over the past few years; see, for example, Vélez-Cordero and Lauga [134] and the references therein, as well as the chapter by Guy and Thomases in this volume. The main issue considered has generally been the effect of fluid viscoelasticity upon single swimmer speeds and efficiencies, while comparatively little has as yet been understood on non-Newtonian effects upon collective behavior. In a first effort, Bozorgi and Underhill [135, 136] have extended the kinetic model of Sect. 2 by adding a non-Newtonian stress tensor to the momentum equation (9.19):

$$-\nabla^2\mathbf{u}+\nabla q=\nabla\cdot(\langle\alpha\mathbf{pp}\rangle+\beta\Sigma^e), \tag{9.39}$$

where the stress tensor Σ^e arises from various viscoelastic constitutive laws and where β captures the nondimensional strength of polymer stress coupling to the momentum balance. Bozorgi and Underhill analyzed the linear stability of various viscoelastic fluid models near the state of isotropy and homogeneity.

One model they studied closely is the Oldroyd-B model [137], which is built upon the assumption that polymer coils respond as Hookean springs to distension by the flow. In this case, the polymer stress obeys the upper-convected evolution equation

$$\mathscr{D}_t\Sigma^e - \left(\nabla\mathbf{u}\cdot\Sigma^e + \Sigma^e\cdot\nabla\mathbf{u}^T\right) = -\mathrm{Wi}^{-1}\left(\Sigma^e - \mathbf{I}\right), \tag{9.40}$$

where Wi is the Weissenberg number relating the strength of flow forcing to polymer relaxation. For Oldroyd-B, they use the analytic reduction developed by Hohenegger and Shelley [86] and study the linearized dynamics when projected to the first azimuthal mode on the unit sphere of orientations. By expanding in associated Legendre polynomials, this yields an infinite-dimensional, but essentially tridiagonal, eigenfunction/eigenvalue problem for the growth rates. From this, they showed that rotational diffusion, in confluence with viscoelasticity, fundamentally alters the nature of collective instabilities, yielding growing oscillations at long wavelengths, and a biased suppression of growth as a function of k that shifts the maximal growth rate from $k = 0$ to intermediate values. One possible weakness of their approach is that viscoelasticity is only felt by the swimmers through the large-scale stresses that produce the background velocity field against which the swimmers move. In particular, viscoelasticity is not introduced in determining the single-particle fluxes of Eqs. (9.17)–(9.18), which assume a Newtonian flow on the scale of the particles.

4 Other Active Fluids

While we have focused on suspensions of micro-swimmers, there are other examples of active fluids where the active stresses devolve from other sources of activity and microstructural displacement. We discuss two here: suspensions of microtubules and bound translocating motor proteins and surface-bound populations of particles whose chemical activity creates Marangoni stresses.

4.1 Microtubules and Motor Proteins

Microtubules (MTs; stiff biological polymers composed of tubulin protein subunits) and motor proteins are the building blocks of self-organized biological structures such as the mitotic spindle and the centrosomal MT array [138]. They are also the ingredients in liquid-crystalline active fluids powered by ATP and driven out of equilibrium by motor-protein activity to display complex flows and persistent defect dynamics. MTs are polar polymers, typically polymerizing and depolymerizing from their "plus-end." The interactions of a motor protein with an MT are also typically polar, with its active motion along the MT being towards either plus- or minus-ends, depending on the motor type.

Active stresses or forces can be created in such systems by the interaction of MTs with immersed motor proteins, often bound to cellular organelles or vesicles, or by motor proteins mechanically coupling together MTs, with their activity inducing their relative displacement. A possible example of the first is the process of nuclear migration in early development, where the "pronuclear complex" containing male and female genetic material is transported to the center of an embryonic cell as shown in Fig. 9.11a. This transport is associated with two dynamic MT arrays emanating from "centrioles," and ends with the nuclear complex rotating so that the centrioles are aligned with the cell's anterior-posterior axis. This is the so-called proper position of the complex so that cell division may proceed smoothly.

While various models of pronuclear migration have been put forward, including interactions of the MT array with the cell cortex, one possible contributing mechanism is the active transport of organelles along MTs towards the centrosomes by dynein motor proteins–minus-end-directed motor proteins–bound to organelle surfaces. Inspired by previous modeling work by Kimura and Onami [139], Shinar et al. [14] investigated this nuclear positioning model as a fluid-structure interaction problem where active agents within the cytoplasm (the cellular fluidic medium) exert minus-end-directed pulling forces upon immersed MTs. To achieve proper force balance–motor proteins can exert no mean force upon the system—these pulling forces upon MTs are balanced by oppositely directed forces acting upon the cytoplasm. Shinar et al. [14] simulated this model using a computational method related to the immersed boundary method [140], and Fig. 9.11b shows the migration

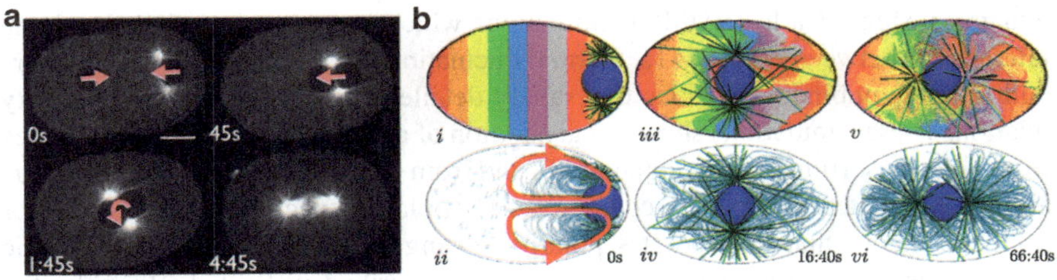

Fig. 9.11 (**a**) MT-based dynamics in a live single-celled *Caenorhabditis elegans* embryo: migration of the male and female pronuclei, pronuclear meeting, centration, and spindle reorientation. (**b**) Numerical simulation of MT-based pronuclear translation, showing transport of a passive scalar (*top*) and flow streamlines (*bottom*). (Adapted with permission from Shinar et al. [14])

and cytoplasmic flows as a model nuclear complex is pulled into the cell center by immersed motor proteins and is then rotated into proper position. We note that the observed cytoplasmic flows along MTs are also observed in vivo and that experiments of Kimura and Kimura [141] showed that positioning and cytoplasmic flows were much slowed by blocking the binding of dynein to organelles. Finally, in unpublished work using a reduced model, Fang and Shelley have shown that the rotation can be explained by a stability calculation that shows that "proper position" is the only mechanically stable orientation for the centriole axis.

Motor proteins can also mediate interactions between MTs by providing a direct and active mechanical coupling of MTs by motor complexes of two or more end-directed motors connected by a molecular tether. Here, the nature of this interaction will depend strongly on whether an MT pair is polar-aligned or anti-aligned. In the latter case, the complex's motors walk in opposite directions on each MT, inducing a relative sliding of the MTs. This process is called "polarity sorting."

The physics of filament sliding and polarity sorting by two-headed molecular motors has been studied experimentally [4, 16, 143]. In early experiments, biofilaments were driven into static self-organized patterns such as vortices and asters, reminiscent of structures observed in vivo. Very recently, in experiments of Sanchez et al. [3], active networks were formed of MTs and synthetic tetrameric kinesin-1 motor complexes with the aid of a depletant. In the presence of ATP, motor complexes can bind pairs of MTs and walk along MTs towards their plus-ends. When suspended in bulk, depletion interactions drove the formation of extended, highly ordered MT bundles characterized by bundle extension and fracture and correlated with spontaneous large-scale fluid flows. When MT bundles were adsorbed onto an oil-water interface, they formed a dense, nematically ordered 2D state and exhibited an active nematic phase characterized by the spontaneous generation and annihilation of disclination defect pairs.

Gao et al. [142] have developed a multiscale model that identifies the possible sources of destabilizing active stresses. They first performed detailed Brownian dynamics-Monte Carlo (BDMC) simulations which incorporate excluded-volume interactions between model MTs, thermal fluctuations, explicit translocating motors with binding/unbinding kinetics that satisfy detailed balance, and a force-velocity relation. These simulations show the generation of activity-driven extensile stresses from polarity sorting of anti-aligned MTs and from "cross-link relaxation" of polar-aligned MTs. It also provides coefficients for polarity-specific active stresses for a kinetic theory that incorporates polarity sorting and long-range hydrodynamic interactions, using a similar approach to that described in Sect. 2. Roughly, the center-of-mass flux in Eq. (9.17) is replaced by

$$\dot{\mathbf{r}} = (\mathbf{n} - \mathbf{p}) + \mathbf{u} - D\nabla_r \ln \Psi, \tag{9.41}$$

where, again, $\mathbf{n}$ is the polar order parameter defined in Eq. (9.6). This new term captures the sliding of an MT (at orientation $\mathbf{p}$) relative to a background of MTs of mixed polarity and can be derived by considering a cluster of polar-aligned and anti-aligned MTs coupled together by translocating cross-links. The active stress is replaced by

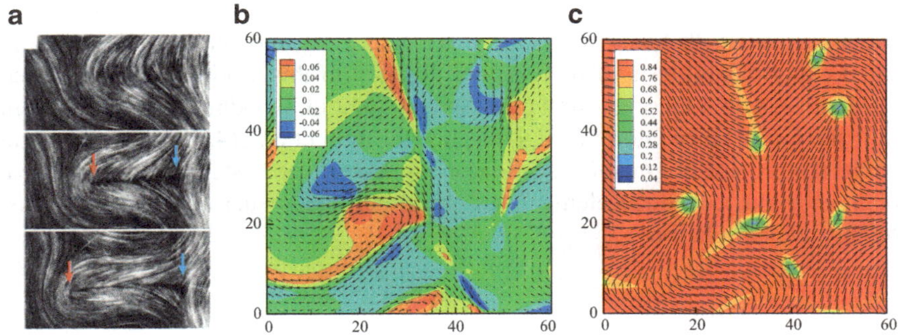

Fig. 9.12 (**a**) Experiment by Sanchez et al. [3] showing the active nematic state of a suspension of MTs and kinesin clusters confined at an interface between two fluids, showing the generation and annihilation of disclination defect pairs. (**b**)–(**c**) Two-dimensional continuum simulation by Gao et al. [142] of a suspensions of MTs and kinesin clusters: (**b**) velocity field overlaying the vorticity, and (**c**) nematic director field and scalar order parameter. (Adapted with permission)

$$\Sigma^a = \frac{\alpha_{aa}}{2} c(\mathbf{Q} - \mathbf{nn}) + \frac{\alpha_{pa}}{2} c(\mathbf{Q} + \mathbf{nn}), \tag{9.42}$$

where α_{aa} and α_{pa} are dimensionless stresslet strengths associated with anti-aligned and polar-aligned interactions, respectively. The BDMC simulations estimate these as being negative, and hence corresponding to destabilizing dipolar stresses, with α_{aa} being the larger. The anti-aligned stresses arise from extensional flows, similar to those for pusher particles, induced by polarity sorting and biases in motor-protein binding and unbinding. The polar-aligned stresses are also extensile but arise from a more subtle statistical mechanical effect associated with temporal relaxation of the motor-protein tether.

Simulations of this polar active nematic model are shown in Fig. 9.12b–c. Simulating in regions of flow instability, Gao et al. [142] find persistently unsteady flows that are correlated with the continual genesis, propagation, and annihilation of $\pm 1/2$ order disclination defect pairs. To wit, Fig. 9.12b shows the background velocity field $\mathbf{u} = (u, v)$ overlaying the vorticity ω. The dynamics are complex and turbulent, and qualitatively very similar to those reported by Sanchez et al. [3]. Also very similar are the MT orientation dynamics. Fig. 9.12c shows the nematic director field and scalar order parameter from the tensor order parameter $\mathbf{Q}$. The local orientation is highly correlated with the flow structures seen in (b). We see also that the plane is littered with $\pm 1/2$ order defects which propagate freely about the system. These defects exist in regions of small nematic order and are born as opposing pairs in elongated low-order regions. These regions are themselves associated with fluid jets, locally decreasing nematic order, and increasing curvature of director field lines. The $+1/2$ order defects propagate away along their central axis and at a much higher velocity than those of $-1/2$ order. The relatively higher velocity in the neighborhood of a $+1/2$ order defect appears as a localized jet, in the direction of defect motion, between two oppositely signed vortices.

Gao et al. [142] also identified the characteristic length scales of this model as those associated with linearized plane-wave models of maximal growth rate, and posited experimental tests of their model. Related to this work are studies based upon **Q**-tensor field theories similar in flavor to that of Woodhouse and Goldstein [106]; see for instance Giomi et al. [144] and Thampi et al. [145]. In these general models, the precise origins of the active stress driving the system are unidentified, though they do reproduce elements of the experiments such as defect genesis, motion, and annihilation.

4.2 Chemically Active Particles

Recent technological advances have enabled the fabrication of synthetic microswimmers that convert chemical energy into directional motion [20, 147]. One widely studied system consists of micron-scale bimetallic gold-platinum rods. When immersed in a hydrogen peroxide solution, the rods show directed motion along their axes [18]. Theoretical studies have proposed that these rods move through a chemically powered electrophoretic mechanism which generates a slip flow along the rod surface from the gold to the platinum portions [148]. Experimental studies show that such particles interact with surfaces by flipping and sliding along walls and being captured into orbits around sedimented colloids as shown in Fig. 9.13a and b. Little if any work has, as yet, studied hydrodynamically mediated collective dynamics. One inhibiting feature of this system is that oxygen is an end product of the chemical reactions driving the rods, and when the rods are at high concentration the dissolved oxygen comes out of solution, forming large bubbles that disrupt the experiment. Work on collective behavior in these systems

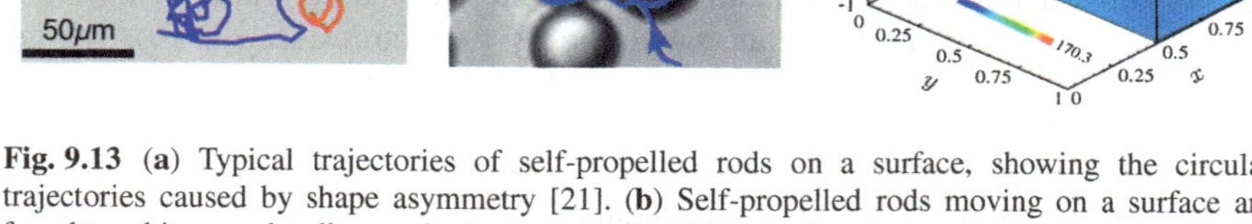

Fig. 9.13 (**a**) Typical trajectories of self-propelled rods on a surface, showing the circular trajectories caused by shape asymmetry [21]. (**b**) Self-propelled rods moving on a surface are found to orbit around sedimented spherical colloids [114]. (**c**) Continuum simulations of the model of Masoud and Shelley [146] for chemotactic collapse of active colloids at an interface, showing both the velocity field and chemical concentration field in the bulk of the liquid. (Adapted with permission)

has tended to focus instead on the role of the fuel concentration field, which diffuses and is consumed by the active particles, and its relations to chemokinetic behaviors [149, 150].

Chemically active particles were recently considered in a very different setting. Masoud and Shelley [146] studied the dynamics of chemically active *immotile* particles that are embedded in a gas-fluid interface. The particles' chemical activity does not produce any phoretic flows, but does create a spatially diffusing chemical concentration field C. On the surface, this chemical field changes the local surface tension, and any consequent surface tension gradients will produce "active" Marangoni shear stresses driving fluid flows that move the particles [151].

Masoud and Shelley [146] considered the case of a flat interface over a incompressible Stokesian fluid of depth H and viscosity η, assuming that diffusion of the chemical species was fast compared with advection and that surface tension depended linearly upon the surface chemical concentration. The surface concentration Ψ of active particles obeys the advection-diffusion equation

$$\partial_t \Psi + \nabla_2 \cdot (\mathbf{U}\Psi) = \frac{1}{\mathrm{Pe}_p} \Delta_2 \Psi, \tag{9.43}$$

where Pe_p is a Péclet number comparing the particle diffusion time scale to advection arising from Marangoni stresses, $\nabla_2 = (\partial_x, \partial_y)$ is the 2D surface gradient operator, $\Delta_2 = \nabla_2^2$, and $\mathbf{U}$ is the 2D surface velocity found by solving the 3D Stokes equations driven by a surface Marangoni stress induced by chemical gradients. Of particular interest is the case where particle activity *raises* the surface tension.

Both the 3D quasi-static diffusion equation for chemical concentration and the 3D Stokes equations for the fluid flow can be solved via Fourier transform in (x, y), and Masoud and Shelley [146] show that the surface velocity's Fourier transform satisfies the relation

$$\tilde{\mathbf{U}}(\mathbf{k}, t) = (i\mathbf{k}/k^2)\Omega(k\delta)\tilde{\Psi}(\mathbf{k}, t) , \tag{9.44}$$

where $\mathbf{k} = (k_x, k_y)$ is the 2D wave vector, $k = |\mathbf{k}|$, and $\delta = H/L$ is the dimensionless layer depth where L is a horizontal system length scale. Equation (9.44) can be interpreted as a nonlocal surface integral operator acting upon the density Ψ. Here $\Omega(\lambda)$ is an explicit monotonically increasing function for which $\Omega = 1/4 + O(\lambda^2)$ for small λ (shallow layers) and $\Omega \to 1/2$ exponentially fast as $\lambda \to \infty$ (deep layers).

Particularly interesting are the limits of shallow and deep layers where Eq. (9.44) reduces to $\tilde{\mathbf{U}} = \nu i\mathbf{k}/k^2\tilde{\Psi}$ with $\nu = 1/4$ (shallow) or $1/2$ (deep) or $\mathbf{U} = -\nu\Delta_2^{-1}\nabla_2\Psi$. Hence, in real space we have

$$\partial_t \Psi - \nu\nabla_2 \cdot \left([\Delta_2^{-1}\nabla_2\Psi]\Psi\right) = \frac{1}{\mathrm{Pe}_p}\Delta_2\Psi, \tag{9.45}$$

and this equation is spatially nonlocal due to the inverse Laplacian.

Very surprisingly, upon rescaling, this PDE recovers the iconic 2D parabolic-elliptic Keller–Segel (KS) model of autotactic aggregation:

$$\partial_t \phi = \nabla \cdot ([\chi \nabla \rho] \phi) + \Delta \phi \quad \text{and} \quad \Delta \rho = \phi, \tag{9.46}$$

which was originally conceived as a model for the aggregation of slime molds [133]. Here ϕ is the concentration field of microorganisms that produces a rapidly diffusing chemoattractant of concentration ρ. Read as a kinetic equation for species number conservation, Eq. (9.46) states that microorganisms move along gradients of the self-generated chemoattractant with speed $\chi \nabla \rho$. The KS model has been the focus of decades of study in PDE analysis (see Horstmann [153] for a comprehensive review), and a great deal is understood about its dynamics. Especially interesting is the 2D case, as is relevant here. For instance, given a sufficient mass of organisms in the plane, the 2D Keller–Segel model suffers *chemotactic collapse* in finite time, with a finite mass of organisms concentrating at a point. The collapse is approximately self-similar, with $\phi \approx \zeta(t)^{-2} \Phi(\mathbf{x}/\zeta(t))$ for some scaling function Φ and a scale ζ whose dominant algebraic behavior is $\sqrt{t_c - t}$ where t_c is the collapse time.

Chemotactic collapse describes very well the aggregation dynamics observed for chemically active particles. Masoud and Shelley [146] simulated Eqs. (9.43)–(9.44) for mean values of Ψ that are large enough to induce two-dimensional instabilities. Figure 9.13c shows the result by plotting the 3D structure of the chemical concentration field and the fluid velocity field. On the surface there has been a rapid accumulation of active particles to the centers and corners of the domain, where the initial particle concentration was peaked. Descriptively, the initial higher concentration of particles yielded a peak in the chemical surface concentration and hence higher surface tensions there. The associated Marangoni stresses created inward flows which concentrated yet more active particles there, leading to yet greater surface tension and stronger flows. Like the KS model, an aggregative finite-time collapse is observed, and Fig. 9.13c is a snapshot right before the collapse time. The particle density field Ψ has a similar structure to that of the surface field of C, but is yet sharper as C is one derivative smoother. In a marked difference from the KS model, here the surface fluid flows towards the aggregation points are associated with 3D flow structures, and Fig. 9.13c shows the formation of a downward jet and encircling vortex ring within the bulk fluid.

The model by Masoud and Shelley [146] is in search of an experiment and originally arose from casual observations of chemically powered motile rods moving on a free surface. However, if such a scenario could be realized, then this dynamics of particle aggregation and 3D flows might prove useful in self-assembly processes and in droplet locomotion.

5 Outlook

There is tremendous ongoing activity in the field of active matter, of which active suspensions are a particular subset. One area which we did not discuss here, because it is yet substantially unformed, is that of flocking or schooling of organisms flying or swimming at high Reynolds number. A complicating factor is that a flocking organism is likely responding to both unsteady fluid forces and sensory information of multiple modalities (and of course, these are not even well separated). There has been great progress in understanding how perception and response may influence ordering and collective behavior, via Vicsek-type models [25, 152]. However, a particular aspect of high Reynolds flows is that the "storage" of shed vorticity into the flow yields a history dependence to body-body interactions that is difficult to capture in a phenomenological model.

Another large area of increasing inquiry is the activity-induced robustness and self-assembly of cellular structures such as the mitotic spindle and the cellular cytoskeleton. Here, theoretical approaches from soft-condensed matter physics, such as generalized hydrodynamics, elasticity, liquid crystals, and polymer dynamics, have proven very useful. Most of the activity in these areas has been carried out by theoretical biophysicists and relatively little by applied mathematicians and engineers. Consequently, tools such as high-performance computing and sophisticated methods from computational fluid dynamics have not as yet made a substantial impact.

Finally, we have heard it remarked that there are many more theoretical models in the field of active matter than there are definitive experiments. This seems patently true and is partly a reflection of the relative ease of coming up with a model with some interesting dynamics (usually agent-based) versus the difficulty of performing experiments using real organisms or synthesizing active materials. Firmly connecting mathematical models to experiments through principled modeling and thorough exploration is difficult and seems best pursued through both many-particle simulations and continuum models.

Acknowledgment We thank our many students and postdocs that have worked with us on the topic of active suspensions and materials. We also much appreciate the editorial patience extended to us during the preparation of this chapter. DS acknowledges the support of NSF grants DMS-0920931 and CBET-1150590 (CAREER). MJS acknowledges the support of NSF grants DMR-0820341 (NYU MRSEC), DMS-0920930, DOE grant DE-FG02-88ER25053, and NIH grant R01 GM104976-03.

References

1. I. Aranson, Physics **6**, 61 (2013)
2. H.P. Zhang, A. Be'er, E.L. Florin, H.L. Swinney, Proc. Natl. Acad. Sci. USA **107**, 13626 (2010)
3. T. Sanchez, D. Chen, S. DeCamp, M. Heymann, Z. Dogic, Nature **491**, 431 (2012)

4. V. Schaller, C. Weber, C. Semmrich, E. Frey, A.R. Bausch, Nature **467**, 73 (2010)
5. S. Thutupalli, R. Seeman, S. Herminghaus, New J. Phys. **13**, 073021 (2011)
6. C.A. Weber, T. Hanke, J. Deseigne, S. Léonard, O. Dauchot, E. Frey, H. Chaté, Phys. Rev. Lett. **110**, 208001 (2013)
7. S. Ramaswamy, Annu. Rev. Condens. Matt. **1**, 323 (2010)
8. M.C. Marchetti, J.F. Joanny, S. Ramaswamy, T.B. Liverpool, J. Prost, M. Rao, R. Aditi Simha. Rev. Mod. Phys. **85**, 1143 (2013)
9. D. Saintillan, M.J. Shelley, C.R. Physique **14**, 497 (2013)
10. C. Dombrowski, L. Cisneros, S. Chatkaew, R.E. Goldstein, J.O. Kessler, Phys. Rev. Lett. **93**, 098103 (2004)
11. L.H. Cisneros, J.O. Kessler, S. Ganguly, R.E. Goldstein, Phys. Rev. E **83**, 061907 (2011)
12. J. Dunkel, S. Heidenreich, K. Drescher, H.H. Wensink, M. Bar, R.E. Goldstein, Phys. Rev. Lett. **110**, 228102 (2013)
13. S. Köhler, V. Schaller, A.R. Bausch, PLoS ONE **6**, e23798 (2011)
14. T. Shinar, M. Mana, F. Piano, M. Shelley, Proc. Natl. Acad. Sci. **108**, 10508 (2011)
15. F. Woodhouse, R.E. Goldstein, Proc. Natl. Acad. Sci. USA **110**, 14132 (2013)
16. F. Nedelec, T. Surrey, A.C. Maggs, S. Leibler, Nature **389**, 305 (1997)
17. Y. Sumino, K.H. Nagai, Y. Shitaka, D. Tanaka, K. Yoshikawa, H. Chaté, K. Oiwa, Nature **483**, 448 (2012)
18. W.F. Paxton, K.C. Kistler, C.C. Olmeda, A. Sen, S.K.S. Angelo, Y. Cao, T.E. Mallouk, P.E. Lammert, V.H. Crespi, J. Am. Chem. Soc. **126**, 13424 (2004)
19. J.R. Howse, R.A.L. Jones, A.J. Ryan, T. Gough, R. Vafabakhsh, R. Golestanian, Phys. Rev. Lett. **99**, 048102 (2007)
20. S.J. Ebbens, J.R. Howse, Soft Matter **6**, 726 (2010)
21. D. Takagi, A. Braunschweig, J. Zhang, M. Shelley, Phys. Rev. Lett. **110**, 038301 (2013)
22. A. Bricard, J.B. Caussin, N. Desreumaux, O. Dauchot, D. Bartolo, Nature **503**, 95 (2013)
23. A. Kudrolli, G. Lumay, D. Volfson, L.S. Tsimring, Phys. Rev. Lett. **100**, 058001 (2008)
24. J. Deseigne, O. Dauchot, H. Chaté, Phys. Rev. Lett. **105**, 098001 (2010)
25. T. Vicsek, A. Zafeiris, Phys. Rep. **517**, 71 (2012)
26. N.H. Mendelson, A. Bourque, K. Wilkening, K.R. Anderson, J.C. Watkins, J. Bacteriol. **181**, 600 (1999)
27. I. Tuval, L. Cisneros, C. Dombrowski, C.W. Wolgemuth, J.O. Kessler, R.E. Goldstein, Proc. Natl. Acad. Sci. USA **102**, 2277 (2005)
28. A. Sokolov, R.E. Goldstein, F.I. Feldchtein, I.S. Aranson, Phys. Rev. E **80**, 031903 (2009)
29. X.L. Wu, A. Libchaber, Phys. Rev. Lett. **84**, 3017 (2000)
30. K.C. Leptos, J.S. Guasto, J.P. Gollub, A.I. Pesci, R.E. Goldstein, Phys. Rev. Lett. **103**, 198103 (2009)
31. G.L. Miño, J. Dunstan, A. Rousselet, E. Clément, R. Soto, J. Fluid Mech. **729**, 423 (2013)
32. J. Orozco, B. Jurado-Sánchez, G. Wagner, W. Gao, R. Vazquez-Duhalt, S. Sattayasamitsathit, M. Galarnyk, A. Cortés, D. Saintillan, J. Wang, Langmuir **30**, 5082 (2014)
33. M.J. Kim, K.S. Breuer, Anal. Chem. **79**, 955 (2007)
34. H. Kurtuldu, J.S. Guasto, K.A. Johnson, J.P. Gollub, Phys. Rev. Lett. **108**, 10391 (2011)
35. A. Sokolov, I.S. Aranson, Phys. Rev. Lett. **103**, 148101 (2009)
36. L. Giomi, T.B. Liverpool, M.C. Marchetti, Phys. Rev. E **81**, 051908 (2010)
37. D. Saintillan, Exp. Mech. **50**, 1275 (2010)
38. J. Gachelin, G. Miño, H. Berthet, A. Lindner, A. Rousselet, E. Clément, Phys. Rev. Lett. **110**, 268103 (2013)
39. E. Lauga, T.R. Powers, Rep. Prog. Phys. **72**, 096601 (2009)
40. E.M. Purcell, Am. J. Phys. **45**, 3 (1977)
41. C. Brennen, H. Winet, Annu. Rev. Fluid Mech. **9**, 339 (1977)
42. K. Drescher, J. Dunkel, L.H. Cisneros, S. Ganguly, R.E. Goldstein, Proc. Natl. Acad. Sci. USA **108**, 10940 (2011)
43. J.S. Guasto, K.A. Johnson, J.P. Gollub, Phys. Rev. Lett. **105**, 168102 (2010)

44. S. Kim, S.J. Karrila, *Microhydrodynamics: Principles and Selected Applications* (Dover, NewYork, 2005)
45. K. Drescher, R.E. Goldstein, N. Michel, M. Polin, I. Tuval, Phys. Rev. Lett. **105**, 168101 (2010)
46. S. Fürthauer, S. Ramaswamy, Phys. Rev. Lett. **111**, 238102 (2013)
47. M. Leoni, T.B. Liverpool, Phys. Rev. Lett. **112**, 148104 (2014)
48. T. Ishikawa, M.P. Simmonds, T.J. Pedley, J. Fluid Mech. **568**, 119 (2006)
49. Q. Liao, G. Subramanian, M.P. DeLisa, D.L. Koch, M. Wu, Phys. Fluids **19**, 061701 (2007)
50. V. Gyrya, I.S. Aranson, L.V. Berlyand, D. Karpeev, Bull. Math. Bio. **72**, 148 (2010)
51. J.P. Hernández-Ortiz, C.G. Stoltz, M.D. Graham, Phys. Rev. Lett. **95**, 204501 (2005)
52. J.P. Hernández-Ortiz, P.T. Underhill, M.D. Graham, J. Phys.: Condens. Matter **21**, 204107 (2009)
53. J.R. Blake, J. Fluid Mech. **46**, 199 (1971)
54. T. Ishikawa, T.J. Pedley, J. Fluid Mech. **588**, 437 (2007)
55. T. Ishikawa, J.T. Locsei, T.J. Pedley, Phys. Rev. E **82**, 021408 (2010)
56. T. Ishikawa, T.J. Pedley, Phys. Rev. Lett. **100**, 088103 (2008)
57. T. Ishikawa, J.T. Locsei, T.J. Pedley, J. Fluid Mech. **615**, 401 (2008)
58. A.A. Evans, T. Ishikawa, T. Yamaguchi, E. Lauga, Phys. Fluids **23**, 111702 (2011)
59. D. Saintillan, M. Shelley, J.R. Soc. Interface **9**, 571 (2012)
60. D. Saintillan, M. Shelley, Phys. Rev. Lett. **99**, 058102 (2007)
61. R.A. Simha, S. Ramaswamy, Phys. Rev. Lett. **89**, 058101 (2002)
62. I.S. Aranson, A. Sokolov, J.O. Kessler, R.E. Goldstein, Phys. Rev. E **75**, 040901 (2007)
63. C. Wolgemuth, J. Biophys.**95**, 1564 (2008)
64. S. Mishra, A. Baskaran, M.C. Marchetti, Phys. Rev. E **81**, 061916 (2010)
65. J. Dunkel, S. Heidenreich, M. Bär, R.E. Goldstein, New J. Phys. **15**, 045016 (2013)
66. D. Saintillan, M. Shelley, Phys. Rev. Lett. **100**, 178103 (2008)
67. D. Saintillan, M. Shelley, Phys. Fluids **20**, 123304 (2008)
68. A. Alizadeh Pahlavan, D. Saintillan, Phys. Fluids **23**, 011901 (2011)
69. B. Ezhilan, A. Alizadeh Pahlavan, D. Saintillan, Phys. Fluids **24**, 091701 (2012)
70. E. Lushi, R.E. Goldstein, M.J. Shelley, Phys. Rev. E **86**, 040902 (2012)
71. B. Ezhilan, M.J. Shelley, D. Saintillan, Phys. Fluids **25**, 070607 (2013)
72. M. Doi, S.F. Edwards, *The Theory of Polymer Dynamics* (Oxford University Press, Oxford, 1986)
73. D.L. Koch, E.S.G. Shaqfeh, J. Fluid Mech. **209**, 521 (1989)
74. G. Subramanian, D.L. Koch, J. Fluid Mech. **632**, 359 (2009)
75. G.B. Jeffery, Proc. R. Soc. Lond. A **102**, 161 (1922)
76. F.P. Bretherton, J. Fluid Mech. **14**, 284 (1962)
77. G.K. Batchelor, J. Fluid Mech. **44**, 419 (1970)
78. M. Garcia, S. Berti, P. Peyla, S. Rafaï, Phys. Rev. E **83**, 035301 (2011)
79. H. Brenner, J. Colloid Interface Sci. **71**, 189 (1979)
80. H. Brenner, PhysicoChem. Hydrodyn. **1**, 91 (1980)
81. H.C. Berg, *Random Walks in Biology* (Princeton University Press, Princeton, 1983)
82. P. Underhill, J. Hernández-Ortiz, M. Graham, Phys. Rev. Lett. **100**, 248101 (2008)
83. E.J. Hinch, L.G. Leal, J. Fluid Mech. **76**, 187 (1976)
84. J.H. Irving, J.G. Kirkwood, J. Chem. Phys. **18**, 817 (1950)
85. G.K. Batchelor, J. Fluid Mech. **41**, 545 (1970)
86. C. Hohenegger, M. Shelley, in *New Trends in the Physics and Mechanics of Biological Systems*, ed. by M. Ben-Amar, A. Goriely, M. Muller, L. Cugliandolo (Oxford University Press, Oxford, 2011)
87. M. Doi, S.F. Edwards, J. Chem. Soc. Faraday Trans. II **74**, 560 (1978)
88. X. Chen, J.G. Liu, J. Differ. Equat. **254**, 2764 (2013)
89. A. Baskaran, M.C. Marchetti, Proc. Natl. Acad. Sci. USA **106**, 15567 (2009)
90. C. Hohenegger, M. Shelley, Phys. Rev. E **81**, 046311 (2010)
91. M. Betterton, A.S. Jhang, M. Shelley, In preparation (2014)

92. J.E. Butler, E.S.G. Shaqfeh, J. Fluid Mech. **468**, 205 (2002)
93. A. Baskaran, M.C. Marchetti, J. Stat. Mech.: Theor. Exp. **2010**, 04019 (2010)
94. L. Onsager, Ann. N. Y. Acad. Sci. **51**, 627 (1949)
95. W. Maier, A. Saupe, Z. Naturforsch. **13**, 564 (1958)
96. P.D. Cobb, J.E. Butler, J. Chem. Phys. **123**, 054908 (2005)
97. E.S.G. Shaqfeh, G.H. Fredrickson, Phys. Fluids A **2**, 7 (1990)
98. P.R. Nott, E. Guazzelli, O. Pouliquen, Phys. Fluids **23**, 043304 (2011)
99. A.P. Berke, L. Turner, H.C. Berg, E. Lauga, Phys. Rev. Lett. **101**, 038102 (2008)
100. G. Li, J.X. Tang, Phys. Rev. Lett. **103**, 078101 (2009)
101. J. Hill, O. Kalkanci, J.L. McMurry, H. Koser, Phys. Rev. Lett. **98**, 068101 (2007)
102. V. Kantsler, J. Dunkel, M. Polin, R.E. Goldstein, Proc. Natl. Acad. Sci. USA **110**, 1187 (2013)
103. E. Altshuler, G. Miño, C. Perez-Penichet, L. del Rio, A. Lindner, A. Rousselet, E. Clément, Soft Matter **9**, 1864 (2013)
104. G. Miño, T.E. Mallouk, T. Darnige, M. Hoyos, J. Dauchet, J. Dunstan, R. Soto, Y. Wang, A. Rousselet, E. Clément, Phys. Rev. Lett. **106**, 048102 (2011)
105. E. Lauga, W.R. DiLuzio, G.M. Whitesides, H.A. Stone, J. Biophys. **90**, 400 (2006)
106. F.G. Woodhouse, R.E. Goldstein, Phys. Rev. Lett. **109**, 168105 (2012)
107. H. Wioland, F.G. Woodhouse, J. Dunkel, J.O. Kessler, R.E. Goldstein, Phys. Rev. Lett. **110**, 268102 (2013)
108. A. Costanzo, R. Di Leonardo, G. Ruocco, L. Angelani, J. Phys.: Condens. Matter **24**, 065101 (2012)
109. S. Chilukuri, C.H. Collins, P.T. Underhill, J. Phys.: Condens. Matter **26**, 115101 (2014)
110. S.E. Spagnolie, E. Lauga, J. Fluid Mech. **700**, 105 (2012)
111. R. Voituriez, J.F. Joanny, J. Prost, Europhys. Lett. **70**, 404 (2005)
112. S.A. Edwards, J.M. Yeomans, Europhys. Lett. **85**, 18008 (2009)
113. M. Ravnik, J.M. Yeomans, Phys. Rev. Lett. **110**, 026001 (2013)
114. D. Takagi, J. Palacci, A. Braunschweig, M. Shelley, J. Zhang, Soft Matter **10**, 1784 (2014)
115. R.L. Schiek, E.S.G. Shaqfeh, J. Fluid Mech. **296**, 271 (1995)
116. B. Ezhilan, D. Saintillan, Submitted (2014)
117. T. Kaya, H. Koser, Biophys. J. **102**, 1514 (2012)
118. Y. Shen, A. Siryapon, S. Lecuyer, Z. Gitai, H.A. Stone, Biophys. J. **103**, 146 (2012)
119. R.W. Nash, R. Adhikari, J. Tailleur, M.E. Cates, Phys. Rev. Lett. **104**, 258101 (2010)
120. A. Jhang, M. Shelley, In preparation (2014)
121. T. Brotto, J.B. Caussin, E. Lauga, D. Bartolo, Phys. Rev. Lett. **110**, 038101 (2013)
122. N. Liron, S. Mochon, J. Eng. Math. **10**, 287 (1976)
123. S. Bhattacharya, J. Bławzdziewicz, E. Wajnryb, J. Comp. Phys. **212**, 718 (2006)
124. A. Lefauve, D. Saintillan, Phys. Rev. E **89**, 021002 (2014)
125. A.C.H. Tsang, E. Kanso, Flagella-induced transitions in the collective behavior of confined microswimmers. Phys. Rev. E **90**, 021001(R) (2014)
126. H.C. Berg, D.A. Brown, Nature **239**, 500 (1972)
127. G. Subramanian, D.L. Koch, S.R. Fitzgibbon, Phys. Fluids **23**, 041901 (2011)
128. T. Kasyap, D. Koch, Phys. Rev. Lett. **108**, 038101 (2012)
129. T. Kasyap, D. Koch, J. Fluid Mech. **741**, 619 (2014)
130. E.O. Budrene, H.C. Berg, Nature **349**, 630 (1991)
131. B.L. Bassler, Cell **109**, 421 (2002)
132. S. Park, P.M. Wolanin, E.A. Yuzbashyan, P. Silberzan, J.B. Stock, R.H. Austin, Science **301**, 188 (2003)
133. E.F. Keller, L.A. Segel, J. Theor. Biol. **30**, 225 (1971)
134. J.R. Vélez-Cordero, E. Lauga, J Non-Newtonian Fluid Mech **199**, 37 (2013)
135. Y. Bozorgi, P. Underhill, Phys. Rev. E **84**, 061901 (2011)
136. Y. Bozorgi, P. Underhill, J. Rheo. **57**, 511 (2013)
137. R.B. Bird, R.C. Armstrong, O. Hassager, *Dynamics of Polymeric Liquids* (Wiley-Interscience, 1987)
138. K.J. Helmke, R. Heald, J.D. Wilbur, Int Rev Cell Mol Biol **306**, 83 (2013)

139. A. Kimura, S. Onami, Dev Cell **8**, 765 (2005)
140. C.S. Peskin, Acta numerica **11**, 479 (2002)
141. K. Kimura, A. Kimura, Proc. Natl. Acad. Sci. USA **108**, 137 (2011)
142. T. Gao, R. Blackwell, M. Glaser, M. Betterton, M. Shelley, Submitted (2014)
143. T. Surrey, F. Nédélec, S. Leibler, E. Karsenti, Science **292**, 1167 (2001)
144. L. Giomi, M. Bowick, X. Ma, M. Marchetti, Phys. Rev. Lett. **110**, 228101 (2013)
145. S. Thampi, R. Golestanian, J. Yeomans, Phys. Rev. Lett. **111**, 118101 (2013)
146. H. Masoud, M. Shelley, Phys. Rev. Lett. **112**, 128304 (2014)
147. J. Wang, *Nanomachines: Fundamentals and Applications* (Wiley, NewYork, 2013)
148. J.L. Moran, P.M. Wheat, J.D. Posner, Phys. Rev. E **81**, 065302 (2010)
149. Y. Hong, N.M. Blackman, N.D. Kopp, A. Sen, D. Velegol, Phys. Rev. Lett. **99**, 178103 (2007)
150. Y. Hong, D. Velegol, N. Chaturvedi, A. Sen, Phys. Chem. Chem. Phys. **12**, 1423 (2010)
151. H. Masoud, H.A. Stone, J. Fluid Mech. **741**, R4 (2014)
152. T. Vicsek, A. Czirók, E. Ben-Jacob, I. Cohen, O. Shochet, Phys. Rev. Lett. **74**, 1226 (1995)
153. D. Horstmann, From 1970 until present: The Keller-Segel model in chemotaxis and its consequences. Jahresber. Deutsch. Math.-Verein. **105**, 103–165 (2003)

Part IV
Computational Methods

Chapter 10
Computational Challenges for Simulating Strongly Elastic Flows in Biology

Robert D. Guy and Becca Thomases

Abstract Understanding the behavior of complex fluids in biology presents mathematical, modeling, and computational challenges not encountered in classical fluid mechanics, particularly in the case of fluids with large elastic forces that interact with immersed elastic structures. We discuss some of the characteristics of strongly elastic flows and introduce different models and methods designed for these types of flows. We describe contributions from analysis that motivate numerical methods and illustrate their performance on different models in a simple test problem. Biological problems often involve the coupled dynamics of active elastic structures and the surrounding fluid. The immersed boundary method has been used extensively for such problems involving Newtonian fluids, and the methodology extends naturally to complex fluids in conjunction with the algorithms described earlier in this chapter. We focus on implicit-time methods because the large elastic stresses in complex fluids necessitate high spatial resolution and long time simulations. As an example to highlight some of the challenges of strongly elastic flows, we use the immersed boundary method to simulate an undulatory swimmer in a viscoelastic fluid using a data-based model for the prescribed shape.

There are many different kinds of complex fluids in biology, and they frequently contain dynamic active or passive structures in the fluid. Numerical simulations of these complex flows can be a powerful tool in understanding these biological systems. Existing techniques in computational fluid dynamics are often sufficient for problems with weak flows and low elasticity. However, when elastic forces become large due to, for example, long relaxation times, extra forces from internal structures, or interactions with complex boundaries, more care must be taken to properly simulate these flows. This chapter is devoted to the challenges that

R.D. Guy (✉) • B. Thomases
Department of Mathematics, University of California, Davis,
1 Shields Ave., Davis, CA 95616, USA
e-mail: guy@math.ucdavis.edu; thomases@math.ucdavis.edu

S.E. Spagnolie (ed.), *Complex Fluids in Biological Systems*, Biological and Medical Physics, Biomedical Engineering,
DOI 10.1007/978-1-4939-2065-5_10

arise when internal elastic forces are modeled in complex fluids with an eye towards recognizing, understanding, and properly treating the features of strongly elastic flows.

We will focus on the Oldroyd-B model as one of the simplest closed continuum models of viscoelastic fluids. In the original derivation [1], Oldroyd set out requirements for constitutive equations so that the material properties would be frame invariant in a coordinate system which convected with the material. This procedure leads to the upper-convected time derivative (also called the Oldroyd derivative, see Eq. (10.5)) which gives the rate of change of a tensor property of a small volume of fluid written in a coordinate system rotating and stretching with the fluid. The most widely used Oldroyd model is the Oldroyd-B model, in part because this model can also be derived from a theory of dilute polymer solutions [2]. The Oldroyd-B model is given below for $\mathbf{u}$ the velocity of the fluid, p the pressure, and $\boldsymbol{\tau}$ the deviatoric stress tensor; see also Chap. 1. From balance of momentum and mass conservation for an incompressible fluid we have

$$\rho \frac{D\mathbf{u}}{Dt} = -\nabla p + \nabla \cdot \boldsymbol{\tau} \tag{10.1}$$

$$\nabla \cdot \mathbf{u} = 0, \tag{10.2}$$

with

$$\boldsymbol{\tau} = \eta_s \dot{\boldsymbol{\gamma}} + \boldsymbol{\tau}_p. \tag{10.3}$$

Here $\eta_s \dot{\boldsymbol{\gamma}}$ is the viscous stress from the Newtonian solvent, with viscosity η_s and rate-of-strain tensor $\dot{\boldsymbol{\gamma}} = \nabla\mathbf{u} + (\nabla\mathbf{u})^T$, and $\boldsymbol{\tau}_p$ is the polymeric stress contribution. In the Oldroyd-B model the polymer stress evolves by

$$\boldsymbol{\tau}_p + \lambda \overset{\nabla}{\boldsymbol{\tau}}_p = \eta_p \dot{\boldsymbol{\gamma}}, \tag{10.4}$$

with polymer viscosity η_p and relaxation time λ. The upper-convected derivative is defined as

$$\overset{\nabla}{\boldsymbol{\tau}}_p \equiv (D/Dt)\boldsymbol{\tau}_p - (\nabla\mathbf{u})^T \cdot \boldsymbol{\tau}_p - \boldsymbol{\tau}_p \cdot \nabla\mathbf{u}. \tag{10.5}$$

The relaxation time characterizes the time it takes for a material to adjust to applied stresses or deformations. For the majority of this chapter we will focus on the low Reynolds number (or creeping flow) regime where inertial forces are small compared with viscous forces.

There are two important dimensionless parameters related to the relaxation time of a fluid used in rheology. The Weissenberg number (Wi) is the ratio of the relaxation time of the fluid and a specific process time. For example, in steady shear, the Weissenberg number is defined as the shear rate ($\dot{\gamma} = \sqrt{\dot{\boldsymbol{\gamma}} : \dot{\boldsymbol{\gamma}}/2}$, see Chap. 1) times the relaxation time $\mathrm{Wi} = \dot{\gamma}\lambda$. The Deborah number (De) is used to characterize flows under specific flow conditions and is defined as the ratio of the relaxation time to the characteristic time scale for fluid deformations, such as an imposed oscillation period, $\mathrm{De} = \frac{\lambda}{T_f}$.

Weakly elastic flows (De or Wi $\ll 1$) can be handled using standard techniques from computational fluid dynamics, and special treatment of the stress tensor is not necessary. By contrast, strongly elastic flows (De or Wi $\gg 1$) create regions of high stress and fine features that require high resolution for accurate flow solutions. Naive implementations of standard CFD techniques may fail.

In Sect. 1 we identify and demonstrate some of the characteristics of strongly elastic flows. These include large stresses and large gradients which require fine meshes for accurate representation. We introduce a host of models and methods designed to overcome these challenges and use a simple test problem to demonstrate some of these techniques and models. In Sect. 2 we briefly present the ideas of the immersed boundary method, a popular technique for simulating problems in biology which typically involve fluid-structure interactions with large deformations and complex flow patterns. One advantage of the immersed boundary method is that it can be easily coupled with a preexisting fluid solver; however, long time simulations and fine meshes necessitate the use of an implicit time stepping method. In Sect. 3 we use the immersed boundary method to simulate an undulatory swimmer with a data-based model for the target shape. This problem highlights some of the additional challenges associated with strongly elastic flows in biology, due, in part, to the large forces which arise in the coupling of fluid to dynamic structures.

1 Strongly Elastic Flows

Simulating strongly elastic flows is difficult and requires care in the choice of model and numerical method to ensure that the chosen technique treats the elastic stresses and corresponding time scale properly. In this chapter we emphasize the importance of recognizing these difficulties and understanding their origin, and we provide some approaches for fixing them. As an example of one difficulty with the Oldroyd-B model, we present a sample simulation of an extensional point flow in Sect. 1.6 which demonstrates fine scales: near-singularities in the stress field and near-jumps in the vorticity. In these simulations a body force drives a flow in which $\mathbf{u} \sim \dot{\varepsilon}(x, -y)$ near the origin, where $\dot{\varepsilon}$ is the strain rate. To demonstrate the flow behavior we examine the vorticity and the trace of the polymer stress tensor, which is proportional to the elastic energy. Figure 10.1a shows a contour plot of the trace of the conformation tensor, $\mathrm{Tr}(\mathbf{C})$, which is related to the polymer stress tensor for the Oldroyd-B model by $\boldsymbol{\tau}_p = \mathrm{Wi}^{-1}(\mathbf{C} - \mathbf{I})$. The vorticity of the flow ($\nabla \times \mathbf{u}$) is plotted in Fig. 10.1b. These plots show the near-steady behavior for a strongly elastic flow where large stress and stress gradients have developed at the extensional stagnation point at the origin. The fine structures can be seen in slices of $\mathrm{Tr}(\mathbf{C})$ and vorticity, which are shown in Figs. 10.1c and d. The details of the simulation are given in Sect. 1.6.

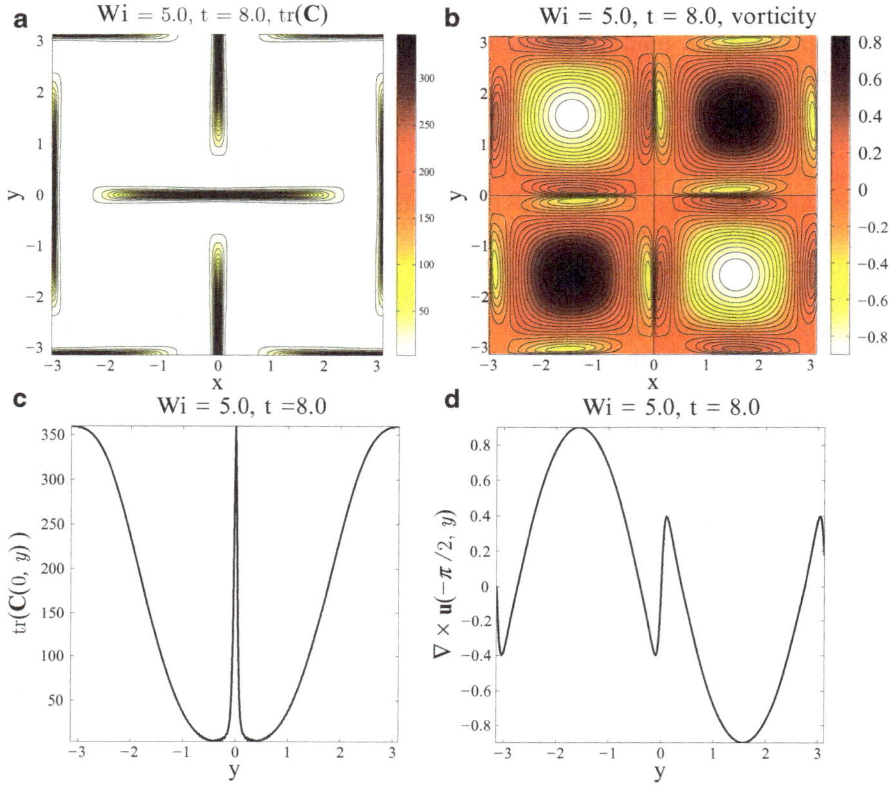

Fig. 10.1 Contour plots of (**a**) Tr(**C**) and (**b**) vorticity for Wi $= 5.0$ at $t = 8$. (**c**) Trace of conformation tensor along line $(0,y)$: $\mathrm{Tr}(\mathbf{C}(0,y))$. (**d**) Vorticity along line $(-\pi/2,y)$: $\nabla \times \mathbf{u}(-\pi/2,y)$

1.1 Historical Perspective

Computational simulations of strongly elastic flows have historically suffered from difficulties not seen in comparable Newtonian flows. These difficulties frequently manifest in numerical methods as a breakdown beyond a critical Deborah (or Weissenberg) number. These computational challenges have been observed since the earliest numerical approximations of complex fluids were attempted in the late 1970s. Standard finite difference and Galerkin finite element approaches that were successful for Newtonian flows were converging only for $\mathcal{O}(1)$ Deborah (or Weissenberg) numbers [3–5]. There are modeling and analytical questions underlying the somewhat mysterious "high-Weissenberg number problem," and so the appropriate choice of numerical method is a delicate and extremely important question which has received much attention over the years. Significant progress has been made in the subsequent decades, but many questions related to the high-Weissenberg number problem are still subjects of active research. We discuss some of the open analytical questions related to these problems in Sect. 1.2 and some current numerical approaches which address these difficulties in Sect. 1.4. We focus on the Oldroyd-B model in Sect. 1.3 and models derived from different molecular assumptions in Sect. 1.5.

Several benchmark problems in engineering have been developed to test proposed constitutive laws and numerical methods. These problems include flow around a cylinder or sphere [6–12] and planar contractions [13–19]. A simplified explanation for why flow around obstacles or near boundaries can cause difficulties in viscoelastic flows which are not seen in corresponding Newtonian flows is that when the velocity of the flow is near zero (due to a no-slip boundary condition or a stagnation point in the flow) the internal elastic structures in the fluid have a long time to get stretched and induce areas of high stress. Large stress gradients then induce more stress on the flow and this leads to sharp boundary layers which require high spatial resolution. Long time simulations or time-dependent simulations for finding steady-state solutions are particularly difficult.

One must be aware of other challenges which arise when modeling viscoelastic fluids such as possible change of type or loss of evolution in the flow. These problems do not arise with the Oldroyd-B and the other models discussed here. We leave further discussions of those problems and more detailed reviews of numerical issues to the following books and review articles, and the citations contained within, [5, 20–26].

1.2 Advances from Analysis

To begin to understand the complicated high-Weissenberg problem we note that fundamental analytical questions about the Oldroyd-B model (Eq. (10.1)–(10.4)) are still open. For example, it is unknown whether global solutions to the Oldroyd-B equations exist. Even for the Stokesian limit in two space dimensions, the question is challenging due to the lack of scale-dependent dissipation in the polymer stress advection equation[1]. These basic questions are important because if a model is well posed, i.e., unique solutions exist and are sufficiently smooth on some time interval, then appropriate numerical approximations of these solutions are reliable. If not, then even convergence of the method cannot guarantee that the correct solution has been chosen.

Though having a well-posedness theory is useful for a numerical study of a particular model, in many cases, notably the 3D Euler or Navier-Stokes equations,

[1] When derived from the kinetic theory of dumbbells [2, 27] there is polymer stress diffusion; however the stress diffusion coefficient is proportional to the square of the ratio of the bead diameter (or polymer radius of gyration) to the flow length scale, and even in the context of microfluidics, it is minute, $\mathcal{O}(10^{-9})$, and is typically ignored.

this is not available. However, there are often mathematical theorems which can be helpful for guiding numerics. One particular example is a well-known condition for the 3D Euler equations, referred to as the Beale-Kato-Majda criteria [28], which states that the maximum norm of the vorticity controls the breakdown of smooth solutions. More specifically this criteria states that the breakup of solutions in any norm will imply the divergence of the supremum norm of the vorticity. This makes investigating the loss of existence of solutions more tractable as you only have to keep track of one quantity. A similar result can be derived for the Oldroyd-B equations and the quantity to track is the supremum norm of the polymer stress tensor [29, 30].

Along with this nonexistence criterion, some progress in well posedness has been made. For example, if the initial data are sufficiently small, solutions to the Oldroyd-B model are globally well posed [29, 31, 32], but this prescribes an unrealistic constraint for many problems of interest. Additionally, these analytical proofs require either an unbounded or periodic domain, while the question of how to treat problems in complicated geometries remains an issue of current research that is of significant importance in many applications.

As mentioned above, one of the main difficulties in obtaining analytical proofs of global existence for the Oldroyd-B model is the lack of scale-dependent dissipation in the stress advection equation, Eq. (10.4). A simple regularization is to include a dissipative term in such as $\alpha\nabla^2\boldsymbol{\tau}_p$. With this type of polymer stress diffusion, some analytical results are available [27], and, in particular, the Oldroyd-B equations in two space dimensions are globally well posed for all initial data [33, 34]. The addition of polymer stress diffusion is not without physical justification as stress diffusion does arise in the physical model [2, 27, 35], but the diffusion is at such small scales that it is typically ignored. However, artificially large polymer stress diffusion can be introduced as a regularization parameter in numerical simulations [36, 37]. The effect of artificially large stress diffusion was studied in [36] where the authors concluded that the stress diffusion had a stabilizing effect, in particular for large Reynolds number calculations. Adding polymer stress diffusion is particularly common in simulations of turbulent drag reduction [38, 39]. Although the polymer stress diffusion may be artificially large in the context of the molecular derivation, if the length scale of the artificial polymer stress diffusion is on the scale of the grid, then the errors due to the regularization are on the order of the spatial discretization errors.

1.3 High-Weissenberg Number Problem in the Oldroyd-B Model

If the simulations presented in Fig. 10.1 are continued (without decreasing the grid size), numerical oscillations will cause the solutions to break down at some point in time. A "local" analytical solution given in [40] indicates that the polymer stress is likely approaching a diverging solution exponentially in time. Hence the breakdown of the numerical solution appears to be due to the fact that at a fixed grid size it becomes impossible to resolve the steeper and steeper gradients of an underlying diverging solution. These oscillations can already be seen if one looks closely in Fig. 10.1c at the polymer stress near $y = 0$. There are many possible reasons for the breakdown of the numerical simulation; in particular it may be the case that the solution itself does not exist for all time, and this is simply not known. However, even if the solution remains smooth in time, other problems including loss of positive-definiteness of the stress tensor, problems with the model (such as infinite extension of polymer coils), or large stress gradients which create even larger forces on the fluid which must be resolved can lead to the breakdown of a numerical method. Problems with the Oldroyd-B model are typically blamed on the linear elastic nature of the model, but as we will see in Sect. 1.6 this is probably not the heart of the problem.

The Oldroyd-B model is attractive because it is the simplest closed continuum model which can be derived from molecular assumptions. One derivation of the Oldroyd-B model comes from representing immersed polymer coils in a Newtonian solvent as two beads connected by a linear spring with a Hookean spring force [2,41]. Additional forces on the beads include drag from the Newtonian solvent and randomly fluctuating Brownian forces. Using statistical mechanics, one can derive an expression for the stress tensor which evolves according Eq. (10.4).

This model predicts that in steady extensional flows, such as uniaxial extension, the extensional viscosity, defined as the ratio of extensional stress to extensional strain rate, will become infinite at finite strain rate. This happens when the frictional drag force that stretches the dumbbell overcomes the spring force. When the strain rate is "small" relative to the relaxation time the spring force dominates and the dumbbell remains coiled. As the strain rate is increased the molecules undergo a "coil-stretch" transition and the steady-state extensional viscosity goes to infinity [41]. This is related to the fact that the linear Hooke's law puts no limit on the length of a dumbbell and has been seen as an underlying cause of the "high Weissenberg number problem." Rallison and Hinch questioned the "physics" of the constitutive model [42], and later it was noted that even below the coil-stretch transition the "smoothness of stresses should be expected to deteriorate with increasing Weissenberg number" [43].

Numerical simulations in [40] found solutions that exhibit the Weissenberg number-dependent smoothness described above. When an extensional flow is posited, namely $\mathbf{u} = \dot{\varepsilon}(x, -y)$, the equation for the polymer stress tensor (10.4) decouples and can be solved exactly via the method of characteristics. In two space dimensions the solution for one component of the stress tensor can be written as

$$S_{11}(x,y,t) = \frac{1}{1 - 2\dot{\varepsilon}\mathrm{Wi}} + e^{(2\dot{\varepsilon}\mathrm{Wi}-1)t} F(xe^{-\dot{\varepsilon}\mathrm{Wi}t}, ye^{\dot{\varepsilon}\mathrm{Wi}t}), \tag{10.6}$$

where the unknown function F must be determined with proper initial and boundary conditions. This solution indicates that for $\dot{\varepsilon}\mathrm{Wi} < 1/2$ the stress should be bounded but for $\dot{\varepsilon}\mathrm{Wi} > 1/2$ the stress is diverging exponentially in time. Furthermore, the solutions have a collapsing inner length scale in y which also depends on $\dot{\varepsilon}\mathrm{Wi}$.

One time-independent solution which was found to be in close agreement with numerical simulations and which demonstrates decreasing regularity in the Weissenberg number is

$$S_{11}^{\infty} = |y|^{(1-2\dot{\varepsilon}\mathrm{Wi})/\dot{\varepsilon}\mathrm{Wi}}. \tag{10.7}$$

It is important to be aware of exponential-in-time stress near-singularities, which are not removed by simply modifying the constitutive model, as we will show in Sect. 1.6. Next we look at a few ways to treat these near-divergent solutions.

1.4 Numerical Approaches

Defeating the "high-Weissenberg number problem" has been the aim of many numerical methods developed over the past several decades. Techniques have been developed to address issues of stability and convergence, for example, the use of upwinding [44,45] has been successfully applied to viscoelastic flow problems using finite element methods [46]. Other variations of finite element methods applied to viscoelastic fluid simulations include using discontinuous Galerkin techniques [47–49] and splitting techniques, such as EVSS (elastic viscous stress splitting) [19,23,50–52], among others [53–55]. Another way to provide local diffusion is by applying ENO schemes (essentially non-oscillatory shock capturing) [56] which use upwinding with a high-order correction [57]. To resolve fine structures in problems like the flow around a cylinder, techniques to improve accuracy include using hp-spectral elements [12] and highly accurate finite volume methods [9]. For a more complete review of the literature we refer the interested reader to a computational rheology book [5] and several review articles that deal with specific aspects of the numerical simulations of viscoelastic fluid flow [22–26]. In what follows we describe in more detail a few ways to handle the near-singularities that arise in the Oldroyd-B model discussed in Sect. 1.3.

1.4.1 Log-Conformation Method

The *log-conformation method* [58,59] is a numerical approach specifically designed to address exponential singularities in the polymer stress tensor. The method was designed for a large class of differential constitutive models (including Oldroyd-B) in which an equation is derived for the matrix logarithm of the conformation tensor or configuration tensor, $\mathbf{C}(\mathbf{x},t)$. $\mathbf{C}$ is the conformational average of the dumbbells,

$$\mathbf{C} = \int \mathbf{R}\mathbf{R}\Psi dR, \tag{10.8}$$

where $\mathbf{R}$ is the end-to-end vector of the dumbbell and $\Psi(\mathbf{x}, \mathbf{R}, t)$ is the probability that a dumbbell at position $\mathbf{x}$ in the flow has orientation and extension $\mathbf{R}$ at time t (see Chaps. 1 and 9). For the Oldroyd-B model, the conformation tensor is related to the polymer stress tensor by

$$\boldsymbol{\tau}_p = \mathrm{Wi}^{-1}(\mathbf{C} - \mathbf{I}), \tag{10.9}$$

and is advected by

$$(\mathbf{C} - \mathbf{I}) + \mathrm{Wi}\overset{\nabla}{\mathbf{C}} = 0. \tag{10.10}$$

From the molecular derivation, the conformation tensor should be symmetric positive definite, and it will remain so according to Eq. (10.10) if it is initially. Loss of positivity of $\mathbf{C}$ is one source of numerical errors. The log-conformation method maintains positivity by definition which can be a source of increased stability.

The log-conformation method replaces Eq. (10.10) with an equation for the matrix logarithm of $\mathbf{C}$:

$$\mathbf{A}(\mathbf{x}, t) = \log \mathbf{C}(\mathbf{x}, t).$$

This is possible because a symmetric positive definite matrix, $\mathbf{S}$, can always be diagonalized, $\mathbf{S} = \mathbf{R}\boldsymbol{\Lambda}\mathbf{R}^T$ and hence $\log \mathbf{S} = \mathbf{R}(\log \boldsymbol{\Lambda})\mathbf{R}^T$. The method relies on the fact that if $\mathbf{u}$ is a divergence-free velocity field and $\mathbf{C}$ is a symmetric positive definite tensor, then there is a decomposition

$$\nabla \mathbf{u} = \boldsymbol{\Omega} + \mathbf{B} + \mathbf{N} \cdot \mathbf{C}^{-1}, \tag{10.11}$$

where $\boldsymbol{\Omega}$ and $\mathbf{N}$ are anti-symmetric and $\mathbf{B}$ is symmetric, traceless, and commutes with $\mathbf{C}$. With this decomposition, the evolution of $\mathbf{A}$ is

$$\frac{\partial \mathbf{A}}{\partial t} + (\mathbf{u} \cdot \nabla)\mathbf{A} - (\boldsymbol{\Omega} \cdot \mathbf{A} - \mathbf{A} \cdot \boldsymbol{\Omega}) - 2\mathbf{B} = \mathrm{Wi}^{-1} e^{-\mathbf{A}}(\mathbf{I} - e^{\mathbf{A}}). \tag{10.12}$$

Under this transformation, the extensional components of the deformation act additively, rather than multiplicatively. Higher Wi values can be achieved than in similar studies without the matrix logarithm, and the log-conformation method has been particularly successful in some standard benchmark problems [10,60–62]. This method has been implemented in many different numerical frameworks including finite difference, finite volume, and finite elements. There is a nontrivial cost associated with implementation of this method, both in obtaining the decomposition in Eq. (10.11) and in computing the matrix exponential to obtain $\mathbf{C}$ from $\mathbf{A}$.

1.4.2 Square-Root Method

A method which is simpler to implement and also maintains positive definiteness of the polymer stress tensor is the *square-root method* [63]. An exact equation for the square root of the conformation tensor is advected rather than the conformation tensor itself, and therefore, the conformation tensor will remain positive. In this method Eq. (10.10) is replaced with an equation for $\mathbf{b}(\mathbf{x},t)$, the unique positive symmetric square root of $\mathbf{C}(\mathbf{x},t)$. The equation for $\mathbf{b}$ is

$$\frac{\partial \mathbf{b}}{\partial t} + (\mathbf{u}\cdot\nabla)\mathbf{b} = \mathbf{b}\cdot\nabla\mathbf{u} + \mathbf{a}\cdot\mathbf{b} + \frac{1}{2\mathrm{Wi}}\left((\mathbf{b}^T)^{-1} - \mathbf{b}\right), \tag{10.13}$$

where $\mathbf{a}$ is any anti-symmetric matrix. Furthermore, $\mathbf{a}$ can be prescribed uniquely so that if $\mathbf{b}$ is initially symmetric it will remain symmetric. In two dimensions the form of $\mathbf{a}$ is

$$\mathbf{a} = \begin{pmatrix} 0 & a_{12} \\ -a_{12} & 0 \end{pmatrix}, \tag{10.14}$$

where

$$a_{12} = \frac{\left(b_{12}\dfrac{\partial u}{\partial x} - b_{11}\dfrac{\partial v}{\partial x}\right) + \left(b_{22}\dfrac{\partial u}{\partial y} - b_{12}\dfrac{\partial v}{\partial y}\right)}{b_{11}+b_{22}},$$

for $\mathbf{u} = (u,v)$. An exact formula is also available in 3 space dimensions, but the details are more complicated [63]. This method was tested in a spectral framework, and it was observed that in practice the square-root method can be applied at higher Wi and for longer time than methods for evolving the conformation tensor directly [63]. This method does not address the exponential nature of the singularities of the polymer stress tensor like the log-conformation method, but the cost of implementation is no different than a direct implementation of the original model.

1.4.3 Polymer Stress Diffusion

As mentioned in Sect. 1.2, adding polymer stress diffusion will regularize the Stokes-Oldroyd-B equations, and in two space dimensions, the problem is well posed. In [37] it was shown that the exponential-in-time singularity obtained in [40] is removed with the addition of polymer stress diffusion, and smooth and bounded steady-state solutions can be found. Consider the polymer stress advection equation with diffusion, which we write in nondimensional form as

$$(\mathbf{C}-\mathbf{I}) + \mathrm{Wi}\,\overset{\nabla}{\mathbf{C}} = \alpha\nabla^2\mathbf{C}. \tag{10.15}$$

If a steady flow, $\mathbf{u} = \mathrm{Wi}^{-1}(x, -y)$, is prescribed, this leads to a decoupling of Eq. (10.15), which is now linear in $\mathbf{C}$. An exact solution can be found which has the form

$$\mathbf{C} = \begin{pmatrix} -1 + Ae^{-y^2/(2\alpha)} & 0 \\ 0 & 1/3 \end{pmatrix}. \tag{10.16}$$

The Gaussian structure of $\mathbf{C}_{11}$ is a regularization of the delta-like singularities seen without diffusion in [40]. When comparing this local analytical solution to the numerical simulations the dependence on α and Wi is

$$\mathbf{C}_{11}(0, y) \approx -1 + C\,\mathrm{Wi}\,\alpha^{-1/2} e^{-y^2/(2\alpha)}. \tag{10.17}$$

These solutions are bounded and smooth for all $\alpha > 0$, and hence polymer diffusion may be used with some confidence as a regularization of the Oldroyd-B equation as long as care is taken to choose the length scale over which diffusion acts to be at or below the grid discretization.

1.5 Molecular Models

Instead of trying to address the singularities of the Oldroyd-B model directly, it is reasonable to criticize the molecular derivation of the Oldroyd-B model and use a model with a bounded extensional viscosity or a model which penalizes infinite extension of polymer coils. Many such models exist, and new molecular models are still being developed to match desired experimental data. Unfortunately, many modifications made at the molecular scale cannot be closed at the macroscopic level, resulting in multiscale or micro-macro models which are extremely computationally expensive. We discuss a few of the macroscopic models which are related to the Oldroyd-B model here and refer the interested reader to [2, 5, 41] for many other models. For various approaches to multiscale modeling of viscoelastic fluids see [26, 64, 65].

1.5.1 Giesekus Model

Like the Oldroyd-B model, the Giesekus model is also derived using a simple dumbbell model [66]. Giesekus proposed introducing an anisotropic drag force on the dumbbell which depends on the stress tensor based on the reasoning that the drag should be lower in the "direction" the fluid has been stressed. This could be true for a polymer solution in which the stress causes the molecules to line up in one direction resulting in lower drag in the direction of alignment. In this model, the drag coefficient becomes a drag tensor and adds a nonlinear term to the Oldroyd-B model. Eq. (10.4) is replaced with

$$\boldsymbol{\tau}_p + \lambda \overset{\nabla}{\boldsymbol{\tau}}_p + \alpha \frac{\lambda}{\eta_p} (\boldsymbol{\tau}_p)^2 = \eta_p \dot{\boldsymbol{\gamma}}. \tag{10.18}$$

This additional nonlinear term leads to physically realistic normal stress differences and bounded extensional viscosity but does not address infinite extension of polymer coils.

1.5.2 PTT Model

A very different way to derive a similar constitutive model comes from transient network model theory. Phan-Thien and Tanner [67] derived a model (called the PTT model) which assumes that polymers are entangled but they can break and reform. If the breaking rate increases with increasing average chain length, then a closed constitutive model can be derived which predicts shear thinning and bounded extensional viscosity. Eq. (10.4) is replaced with

$$\boldsymbol{\tau}_p + \lambda \overset{\nabla}{\boldsymbol{\tau}}_p + \alpha \frac{\lambda}{\eta_p} \mathrm{Tr}(\boldsymbol{\tau}_p) \boldsymbol{\tau}_p = \eta_p \dot{\boldsymbol{\gamma}}. \tag{10.19}$$

1.5.3 Finite Extension Models

Another modification at the molecular level involves enforcing finite extensibility of polymer coils. The following modification to the linear Hooke's law was proposed by Warner [68]. The force is penalized if the polymers stretch beyond some given maximum stretch length, R_0, which results in a force law

$$\mathbf{F} = \frac{H\mathbf{R}}{\left(1 - \mathrm{Tr}\left(\mathbf{RR}\right)/R_0^2\right)}, \tag{10.20}$$

where H is a spring constant and $\mathbf{R}$ is the end-to-end vector of the dumbbell. The main drawback with this force law is that one cannot obtain a closed continuum model for the polymer stress tensor. There have been many closure approximations suggested, see for example [69–72]. The simplest and most commonly used approximation is to assume that the force depends on the average extension of the distribution of springs, and this leads to the FENE-P model [73]; see also Chap. 1. Including the full form for the force in Eq. (10.20) leads to micro-macro models which are numerically expensive and beyond the scope of this chapter.

1.6 Extensional Flow Simulations

Simulations with these models in place of the Oldroyd-B model show that neither finite extensibility nor maintaining positive definiteness of the polymer stress tensor will solve the high-Weissenberg number problem. An unavoidable cause of the problem seems to be that advection of the polymer stress near extensional points causes large stress gradients which create large forces on the fluid. In [40] it was shown that although the polymer stress was bounded for the FENE-P model, sharp gradients and corner singularities were still found in the simulations, perhaps as a consequence of the force penalization. Here we show new simulations using the Giesekus and PTT models, which are more stable than the Oldroyd-B model, but they still break down in long time simulations beyond a critical Wi. These simulations illustrate some of the ideas mentioned in Sects. 1.4 and 1.5 and highlight the need for careful consideration of the polymer stress tensor. We compare the PTT, Giesekus, and polymer stress diffusion models to the (Stokes) Oldroyd-B model in a simple 2D (periodic) extensional flow. We repeat the numerical experiment performed in [40] which involved solving the Stokes-Oldroyd-B equations in two space dimensions with a background force prescribed to enforce an extensional flow.

The various models used for the following example all have the same form, given non-dimensionally as

$$\nabla^2\mathbf{u} - \nabla p + \xi\nabla\cdot\boldsymbol{\tau}_p + \mathbf{f} = 0, \tag{10.21}$$

$$\nabla\cdot\mathbf{u} = 0, \tag{10.22}$$

$$(\mathbf{C}-\mathbf{I}) + \mathrm{Wi}\overset{\nabla}{\mathbf{C}} + \alpha\mathscr{R}(\mathbf{C}) = 0, \tag{10.23}$$

where the conformation tensor is related to the polymer stress tensor as $\boldsymbol{\tau}_p = \mathrm{Wi}^{-1}(\mathbf{C}-\mathbf{I})$. For the Stokes-Oldroyd-B equations we set $\mathscr{R} \equiv 0$; the other models are defined below.

Giesekus	$\mathscr{R}(\boldsymbol{\tau}_p) = (\boldsymbol{\tau}_p)^2$
PTT	$\mathscr{R}(\boldsymbol{\tau}_p) = \mathrm{Tr}(\boldsymbol{\tau}_p)\boldsymbol{\tau}_p$
polymer stress diffusion	$\mathscr{R}(\boldsymbol{\tau}_p) = -\nabla^2\boldsymbol{\tau}_p$

$\mathrm{Wi} = \lambda/T_f$ is the *Weissenberg number*, with λ the polymer relaxation time and T_f the time scale of the fluid flow. The dimensional scaling F of the forcing $\mathbf{f}$ is used to set the flow time scale as $T_f = \eta_s/\rho LF$, where η_s is the solvent viscosity, ρ the fluid density, and L the system size. This sets the dimensionless force and the time scale of transport to be order one. The parameter $\xi = GT_f/\eta_s$ measures the relative contribution of the polymer stress to momentum balance, where G is the isotropic stress in the polymer field in the absence of flow. Note that the parameter ξ is the ratio of the polymer viscosity to the solvent viscosity and in what follows we set $\xi = 0.5$.

The Giesekus and PTT models both have bounded extensional viscosity, unlike Oldroyd-B, although we will see below that they still have large stress gradients near extensional points in the flow for sufficiently large Wi. In what follows, the parameter α is fixed at 0.001. We note that for this value, the resulting length scale for the polymer stress diffusion is smaller than the grid spacing.

The numerical experiment in [40] involved analyzing the stress near hyperbolic extensional points in the flow. The background force

$$\mathbf{f} = \begin{pmatrix} 2\sin x \cos y \\ -2\cos x \sin y \end{pmatrix}. \tag{10.24}$$

sets up a four vortex "mixer" in each $[-\pi,\pi]^2$ cell, which in a purely Newtonian Stokes flow (Wi $= 0$) has solution $\mathbf{u} = -\mathbf{f}/2$.

This 2D periodic "4-roll mill" geometry provides an opportunity to compare solutions to these different models in an extensional flow with no boundary effects. Given the regular domain, this problem is well suited to a pseudo-spectral method. Furthermore, as the local analytical solution in Eq. (10.6) suggests, beyond a critical Wi the polymer stress $\boldsymbol{\tau}_p$ grows exponentially near extensional points, and using a pseudo-spectral method allows one to analyze the regularity and evolution of the singularity in Fourier space [74–76].

The algorithm used in [40] for solving (10.21)–(10.23), which is similar here for $\mathscr{R} \neq 0$, was to prescribe initial data for $\boldsymbol{\tau}_p$, invert the Stokes equation to find the velocity, and with that update the polymer stress via any appropriate time stepping method (second-order Adams-Bashforth was used in [40] and will be used here). We set $\mathbf{C}(0) = \mathbf{I}$ for isotropic initial data. Inverting the Stokes equation amounts to solving for u_j in Fourier space ($j = 1,2$)

$$\hat{u}_j = \frac{1}{|\mathbf{k}|^2}\left[ik_j\hat{p} + i\xi k_\ell(\hat{\boldsymbol{\tau}}_p)_{\ell j} + \widehat{f_j}\right], \tag{10.25}$$

where the pressure is found using the incompressibility constraint

$$\hat{p} = \frac{1}{|\mathbf{k}|^2}\left[\xi k_j k_\ell(\hat{\boldsymbol{\tau}}_p)_{\ell j} - ik_j\widehat{f_j}\right], \tag{10.26}$$

with summation convention applied for all repeated indices. The stress update is also performed in Fourier space but care must be taken as the stress advection equation is nonlinear. In Eq. (10.23) the quadratic nonlinearities are computed using de-aliasing techniques [77]. We use a filter to zero the high wave numbers before inverting and multiplying the terms in real space to avoid aliasing errors. The filter we apply is similar to a simple 2/3 cutoff, but instead we apply a smooth rapidly decaying exponential cutoff [78] which helps stabilize the simulations. The Fourier transform is applied again and the nonlinear terms are used to update the polymer stress.

Simulations of the Stokes-Oldroyd-B model ($\mathscr{R} = 0$) were performed with this algorithm for $n^2 = 1024^2$ grid points in the $[-\pi,\pi]^2$ domain. In [40] two critical

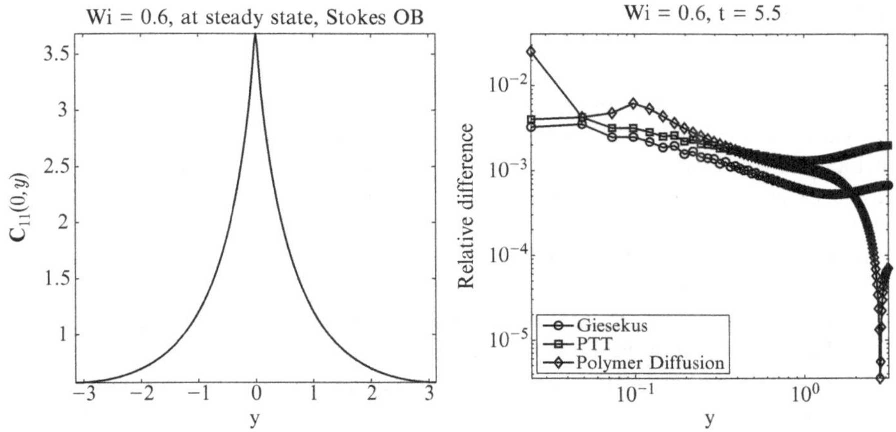

Fig. 10.2 Wi $= 0.6$ at $t = 5.5$: (*left*) $\mathbf{C}_{11}(0,y)$ the first component of the conformation tensor along the axis of compression where solution is at steady state. (*right*) Relative difference between models (Giesekus, PTT, and Polymer diffusion) and the Stokes-OB solution

Weissenberg numbers were identified from the numerical simulations: for Wi < 0.5 it was seen that the polymer stress was smooth, for $0.5 <$ Wi < 1 the stress approached a finite-valued cusp exponentially in time and for Wi > 1 the stress was diverging exponentially in time. These different polymer stress solutions correspond to different modifications to the velocity field. For sufficiently small Weissenberg number, Wi $\lesssim 1$, the stress perturbation only modifies the amplitude of the flow, namely $\mathbf{u} \approx C\mathbf{f}$ where C depends on Wi. The dependence of C on Wi is described in [40]. However for Wi $\gtrsim 1$ the 4-roll mill structure persists, but oppositely signed vortices arise along the axis of extension and compression. Figure 10.1a shows a contour plot of $\mathrm{Tr}(\mathbf{C})$ at $t = 8$ for Wi $= 5.0$. At $t = 8$ the solution is at "near-steady state." Although the polymer stress is still increasing in a neighborhood of the axis of compression and extension, the size of that neighborhood is decreasing, and the L^2 norm of the stress is nearly constant as is the L^2 norm of the velocity. The stress has concentrated along the stable and unstable manifolds of the extensional stagnation points in the flow. These stress regions are localized and need to be well resolved for accurate simulations. Figure 10.1b shows a contour plot of the vorticity at the same time. The near-delta-function stress creates near-cusps in the velocity field which yield near-jumps in the vorticity. These features are seen in detail when looking at slices of the polymer stress and vorticity in Figs. 10.1c and d.

In what follows, to compare the different models, we look at two singular cases separately, namely Wi $= 0.6$ which has a cusp solution and Wi $= 2.0$ which has diverging solutions. Simulations were done for the Stokes-Oldroyd-B model ($\mathscr{R} \equiv 0$) for Wi $= 0.6$ and Wi $= 2.0$ with $n^2 = 1024^2$ grid points in the $[-\pi, \pi]^2$ domain. These "exact solutions" are compared with coarser solutions of the various models: Giesekus, PTT, polymer diffusion, at $4\times$ coarser resolution, $n^2 = 256^2$, and $\alpha = 0.001$.

1.6.1 Cusp Solution: Wi=0.6

Figure 10.2 (left) shows the solution, $\mathbf{C}_{11}(0,y)$, for Stokes OB with Wi $= 0.6$ at $t = 5.5$. At this time, the solution has reached steady state, the maximum difference in the stress over 0.1 time units is $\mathscr{O}(10^{-3})$, and the maximum difference in the velocity is 2 orders of magnitude smaller. The cusp in the polymer stress is located at the extensional point in the flow $(0,0)$. The other constitutive models were run at $4\times$ coarser resolution, $n^2 = 256^2$ grid points, and are compared with the Stokes-OB solution in Fig. 10.2 (right). The different models agree well with Stokes OB. As expected, the polymer stress diffusion model fails to capture the cusp. However, outside an $\mathscr{O}(10^{-1})$ region near the extensional point all three models are accurate to 2 digits. Examining the Fourier spectrum for each of these models (not shown) reveals that the polymer diffusion model has a decaying spectrum similar to that of the higher resolution Stokes-OB solution. The spectra of the Giesekus and PTT models decay much less rapidly, which indicates the approach of a singularity. Long time simulations with polymer diffusion go to steady state, while the other models and the Stokes-OB simulations will eventually break down from oscillations as the solutions become more singular over time.

1.6.2 Diverging Solution: Wi=2.0

When Wi $= 2.0$ the results from [40] and the local analytical solution indicate that the polymer stress should be diverging near the extensional point at $(0,0)$. Although the stress is diverging, the set on which the stress is growing diminishes in time so that the resultant velocity field approaches a steady state. In these simulations between $t = 6.9$ and $t' = 7.0$ we see $\sup_{(x,y)} |\mathbf{u}(t) - \mathbf{u}(t')| = \mathscr{O}(10^{-4})$, although the polymer stress is diverging. If we revisit Fig. 10.1, we see that the effect of the concentration of polymer stress is to create recirculation cells in the vorticity. Figure 10.3a shows a slice of the $x-$component of the velocity $\mathbf{u}_1(\pi/2,y)$, at $t = 7$ (near-steady state) for Wi $= 2$. This value of Wi is beyond the coil-stretch transition and we see that the diverging stress leads to a more significant modification to the flow than the near-cusp stresses for Wi $= 0.6$. In Fig. 10.3a the flow perturbation occurs near $u_1(\pi/2,0)$ and as $t \to \infty$ this becomes a near-corner singularity. These near-corners lead to the near-jumps in the vorticity seen in Fig. 10.1b.

Figure 10.3b compares the other constitutive models (at $4\times$ coarser resolution) to the Stokes-OB solution for the velocity $u_1(\pi/2,y)$ for Wi $= 2$, at $t = 7$. The results are plotted on a log-log scale for $0 < y < \pi/2$. Here the polymer stress diffusion model captures the near-corner singularity almost one order of magnitude better than the Giesekus and PTT models (whose data lies practically on top of one another in the figure). Figures 10.3c and d show the Fourier spectra of the spatial data in Fig. 10.3a and b, respectively. The Stokes-OB solution is well resolved at this time with $n^2 = 1024^2$ grid points, but the polymer diffusion model captures the essential features of the flow and is well resolved with $n^2 = 256^2$ grid points. The Giesekus and PTT models are both beginning to lose accuracy in the high frequencies which

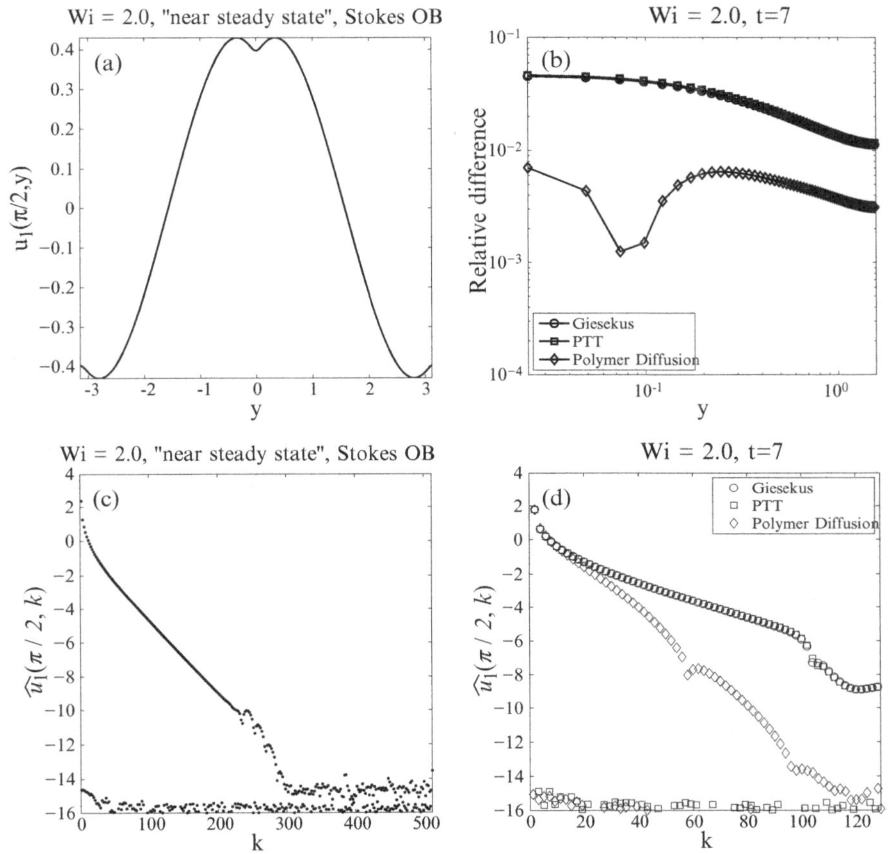

Fig. 10.3 Wi = 2.0 at $t = 7$: (**a**) $\mathbf{C}_{11}(0,y)$ the first component of the conformation tensor along the axis of compression where solution is "near-steady state." (**b**) Difference between models (Giesekus, PTT, and polymer diffusion) and the Stokes-OB solution. (**c**) and (**d**) Fourier spectra $\hat{\mathbf{u}}_1(\pi/2,k)$ for different models

may indicate that the solutions are approaching a singularity. Long time simulations of this problem break down for the Stokes-OB, PTT, and Giesekus model, whereas the polymer stress diffusion smooths the singularity and the solutions converge to steady state.

Simulations of FENE-P in this simple framework were done in [40]. It was shown that simply adding a cutoff to the polymer stress does not solve the problem of large stress gradients in extensional flows. The cusp-type singular solutions for $1/2 < \mathrm{Wi} < 1$ were still found for a sufficiently large maximum extension parameter ℓ and for $\mathrm{Wi} > 1$ the diverging solutions became bounded for finite ℓ, but numerical evidence indicated an exponential approach to a pair of corner singularities. Some numerical smoothing appears to be necessary even for this finite extension model for accurate long time simulations.

1.6.3 Simulation Conclusions

We see from this simple numerical test that long time simulations for strongly elastic flows require care in choice of model and method. These high-stress regions arise in many continuum models based on the upper-convected derivative (Eq. (10.5)) and care must be taken to resolve these regions for accurate information about the polymer stress contribution to the flow. Adding polymer stress diffusion leads to a nice balance of smoothing the stress locally while closely matching the flow outside the small smoothing region.

It is important to be aware of underlying near-singularities in these complex fluid models when designing numerical experiments. Model modifications which add finite extension (FENE-P) or which produce finite steady-state extensional viscosity (PTT and Giesekus) do not entirely get away from the underlying difficulties with the polymer stress advection equation given in Eq. (10.4). When long time simulations are required, these stress near-singularities can lead to loss of accuracy in numerical simulations unless care is taken to identify and address those features. The addition of polymer stress diffusion as a numerical smoothing, carefully chosen, does a good job capturing the fine structures while maintaining accuracy over long times.

Biological applications nearly always involve complex flows near walls and around obstacles and these flows contain extensional stagnation points which are subject to high-Weissenberg number problems. Near these regions, the flow has to be sufficiently well resolved to incorporate the fine stress structures which arise and feedback to create nontrivial flow patterns. The relaxation time of the elastic stress also introduces a new time scale which needs to be considered. Coupling fluids with structures adds further challenges which need to be addressed separately.

2 Immersed Boundary Methods

A common challenge of computational fluid mechanics is solving the equations of motion in complex geometry. Biological systems are particularly challenging because they often involve the coupled dynamics of active elastic structures and the surrounding fluid. There are many examples of such systems involving complex fluids, some of which were discussed in earlier chapters: beating cilia in mucus in the respiratory system, swimming sperm in the mucus of the female reproductive tract, and peristaltic pumping in the reproductive and digestive systems. In each of these systems, the collective dynamics of the material is an emergent phenomenon and is the product of the interactions between the fluid and the elastic material.

A popular method for biological fluid dynamics problems involving large deformations is the immersed boundary (IB) method [79]. The IB method was originally developed to study blood flow in the heart [80], and it has been applied to a large range of biological and nonbiological systems over the past thirty years. The IB method uses two coordinate systems: a moving Lagrangian coordinate system for the immersed structures and a fixed Eulerian coordinate system for the fluid. The Eulerian domain is discretized with a regular, Cartesian mesh which permits the use of fast methods for solving the equations of fluid mechanics. A key feature of the method is that it does not require conforming discretizations of the fluid and structure; instead, the curvilinear mesh is free to cut through the background Cartesian grid in an arbitrary manner. Consequently, IB simulations do not require dynamic grid generation, even for problems involving very large structural deformations.

The popularity and longevity of the immersed boundary method are partly due to its robustness and simplicity. The method is highly adaptable, and codes require few changes to be modified for different applications. Typical implementations of the IB method generally require only solvers for the fluid equations along with routines to compute elastic forces and to transfer data between the Lagrangian and Eulerian grids. IB codes can be built on top of existing codes for solving the equations of the fluid. In fact, the computational examples presented in Sect. 3 use the same viscoelastic fluid solver for Stokes Oldroyd-B with polymer diffusion that was used to generate the results in Sect. 1. In Chap. 11, a boundary integral method is presented to simulate the dynamics of a suspension. We note that boundary integral methods can only be used for linear equations, and they do not extend to the Stokes Oldroyd-B model. In this way, immersed boundary methods are more general.

Recently, the IB method has been used to investigate several classical low Reynolds number bio-fluid problems involving complex fluids in place of Newtonian fluids. For example, [57, 81] use the IB method to investigate peristaltic pumping of viscoelastic fluids, and [82–84] use the IB method to explore the swimming of microorganisms in complex fluids. The goal of this chapter is not to review the many variants and applications of immersed boundary methods, but rather to highlight key considerations when using immersed boundary methods for applications involving strongly elastic fluids at low Reynolds number. In the remainder of this section, we briefly introduce the key ideas of the immersed boundary method. We pay particular attention to time stepping algorithms. As we show in the later sections, traditional explicit-time methods are severely limited for strongly elastic flows with stiff boundaries.

As discussed in Sect. 1, viscoelastic flows tend to form regions of high stress with sharp velocity gradients near extensional points. These regions of highly concentrated stress can be generated by active immersed objects in the fluid, and resolving these stresses requires finer grid resolution than similar problems of Newtonian fluid mechanics. Viscoelastic stresses introduce additional time scales into the problem, and at moderate to high-Weissenberg numbers, the stress evolves on very long time scales. Capturing these slow dynamics requires performing long time simulations. In Sect. 3, we illustrate these ideas using an example problem of a swimming organism in a viscoelastic fluid.

2.1 Immersed Boundary Equations

Let $\mathbf{x} \in \boldsymbol{\Omega}$ denote fixed physical coordinates, with $\boldsymbol{\Omega} \subset \mathbb{R}^{n_\mathrm{f}}$ being the physical domain where $n_\mathrm{f} = 2$ or 3 is the dimension of the fluid. Let $\mathbf{s} \in \Gamma$ denote material coordinates attached to the immersed structure, with $\Gamma \subset \mathbb{R}^{n_\mathrm{s}}$ denoting the Lagrangian coordinate domain and n_s denoting the dimension of the structure. The physical location of material point $\mathbf{s}$ at time t is given by $\mathbf{X}(\mathbf{s},t) \in \boldsymbol{\Omega}$. The name *immersed boundary method* suggests that the elastic structure is a thin interface, i.e., an object of codimension one with respect to the fluid ($n_\mathrm{s} = n_\mathrm{f} - 1$). While this is the case in many applications of the IB method, this formulation applies equally well to immersed structures that have nonzero thickness.

In many biological problems, the structure has the same density as the fluid, and thus the combined fluid and structure can be described by a single momentum balance equation, and their motion can be described by a single velocity field. In the absence of other loading, the forces generated by the deformations of the structure drive the motion of the fluid through a body force term in the balance of momentum equation. In this chapter we consider the boundary to be immersed in an Oldroyd-B fluid (with diffusion coefficient α) at zero Reynolds number. The system of equations describing the fluid and structure is

$$\eta_s \Delta \mathbf{u} - \nabla p + \nabla \cdot \boldsymbol{\tau}_p + \mathbf{f} = 0, \tag{10.27}$$

$$\nabla \cdot \mathbf{u} = 0, \tag{10.28}$$

$$\boldsymbol{\tau}_p + \lambda \overset{\nabla}{\boldsymbol{\tau}}_p = \mu_p \dot{\gamma} + \lambda \alpha \Delta \boldsymbol{\tau}_p, \tag{10.29}$$

$$\mathbf{f}(\mathbf{x},t) = \int_\Gamma \mathbf{F}(\mathbf{s},t)\, \delta(\mathbf{x} - \mathbf{X}(\mathbf{s},t))\, \mathrm{d}\mathbf{s}, \tag{10.30}$$

$$\frac{\partial \mathbf{X}(\mathbf{s},t)}{\partial t} = \mathbf{U}(\mathbf{s},t) = \int_{\boldsymbol{\Omega}} \mathbf{u}(\mathbf{x},t)\, \delta(\mathbf{x} - \mathbf{X}(\mathbf{s},t))\, \mathrm{d}\mathbf{x}, \tag{10.31}$$

in which $\mathbf{u}(\mathbf{x},t)$ is the velocity field, $\mathbf{U}(\mathbf{s},t)$ is the velocity of the structure, $p(\mathbf{x},t)$ is the pressure, $\boldsymbol{\tau}_p(\mathbf{x},t)$ is the viscoelastic stress, $\mathbf{f}(\mathbf{x},t)$ is the Eulerian elastic force density generated by the immersed structure, $\mathbf{F}(\mathbf{s},t)$ is the Lagrangian elastic force density generated by the immersed structure, and δ denotes the Dirac delta function.

The integral operator in (10.30) that determines the Eulerian force density from the Lagrangian force density is called the *spreading* operator, which we denote by S. The *interpolation* operator that transfers the velocity to the structure is the adjoint of the spreading operator. Using this notation, equations (10.30) and (10.31) can be compactly expressed as

$$\mathbf{f} = S\mathbf{F}, \tag{10.32}$$

$$\frac{\partial \mathbf{X}}{\partial t} = \mathbf{U} = S^* \mathbf{u}, \tag{10.33}$$

respectively. Note that the spreading operator depends on the current position of the structure. For a discussion of discretizing these operators see [79].

A constitutive law which relates the boundary configuration to the force is needed to complete the description of the system. The form of the constitutive law depends on the application being considered. In Sect. 3 we present an example in which penalty forces are used to actively drive the immersed boundary to follow a prescribed shape to mimic the undulatory stroke of a swimming worm.

2.2 Explicit-Time Stepping

Typical implementations of the IB method use a fractional time stepping approach to solve the equations of motion. In the simplest version of such a scheme, the fluid velocity, pressure, and viscoelastic stress are updated while keeping the position of the structure fixed, and then the structural position is updated using the newly computed velocity. In this section, we consider that the viscoelastic stress is known at the beginning of a time step and thus can be described by its force density $\mathbf{f}_{\text{ve}} = \nabla \cdot \boldsymbol{\tau}_p$. For the model equations considered here, the explicit-time method advances the solution variables from time $t^n = n\Delta t$ to time $t^{n+1} = (n+1)\Delta t$ via

$$\eta_s \Delta_h \mathbf{u}^{n+1} - \nabla_h p^{n+1} + \mathbf{f}_{\text{ve}}^{n+1} + S_h^n \mathbf{F}(\mathbf{X}^n) = 0, \tag{10.34}$$

$$\nabla_h \cdot \mathbf{u}^{n+1} = 0, \tag{10.35}$$

$$\mathbf{X}^{n+1} = \mathbf{X}^n + \Delta t \, (S_h^n)^* \, \mathbf{u}^{n+1}. \tag{10.36}$$

Notice that the explicit-time method effectively decouples the computation of the boundary mechanics from the computation of the fluid velocity, pressure, and viscoelastic stress. This decoupling makes it easy to use existing codes for solving the fluid mechanics for IB simulations. In this way, the IB method for complex fluids is identical to the IB method for Newtonian fluids.

It is well known that for applications involving stiff elastic structures, the explicit-time method requires very small time steps to maintain stability. We let $\mathscr{L}^{-1}$ denote the inverse Stokes operator which maps force densities to velocity fields. The boundary update equation (10.36) can be expressed as

$$\mathbf{X}^{n+1} = \mathbf{X}^n + \Delta t \, (S_h^n)^* \, \mathscr{L}^{-1} S_h^n \mathbf{F}(\mathbf{X}^n), \tag{10.37}$$

where we have suppressed the viscoelastic forces for simplicity. This expression shows that the explicit-time scheme is essentially a forward-Euler method for the boundary positions, which explains the origin of the stability restriction. For many constitutive laws, as the grid is refined the problem becomes increasingly stiff. Because viscoelastic fluids require high grid resolution and sometimes involve long time integration, the stability restriction may be more limiting than for Newtonian fluids.

2.3 Implicit-Time Stepping

Much effort has been devoted both to understanding and to alleviating the severe time step restriction of fractional step IB methods [85–87]. The solution methods used in early implicit IB methods were not efficient and were not competitive with explicit methods [88], and some semi-implicit methods intended to allow for large time steps still suffered from significant time step restrictions [89, 90]. Newren *et al.* [86] analyzed the origin of instability in semi-implicit IB methods using energy arguments, and they gave sufficient conditions to achieve unconditional stability in the sense that the total energy is bounded independent of the time step size. An important result by Newren *et al.* [86] is that it is not necessary to employ a *fully* implicit-time discretization to achieve unconditional stability, but the stable time stepping schemes proposed therein do simultaneously solve for both the Eulerian velocity field and the Lagrangian structural configuration. As indicated by the early experience with implicit IB methods, however, developing efficient solvers for the coupled equations is challenging.

More recently, a number of stable semi-implicit [91–94] and fully implicit [95, 96] IB methods have been developed. The efficiency of these methods is generally competitive with explicit methods, and in some special cases, these implicit schemes can be faster than explicit methods by several orders of magnitude. Many implicit methods use a Schur complement approach to reduce the coupled Lagrangian-Eulerian equations to purely Lagrangian equations [94, 95, 97]. These methods achieve a substantial speedup over explicit methods especially when there are relatively few Lagrangian mesh nodes [94]. An open question is whether there exist robust, general-purpose implicit methods that are more efficient than explicit methods or whether specialized methods must be developed for specific problems.

We present an example of a semi-implicit method which is very similar to methods presented in [91, 94, 95]. In our implicit-time method, the fluid velocity at time t^{n+1} depends on structure forces at time t^{n+1}, rather than on the forces at time t^n as in the explicit method. Again we consider the viscoelastic forces as given. The system of equations that must be solved for the fluid velocity, pressure, and structure position is

$$\eta_s \Delta_h \mathbf{u}^{n+1} - \nabla_h p^{n+1} + \mathbf{f}_{\text{ve}}^{n+1} + S_h^n \mathbf{F}(\mathbf{X}^{n+1}) = 0, \tag{10.38}$$

$$\nabla_h \cdot \mathbf{u}^{n+1} = 0, \tag{10.39}$$

$$\mathbf{X}^{n+1} = \mathbf{X}^n + \Delta t \, (S_h^n)^* \, \mathbf{u}^{n+1}. \tag{10.40}$$

Notice that in this time stepping scheme, the structural positions used to define the spreading and interpolation operators are lagged in time. As shown by Newren *et al.* [86], this scheme is unconditionally stable, despite the fact that the positions of the spreading and interpolation operators are treated explicitly rather than implicitly. This type of scheme is often termed semi-implicit to emphasize that not all of the terms are treated implicitly.

Rather than solve for the velocity, pressure, and boundary position simultaneously, we reduce the equations to a single equation for the unknown boundary positions. Using the inverse operator for the Stokes equations, $\mathscr{L}^{-1}$, the boundary update equation (10.40) can be written as

$$\mathbf{X}^{n+1} = \mathbf{X}^n + \Delta t\,(S_h^n)^*\,\mathscr{L}^{-1} S_h^n \mathbf{F}(\mathbf{X}^{n+1}), \tag{10.41}$$

which resembles a backward-Euler scheme for advancing the boundary positions. This is a nonlinear equation, which we solve by applying Newton's method to

$$\mathbf{G}(\mathbf{X}) = \mathbf{X} - \mathbf{X}^n - \Delta t\,(S_h^n)^*\,\mathscr{L}^{-1} S_h^n \mathbf{F}(\mathbf{X}) = 0. \tag{10.42}$$

Let $\mathbf{X}^k$ represent the approximate solution at the kth step of the Newton iteration. Each Newton step involves the update

$$\mathbf{J}\delta\mathbf{X} = -\mathbf{G}(\mathbf{X}^k) \tag{10.43}$$

$$\mathbf{X}^{k+1} = \mathbf{X}^k + \delta\mathbf{X}, \tag{10.44}$$

where $\mathbf{J}$ is the Jacobian of $\mathbf{G}$. We do not explicitly form $\mathbf{J}$. Rather, we perform multiplication by $\mathbf{J}$ as described below, and we solve equation (10.43) using the generalized minimum residual method (GMRES).

Let $\mathbf{J}_F$ represent the Jacobian of the force function, $\mathbf{F}$. The Jacobian of $\mathbf{G}$ may be expressed as

$$\mathbf{J} = \mathbf{I} - \Delta t\,(S_h^n)^*\,\mathscr{L}^{-1} S_h^n \mathbf{J}_F. \tag{10.45}$$

The product $\mathbf{J}\delta\mathbf{X}$ is accomplished by first explicitly multiplying by $\mathbf{J}_F$ (which is sparse), then spreading the result to the grid, solving the Stokes equations, and interpolating back to the structure. This procedure avoids the need to explicitly form $\mathbf{J}$. The GMRES solver requires a good preconditioner for efficient performance. As a preconditioner, we ignore the Stokes solve and use

$$\mathbf{M} = \mathbf{I} - \Delta t\,(S_h^n)^*\,S_h^n \mathbf{J}_F. \tag{10.46}$$

This matrix is sparse and relatively small (number of the IB points) and can be factored quickly. The performance of the method is discussed in Sect. 3.5.

Note that each evaluation of the objective function, $\mathbf{G}$, as well as each application of the Jacobian, J, involves the application of $\mathscr{L}^{-1}$. This operator is never explicitly constructed. Its application is achieved by solving the Stokes equations for the fluid velocity. This implicit-time method uses the same code as the explicit-time method for finding the fluid velocity. Although the immersed boundary code for the implicit-time method is significantly more involved than the explicit-time method, this method retains the appealing modularity of the original IB method and allows the algorithms presented in Sect. 1 to be used without modification.

3 Locomotion of Undulatory Swimmers

Locomotion of microorganisms at low Reynolds number occurs in numerous biological processes, and swimming in a Stokesian Newtonian fluid has been extensively studied and the underlying physics is well understood. See [98] for a review of low Reynolds number locomotion. There have been many recent theoretical studies on locomotion in complex fluids [83, 84, 99–106]. Asymptotic analyses of infinitely long, small-amplitude, undulatory swimmers in a viscoelastic fluid showed that swimming is hindered by the addition of elastic stresses [100, 101]. However, numerical simulation of finite-length large-amplitude swimmers in a viscoelastic fluid showed that under some conditions, the swimming speed may be enhanced by the elastic stresses [83]. A similar enhancement was shown for numerical simulations of infinite-length helical swimmers with large pitch angles [105]. The results from these papers highlight the importance of computational methods in exploring problems that are beyond the reach of asymptotic analysis.

In the remainder of this chapter, we use the problem of a finite-length free swimmer to illustrate the ideas of the immersed boundary method. We pay particular attention to the additional complications introduced from the viscoelastic fluid, namely, the need for high grid resolution and the presence of long time dynamics at high-Weissenberg numbers. The problem we explore is very similar to that presented in [83], except that the stroke pattern of the swimmer is based on data for a swimming nematode presented in Chap. 7 of this book and in [107]. In this chapter, we primarily use this problem as a computational example of the methodology. For a more in-depth analysis of how the fluid elasticity, body elasticity, and stroke kinematics affect swimming speed in this problem, see [106].

3.1 Swimmer Model

The worm is modeled as an inextensible infinitely thin sheet, which in two dimensions is a curve in the plane. The position of the worm is given by $\mathbf{X}(s,t)$, where $s \in [0,L]$ is the Lagrangian coordinate. The swimming is driven using a prescribed target curvature $\kappa_0(s,t)$, which in the absence of resistance from the surrounding medium represents the shape of the worm.

3.1.1 Immersed Boundary Forces

Both the inextensibility and the shape are enforced by forces that are designed to penalize extension and deviation from the prescribed curvature. These forces are derived from expressions for the bending and extension (stretching) energy, which are given below. For a given configuration of the worm the energy from stretching is

$$E_{\mathrm{s}} = \frac{k_s}{2} \int_{\Gamma} \left(|\mathbf{X}_s| - 1 \right)^2 ds, \tag{10.47}$$

where k_s is a stiffness coefficient. The bending energy is

$$E_{\mathrm{b}} = \frac{k_b}{2} \int_{\Gamma} \left(\kappa - \kappa_0(s,t) \right)^2 ds, \tag{10.48}$$

where k_b is the bending stiffness, κ is the curvature of the worm, and κ_0 is the prescribed target curvature. The total energy is the sum of the bending and stretching energy:

$$E = E_{\mathrm{s}} + E_{\mathrm{b}}. \tag{10.49}$$

The Lagrangian force densities come from the variational derivative of the total energy:

$$\frac{\delta E}{\delta \mathbf{X}} \tilde{\mathbf{X}} = -\int_0^L \mathbf{F} \tilde{\mathbf{X}} ds = -\int_0^L (\mathbf{F}_s + \mathbf{F}_b) \tilde{\mathbf{X}} ds, \tag{10.50}$$

where $\mathbf{F}_s$ and $\mathbf{F}_b$ are the force densities corresponding to stretching and bending, respectively. The expressions for the force densities are derived by first discretizing the structure (and hence the energy functional) and then taking the variational derivative of the discrete energy. The advantage of this approach is that it guarantees that the total forces discretely sum to zero, which is a requirement to be able to find a solution to Stokes equations in a periodic domain.

The signed curvature is computed using a discretized version of

$$\kappa = \hat{\mathbf{n}} \cdot \frac{\partial \hat{\mathbf{t}}}{\partial s} = \hat{\mathbf{n}} \cdot \frac{\partial^2 \mathbf{X}}{\partial s^2}, \tag{10.51}$$

where $\hat{\mathbf{t}}$ is the tangent vector and $\hat{\mathbf{n}}$ is the normal vector. We are assuming that the material is inextensible and the tangent vector is expressed as $\hat{\mathbf{t}} = \mathbf{X}_s$. We compute the discrete curvature at an interior point using the expression

$$\kappa_j = \left(\frac{\hat{\mathbf{n}}_{j+1/2} + \hat{\mathbf{n}}_{j-1/2}}{2} \right) \cdot \left(\frac{\hat{\mathbf{t}}_{j+1/2} - \hat{\mathbf{t}}_{j-1/2}}{\Delta s} \right), \tag{10.52}$$

where the discrete tangent vector is

$$\hat{\mathbf{t}}_{j+1/2} = \frac{\mathbf{X}_{j+1} - \mathbf{X}_j}{\Delta s} \tag{10.53}$$

and the discrete normal is the $\pi/2$ rotation of the tangent. The discretized expression for the total energy is

$$E = \frac{k_s}{2} \sum_{j=1}^{N-1} \left(\left| \frac{\mathbf{X}_{j+1} - \mathbf{X}_j}{\Delta s} \right| - 1 \right)^2 \Delta s + \frac{k_b}{2} \sum_{j=2}^{N-1} \left(\kappa_j - \kappa_{0j} \right)^2 \Delta s. \tag{10.54}$$

3.1.2 Curvature Model

We take shape data of a nematode swimming in the water to identify the target curvature.[2] The measured curvature as a function of time and position along the worm are shown in Fig. 10.4. The body coordinate (in units of mm) runs from $s = 0$ at the head to $s = 1.2$ at the tail. The worm shows a periodic motion with a dominant frequency around 2 Hz (period of 0.5 s), which is evident in the curvature data.

For the model swimmer, we use a curvature function of the form

$$\kappa_0 = A(s) \cos \left(\frac{2\pi}{T} (t + \phi(s)) \right), \tag{10.55}$$

where A is the amplitude, ϕ is the phase, and $T = 0.5$ s is the period. We identify the phase function, ϕ, by finding the peak cross-correlation between the head and points along the body. This phase function and a linear fit to it are shown in Fig. 10.5. The slope of the linear fit is -0.250 s/mm, and so we use a phase function of $\phi(s) = -s/4 + \phi_0$, where ϕ_0 is the value of the phase assigned to the head.

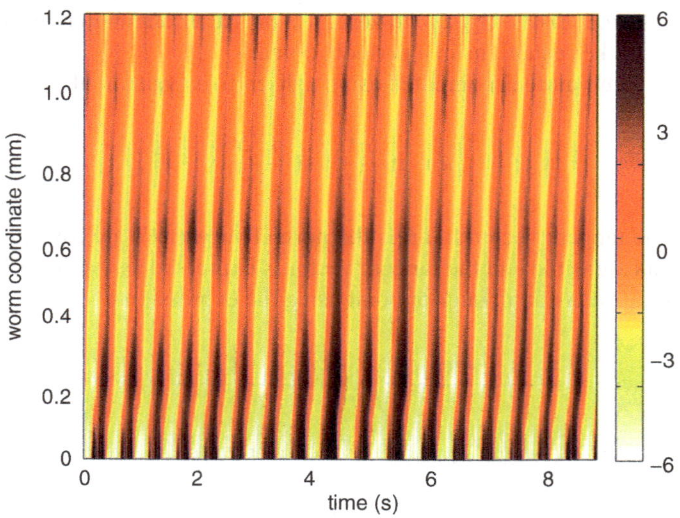

Fig. 10.4 Color field of the curvature (in mm^{-1}) data of the swimming nematode as a function of time and body coordinate along the worm. $s = 0$ is the head and $s = 1.2$ is the tail

[2]The data are kindly provided by Paulo Arratia and Xiaoning Shen.

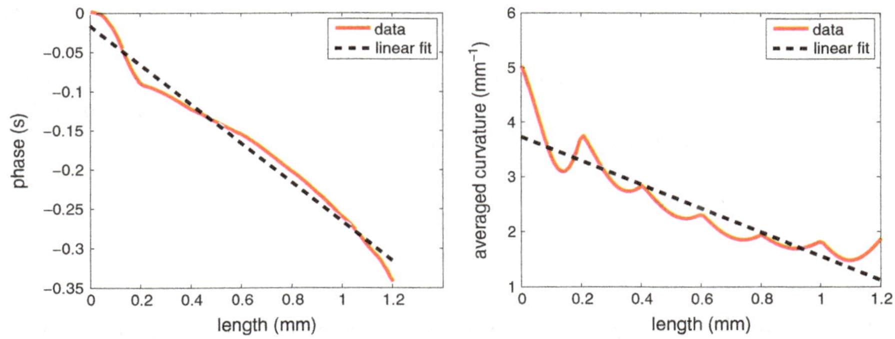

Fig. 10.5 Phase (*left*) and amplitude (*right*) functions based on the curvature data of a swimming nematode and their linear fits that are used to define the curvature in the model computations

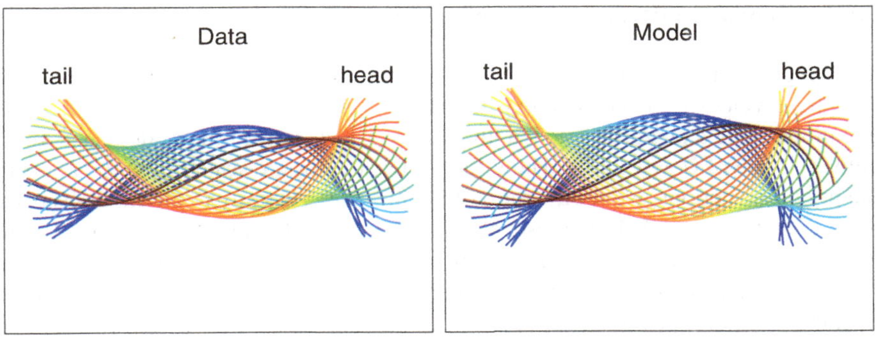

Fig. 10.6 Time sequence of worm shapes from the data (*left*) and model (*right*). The colors correspond to the time progressing from blue to red. The body is positioned with a fixed center of mass and horizontal end-to-end vector

We compute the time-averaged curvature of the data by

$$\bar{\kappa}(s) = \sqrt{\frac{1}{N}\sum_{t_j} \kappa(s,t_j)^2} \tag{10.56}$$

and fit this with a linear function. The averaged curvature and the linear fit are displayed in Fig. 10.5. We choose the amplitude of the curvature in the model by $A(s) = \sqrt{2}\bar{\kappa}_{\mathrm{fit}}$, so that the model matches the data in the mean squared curvature.

Figure 10.6 shows a time sequence of the actual shapes and the shapes produced by the model fit. Changes in curvature propagate as phase waves from the head ($s = 0$), where the amplitude of the curvature is largest, to the tail. In what follows we call this type a swimmer a "burrower." The exact form of the curvature function used to drive the swimming worm is

$$\kappa_0^{\text{burrower}} = (5.3 - 3.1s)\cos\big(4\pi(t - s/4 + \phi_0)\big), \tag{10.57}$$

where $\phi_0 = L/4$ is the phase shift chosen to have zero phase at the tail. We note that in [83] the desired motion of the swimmer involved waves of curvature that propagated with an increasing amplitude along the direction of the wave. We call this type of swimmer a "kicker." For comparison we simulate a kicker using the change of coordinates

$$\kappa_0^{\text{kicker}}(s,t) = \kappa_0^{\text{burrower}}(L - s, t_0 - t), \tag{10.58}$$

where L is the length of the swimmer and t_0 is a phase shift that keeps the head at the same phase as the burrower.

3.1.3 Simulation Parameters/Nondimensionalization

The length of the swimmer is 1.2 mm, the period of the oscillation is 0.5 s, and the phase velocity of the bending motion is 4 mm/s. We nondimensionalize the equations using a characteristic length scale of $L = 1$ mm, a time scale of 1 s, and a velocity scale of $U = 1$ mm/s. The viscoelastic stresses are scaled by $\eta_p U/L$. As in the previous sections, we set the ratio of the polymer viscosity to the fluid viscosity to be $\xi = \eta_p/\eta_s = 0.5$. In the viscoelastic fluid we use the Deborah number, De, to characterize the ratio of the relaxation time of the polymers to the time scale of the flow. We note that we use the characteristic time scale of 1 s to define the De rather than the period of swimmer.

For the bending stiffness we choose a characteristic bending stiffness of nematodes: $k_b = 2 \times 10^{-15}$ Nm^2 [108], which when nondimensionalized becomes $k_b = 2$. For the stretching stiffness we choose a nondimensional value of $k_s = 2500$. We note that with this value of the bending stiffness, the actual curvature may be significantly different from the target curvature. We return to examining the effect of changing the bending stiffness in Sect. 3.4.

The computations are performed in a 2 mm by 1 mm doubly periodic domain, with the worm initially aligned in the x-direction. The domain is discretized using a regular $N_x = 256$ by $N_y = 128$ grid, and the points on the swimmer are spaced so that $\Delta s \approx \Delta x$. We use the implicit-time stepping scheme with a time step of $\Delta t = 10^{-3}$ and a stopping tolerance of 5×10^{-5} for the nonlinear solver.

The viscoelastic fluid solver used here is the same solver described in Sect. 1.6. The dimensionless polymer stress diffusion coefficient is fixed at $\alpha = 0.01$. The length scale associated with polymer stress diffusion over a time step is smaller than the grid spacing ($\sqrt{\Delta t \alpha} \approx 0.0032 \lesssim \Delta x = 1/2^7 \approx 0.0078$).

3.2 Swimming Speed

We simulate the swimming of both the burrower and the kicker until time $t = \max(10, 10\,\mathrm{De})$ which allows sufficient time to establish a periodic motion independent of the initial transients. We measure their steady-state swimming speed by computing the displacement of the center of mass over the last period of the simulation. In Fig. 10.7 we show the swimming speeds for both the burrower and the kicker as a function of De, scaled by the swimming speeds for De $= 0$. While the burrower always swims slower in a viscoelastic fluid, the kicker swims up to 25% faster in a viscoelastic fluid. The swimming speed changes non-monotonically as a function of De for both swimmers, with a local maximum a little beyond De $= 1$. The relative swimming speed for the kicker is consistent with the result reported in [83].

Figure 10.8 shows contours of the elastic energy, $\mathrm{Tr}(\boldsymbol{\tau}_p)$, for both the burrower and kicker at the ending time of the simulation for De $= 0.1, 0.5, 1.0, 2.0$. These plots demonstrate the significant differences between both the size and location of the elastic stresses for the burrower and the kicker. Generally, the elastic stresses are larger for the kicker, and much more concentrated at the tail.

In Sect. 1, we showed that large, highly concentrated elastic stresses were produced around extensional points. It is not a priori clear whether the swimmer problem is prone to exhibit the flow characteristics associated with high-Weissenberg number problems. In Fig. 10.9 we show the extensional points, centers, and streamlines at four representative times in a steady frame translating at the worm's average swimming speed for the burrower at De $= 0$. Periodically, a new extensional point and center are generated at the head, and shortly after this, an extensional point and center coalesce at the tail. In between these events, there are a pair of extensional points and a pair of centers traveling from the head to the tail. In the plots shown, the strengths of the extensional points range from 0.9 to 2.8, which is in the range subject to high-Weissenberg number phenomena.

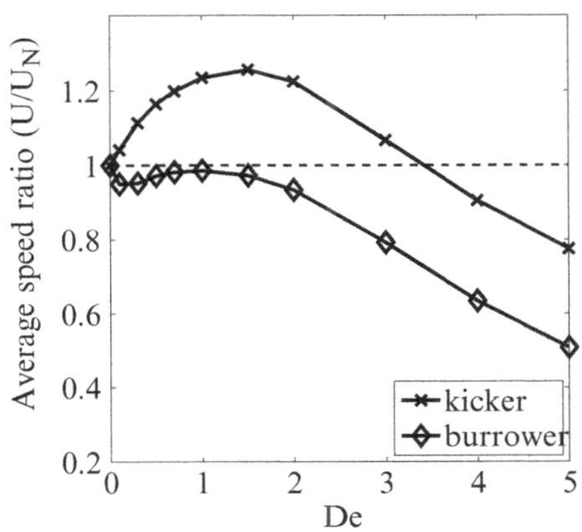

Fig. 10.7 Scaled swimming speed of both the burrower and the kicker as a function of De. The swimming speed is scaled by the speed at De $= 0$

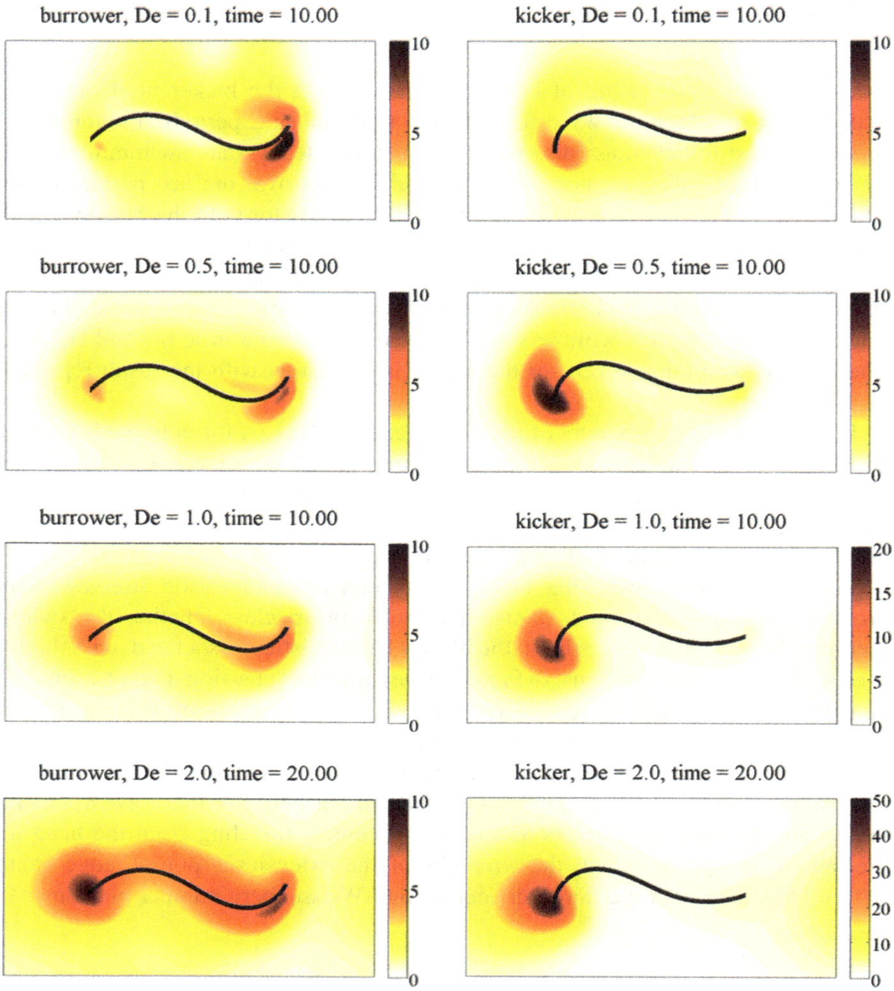

Fig. 10.8 Contours of the elastic energy at the end of the simulation for both the burrower and kicker for different De. The head is on the right

However, examining streamlines provides only limited information. It remains an open question whether these moving extensional points are related to high-Weissenberg number phenomena. See Chap. 7 for a similar discussion, and see [106] for a more detailed study of the effects of fluid elasticity, body mechanics, and stroke kinematics on swimming speed for this problem.

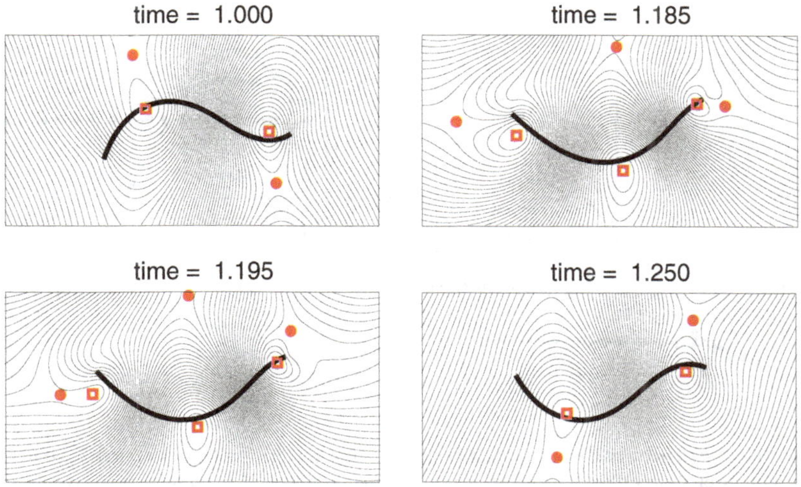

Fig. 10.9 Streamlines in the frame translating with the average swimming speed of the worm for the burrower in a Newtonian fluid (De = 0). The extensional points are marked with *dots* and centers are marked with *squares*. Four representative times are shown including the creation of a new extensional point at the head and the disappearance of an extensional point at the tail. The head is on the right

3.3 Time and Space Resolution

The results of the previous section show that the swimmer problem involves strong extensional points and highly concentrated elastic stresses that are associated with high-Weissenberg number problems. Here we demonstrate the importance of long time scales and fine spatial scales for the swimmer problem.

Figure 10.10 (left) shows the polymer elastic energy, measured as the trace of the polymer stress tensor, for three different De up to time 20. These data demonstrate that for high De the elastic stresses evolve very slowly. For De = 5 the elastic energy is still growing substantially at time $t = 20$, while the energy is near-steady state for the lower values of De. The elastic energy for De = 5 up to time 50 is shown in the inset. The effect of this slow evolution of stress on the swimming speed is demonstrated in Fig. 10.10 (right) which shows the swimming speed (averaged over the previous period) as a function of time for different De. For De = 1, the swimming speed rapidly approaches a steady state, but for high De the swimming speed increases on a time scale on the order of De, but on long time scales it slowly decays to a much smaller value. These data demonstrate that understanding the effects of elastic stresses in systems with large De (or large Wi) requires performing long-time simulations.

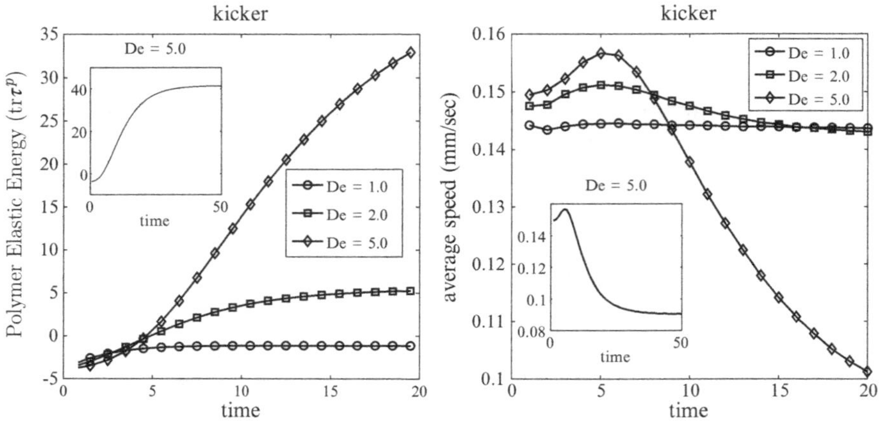

Fig. 10.10 (*left*) Time course of the elastic energy in the kicker simulations up to $t = 20$ for De = 1,2,5. The inset shows the data for De = 5 up to time 50. (*right*) Swimming speed averaged over previous period as a function of time for the same parameters

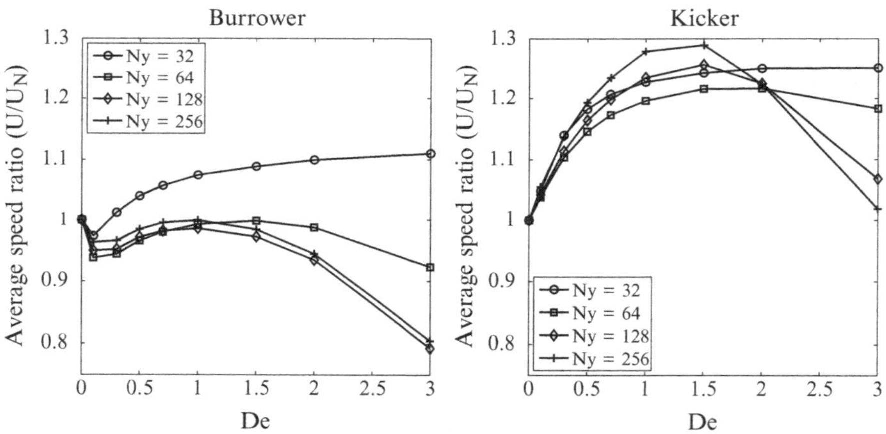

Fig. 10.11 Swimming speed for burrower and kicker for $Ny = 32, 64, 128, 256$ and varying De

As Fig. 10.8 shows, highly localized stresses develop around the swimmer for large De. Here we demonstrate the importance of adequately resolving these large stress gradients. Figure 10.11 shows the swimming speeds as a function of De for both the burrower and the kicker as the grid is refined. The results from the two coarsest meshes and the two finest meshes are notably different. In fact, on the coarsest mesh, the simulations predict the wrong dependence of swimming speed on De.

3.4 Effect of Increasing Bending Stiffness

In these simulations, the active forces of the worm are driven by changes in the desired curvature, κ_0. In the limit that the bending stiffness goes to infinity, the curvature of the worm approaches the desired curvature, but for any finite value of the bending stiffness, the curvature of the worm and the desired curvature are different. In the simulations presented in the previous sections, the bending stiffness was chosen based on the measured bending stiffness of nematodes. In this section we examine the effect of changing the bending stiffness on the swimming speed.

Figure 10.12 shows the swimming speed at De $= 0.5$ for eight different bending stiffness values from four times lower than that used in the previous sections to 50 times higher. As the bending stiffness increases, the curvature of the simulated worm more closely follows the prescribed curvature, and the swimming speed approaches a constant. It is notable that the swimming speed for $k_b = 100$ is about 80 % larger than for $k_b = 2$, which was the value used in the previous sections.

We repeat the calculations of the steady-state swimming speed from Sect. 3.2 for a range of De for a larger bending stiffness. Figure 10.13 shows the relative swimming speeds for both the burrower and the kicker for De $= 0 - 5$ for both $k_b = 2$ and $k_b = 40$. For both swimmers, the results are very different between the two stiffness values. In contrast to the softer worm, with the larger bending stiffness, the burrower's swimming speed decreases monotonically with De, and the kicker's swimming speed is always slower than in a Newtonian fluid.

For the softer bending stiffness, there is a significant difference between the prescribed curvature and the realized curvature. The target curvature represents the equilibrium shape that would be achieved in the absence of resistance from the surrounding medium. The product $k_b \kappa_0$ can be related to the internal torque density resulting from muscular contraction. In the soft regime, the realized shape

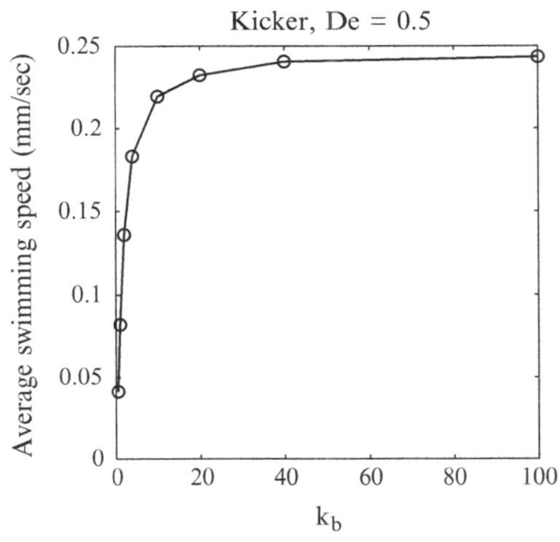

Fig. 10.12 Average swimming speed at De $= 0.5$ for different values of the bending stiffness

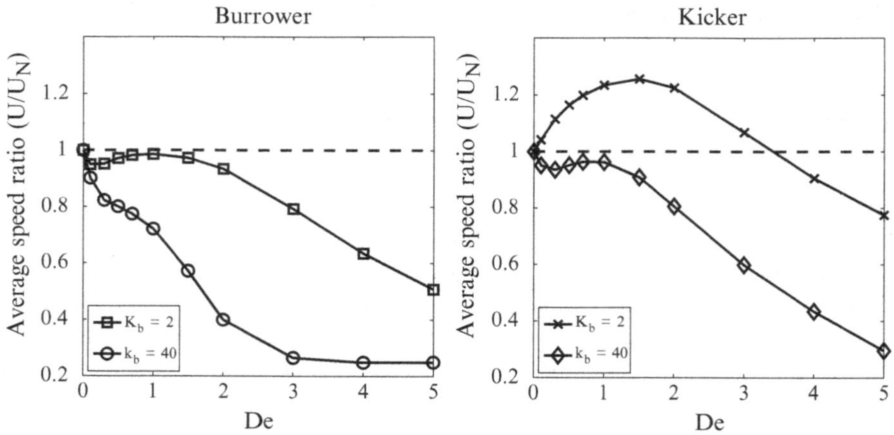

Fig. 10.13 Scaled swimming speed of both the burrower and the kicker as a function of De for two different bending stiffnesses. The swimming speed is scaled by the speed at De = 0. The data for $k_b = 2$ is the same as that from Fig. 10.7

results from the complex interaction between the fluid, passive body mechanics, and actively generated torques. For large elastic stresses, the fluid offers significant resistance to deformation, and the bending forces must be stiff enough to maintain the prescribed shape if the intention is to simulate that prescribed shape.

These results demonstrate that care must be exercised when using penalty methods to ensure that the forces are sufficiently stiff so that the choice of numerical parameters does not affect the predicted results. Even with the lower bending stiffness, the equations are extremely numerically stiff, and explicit-time methods would be very inefficient (see next section). The implicit-time method is necessary to be able to simulate on long time scales, with a fine grid, and with sufficiently large stiffness to capture the correct dynamics of the prescribed shape swimmers.

3.5 *Efficiency of the Implicit-Time Method*

For given stiffness coefficients and grid spacing, we let Δt_{exp} represent the largest time step that gives a stable simulation in the explicit-time method. We identify Δt_{exp} using a bisection algorithm. The worm simulation is run until time 0.1, and considered unstable if the elastic energy rises above a prescribed threshold. In Table 10.1, we report Δt_{exp} for two different bending stiffnesses and four different grid resolutions, for De = 0.5. The table shows that when the grid is refined by a factor of 2, the time step shrinks by a factor of about 8, which is as expected for bending forces at low Reynolds number [109]. On the finest grid, the reported time step restriction is an estimate from the coarse grid values.

Each time step of the implicit method requires repeated solution of the Stokes equations. Each evaluation of the objective function in the Newton's method and

Table 10.1 Maximum stable time step for the explicit method. The value reported on the finest grid is estimated from the coarser grids

Δx	$k_b = 2$	$k_b = 20$
2^{-5}	2.76e-4	2.65e-5
2^{-6}	3.28e-5	3.14e-6
2^{-7}	4.01e-6	3.86e-7
2^{-8} (est)	4.81e-7	3.49e-8

Table 10.2 Average number of Stokes solves over 1,000 time steps of the implicit code with $\Delta t = 10^{-3}$

	tol = 5×10^{-5}		tol = 1×10^{-5}	
Δx	$k_b = 2$	$k_b = 20$	$k_b = 2$	$k_b = 20$
2^{-5}	17.82	21.70	22.21	32.34
2^{-6}	20.99	33.45	33.10	43.02
2^{-7}	24.23	44.11	39.48	67.44
2^{-8}	39.58	–	64.41	102.93

each application of the Jacobian require solving the Stokes equations. In Table 10.2 we report the number of times the Stokes equations are solved per time step for 1,000 time steps of the implicit method with $\Delta t = 10^{-3}$ (two periods of the swimming motion). We report these values for two different stiffnesses, four different grid resolutions, and two different stopping tolerances for the Newton's method.

Note that on the finest grid with the largest stiffness, the solver failed with the larger tolerance during the simulation. With a smaller tolerance, no failure occurred. A more sophisticated nonlinear solver, e.g., including line searching or trust region methods, would likely be able to compute with the larger tolerance [110, 111]. This failure illustrates one of the many added difficulties of working with implicit-time methods in place explicit-time methods.

To estimate the efficiency of the implicit method we compare the number of times the Stokes equations are solved per time unit with the explicit-time method. We measure the efficiency as the ratio of Stokes solves in the explicit method to the implicit method per unit time. This efficiency estimate ignores the extra work involved in the GMRES iteration in the implicit method. However, it also ignores the extra updates of the stress equation in the explicit method, and the fact that one is not likely to run the explicit method exactly at the stability limit. With this in mind, one may loosely interpret the efficiency measurement as the expected speedup gained by using the implicit method in place of the explicit method.

The efficiency is reported in Table 10.3 for two different bending stiffnesses, four different grid resolutions, and two different stopping tolerances for the Newton's method. Only on the coarsest grid is the explicit method ever significantly more efficient than the implicit method. However, as noted previously, the solutions on the two coarsest grids have very large errors. The coarsest grid spacing which we consider usable is $\Delta x = 2^{-7}$. At this resolution and above, the implicit solver significantly outperforms the explicit method.

As the grid is refined, the problem gets significantly stiffer, but the work involved in the nonlinear solver grows only mildly. The number of Stokes solves required per

Table 10.3 Efficiency of the implicit method measured as the ratio of the number of times the Stokes equations are solved per time unit in the explicit method to the implicit method

	tol = 5×10^{-5}		tol = 1×10^{-5}	
Δx	$k_b = 2$	$k_b = 20$	$k_b = 2$	$k_b = 20$
2^{-5}	0.20	1.15	0.16	1.16
2^{-6}	1.45	6.84	0.92	7.39
2^{-7}	10.28	58.68	6.31	38.38
2^{-8}	52.42	–	32.27	278.23

Numbers greater than one indicate that the implicit method is more efficient than the explicit method

time step increases by only a factor between about 2 and 3 as the grid is refined by a factor of 8, while in the explicit method, the number of Stokes solves would increase by a factor of over 500. Similarly, when the bending stiffness is increased by a factor of 10, the work increases by less than a factor of 2 in the implicit method and by about a factor of 10 in the explicit method. These results illustrate that implicit methods far outperform explicit methods for very stiff problems on fine grids.

4 Conclusions

Strongly elastic flows at low Reynolds numbers share some characteristics of high Reynolds number flows, namely, regions of highly localized stress and sharp gradients in the velocity. One does not expect to be able to use the same algorithms for low and high Reynolds number, and similarly, special care must be used to simulate flows at high-Weissenberg numbers. We have highlighted some recent analytical work on low Reynolds number viscoelastic fluids which has led to new algorithms for successfully simulating flows at high-Weissenberg numbers. The unbounded stress growth exhibited by the Oldroyd-B model is not present in many other models with nonlinear relaxation rates. However, these models also exhibit large stress gradients at high-Weissenberg numbers, and the algorithms developed for Oldroyd-B are often still necessary to mitigate these numerical challenges. In particular, we have demonstrated that the polymer stress diffusion modification regularizes the Oldroyd-B model so that accurate, smooth, and bounded solutions are obtained in periodic extensional flow.

It is not obvious how the standard benchmark problems for the high-Weissenberg number phenomenon such as flow in a contracting channel or steady elongational flow are directly related to biological problems. We used the classical problem of an undulatory swimmer at low Reynolds number to demonstrate that when elastic stresses are under-resolved, the predicted relative swimming speed is qualitatively different from simulations with resolved stresses. We note that the immersed boundary method is first-order accurate near structures, and this low accuracy may make the resolution of stresses more difficult.

The popularity of the immersed boundary method comes from its simplicity and robustness which make it more attractive than using a significantly more complicated high-order method, in particular for problems in biology where high-order accuracy is often not paramount. Any algorithm for solving the forced equations of motion for the fluid on a structured grid can be used in an immersed boundary simulation. The methodology extends to complex fluids without modification. The method is not without its drawbacks; notably, its low accuracy and the severe stability restriction imposed by explicit-time stepping schemes in problems involving stiff elastic structures. Accurately resolving the stress in strongly elastic flows requires high grid resolution and long time integration of the equations. The implicit-time method presented in this chapter makes it possible to perform simulations on long time scales, with a fine grid, and with large elastic stiffness.

Acknowledgment The authors would like to thank Lisa Fauci and Michael Shelley for interesting discussions and suggestions on this work. We also acknowledge Paulo Arratia and Xiaoning Shen for allowing us to use their data. This work was supported in part by NSF grants DMS-1160438 and DMS-1226386 to RDG.

References

1. J. Oldroyd, Proceedings of the Royal Society of London. Series A. Math. Phys. Sci. **200**(1063), 523 (1950)
2. R.B. Bird, O. Hassager, R. Armstrong, C. Curtiss, *Dynamics of Polymeric Liquids, Vol. 2: Kinetic Theory* (John Wiley and Sons, London 1980)
3. M. Mendelson, P.W. Yeh, R. Brown, R. Armstrong, J. Non-Newtonian Fluid Mech. **10**(1), 31 (1982)
4. R. Brown, R. Armstrong, A. Beris, P. Yeh, Comput. Meth. Appl. Mech. Eng. **58**(2), 201 (1986)
5. R.G. Owens, T.N. Phillips, *Computational Rheology*, vol 2 (World Scientific, Singapore 2002)
6. G.H. McKinley, R.C. Armstrong, R.A. Brown, Philos. Trans. R. Soc. Lond. Ser. A: Phys. Eng. Sci. **344**(1671), 265 (1993)
7. M. Chilcott, J. Rallison, J. Non-Newtonian Fluid Mech. **29**, 381 (1988)
8. A.W. Liu, D.E. Bornside, R.C. Armstrong, R.A. Brown, J. Non-Newtonian Fluid Mech. **77**(3), 153 (1998)
9. M. Alves, F. Pinho, P. Oliveira, J. Non-Newtonian Fluid Mech. **97**(2), 207 (2001)
10. M.A. Hulsen, R. Fattal, R. Kupferman, J. Non-Newtonian Fluid Mech. **127**(1), 27 (2005)
11. H.S. Dou, N. Phan-Thien, Chem. Eng. Sci. **62**(15), 3909 (2007)
12. S. Claus, T. Phillips, J. Non-Newtonian Fluid Mech. **200**, 131 (2013)
13. H. Nguyen, D. Boger, J. Non-Newtonian Fluid Mech. **5**, 353 (1979)
14. D. Boger, Annual Rev. Fluid Mech. **19**(1), 157 (1987)
15. G.H. McKinley, W.P. Raiford, R.A. Brown, R.C. Armstrong, J. Fluid Mech. **223**, 411 (1991)
16. S.C. Xue, N. Phan-Thien, R. Tanner, J. Non-Newtonian Fluid Mech. **74**(1), 195 (1998)
17. M.A. Alves, P.J. Oliveira, F.T. Pinho, J. Non-Newtonian Fluid Mech. **122**(1), 117 (2004)
18. M. Webster, H. Tamaddon-Jahromi, M. Aboubacar, J. Non-Newtonian Fluid Mech. **118**(2), 83 (2004)
19. D. Trebotich, P. Colella, G. Miller, J. Comput. Phys. **205**(1), 315 (2005)
20. D.D. Joseph, *Fluid Dynamics of Viscoelastic Liquids*, vol. 84 (Springer, New York, 1990)
21. M. Renardy, *Mathematical Analysis of Viscoelastic Flows*, vol 73 (SIAM, Philadelphia, 2000)

22. M.J. Crochet, Rubber Chem. Technol. **62**(3), 426 (1989)
23. F. PT Baaijens, J. Non-Newtonian Fluid Mech. **79**(2), 361 (1998)
24. R. Keunings, in *Proceedings of the XIIIth International Congress on Rheology*, vol 1 (British Soc. Rheol, 2000), vol 1, pp. 7–14
25. R. Keunings, Rheology Rev. pp. 167–196 (2003)
26. R. Keunings, Rheology Rev. pp. 67–98 (2004)
27. A.W. El-Kareh, L.G. Leal, J. Non-Newton. Fluid Mech. **33**, 257 (1989)
28. J.T. Beale, T. Kato, A. Majda, Comm. Math. Phys. **94**(1), 61 (1984)
29. J.Y. Chemin, N. Masmoudi, SIAM J. Math. Anal. **33**(1), 84 (2001)
30. R. Kupferman, C. Mangoubi, E.S. Titi, arXiv preprint arXiv:0709.1455 (2007)
31. F.H. Lin, C. Liu, P. Zhang, Comm. Pure Appl. Math. **58**(11), 1437 (2005)
32. T.C. Sideris, B. Thomases, Comm. Pure Appl. Math. **58**(6), 750 (2005)
33. P. Constantin, Remarks on complex fluid models. Mathematical aspects of fluid mechanics. London Mathematical Society Lecture Note Series (No. 402) (Cambridge University Press, Cambridge, 2012), pp. 70–87
34. P. Constantin, M. Kliegl, ARMA **206**(1), 725 (2012)
35. R. Larson, *The Structure and Rheology of Complex Fluids* (Oxford University Press, Oxford, 1998)
36. R. Sureshkumar, A.N. Beris, J. Non-Newtonian Fluid Mech. **60**, 53 (1995)
37. B. Thomases, J. Non-Newt. Fluid Mech. **166**, 1221 (2011)
38. C.D. Dimitropoulos, R. Sureshkumar, A.N. Beris, J. Non-Newt. Fluid Mech. **79**, 433 (1998)
39. K.D. Housiadas, A.N. Beris, Phys. Fluids **15**, 2369 (2003)
40. B. Thomases, M. Shelley, Phys. Fluids **19**, 103103 (2007)
41. R.G. Larson, *The structure and rheology of complex fluids*, vol. 2 (Oxford university press, New York, 1999)
42. J.M. Rallison, E.J. Hinch, J. Non-Newt Fluid Mech. **29**, 37 (1988)
43. M. Renardy, J. Non-Newt. Fluid Mech. **138**, 204 (2006)
44. A.N. Brooks, T.J. Hughes, Comput. Meth. Appl. Mech. Eng. **32**(1), 199 (1982)
45. M.D. Gunzburger, *Finite Element Methods for Viscous Incompressible Flows: A Guide to Theory, Practice, and Algorithms* (Elsevier, Amsterdam, 2012)
46. J. Marchal, M. Crochet, J. Non-Newtonian Fluid Mech. **26**(1), 77 (1987)
47. M. Fortin, A. Fortin, J. Non-Newtonian Fluid Mech. **32**(3), 295 (1989)
48. H.P. Baaijens, G.W. Peters, F.P. Baaijens, H.E. Meijer, J. Rheol. (1978–present) **39**(6), 1243 (1995)
49. B. Cockburn, G.E. Karniadakis, C.W. Shu, *The Development of Discontinuous Galerkin Methods* (Springer, NewYork 2000)
50. D. Rajagopalan, R.C. Armstrong, R.A. Brown, J. Non-Newtonian Fluid Mech. **36**, 159 (1990)
51. F. Baaijens, J. Non-Newtonian Fluid Mech. **75**(2), 119 (1998)
52. H. Matallah, P. Townsend, M. Webster, J. Non-Newtonian Fluid Mech. **75**(2), 139 (1998)
53. R.C. King, M.R. Apelian, R.C. Armstrong, R.A. Brown, J. Non-Newtonian Fluid Mech. **29**, 147 (1988)
54. R. Guénette, M. Fortin, J. Non-Newtonian Fluid Mech. **60**(1), 27 (1995)
55. I. Keshtiban, F. Belblidia, M. Webster, J. Non-Newtonian Fluid Mech. **126**(2), 123 (2005)
56. C.W. Shu, S. Osher, J. Comput. Phys. **77**(2), 439 (1988)
57. H.D. Ceniceros, J.E. Fisher, J. Non-Newtonian Fluid Mech. **171**, 31 (2012)
58. R. Fattal, R. Kupferman, J. Non-Newtonian Fluid Mech. **123**(2), 281 (2004)
59. R. Fattal, R. Kupferman, J. Non-Newtonian Fluid Mech. **126**(1), 23 (2005).
60. A. Afonso, M. Alves, F. Pinho, P. Oliveira, in *3rd Annual European Rheology Conference* (2006), pp. 27–29
61. A. Afonso, P. Oliveira, F. Pinho, M. Alves, J. Non-Newtonian Fluid Mech. **157**(1), 55 (2009)
62. H. Damanik, J. Hron, A. Ouazzi, S. Turek, J. Non-Newtonian Fluid Mech. **165**(19), 1105 (2010)
63. N. Balci, B.Thomases, M. Renardy, C. Doering, J. Non-Newt. Fluid Mech. (2011)
64. M. Laso, H. Öttinger, J. Non-Newtonian Fluid Mech. **47**, 1 (1993)

65. A. Lozinski, C. Chauvière, J. Comput. Phys. **189**(2), 607 (2003)
66. H. Giesekus, J. Non-Newtonian Fluid Mech. **11**(1), 69 (1982)
67. N.P. Thien, R.I. Tanner, J. Non-Newtonian Fluid Mech. **2**(4), 353 (1977)
68. H.R. Warner Jr, Ind. Eng. Chem. Fundamentals **11**(3), 379 (1972)
69. M. Chilcott, J. Rallison, J. Non-Newtonian Fluid Mech. **29**, 381 (1988)
70. G. Lielens, P. Halin, I. Jaumain, R. Keunings, V. Legat, J. Non-Newtonian Fluid Mech. **76**(1), 249 (1998)
71. G. Lielens, R. Keunings, V. Legat, J. Non-Newtonian Fluid Mech. **87**(2), 179 (1999)
72. Q. Du, C. Liu, P. Yu, Multiscale Modeling and Simulation **4**(3), 709 (2005)
73. A. Peterlin, J. Polymer Sci. Part B: Polymer Lett. **4**(4), 287 (1966)
74. C. Sulem, P.L. Sulem, H. Frisch, J. Comput. Phys. **50**(1), 138 (1983)
75. R. Krasny, J. Fluid Mech. **167**, 65 (1986)
76. M. Shelley, J. Fluid Mech. **244**(1), 493 (1992)
77. R. Peyret, *Spectral Methods for Incompressible Viscous Flow*, vol 148 (Springer, Newyork, 2002)
78. T.Y. Hou, R. Li, J. Comput. Phys. **226**(1), 379 (2007)
79. C.S. Peskin, Acta Numer. **11**(-1), 479 (2002)
80. C.S. Peskin, J. Comput. Phys. **25**(3), 220 (1977)
81. J. Chrispell, L. Fauci, Math. Modelling of Nat. Phenomena **6**(05), 67 (2011)
82. J.C. Chrispell, L.J. Fauci, M. Shelley, Phys. Fluids **25**(1), 013103 (2013)
83. J. Teran, L. Fauci, M. Shelley, Phys. Rev. Lett. **104**, 038101 (2010).
84. J. Du, J.P. Keener, R.D. Guy, A.L. Fogelson, Phys. Rev. E **85**, 036304 (2012)
85. J.M. Stockie, B.R. Wetton, J. Comput. Phys. **154**(1), 41 (1999)
86. E.P. Newren, A.L. Fogelson, R.D. Guy, R.M. Kirby, J. Comput. Phys. **222**(2), 702 (2007)
87. Z. Gong, H. Huang, C. Lu, Commun. Comput. Phys. **3**, 704 (2008)
88. C. Tu, C.S. Peskin, SIAM J. Sci. Stat. Comput. **13**(6), 1361 (1992).
89. A.A. Mayo, C.S. Peskin, in *Fluid Dynamics in Biology (Seattle, WA, 1991), Contemp. Math.*, vol. 141 (Amer. Math. Soc., Providence, RI, 1993), pp. 261–277
90. L. Lee, R.J. LeVeque, SIAM J. Sci. Comput. **25**(3), 832 (2003).
91. E.P. Newren, A.L. Fogelson, R.D. Guy, R.M. Kirby, Comput. Methods Appl. Mech. Engrg. **197**(25–28), 2290 (2008).
92. T.Y. Hou, Z. Shi, J. Comput. Phys. **227**(20), 8968 (2008)
93. T.Y. Hou, Z. Shi, J. Comput. Phys. **227**(21), 9138 (2008)
94. H.D. Ceniceros, J.E. Fisher, A.M. Roma, J. Comput. Phys. **228**(19), 7137 (2009).
95. Y. Mori, C.S. Peskin, Comput. Methods Appl. Mech. Engrg. **197**(25–28), 2049 (2008).
96. D. Le, J. White, J. Peraire, K. Lim, B. Khoo, J. Comput. Phys. **228**(22), 8427 (2009).
97. H.D. Ceniceros, J.E. Fisher, J. Comput. Phys. **230**(12), 5133 (2011).
98. E. Lauga, T. Powers, Rep. Prog. Phys. **72**, 096601 (2009)
99. T. Normand, E. Lauga, Phys. Rev. E **78**, 061907 (2008)
100. E. Lauga, Phys. Fluids **19**, 083104 (2007)
101. H.C. Fu, T.R. Powers, C.W. Wolgemuth, Phys. Rev. Lett. **99**, 258101 (2007)
102. E. Lauga, Europhys. Lett. **86**, 64001 (2009)
103. A.M. Leshansky, Phys. Rev. E **80**, 051911 (2009)
104. H. Fu, V. Shenoy, T. Powers, Europhys. Lett. **91**, 24002 (2010)
105. S.E. Spagnolie, B. Liu, T.R. Powers, Phys. Rev. Lett. **111**(6), 068101 (2013)
106. B. Thomases, R.D. Guy, Phys. Rev. Lett. **113**(9), 098102 (2014)
107. X.N. Shen, P.E. Arratia, Phys. Rev. Lett. **106**(20), 208101 (2011)
108. M. Backholm, W.S. Ryu, K. Dalnoki-Veress, Proc. Natl. Acad. Sci. **110**(12), 4528 (2013)
109. R.D. Guy, B. Philip, Commun. Comput. Phys. **12**, 378 (2012)
110. C. Kelley, *Iterative Methods for Linear and Nonlinear Equations* (SIAM, Philadelphia, 1995)
111. J. Nocedal, S.J. Wright, *Numerical Optimization* (Springer, New York, 1999)

Chapter 11
Cell Distribution and Segregation Phenomena During Blood Flow

Amit Kumar and Michael D. Graham

Abstract Blood is the archetype of a biological complex fluid. It is complex in the microstructural and mechanical sense, as a multiphase non-Newtonian viscoelastic fluid, and also in the biological sense, as a tissue that has a wide range of functions from delivery of oxygen and nutrients to response to injury and inflammation. These forms of complexity are interconnected, as the physical nature of blood as a multiphase fluid is intimately related to its biological functions. In the present chapter, we summarize basic features of the structure and biology of blood as well as observations of its dynamics during flow in the body. Emphasis will be put on flow at small scales, where the particulate nature of blood as a suspension of many different types of cells becomes important both physically and functionally. The first part of the chapter describes the nature and biological functions of the various components of blood, as well as the distribution of these components in blood vessels. In particular, it has long been observed that the various cellular components of blood are distributed very nonuniformly, a phenomenon that is physiologically important as well as fascinating from the fluid-dynamical point of flow. The second part of the chapter focuses on computational and theoretical approaches for predicting and understanding the distribution and segregation of blood cells in flow. Various numerical methods are described, with a focus on one of the most widely used for multiphase small-scale flows, the boundary integral method. Computational results for model suspensions are presented that allow careful study of the basic mechanisms underlying segregation phenomena and a model framework is introduced that incorporates these mechanisms in an idealized way. This framework is a stepping stone toward a unified understanding of these segregation phenomena that will hopefully be useful in aiding the development of therapies to modify and exploit blood flow phenomena for disorders as varied as cancer, sickle cell disease, and hemorrhage.

A. Kumar • M.D. Graham (✉)
University of Wisconsin-Madison, Madison, WI 53706, USA
e-mail: kumar20@gmail.com; graham@engr.wisc.edu

S.E. Spagnolie (ed.), *Complex Fluids in Biological Systems*, Biological and Medical Physics, Biomedical Engineering,
DOI 10.1007/978-1-4939-2065-5_11

Blood is the archetype of a biological complex fluid. It is complex in the microstructural and mechanical sense, as a multiphase non-Newtonian viscoelastic fluid, and also in the biological sense, as a tissue that has a wide range of functions from delivery of oxygen and nutrients to response to injury and inflammation. These forms of complexity are interconnected, as the physical nature of blood as a multiphase fluid is intimately related to its biological functions. In the present chapter, we summarize basic features of the structure and biology of blood as well as observations of its dynamics during flow in the body. Emphasis will be put on flow at small scales, where the particulate nature of blood as a suspension of many different types of cells becomes important both physically and functionally. The first part of the chapter describes the nature and biological functions of the various components of blood, as well as the distribution of these components in blood vessels. In particular, it has long been observed that the various cellular components of blood are distributed very nonuniformly, a phenomenon that is physiologically important as well as fascinating from the fluid-dynamical point of flow. The second part of the chapter focuses on computational and theoretical approaches for predicting and understanding the distribution and segregation of blood cells in flow. Various numerical methods are described, with a focus on one of the most-widely used for multiphase small-scale flows, the boundary integral method. Computational results for model suspensions are presented that allow careful study of the basic mechanisms underlying segregation phenomena and a model framework is introduced that incorporates these mechanisms in an idealized way. This framework is a stepping stone toward a unified understanding of these segregation phenomena that will hopefully be useful in aiding development of therapies to modify and exploit blood flow phenomena for disorders as varied as cancer, sickle cell disease and hemorrhage.

1 Background

1.1 Blood: Components and Physiological Functions

Blood is a suspension of various kinds of cells and cell fragments in an aqueous background fluid known as plasma [1]. Red blood cells (RBCs) or erythrocytes are the predominant cellular component; in humans the overall volume percent of RBCs, or hematocrit, is about 40 % in females and 45 % in males [2]. A mature RBC is a biconcave discoid about 8 μm across and 2 μm high. It has no nucleus or cytoskeleton and to a first approximation can be viewed mechanically as a deformable fluid-filled capsule with an elastic membrane [3]. The primary function of red blood cells is oxygen transport—they contain the oxygen-carrying protein hemoglobin. The other cellular components of blood are white blood cells (WBCs) or leukocytes and platelets or thrombocytes . WBCs are outnumbered by the RBCs by about 500:1 and platelets by 12–14:1 [5]. Figure 11.1 shows a schematic of blood flow in a small vessel, qualitatively illustrating the three cell types.

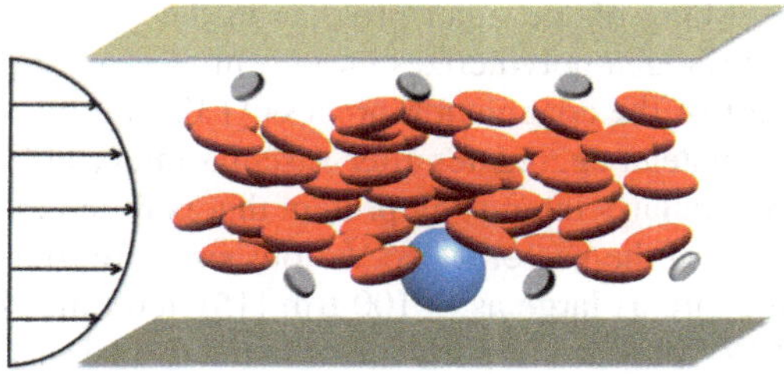

Fig. 11.1 A schematic depicting blood flow in a small channel or blood vessel, showing RBCs (*red*), a leukocyte (*large sphere*), and several platelets (*small disks*) [4]

There are various kinds of WBCs, ranging in size from about 7 μm to 15 μm. The majority, 60–70 %, are neutrophils, which are between 10 and 12 μm across and approximately spherical [6]. In contrast to RBCs, WBCs are intact cells with a cytoskeleton and nucleus and as such are best viewed mechanically as soft solid objects [5]. White blood cells are part of the immune system. For example, neutrophils and monocytes extravasate (pass through blood vessel walls) in response to infection. These cells can then capture and phagocytose the bacteria or the foreign body, thereby killing them [5, 7].

Platelets are 2–3 μm cell fragments that form in the bone marrow by pinching off from precursor cells called megakaryocytes that remain in the marrow. They are discoidal in blood under normal conditions and are central to blood clotting (coagulation). Blood vessel walls are lined with endothelial cells, to which platelets do not normally adhere. An injury to a blood vessel exposes the extracellular matrix, which is predominantly collagen; platelets adhere to the exposed collagen and become activated [8–12], initiating the coagulation process. Once activated, platelets develop fingerlike projections called pseudopodia [9]. In case of injury, coagulation is normal. However, platelets also play a substantial role in the pathological conditions of thrombosis, the formation of a thrombus (clot) within a blood vessel, and atherosclerosis, the thickening of arterial walls. These are both central elements of cardiovascular and cerebrovascular diseases [11].

Functionally, plasma is equally important as the cellular components of blood. Dissolved within it are a great many small molecules–gases, salts, urea, hormones, amino acids, lipids, glucose, vitamins—as well as a variety of proteins. The most abundant of these, making up several percent of the total volume of blood, is albumin. Albumin is a compact globular protein, and aside from its role as a carrier of many small molecules in blood, it is a primary contributor to the osmotic pressure of the blood. Albumin is too large to pass through the vessel walls and thus maintains an osmotic driving force that resists the tendency of water to leak out from the blood vessels. In the physiology literature, the contribution of albumin and other proteins to the osmotic pressure is called the *oncotic* pressure, to distinguish it from the contribution of small molecules like salts that pass much more easily between the blood and surrounding tissues. Globulins are also abundant proteins, transporting other proteins and playing a role in the immune response: antibodies, molecules that bind to foreign bodies as part of the immune response, are members of this family.

Many of the other proteins in the plasma are involved in coagulation [8, 12]. Fibrinogen polymerizes into fibrin, the major protein component of blood clots, as part of the coagulation process [13]. Fibronectin is important for the stabilization of platelet aggregates after vascular injury [14]. Von Willebrand factor (VWF) is a particularly interesting protein that is found in plasma as well as in the endothelial cells of the vessel walls. Various sizes are found, with lengths when fully extended that are as large as $\sim 100\ \mu$m [15]–ten times longer than the size of a blood cell. VWF plays an important role in the formation of clots. Recall that platelet adhesion to collagen initiates the coagulation response. During blood flow with low shear rates at the vessel walls ($\dot{\gamma}_w \lesssim 100\ \mathrm{s}^{-1}$) platelet adhesion is not affected by the presence or absence of VWF. At high shear rates ($\gtrsim 1000\ \mathrm{s}^{-1}$), however, VWF is essential for adhesion [16–18]. Indeed the molecule is named after the physician who identified the clotting disorder that arises when it is deficient [19]. Strong evidence suggests that shear-induced unfolding of VWF at high shear rates leads to its increased efficacy at initiation of coagulation [15, 20].

1.2 Rheology and Nonuniform Flow Phenomena in Blood

In the larger arteries and veins, blood can often be satisfactorily treated as a single-phase fluid. There are many important high-Reynolds number flow phenomena including turbulence, vortices, and recirculation zones in these large vessels, many of which play an important role in cardiovascular health and disease [21]. The focus of the present work, however, is flow at smaller scales, where blood flow is laminar [21] and the multiphase nature of blood comes into play. Typical values [10, 22] of the wall shear rate $\dot{\gamma}_w$ in various vessels are $\sim$ 100–500 s^{-1} in larger veins and arteries such as carotid, femoral, and coronary, $10^4\ \mathrm{s}^{-1}$ in arterioles, $10^3\ \mathrm{s}^{-1}$ in capillaries, and slightly lower in venules. In arteries with severe stenosis, shear rates as high as $10^5\ \mathrm{s}^{-1}$ can occur [10].

Macroscopic viscosity measurements of blood show that it is a shear-thinning fluid with a yield stress [23]. At 37 °C, the viscosity becomes constant at about 3–5 mPa s once $\dot{\gamma}_w \gtrsim 100\ \mathrm{s}^{-1}$ and the yield stress is about 4 mPa. At 20 °C, the viscosity reaches a high-shear-rate plateau of about 10 mPa s. These numbers depend somewhat on temperature and hematocrit. Plasma itself is generally considered to be Newtonian with a viscosity of about 1.3–2 mPa s, although very recent evidence suggests that it does have some viscoelasticity, with a relaxation time of about 2 ms [24]. The shear-thinning of blood is largely attributable to the deformability of the RBCs under flow; suspensions of chemically stiffened RBCs do not shear-thin [25]. Blood is also somewhat viscoelastic, again due to the deformability of the RBCs [26].

The yield stress of blood is small—any flow with $\dot{\gamma} \gtrsim$ 1–10 s^{-1} is strong enough to exceed it. Situations with very weak flow may occur near recirculation regions and stagnation points but will not be present in the bulk of the circulatory system. Nevertheless, the origin of this yield stress is interesting. It has long been known

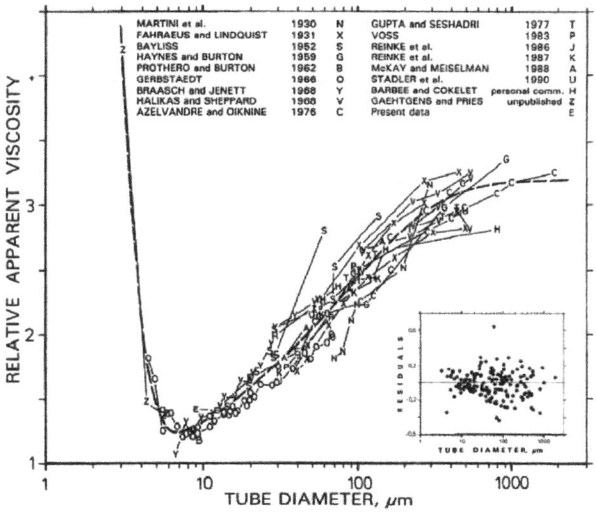

Fig. 11.2 Apparent viscosity vs. tube diameter for blood as measured by a large number of separate investigators [31]

that RBCs in stagnant blood form rouleaux, aggregates in which the RBCs stack like coins. Rouleau formation does not seem to be completely understood, but classical depletion flocculation (which is osmotic in origin [27]) due to the presence of the large number of albumin and globulin molecules seems to play a substantial role [28, 29]. In any case the rouleaux are broken up in any appreciable flow field.

The rheological properties just described are relevant for macroscopic flow geometries, i.e., those with scales much larger than that of a blood cell. Fåhræus and Lindqvist [30] made the important observation that the apparent viscosity at shear rates $\gtrsim 100\ \mathrm{s}^{-1}$ (i.e., in the constant viscosity plateau beyond the shear-thinning region) determined in a capillary viscometer using Poiseuille's law is a strong function of the capillary tube diameter, as shown in Fig. 11.2 [31]. This phenomenon is called the Fåhræus-Lindqvist effect. For large tubes, there is a plateau, but for diameters between 10 and 1000 μm, the apparent viscosity decreases substantially with decreasing tube size, before increasing sharply for tubes smaller than 10 μm. The increase for very small tubes is easily understood by recalling that the RBCs are about 8 μm across, but the behavior at larger tube diameters is more subtle.

The origin of this phenomenon is related to another classical observation in blood physiology called the Fåhræus effect [32, 33]. This effect is the observation that the measured hematocrit of blood *in* a tube—i.e., the true average volume fraction–can be less than the hematocrit inferred from measurements of the blood discharged by flow out *from* a tube. The ratio of "tube hematocrit" to "discharge hematocrit" displays a similar tube size dependence as the apparent viscosity. In particular, in a tube of 20 μm diameter, the tube hematocrit (i.e., the true hematocrit within the tube) is about half the discharge hematocrit [22, 32, 33].

The resolution to this apparent paradox lies in the fact that the discharge hematocrit is weighted by the nearly parabolic velocity profile in the tube while the tube hematocrit is not. If the local concentration of RBCs in the tube were uniform, then the two measures would yield the same result, so the Fåhræus effect indicates that the concentration is not uniform and that the concentration of RBCs is higher near the center of the channel where the velocity is high than near the walls where the velocity is low. One reason for this concentration variation is the simple volume exclusion effect–cells cannot pass through the wall so there is always a region near the wall with a cell volume fraction lower than the nominal value. The other, more interesting reason is that RBCs and indeed all deformable particles migrate away from walls during shear due to hydrodynamic interaction effects with the walls. This is further discussed below. The cell-depleted region near vessel walls is called a "cell-free layer" or "marginal layer" [34, 35].

A further consequence of the nonuniform concentration of RBCs during flow is "plasma-skimming." The circulatory system is highly branched. The cell-depleted wall region causes the hematocrit in a daughter vessel to be smaller than in the parent vessel, as the daughter disproportionately draws fluid from near the wall of the parent [34, 36]. Related to plasma-skimming is the Zweifach-Fung effect: the observation that when a suspension of particles encounters an asymmetric bifurcation, the RBC volume fraction in the outlet branch with the higher flow rate is higher than in the branch with the lower [37, 38]. Recent work suggests that this effect is primarily due to the cell distribution in the inlet branch: just as in the plasma-skimming effect, the presence of the cell-free layer strongly affects the distribution in the outlet branches [39]. In vivo measurements in small blood vessels confirm the substantial drop in tube hematocrit that arises from the combination of the Fåhræus and plasma-skimming effects [22].

1.3 Distribution of Blood Cells During Flow: Cell-Free Layer and Margination

The Fåhræus-Lindqvist, Fåhræus, plasma-skimming, and Zweifach-Fung effects all indicate that the concentration of RBCs is not uniform in blood vessels during flow. We can qualitatively understand this result through consideration of two basic phenomena in flowing suspensions: cross-stream migration and shear-induced diffusion [40]. Even at zero Reynolds number, any deformable object in shear (capsule, drop, vesicle, flexible macromolecule) migrates away from the wall [41–44] with migration velocity v_{mig}. Roughly speaking, this migration is driven by the flow generated by a deformable particle as it tries to relax to its rest shape and how that flow is modified in the presence of a wall. It has been described in a number of works, both in the context of deformable particles [42, 45] as well as for flexible macromolecules [46–48]. Additionally, if the velocity gradient is not constant, as in Poiseuille flow, a single deformable particle can undergo lateral migration even in the absence of hydrodynamic wall effects [42]. The deformability of a suspended object in flow is characterized by the nondimensional capillary number $\mathrm{Ca} = \eta\dot{\gamma}a/G$, which measures the ratio of viscous to elastic and/or stresses. Here a is a characteristic size of the object (say 4 μm for an RBC) and G is an interfacial modulus: for an RBC, $G \approx 4 \cdot 10^{-6}$ N/m [34, 49–52]. In the circulation the capillary number can be in the range $0.1 \lesssim \mathrm{Ca} \lesssim 1$. For an isolated suspended object flowing far above a plane wall, the migration velocity is given by

$$v_{\mathrm{mig}} = \frac{a^3\dot{\gamma}g(\mathrm{Ca})}{y^2}, \tag{11.1}$$

where y is the distance from the wall and $g(\mathrm{Ca})$ depends on the deformability and shape of the particles. Theory and simulations [53, 54] for deformable cells or capsules indicate that $g \sim \mathrm{Ca}$ at low Ca and $g(\mathrm{Ca}) \sim \mathrm{Ca}^{0.6}$ when $\mathrm{Ca} = O(1)$. If the Reynolds number is finite, there is also an inertial lift force [55–58]. This effect can be important in microfluidic applications at relatively high flow rates [59, 60], but is small under the conditions present close to vessel walls [54].

In a suspension, migration is counterbalanced by shear-induced diffusion, the random motion of particles in a suspension that is driven by the fluid motion generated by other particles in the suspension. In particular, the "collisions" between neighboring particles contribute significantly to this effect. The shear-induced diffusivity D of particles in a dilute suspension is expected to be proportional to the product of collision frequency (proportional to $\dot{\gamma}\phi$, where ϕ is the local volume fraction) and the mean squared displacement per collision (proportional to a^2):

$$D = \dot{\gamma}\phi a^2 f_s, \tag{11.2}$$

where $f_s(Ca)$ depends on the deformability and shape of the capsules. A simple mathematical model [54] containing these two phenomena yields a prediction that the cell-free-layer thickness δ is given by

$$\frac{\delta}{a} = \frac{g(\mathrm{Ca})}{f_s(\mathrm{Ca})\phi_b}, \tag{11.3}$$

where ϕ_b is the bulk volume fraction. An important implication of the above equation is that the thickness of the cell-depleted layer scales inversely with ϕ. It also shows a direct relationship of δ with the wall-induced migration, $g(\mathrm{Ca})$, and an inverse relationship with the magnitude of shear-induced diffusion ($\sim f_s^{-1}(\mathrm{Ca})$)—indicating that wall-induced migration favors the formation of capsule-depleted layer whereas shear-induced diffusion opposes it. This result provides a simple expression that at least qualitatively captures the experimentally observed trends.

White blood cells and platelets are also distributed nonuniformly in flow, but in contrast to RBCs they are preferentially found near the walls of the flow channel, a phenomenon called margination [61–63]; this is schematically shown in Fig. 11.1. Physiologically, margination of WBCs and platelets is desirable in general, because the natural functions of these cells are closely associated with their interactions with blood vessel walls as noted earlier. A number of recent "blood on a chip" devices have relied on this phenomenon to effect separations between different components of normal blood (e.g., [64, 65]) or between healthy and diseased cells [66, 67].

Experimental studies characterizing the margination behavior of leukocytes, both in vivo and in vitro, find a stronger dependence on the shear rate than on the hematocrit [61, 68–70]. At least at shear rates $\gtrsim 100\ \mathrm{s}^{-1}$, the margination of leukocytes is found to decrease with increasing shear rate. Inducing aggregation of RBCs by the addition of dextran (and thus promoting depletion flocculation) increases margination [61, 69], so this trend with shear rate may be a consequence of the decrease in aggregation as shear rate increases [61, 69, 70]. The effect of hematocrit on margination is less clear. Some studies indicate that margination is either unaffected (within experimental errors) with hematocrit or increases slightly [61, 70]. However, a recent study found margination to be a non-monotonic function of hematocrit with a maximum in margination observed at $\phi = 0.2$ [71]. This non-monotonic dependence on hematocrit was also observed in a recent computational effort [72].

The effect of channel width on the margination of leukocytes has also been characterized, with the degree of margination usually found to decrease with increasing channel width [69, 71]. Margination behavior at sudden expansions in in vitro studies suggest that a 25–50 μm sudden expansion results in an optimal margination. This optimal expansion size range is in good agreement with the capillary-postcapillary expansion of 30–40 μm observed during inflammation [71].

Computational studies have shed some light on the mechanism of leukocyte margination. These studies usually find a stack of RBCs leaning upstream on the marginated leukocyte [73, 74], which is believed to stabilize their position near the walls [74]. Furthermore, the computational studies show that, while aggregation of RBCs increases leukocyte margination [72], it is not a necessary condition [72, 74].

The margination characteristics of platelets in blood flow has also been investigated in many in vivo and in vitro experimental studies [62, 75–78]. It is typically found to increase with increasing hematocrit [76–78] as well as with increasing shear rate [76, 77]. However, some studies show a more complicated non-monotonic dependence on shear rate [78]. Nonetheless, the trends in the margination behavior of platelets with increasing shear rate is in direct contrast with that of leukocytes, as the latter exhibit decreasing margination with increasing shear rate. It certainly appears that the RBC aggregation is not as important a factor for platelet margination as it is for leukocyte margination, but the cause of this difference is unknown.

Several computational studies have also focused on the platelet margination behavior. Some authors have suggested that the platelet drift toward the wall at the edge of the red-cell-free layer due to one-sided collisions with the RBCs contributes significantly to the margination of platelets [79]. Other authors have argued that higher-velocity fluctuation near the center of the vessel, due to the higher hematocrit there, drive the platelets toward the walls where these fluctuations are negligible [80]. Lastly, it has been shown that the platelet margination is correlated with the difference in the size of RBCs and platelets–simulations performed with RBCs replaced by platelet-sized (but still deformable) particles did not exhibit platelet margination [81].

There is also significant current interest in the field of targeted drug delivery via the bloodstream. Targeted delivery of drugs has been explored in a variety of diseases including cancer, cardiovascular, pulmonary, and inflammation diseases [82–85]. Drug delivery for cancer treatment is easily the most widely studied problem among these. In fact, efficient delivery of drugs to tumors is one of the major goals of current cancer research as it can lead to a reduction in dosages and side effects [84, 85].

To achieve targeted drug delivery, a variety of nanocarrier-based drug-delivery systems are being developed [86]. These carriers are typically based on polymers (including micelles, vesicles, dendrimers, nanogels), lipids (including liposomes, micelles, niosomes), metals (including iron oxide, gold nanoparticles, quantum dots), and carbon (including carbon nanotubes) among others [84]. In addition to compositions, the carriers can also have varied shapes and sizes as well as surface modifications [86]. The targeting of tumors by these carriers can either be passive or active [82]. In passive targeting, the enhanced permeability and retention effect (EPR) of the tumors is exploited for drug delivery. The EPR effect of tumors is due to their leaky blood vessels and an impaired lymphatic drainage system [85]. In active targeting, the carrier surface is modified with ligands, peptides, or counter-receptors specific to molecules expressed on the tumor cells [82, 87]. Irrespective of targeting mechanism, an efficient drug delivery particle must possess an inherent propensity to marginate so as to maximize its interaction with the tumor vascular walls. Optimization of the physical design parameters of the drug delivery particle, such as shape, size, and deformability, to maximize delivery efficiency is therefore a very important area of research.

Margination of drug delivery particles has not been extensively studied, although there have been a few recent efforts. Charoenphol et al. [88] examined the binding efficiency of vascular-targeted spherical carriers in parallel plate reconstituted blood flow assays. Investigating the effect of particle size, the authors found that the binding of particles to the walls decreased when the particle diameter decreased from 10 μm to 100 nm at a shear rate of 200 s^{-1}. In these studies the hematocrit was varied between $\phi = 0.3$ and 0.5 without a significant variation in the results. Another feature of the particle adhesion behavior was that it exhibited a maximum at a critical wall shear rate, i.e., the adhesion decreases at shear rates lower or higher than the critical shear rate; the critical shear rate was lower for the larger particles. The authors concluded that the particle margination did not change appreciably with changing shear rate–the maximum in the adhesion results from the balance between the increased particle flux with increasing shear rate and the increasingly disruptive effect of the shear stress on the adhered particle. Based on this study, the authors concluded that spheres with diameter in the 2 to 5 μm range display significantly better margination to the wall at intermediate to high wall shear rates and channel heights than the nanometer-sized spheres. Recently, Tan et al. [89] numerically investigated the influence of red blood cells on the drug delivery particle distribution in the vessel. These authors found the particle distribution profile to be nonuniform with a higher particle concentration near the vessel wall. The authors observed that the tumbling motion of RBCs in the core region of the capillary enhances the radial dispersion of the particles. This effect, in addition to the volume exclusion of the RBCs, was suggested to result in the enhanced margination of drug delivery particles.

1.4 Effect of Plasma Rheology on Cell Distribution

Blood plasma is generally regarded as Newtonian, though a subtle viscoelastic effect may be present [24]. A number of studies have addressed how blood flow behavior changes if the plasma rheology is changed by the addition of long-chain water-soluble polymers such as poly(ethylene oxide) (PEO) or hyaluronic acid (HA). These studies were originally motivated by the knowledge that small amounts of long-chain polymer additives—"drag-reducing polymers" (DRPs) —can dramatically reduce drag in turbulent flows while having negligible effect on laminar flows [94]. Studies with laboratory animals indicate that the addition of DRPs to blood leads to many beneficial effects, including (1) reduced atherosclerotic plaque deposition (Figure 11.3(a)) [90, 95]; (2) increased cardiac output [96] and reduced severity of cardiac stenosis or myocardial infarction [97–100]; (3) improved survival probability after severe hemorrhage with use of polymer-containing resuscitation fluids [91, 93, 101] (see Fig. 11.3(b)); (4) reduced immune attack of implanted polymeric tissue scaffolds [92] (see Fig. 11.3c). Blood flow in small vessels is laminar, so turbulent drag reduction per se is not the source of these observations. Chemically different polymers (e.g., PEO, HA, polyacrylamide, various polysaccharides) yield very similar results as long as the molecular weight is sufficiently high (typically $\gtrsim 10^6$ Da). These observations indicate that the origin of the physiological effects is not chemical but physical.

Complementing the in vivo studies are a small number of experimental studies of the effects of DRPs on flow of blood or blood cell suspensions in tubes or fabricated microchannels. Many studies of polymer solutions in channels with complex geometries [102, 103] in the Reynolds number regime relevant for arteries have shown that recirculation regions and the corresponding stagnation zones near contractions, expansions, and branches are reduced in size by polymer additives. Since these zones are at increased risk for atherogenesis [104], their reduction is a likely reason for the observed in vivo reduction in plaque deposition and atherosclerotic lesions. On the other hand, the effect of polymers on the macroscopic fluid dynamics does not seem to explain the other observations described above.

Therefore, examination of the effects of polymer additives on flows at the scale of the microcirculation is important. Experiments in microfluidic devices [93, 105] have shown that the addition of a small amount of DRPs to suspension of RBCs results in a redistribution of RBCs with a significant reduction in the

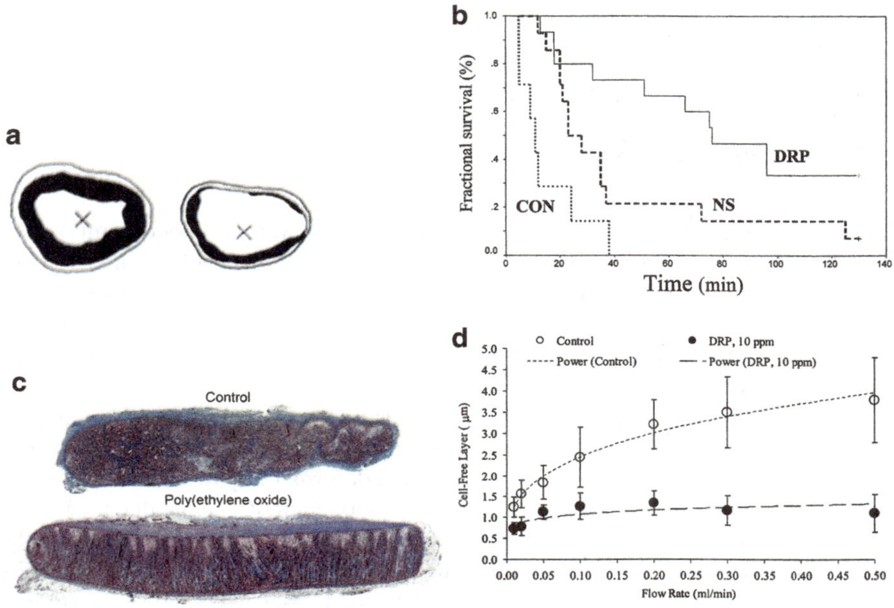

Fig. 11.3 Examples of physiological effects of addition of drag-reducing polymers. (**a**) Processed images of cross sections of aortas from rabbits fed a high-cholesterol diet: left, control animal, right, animal intravenously injected with poly(acrylamide) [90]. (**b**) Survival vs. time for rats subjected to hemorrhage and not treated (CON), treated by injection of normal saline (NS) or intravenously injected with saline containing a drag-reducing polymer [91]. (**c**) Poly(L-lactate) tissue scaffolds recovered after 7 weeks of implantation in a rat. *Top*: from control animal. *Bottom*: from animal treated with intravenous injections of poly(ethylene oxide) [92]. (**d**) Cell-free layer thickness vs. flow rate for microchannel experiments with RBCs in saline or saline with containing a drag-reducing polymer [93]

thickness of the cell-free layer. Direct simulations of a suspension of model cells and polymer molecules corroborates these experimental results [54]. Additionally, the thinning of the cell-free layer is observed to attenuate the plasma-skimming effect at microchannel branches so these branches display a larger RBC concentration than would arise otherwise. By increasing the hematocrit in the smallest arterioles and capillaries, this effect may contribute to the improvements in oxygen transport observed in vivo.

In addition to reducing the thickness of the cell-free layer, the addition of these DRPs is experimentally observed to reduce the concentration of platelets in this region, i.e., it attenuates margination [106]. This likely occurs because the increased number of RBCs near the walls leads to an increased effective (shear-induced) diffusivity of the platelets, leading them to be more uniformly distributed. This phenomenon may alter platelet aggregation and thrombus formation and may provide the explanation for the reduced deposition rate of atherosclerotic plaques. Margination of leukocytes is probably disrupted as well, reducing their ability to adhere to vessel walls and extravasate into surrounding tissue. This phenomenon might explain the reduced immune response to implantation of a polymer scaffold [92].

1.5 Motivation and Goals

The above discussion provides a brief background on various blood components and their distribution during blood flow. Having provided this background, we now turn to the main goal of this chapter, which is to describe efforts to *mechanistically* understand the margination phenomenon observed in blood flow using both simulations and theory. It will be helpful to first critically examine the status of the available literature on this topic.

As noted above, a variety of studies have focused on margination of leukocytes and platelets in blood. A number of physical parameters may contribute to this margination behavior including size, shape, and deformability differences between the leukocytes and platelets and the majority component RBCs. Note that the leukocytes are larger than RBCs and the platelets smaller, while both are considerably stiffer than RBCs. Their shapes differ as well: RBCs have a biconcave discoid shape, leukocytes are nearly spherical, and platelets are disk shaped. In studies on whole blood, on which most prior works have focused [72, 74, 80, 107, 108], the individual components–RBCs, leukocytes, and platelets–differ simultaneously in size, shape, and deformability. As such, it is not clear from these studies how the above determinants affect the margination behavior. To clarify the effects of these determinants, well-controlled studies that focus on changing only one property at a time, for example, the stiffness of the particle, are necessary. Such controlled studies are lacking in the literature.

On the theoretical front, most attempts on elucidating the mechanism of the margination have relied on simple transport models, the most common of which is the drift-diffusion equation [79,109]. These models are, however, phenomenological with the drift and diffusion terms extracted from experimental or computational results. These models by themselves do not provide any mechanistic understanding of the phenomenon, i.e., how the size, shape, and deformability affect the margination behavior. To enable progress, more mechanistic models are necessary, for example, models that estimate the drift and diffusion of the species from fundamental transport processes occurring in the suspension.

The present authors have sought to address the above shortcomings by performing well-controlled computational studies in which the components in a multicomponent suspension differ only in their rigidity [110,111] or their size [112], thereby clarifying their roles on the margination behavior. Furthermore, we have also developed a simple master equation model of suspension dynamics capable of reproducing results from detailed numerical simulations [111]. This master equation model is based on two key transport mechanisms in confined suspensions: the wall-induced migration and hydrodynamic pair collisions. These fundamental transport processes account for the deformability and size of the species and therefore clarify how these determinants contribute to the margination behavior.

The remainder of the article is organized as follows. First, in Sect. 2, we formulate the multiphase flow problem relevant to blood and discuss procedures for its numerical solution. In particular, this section describes the boundary integral method commonly used for simulating blood flow. In Sect. 3, we characterize the effect of rigidity on segregation and margination behavior in binary capsule suspensions. In this section, results are presented both from the detailed boundary integral method as well as from the master equation model discussed above. The master equation model enables a mechanistic understanding of the segregation behavior. In Sect. 4, we briefly discuss results from recent simulation efforts investigating the effect of long-chain polymer additives to blood. Finally, concluding remarks and an outlook are provided in Sect. 5.

2 Problem Formulation and Implementation

We present this section in two parts. We first formulate the multiphase fluid flow problem and subsequently describe its numerical solution procedures (Sec. 2.1). Following this, we introduce the framework employed for computing the elastic membrane forces (Sect. 2.2). These two problems are obviously coupled—the fluid flow leads to membrane deformation and hence forces, while the forces themselves appear as a forcing term in the fluid flow problem. These aspects are discussed further in the sections below.

2.1 *Fluid Flow Problem*

Consider a suspension of fluid-filled deformable particles, such as RBCs or capsules, in a three-dimensional domain as shown in Fig. 11.4. The domain boundary is denoted by S_E, while the interface of the mth particle is denoted by S_m. Both the suspending and the enclosed fluid are Newtonian and incompressible. The viscosity of the suspending fluid is μ, while the viscosity of enclosed fluid is $\lambda\mu$. Our interest is in blood flow in the microcirculation where the Reynolds number is typically small and inertial forces can be neglected [113]. In this case, the momentum balance for the fluid reduces to the Stokes equation :

$$-\nabla P_1 + \mu \nabla^2 \mathbf{u}_1 = 0\,, \quad -\nabla P_2 + \lambda \mu \nabla^2 \mathbf{u}_2 = 0, \tag{11.4}$$

where $\mathbf{u}_1$ and P_1 refer to the velocity and pressure fields in the fluid outside the particles, while $\mathbf{u}_2$ and P_2 refer to the velocity and pressure fields in the fluid inside the particles. The mass balance and incompressibility require that

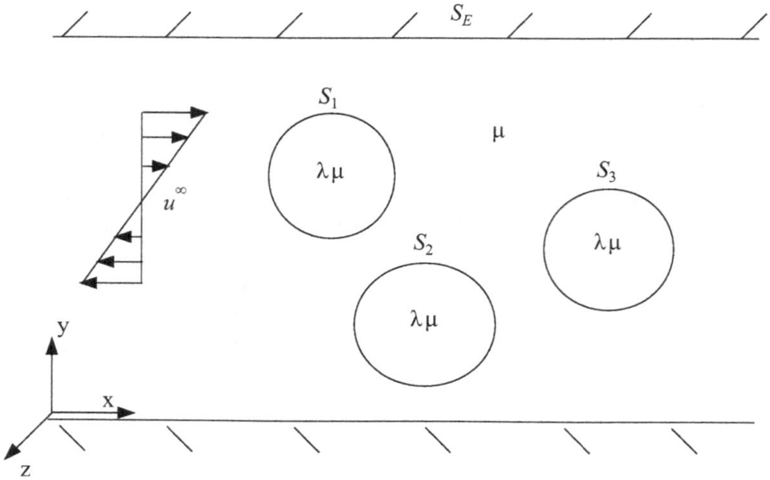

Fig. 11.4 Problem schematic: a dispersed phase with viscosity μ inside the domain boundaries denoted by S^E, containing, for example, three particles with internal viscosity $\lambda\mu$. The surfaces of the particles are denoted by S_1, S_2, S_3, respectively. The undisturbed flow is denoted by $\mathbf{u}^\infty$

$$\nabla \cdot \mathbf{u}_1 = 0\,, \quad \nabla \cdot \mathbf{u}_2 = 0. \tag{11.5}$$

This set of equations are closed with appropriate boundary conditions at the particle interfaces (S_m) as well as at the domain boundary (S_E). The boundary conditions on the particle interface are the continuity of the velocity and zero net force on an infinitesimal area. These are respectively expressed by the following [114]:

$$\mathbf{u}_1 = \mathbf{u}_2\,, \quad (\boldsymbol{\sigma}_1 - \boldsymbol{\sigma}_2) \cdot \mathbf{n} = -\nabla_s \cdot \boldsymbol{\tau}\,, \tag{11.6}$$

where $\boldsymbol{\sigma}$ is the stress tensor in the fluid, $\mathbf{n}$ is the outward normal at the interface, ∇_s is the surface divergence operator, and $\boldsymbol{\tau}$ is the tension tensor in the membrane. We will discuss the membrane tension in detail in Sect. 2.2. The term $(\boldsymbol{\sigma}_1 - \boldsymbol{\sigma}_2) \cdot \mathbf{n}$ expresses the hydrodynamic traction jump across the interface and will be denoted as $\Delta\mathbf{f}^H$; hence, $\Delta\mathbf{f}^H = -\nabla_s \cdot \boldsymbol{\tau}$. As noted above, the solution must also satisfy the appropriate boundary conditions at the domain boundary S_E [115]. Some common examples are periodic boundary conditions and no-slip conditions at solid walls.

There are several commonly employed techniques for solving the multiphase flow equations. A popular technique for multiphase Stokes flow is the boundary integral (BI) method [114]. In this method the fluid velocity at any point in the domain is expressed as an integral over all the boundaries in the system, which includes the particle interface as well as the domain boundary. The BI method has been used successfully for a variety of multiphase Stokes flow problems including blood flow, drops, capsules, and vesicles among others [115–120]. In this chapter, we discuss the BI method in detail. Another commonly used technique for multiphase flows is the immersed boundary (IB) method [121] (see Chap. 10). In this method, the particle interface appears as forcing term–typically known–in the momentum balance for the fluid, which can then be solved by a Navier-Stokes solver on an Eulerian grid. The particle interface in the IB method is explicitly tracked by a separate Lagrangian mesh. An important consideration in the IB method is the regularization of the Dirac-delta function, which is employed to distribute the interfacial forces on the Eulerian grid [121]. The immersed boundary and related techniques like the front tracking method have also been used in many works on blood flow and other multiphase flow problems [122–125]. Besides these methods, there are also various non-continuum-based approaches. These include, among others, the lattice-Boltzmann method where a discrete analog of the Boltzmann equation is solved [126, 127], and the dissipative particle dynamics method where coarse-grained fluid particles are considered [128, 129]. These techniques have also been used in various studies on multiphase flow problems including blood flow [79, 130, 131].

2.1.1 Boundary Integral Method

As preliminaries for the boundary integral equation, we introduce the Green's function **G** and its associated stress tensor **T**. By definition, **G** and **T** give the velocity and stress fields due to a point force F as follows [114]:

$$u_i(\mathbf{x}_0) = F_j G_{ij}(\mathbf{x}_0,\mathbf{x}), \quad \sigma_{ik}(\mathbf{x}_0) = F_j T_{ijk}(\mathbf{x}_0,\mathbf{x}), \tag{11.7}$$

where $\mathbf{x}$ is the location of the point force (pole) and $\mathbf{x}_0$ is the location of the target (field) point. In the present case, it is assumed that the velocity and stress fields above satisfy the appropriate boundary conditions of the domain, i.e., **G** and **T** are geometry-specific. There are several different formulations of the boundary integral equation employed in the literature. The most commonly used form is the following [114]:

$$\begin{aligned} u_j(\mathbf{x}_0) &= \frac{2}{1+\lambda} u_j^{\infty}(\mathbf{x}_0) - \frac{1}{4\pi\mu(1+\lambda)} \sum_{n=1}^{N_p} \int_{S^n} \Delta f_i^H(\mathbf{x})\, G_{ji}(\mathbf{x}_0,\mathbf{x})\, dS(\mathbf{x}) \\ &+ \frac{(1-\lambda)}{4\pi(1+\lambda)} \sum_{n=1}^{N_p} \int_{S^n} u_i(\mathbf{x})\, T_{ijk}(\mathbf{x},\mathbf{x}_0)\, n_k(\mathbf{x})\, dS(\mathbf{x}), \end{aligned} \tag{11.8}$$

where $\mathbf{x}_0$ lies on the interface of particle m (i.e., $\mathbf{x}_0 \in S^m$), $\mathbf{u}^{\infty}(\mathbf{x}_0)$ is the undisturbed fluid velocity at $\mathbf{x}_0$, and the sums are over all the N_p particles in the system. The integrals involving **G** and **T** as kernels are typically referred to as the single-layer and the double-layer integral, respectively [114, 132]. From here onwards, a principal value of the double-layer integral over a part of the boundary is assumed whenever the target point $\mathbf{x}_0$ lies on that boundary. For example, in the above equation, the double-layer integral over S^n is assumed to denote the principal value when $n = m$. A crucial aspect of the above formulation is that the integrals only involve the internal (interfacial) boundaries and not the external domain boundaries. This is a direct consequence of using the geometry-specific **G** and **T**; if **G** and **T** for any other geometry is employed (e.g., free space), additional integrals over the domain boundaries arise in Eq. (11.8).

An alternative formulation of the BI equation is the following where the velocity field is expressed solely in terms of the single-layer integral [114]:

$$u_j(\mathbf{x}_0) = u_j^{\infty}(\mathbf{x}_0) + \sum_{n=1}^{N_p} \int_{S^n} q_i(\mathbf{x})\, G_{ji}(\mathbf{x}_0, \mathbf{x})\, dS(\mathbf{x}), \tag{11.9}$$

where $\mathbf{q}(\mathbf{x})$ is the single-layer density and other quantities are as defined for Eq. 11.8. The single-layer density $\mathbf{q}(\mathbf{x}_0)$ itself satisfies

$$\begin{aligned} q_j(\mathbf{x}_0) + \frac{\kappa}{4\pi} n_k(\mathbf{x}_0) \sum_{n=1}^{N_p} \int_{S^n} q_i(\mathbf{x})\, T_{jik}(\mathbf{x}_0, \mathbf{x})\, dS(\mathbf{x}) \\ = -\frac{1}{4\pi\mu} \left(\frac{\Delta f_j^H(\mathbf{x}_0)}{\lambda_m + 1} + \kappa f_j^{\infty}(\mathbf{x}_0) \right), \end{aligned} \tag{11.10}$$

where $\kappa = (\lambda - 1)/(\lambda + 1)$ and $\mathbf{f}^{\infty}$ is the traction at a given point (computed with the suspending fluid viscosity μ) due to the stress generated in the fluid corresponding to the undisturbed flow $\mathbf{u}^{\infty}$. An important distinction of the second BI formulation over the first formulation is that (a) the target point of both the kernels **G** and **T** is the same as the target point of the overall BI equation and (b) the source point of the kernels **G** and **T** are same as the location of its multiplicand. These properties are critical for employing certain acceleration schemes, especially the general-geometry-Ewald-like method (GGEM) described below [115].

2.1.2 General Considerations

A key step in the numerical evaluation of the integrals is the discretization of the surface into elements and the corresponding choice of basis functions over each element. The surface discretization employed in the literature includes both finite-element-like discretizations with low-order basis functions [116–118] as well as global spectral discretization schemes [119, 133]. A widely used discretization in the former category is the choice of triangular elements with linear basis functions [115, 116]. An example of spectral discretization is the choice of spherical harmonics basis functions [119, 134].

Another point of interest is the numerical evaluation of singular integrals. Singular integrals arise naturally in the BI method because both the single-layer and the double-layer integrals become singular as $\mathbf{x} \to \mathbf{x}_0$, with the singularity scaling as $1/r$, where $r = |\mathbf{x} - \mathbf{x}_0|$ [133]. These singular integrals require special care and a variety of schemes have been developed in the literature to evaluate them accurately. A common approach is to transform the surface integral to polar coordinates centered at $\mathbf{x}_0$, which cancels the $1/r$ singularity automatically with the Jacobian of the polar coordinate transformation [117]. Other approaches include a stretched coordinate mapping that clusters points near the singularity locations and enables accurate evaluation of the integrals [133]. The singularity subtraction technique is also a widely used approach which renders the integral non-singular [116, 118].

2.1.3 Computational Cost

Traditional implementations of the BI equation typically scale as $O(N^2)$, where N is proportional to the number of degrees of freedom in the system. For a system with N_p particles, each of which have been discretized into N_Δ elements, the number of degrees of freedom in the system scales as $N \sim N_p N_\Delta$. The $O(N^2)$ scaling essentially arises as the contribution from every element has to be evaluated at all elements in the system and assumes an iterative solution of the discretized system of equations, where the number of iterations is independent of N; a direct solution will result in a scaling of $O(N^3)$, while a system-size-dependent number of iterations with an iterative solution results in a scaling higher than $O(N^2)$. The $O(N^2)$ scaling is usually prohibitive, such that it precludes a numerical study of large system sizes. It is therefore not surprising that many of the past studies have been limited to an $O(1)$ number of particles.

To overcome these limitations, there have been several efforts to develop *accelerated* techniques giving a scaling closer to the ideal $O(N)$. These techniques employ either some variant of the particle-particle-particle-mesh (P^3M) method [135], or the fast multipole method (FMM) [136]. In the P^3M-based techniques, like the particle-mesh-Ewald (PME) method, the periodic Green's function given by Hasimoto [137] is employed. This periodic Green's function can be expressed as the sum of a real-space sum and a wave-space sum. The real-space sum decays exponentially from the origin of the singularity and hence only the nearest neighbors within a cutoff distance need to be considered when evaluating this contribution. The wave-space sum can be accurately and quickly computed on a uniform mesh using the fast Fourier transforms (FFT). When this acceleration method is applied to the boundary integral method, the real-space sum leads to a computational cost of $O(N)$, while the wave-space sum has a cost of $O(N \log N)$, thereby leading to the overall scaling of $O(N \log N)$ [119]. In the FMM method, the basic idea involved is to calculate the interactions between distant elements using truncated multipole expansions and those between nearby elements directly. The overall approach is developed in such a way that the computational cost is $O(N)$, while maintaining a given level of accuracy. Both of the above methods can be adapted to confined geometries [119, 138].

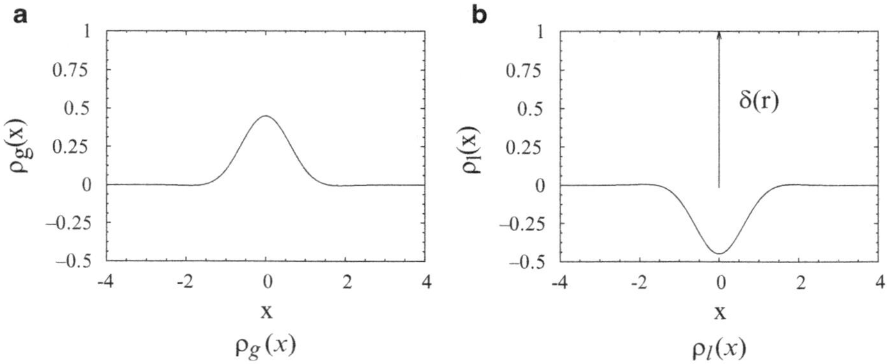

Fig. 11.5 Variation of the global $\rho_g(x)$ and the local $\rho_l(x)$ force density along the x-axis, given the center of the force density is at the origin [115]. Note that $\rho_g(\mathbf{x}) + \rho_l(\mathbf{x}) = \delta(\mathbf{x})$. For plotting $\rho_g(x)$ and $\rho_l(x)$ here, we set $\alpha = 1$. **(a)** $\rho_g(x)$ **(b)** $\rho_l(x)$

An alternative acceleration technique is the general-geometry-Ewald-like method (GGEM) [139]. As the name suggests, it bears similarity to PME-based acceleration technique for periodic geometries. However, the GGEM method is more naturally suited to nonperiodic domains with several distinct advantages over PME-based methods [115]. A particular advantage is that the pressure drop in a simulation can be directly specified in GGEM-accelerated methods, while it cannot be in a PME-based method. In addition, the domain of interest has to be embedded in a periodic domain for it to be amenable for acceleration by PME-based methods, which can lead to unwarranted inefficiencies due to discretization of extra domain not central to the problem. There is no such drawbacks of the GGEM methodology.

The key idea in the GGEM methodology is to express a Dirac-delta force density as a sum of a smooth quasi-Gaussian global density $\rho_g(\mathbf{x})$ and a second local density $\rho_l(\mathbf{x})$; these are respectively given by the following expressions:

$$\rho_g(\mathbf{x}) = \frac{\alpha^3}{\pi^{3/2}} e^{-\alpha^2 r^2} \left(\frac{5}{2} - \alpha^2 r^2 \right), \quad \rho_l(\mathbf{x}) = \delta(\mathbf{x}) - \rho_g(\mathbf{x}), \tag{11.11}$$

where α^{-1} represents a length scale over which the delta-function density $\delta(x)$ has been smeared using the quasi-Gaussian form above, while for the moment $\mathbf{x}$ is position vector relative to the pole of the singularity. See Fig. 11.5 for pictorial representations of ρ_g and ρ_l. It is important to emphasize that the total density remains a delta function; i.e., $\rho_g(x) + \rho_l(x) = \delta(x)$. The solution associated with a local force density, which is known analytically, is short ranged and is neglected beyond a length scale of $O(\alpha^{-1})$ from its pole. We note that the local solution is obtained assuming free-space boundary conditions. The solution associated with the global force density is numerically computed, while ensuring that the boundary conditions associated with the overall problem are satisfied. The cost associated with the local solution is $O(N)$. The cost associated with the global solution depends on the exact choice of numerical scheme and virtually any numerical method developed for the Stokes equation is valid here. When employing a finite difference or a finite-element scheme with proper preconditioners, the global solution can be obtained at a cost of $O(N)$.

2.2 *Membrane Mechanics*

The hydrodynamic traction jump across the interface, $\Delta \mathbf{f}^H$, is an unknown in the fluid flow problem and must be computed independently. This is achieved by writing a force balance on an infinitesimal patch of the membrane, which upon simplification yields: $\Delta \mathbf{f}^H = -\nabla_s \cdot \boldsymbol{\tau}$. This is exactly the same as the second boundary condition in Eq. 11.6. In this section, we will describe the procedure commonly employed to compute the membrane tension (Sect. 2.2.2). However, it is appropriate to first provide some background on the RBC membrane structure and some other relevant properties.

2.2.1 RBC Membrane

We noted in Sect. 1 that RBCs are a nucleate cells and essentially consist of a plasma membrane enclosing the cytoplasm. The plasma membrane has a complex composite structure consisting of a lipid bilayer envelope anchored to a network of skeletal protein [34, 140, 141]. The lipid bilayer essentially behaves as a two-dimensional incompressible fluid that strongly resists surface area changes and endows the membrane with a small bending resistance, but lacks a shear resistance. The protein skeletal network behaves as a two-dimensional elastic solid resisting deformation of the RBCs. The structural properties of RBCs (low elastic modulus coupled with large excess area due to their shape) allow them to undergo large deformation while in circulation, thereby enabling them to flow through capillaries smaller than their size. RBCs have been observed to withstand linear extensions of about 250 %; however, even a small increase in its surface area (3–4 %) can lead to their lysis [141]. A brief comment is also in order about the cytoplasm. The cytoplasm is essentially a concentrated solution of hemoglobin [141]. The flow behavior of cytoplasm is approximately Newtonian with a viscosity in the range 6–10 mPa s [142]. The viscosity of the cytoplasm is strongly dependent on the hemoglobin concentration, which is tightly regulated in the range 27–37 g/dL [141]. See Chaps. 4 and 5 for more detailed discussions of the cell membrane and cytoskeleton.

2.2.2 Membrane Tension Tensor

The tensions in the membrane are typically computed using finite deformation theory for thin continuum elastic shells [53, 143]. Alternative mesoscopic models based on a network of springs have also been developed, e.g., [144], though here we will focus only on continuum models. The membrane bending modulus is very small [34, 145] and, because it adds complexity to the formulation as well as substantial computational expense, is often neglected in simulations. However, its absence can lead to spurious wrinkling instabilities as well as development of artificially sharp corners in the interface shape [143, 146]. These difficulties can be circumvented to some extent by introducing a small prestress into the membrane [147]. While we neglect the membrane bending modulus in the present discussion, this can play a role in determining the detailed dynamics of an RBC in flow [148, 149]. Additionally, recent evidence suggests that the biconcave RBC shape exhibits residual stress and that the spontaneous curvature of the cell actually favors an oblate spheroid [150, 151].

It is common to assume that the membrane is isotropic and hyperelastic, with a strain energy potential that is a function of its deformation. The elastic tension in the membrane can be obtained from the knowledge of this strain energy potential. We next provide a brief overview of the finite deformation theory and connect it to the computation of the membrane tension tensor.

We begin by denoting the coordinates of a material point in the initial unstressed condition by $\mathbf{X}$ and at a later time t by $\mathbf{x}(\mathbf{X},t)$. The surface deformation gradient tensor $\mathbf{F}$ is then defined as follows [53]:

$$\mathbf{F} = (\mathbf{I}-\mathbf{nn}) \cdot \frac{\partial \mathbf{x}}{\partial \mathbf{X}} \cdot (\mathbf{I}-\mathbf{NN}), \tag{11.12}$$

where $\mathbf{N}$ and $\mathbf{n}$ are normals to the surface in the initial and the deformed state, respectively. The symmetric surface left Cauchy-Green strain tensor is then defined as [143]

$$\mathbf{b} = \mathbf{F} \cdot \mathbf{F}^T \tag{11.13}$$

The tensor $\mathbf{b}$ has two nonzero eigenvalues λ_1^2, λ_2^2, where λ_1 and λ_2 are called the principal extension ratios [143]; the third eigenvalue is 0. The local area dilation is denoted by $J = \lambda_1 \lambda_2$. For an isotropic hyperelastic membrane, the strain energy potential W is solely expressible in terms of invariants of the tensor $\mathbf{b}$ [143]. For example, Skalak et al. [152] introduced the following invariants:

$$I_1 = \lambda_1^2 + \lambda_2^2 - 2\,, \quad I_2 = \lambda_1^2 \lambda_2^2 - 1. \tag{11.14}$$

Using standard relations, the Cauchy tension tensor $\boldsymbol{\tau}$ can be expressed in terms of the strain energy potential W as follows [143]:

$$\boldsymbol{\tau} = \frac{1}{J}\mathbf{F}\cdot\frac{\partial W}{\partial \mathbf{F}} = \frac{1}{J}\mathbf{F}\cdot\frac{\partial W}{\partial \mathbf{E}}\cdot\mathbf{F}^T, \tag{11.15}$$

where $\mathbf{E} = (\mathbf{F}^T \cdot \mathbf{F} - \mathbf{I} - \mathbf{NN})/2$ is the Green-Lagrange strain tensor. From the preceding equation, it can be shown that the principal tensions in the membrane are given by [143]

$$\tau_1^p = \frac{1}{\lambda_2}\frac{\partial W}{\partial \lambda_1}, \quad \tau_2^p = \frac{1}{\lambda_1}\frac{\partial W}{\partial \lambda_2}. \tag{11.16}$$

Note that the principal directions of tensions and deformation are collinear in view of the isotropy of the membrane. In terms of Skalak invariants introduced above, the tension tensor takes the form

$$\boldsymbol{\tau} = \frac{2}{\lambda_1\lambda_2}\frac{\partial W}{\partial I_1}\mathbf{b} + 2\lambda_1\lambda_2\frac{\partial W}{\partial I_2}(\mathbf{I} - \mathbf{nn}). \tag{11.17}$$

Therefore, once the strain energy function $W(I_1, I_2)$ or $W(\lambda_1, \lambda_2)$ is known, then the membrane tensions can be computed using the above equations. We discuss several popular strain energy potential constitutive laws next.

2.2.3 Membrane Constitutive Laws

A popular constitutive law is the neo-Hookean (NH) law for which the strain energy function has the following form [53]:

$$W = \frac{G}{2}\left(I_1 + 2 + \frac{1}{I_2 + 1}\right), \tag{11.18}$$

where G is the two-dimensional shear modulus and has units of force per unit length. The neo-Hookean model has a strain-softening response and is a good model for volume-incompressible rubberlike materials like artificial polymeric capsules [53]. The Skalak model for the strain energy function, which is a good model for RBCs, is the following [53]:

$$W = \frac{G}{4}\left((I_1^2 + 2I_1 - 2I_2) + CI_2^2\right) \tag{11.19}$$

where the parameter C is the energy penalty due to area change. The area dilation modulus for the Skalak law can be shown to equal $K = 2G(1 + 2C)$. A variety of experiments have attempted to characterize the RBC membrane properties. These include micropipette aspiration experiments [153] and experiments with optical tweezers [50]. These experiments usually provide values for the shear modulus in the range 2–12 μN/m. The value of the parameter C is typically very high ($C \gg 1$), signifying the near area incompressibility of the membrane [143].

3 Segregation by Membrane Rigidity: Simulations and Theory

The goals of this section are twofold: first is to characterize the effect of rigidity on margination and segregation behavior and second is to establish the mechanism underlying the observed segregation. We do this in two parts. First, in Sect. 3.1, we consider detailed boundary integral simulations of a model problem that isolates the effect of stiffness on the segregation behavior. Following this, in Sect. 3.2, we consider an idealized master equation model of the suspension dynamics that will enable a mechanistic understanding of the segregation phenomenon.

3.1 Boundary Integral Simulations of Binary Suspensions

The effect of rigidity alone on the segregation and margination behavior was recently characterized by Kumar and Graham [110, 111]. These authors considered a binary suspension of neo-Hookean capsules subjected to simple shear or pressure driven flows in a planar slit, i.e., between two planar walls. The two components of the binary suspension had the same spherical rest shape and size, but differed in their membrane rigidity. The rigidity was characterized as usual by the nondimensional capillary number, $\mathrm{Ca} = \mu\dot{\gamma}a/G$, where μ is the suspending fluid viscosity, $\dot{\gamma}$ is the characteristic shear rate, a is the capsule rest radius, and G is the shear modulus. The component with the higher capillary number in these studies was termed floppy, while the component with the lower capillary number was termed stiff. The authors systematically explored the effect of a variety of parameters on the segregation behavior including the capillary number Ca_f of the floppy particle, capillary number of the stiff particle Ca_s ($\mathrm{Ca}_s < \mathrm{Ca}_f$), the overall volume fraction ϕ, the confinement ratio $2a/H$ (H is the channel height), and the number fraction of the floppy particles in the mixture X_f.

Figure 11.6 shows some typical results for the normalized number density profile for both the species as a function of X_f [111]. Note that the number density profile was normalized in such a way that a uniform distribution will result in $\hat{n}_\alpha(y) = 1$ throughout. The parameters in this study were $\mathrm{Ca}_f = 0.5$, $\mathrm{Ca}_s = 0.2$, $2a/H = 0.197$, and $\phi = 0.12$. Also, the bulk flow was a simple shear flow. Focusing first on pure suspensions of stiff and floppy particles, it can be seen that the number density profiles are similar (Figs. 11.6a and 11.6d). Obvious in both of these plots is a particle-depleted layer next to the wall, known as the cell-free layer in the blood literature; see Sect. 1. Also evident is a substantial near-wall peak just beyond the cell-free layer. The cell-free layer arises due to the fact that a deformable particle such as a capsule has a tendency to migrate away from the wall [45]. The interparticle collisions, which are the other key source of particle motion, generally displace the particles toward the wall on average. The near-wall peak is a consequence of the competition between these two opposing effects.

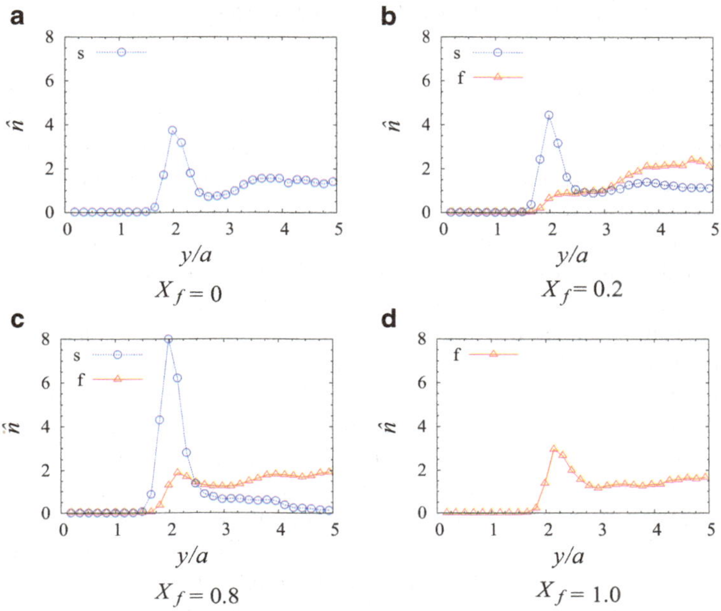

Fig. 11.6 Normalized number density profile $\hat{n}$ for both the stiff and the floppy particles at several values of X_f, the number fraction of floppy capsules [111]. Simulation parameters are $Ca_s = 0.2$, $Ca_f = 0.5$, $\phi = 0.12$, and $a/h = 0.197$. One of the channel walls is at the *left* end, while the channel center is at the *right* end of the plots (only half the channel is shown). Note that a uniform distribution of the species will yield a value of $\hat{n} = 1$ throughout

We next focus on the number density profiles in binary suspensions. Figure 11.6b shows $\hat{n}$ for both the stiff and the floppy particles at $X_f = 0.2$, a suspension which is relatively dilute in floppy particles. In comparison to the pure suspension results, $\hat{n}$ for the stiffer particles indicates a slight enrichment in the near-wall region; in contrast, there is a substantial depletion of the floppy particles in the near-wall peak accompanied by their enrichment near the centerline. The latter behavior is clearly opposite of the margination behavior and was termed demargination by the authors. Consider next a suspension with $X_f = 0.8$, which has the stiff particles as the dilute component. Figure 11.6c shows $\hat{n}$ for both the species in this case. In comparison to the pure suspension results, the floppy particles reveal a slight depletion in the near-wall region. In comparison, there is a significant enrichment of the stiff particles in the near-wall region, indicating that the stiff particles undergo substantial margination. Thus, to summarize, it can be said that stiff particles undergo margination when they are the dilute component in the suspension, while floppy particles undergo demargination when they are the dilute component in the mixture. The authors observed a qualitatively similar behavior over the entire parameter range they investigated.

3.2 Master Equation Model for Binary Suspensions

Kumar and Graham [111] developed a very simple theoretical model for describing particle transport in confined multicomponent suspensions of deformable particles. They subsequently employed this model for providing a mechanistic understanding of the segregation phenomena described above. For making theoretical progress, the authors assumed that the suspension is dilute. Further details of their model are described next. Closely related models for single-component suspensions of rigid and deformable particles appear in [154] and [155], respectively.

In a dilute suspension of particles, the particle interactions can be treated as a sequence of uncorrelated pair collisions [154, 156]. In addition, since the capsules are deformable, they also have a wall-induced migration velocity $v_{\mathrm{mig}}(y)$ away from the wall [45, 48]. The effect of the pair collisions and the wall-induced particle migration can be consistently described by a kinetic master equation (cf. [154]). The authors focused on a model of a monolayer in the $x-y$ (flow-gradient) plane for simplicity–its extension to three dimensions is straightforward [154]. The mean area number density of all the particles in the monolayer is denoted n_0, such that the areal fraction in the $x-y$ plane at rest is $\phi_a = \pi a^2 n_0$. The mean area number density of each of the species α in the mixture is denoted n_0^α, while its distribution in the y direction is denoted $n^\alpha(y)$. The master equation for this case is given by

$$\begin{aligned}\frac{\partial n^\alpha(y)}{\partial t} = & -\frac{\partial\left(v_{\mathrm{mig}}^\alpha n^\alpha\right)}{\partial y} + \sum_{S=1}^{N_s}\left(\int_{-(H-y)}^{y}\left\{n^\alpha(y-\Delta^{\alpha\beta})\, n^\beta(y-\Delta^{\alpha\beta}-\delta)\right.\right.\\ & \left.\left. - n^\alpha(y)\, n^\beta(y-\delta)\right\}\dot{\gamma}|\delta|\,d\delta\right), \qquad (11.20)\end{aligned}$$

where δ is the pre-collision pair offset in the y direction, $\Delta^{\alpha\beta}(\delta)$ is the cross-stream displacement of particle of type α after collision with another particle of type β, while the sum is over all the species N_s in the suspension; see (Fig. 11.7b) for a schematic of a pair collision. The first term on the right-hand side arises from the wall-induced migration, while the integral term represents the effect of pair collisions [154].

The above equation is analogous to the Boltzmann equation for rarefied gases [157, 158]. The dynamic simulation Monte Carlo (DSMC) approach is a popular technique to obtain solutions of the Boltzmann equation [157, 158]. The authors developed a similar technique for obtaining steady-state solutions from the master equation model (Eq 11.20) and named their method as the hydrodynamic Monte Carlo (HMC) method. Similar approaches have also found application in recent works on colloidal suspensions of rigid spheres [154]. As in the case of DSMC [157, 159], the HMC approach is appropriate in the dilute limit and requires the assumption of chaotic particle dynamics, an assumption that is valid for particulate flows [160].

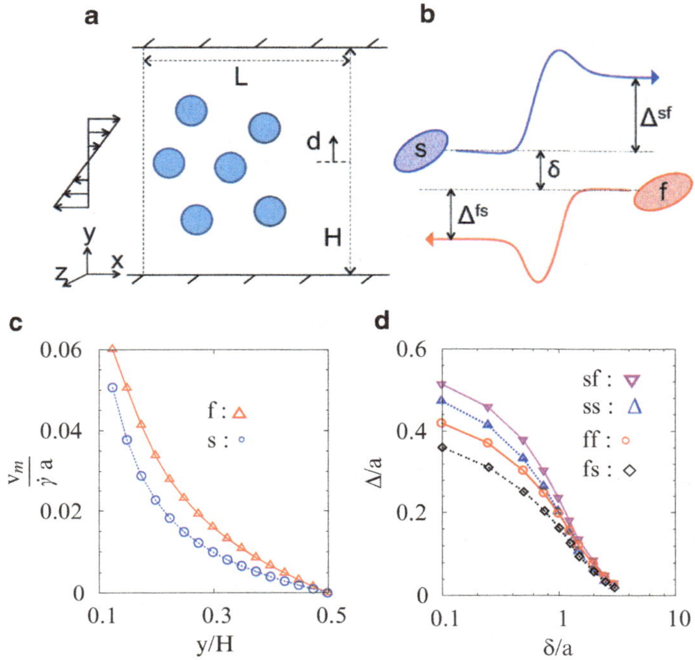

Fig. 11.7 (**a**) System geometry in HMC simulations, (**b**) schematic of particle trajectories in a pair collision, (**c**) isolated particle migration velocity at confinement ratio $2a/H = 0.197$, and (**d**) cross-stream displacement Δ in various types of pair collisions as a function of the initial offset δ. Here "s" refers to the stiffer particle ($\mathrm{Ca}_s = 0.2$), while "f" refers to the floppier particle ($\mathrm{Ca}_f = 0.5$)[111]

In the HMC approach, the y positions of N_p particles are followed in time. Each particle is assumed to represent an infinite number of particles at the same y position randomly distributed in the flow direction with an average spacing of L, where L is given by $L = N_p/(n_0 H)$. A distinguishing feature of this method is that the collisions between particles are treated probabilistically. The authors neglected pair collisions with large initial offsets $\delta > \delta_{\text{cut}}$, as their effect on cross-stream displacement is weak (e.g., $\delta_{\text{cut}} = 2.5a$ in [111]). A time step of the simulation involves choosing a pair of particles which satisfies the condition $\delta \le \delta_{\text{cut}}$. The pair is subsequently selected or rejected for collision with a probability proportional to their relative velocity of approach $\dot{\gamma}|\delta|$ [154, 157, 161]. An important aspect of the simulation is the time interval between collisions Δt, because the wall-induced particle migration occurs simultaneously with the collisions. In order to determine Δt at each time step, the authors assumed that the number of particle collisions with time follows a Poisson process with a mean collision frequency ν, such that the time interval between collisions is distributed with probability $P(\Delta t) = \nu e^{-\nu \Delta t}$ [161]. The collision frequency is estimated as

$$\nu = 0.5 \int_0^L \int_0^H n(y) \left(\int_{y-\delta_{\rm cut}}^{y+\delta_{\rm cut}} n(y+\delta)\,\dot{\gamma}|\delta|\,d\delta \right) dy\,dx. \tag{11.21}$$

Once the collision pair and time interval are set, the positions of all particles are updated. For a particle k this reads as $y_k(t+\Delta t) = y_k(t) + v_{\rm mig}\Delta t + \Delta_k$, where Δ_k is nonzero only for the colliding particle pair (i,j). It is important to emphasize that in a mixture, a collision could occur between two species of the same type, termed as *homogeneous* collisions, or between two species of different type, termed as *heterogeneous* collisions. For the current time step, if the collision is homogeneous, then $\Delta_k = \Delta^{ss}$ or Δ^{ff} for stiff and floppy particles, respectively; if the collision is heterogeneous, then $\Delta_k = \Delta^{sf}$ for the stiff particle and $\Delta_k = \Delta^{fs}$ for the floppy one. The procedure outlined above for a time step is repeated until a statistical steady state is obtained.

3.2.1 Pair Collisions and Wall-Induced Migration

The HMC method requires as inputs the cross-stream displacements $\Delta^{\alpha\beta}$ in pair collisions and the wall-induced migration velocity $v^{\alpha}_{\rm mig}$. The authors computed these using the boundary integral method [115]. Figure (11.7c) shows the isolated particle migration velocity as a function of y/H for $\mathrm{Ca}_s = 0.2$ and $\mathrm{Ca}_f = 0.5$ capsules at a confinement ratio $2a/H = 0.197$. Results for $\Delta^{\alpha\beta}(\delta)$ for the same two species (in an unbounded domain) are shown in Fig. (11.7d) for offsets $\delta \lesssim a$, $\Delta^{ff} < \Delta^{ss}$. Furthermore, an important feature in this plot is that the displacement of the stiffer particle in heterogeneous collisions is higher than that of the floppy particle ($\Delta^{sf} > \Delta^{fs}$), while the cross-stream displacements in homogeneous collisions are between these two limits, i.e., $\Delta^{fs} < \Delta^{ff}, \Delta^{ss} < \Delta^{sf}$. This ordering will turn out to be crucial in determining the segregation behavior.

The above particle migration velocity and pair collision results were determined in idealized systems, namely, by considering an isolated particle and an unconfined system, respectively. Since the system of interest is a *confined suspension*, corrections are necessary for both the migration as well as the pair collision results. The wall-induced migration velocity $v_{\rm mig}$ arises due to the disturbance velocity created by the particle's image and is a far-field effect [45, 48]. As a result of its far-field nature, in a suspension of particles, it can be expected that the wall-induced migration of a particle will result not only due to interaction with its own images, but also due to images of other particles–this can be expected to introduce an averaging effect in a suspension of particle mixtures. To model this, the authors modified the migration velocity of a particle as $v^{\alpha,s}_{\rm mig} = \xi\, v^{\alpha}_{\rm mig} + (1-\xi)\sum_1^{N_s} X_\beta v^{\beta}_{\rm mig}$, where $v^{\alpha,s}_{\rm mig}$ is the migration velocity of the species α in the suspension and ξ is an adjustable parameter. By taking $\xi = 1$, the isolated particle migration velocity is recovered for each species, while for $\xi = 0$, each of the species in the suspension has the same migration velocity.

The authors adopted a similarly simple model to account for the confinement effects on cross-stream displacement Δ in pair collisions. For a spherical particle with center at $y = a$, its Δ in a pair collision is expected to be zero as it will be touching the wall, while its Δ will approach the unconfined result at large particle-wall separations. To account for this effect, the Δ of the particle in an unconfined system was multiplied by a factor $\eta = 1 - e^{-(d_w - a)/a}$, where d_w is the distance of the particle from the nearest wall, assumed to satisfy $d_w > a$.

3.2.2 Evaluation of n_0 and ξ

The HMC model has two adjustable parameters, n_0 and ξ, which can be tuned to obtain good agreement with BI simulations for a given suspension. The authors sought the agreement of the mean normalized distance of a species from the centerline $\hat{d} = 2\bar{d}/H$ (see Fig. 11.7a) between the two methods.

The parameters n_0 and ξ in the HMC method are expected to depend on ϕ and $2a/H$, while their dependence on Ca_s, Ca_f, and X_f is expected to be weak – this assumption was corroborated by the results. To estimate n_0 and ξ corresponding to a suspension with a given (ϕ, $2a/H$, Ca_s, Ca_f), the following two-step procedure was adopted: (i) consider the pure suspension of stiff particles and tune the value of n_0 to obtain a good match in $\hat{d}$ from HMC and BI simulations, and (ii) consider the binary suspension with $X_f = 0.5$ and tune the value of ξ, while using the value of n_0 obtained in (i), to obtain a good match in $\hat{d}$ of both the species between the two methods. As an example, the authors considered a binary suspension with $\phi = 0.04$ and $\mathrm{Ca}_s = 0.2$, $\mathrm{Ca}_f = 0.5$, and $2a/H = 0.197$. For this system, the above procedure to estimate n_0 and ξ yields $n_0 = 0.026a^{-2}$ and $\xi = 0.23$. The HMC method can then be used to *predict* results for other suspensions at the same ϕ and $2a/H$. To show this, consider the same binary suspension as above ($\mathrm{Ca}_s = 0.2$, $\mathrm{Ca}_f = 0.5$) and predict $\hat{d}$ for both the species for a range of X_f and compare them with the corresponding BI results (Fig. 11.8a). Excellent agreement can be observed at all values of X_f. A similar close agreement was also observed for different sets of (Ca_s,Ca_f): (0.1, 0.5) and (0.3, 0.4) with no adjustment of n_0 and ξ (see [111]). The HMC method described above when repeated for higher volume fraction suspensions gave a similar good agreement with BI simulations. Note that the highest volume fraction considered by the authors was $\phi = 0.2$.

Besides the averaged measure $\hat{d}$, the authors also compared the particle number density distribution in the wall normal direction in the HMC and BI simulations. Results for the normalized number density distribution $\hat{n}_\alpha(y)$ are shown in Figs. (11.8c) and (11.8d) for the $\mathrm{Ca}_s = 0.2$ and $\mathrm{Ca}_f = 0.5$ mixture at $\phi = 0.04$, $2a/H = 0.197$, and $X_f = 0.5$ (see [111] for a complete set of plots). Very good agreement of $\hat{n}$ with the BI results is observed in all cases. The agreement is remarkably good in the region around the centerline, though the peak near the wall is usually smeared in the HMC results in comparison to the BI results. Nonetheless,

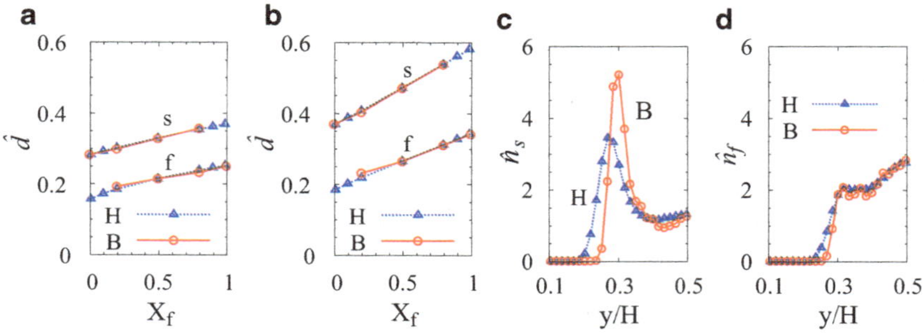

Fig. 11.8 (**a**) and (**b**): Mean normalized distance of a species $\hat{d} = 2\bar{d}/H$ from the centerline in HMC (H) and BI (B) methods in (**a**) $(\mathrm{Ca}_s, \mathrm{Ca}_f, \phi_v) = (0.2, 0.5, 0.04)$ mixture as a function of X_f and in (**b**) $(0.2, 0.5, 0.12)$ mixture. (**c**) and (**d**): Normalized number density profile $\hat{n}$ for (**c**) stiff and (**d**) floppy particles at $X_f = 0.5$ for the suspension in (**a**). In all cases $2a/H = 0.197$ [111]

given the broad agreement of $\hat{d}$ as well $\hat{n}$ in HMC and BI simulations in various suspensions, it is apparent that the HMC model captures the key aspects of the particle distributions in these suspensions.

3.2.3 Mechanisms of Flow-Induced Segregation

A key benefit of the HMC approach is that it allows estimation of independent contributions from various sources of particle dynamics on the margination behavior. To illustrate this, let us consider the same suspension as above with parameters $\mathrm{Ca}_s = 0.2$, $\mathrm{Ca}_f = 0.5$, $\phi = 0.04$, and $2a/H = 0.197$. The HMC results for this system can be generated by setting $n_0 = 0.026a^{-2}$ and $\xi = 0.23$ (see Sect.). Consider first the prediction from the *full* HMC model—for this case, Figs. (11.9a) and (11.9d) show $\hat{n}$ for both the stiff and floppy particles at $X_f = 0.01$ (dilute in floppy) and $X_f = 0.99$ (dilute in stiff), respectively. It is clear from these plots that the stiff particles accumulate in the particle layer formed nearest to the wall as they become dilute in the suspension, i.e., they marginate with increasing X_f. In contrast, the floppy particles do the opposite as they become dilute (X_f decreases), accumulating near the centerline and thus demarginating.

To disentangle the effects of wall-induced migration and pair collisions, a number of control cases were considered. First, the impact of heterogeneous collisions was estimated by (i) setting the particle migration velocities of both the species to the simple average migration velocity of these two species and (ii) setting Δ^{ss} and Δ^{ff} to the average value for the two species. Therefore, the only difference between the two species is their behavior in heterogeneous collisions: $\Delta^{sf} > \Delta^{fs}$. Plots of $\hat{n}(y)$ for these simulations are shown in Figs. (11.9b) and (11.9e). The difference between Δ^{sf} and Δ^{fs} is sufficient to lead to a segregation between the two species. In fact, as will be quantified shortly, most of the segregation appears to result from heterogeneous collisions.

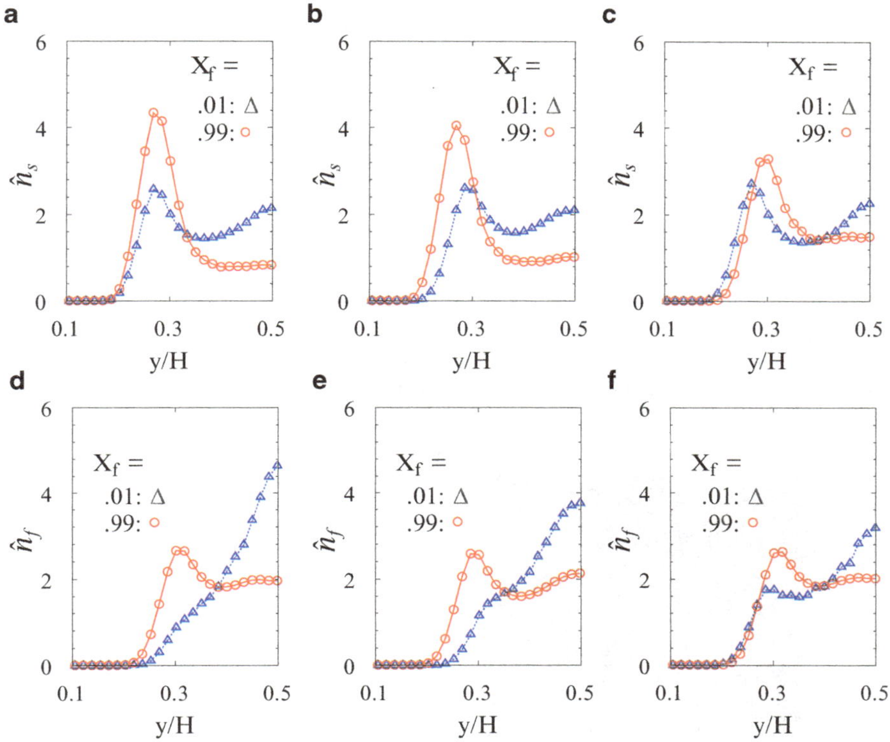

Fig. 11.9 Number density $\hat{n}$ for stiff (*top row*) and floppy (*bottom row*) particles in simulations at $X_f = 0.01$ and $X_f = 0.99$, from: (**a**) and (**d**), full model (case A); (**b**) and (**e**), only difference between the two species being heterogeneous collisions (case B); (**c**) and (**f**), only difference between the two species being migration velocities (case C). These plots are for $(\mathrm{Ca}_s, \mathrm{Ca}_f, \phi_v) = (0.2, 0.5, 0.04)$ [111]

The effect of differences in migration velocity on the segregation behavior was estimated by setting the cross-stream displacement in all types of collisions for both the species to the simple average of the four curves on Fig. (11.7d), yielding $\Delta^{sf} = \Delta^{ss} = \Delta^{fs} = \Delta^{ff}$. Plots for $\hat{n}(y)$ in this case are shown in Figs. (11.9c) and (11.9f). Here too some segregation is observed, though the degree of segregation is considerably smaller than in the full model.

The degree of segregation between the two species is more quantitatively characterized by computing the difference in $\hat{d}$ of each of the species from the corresponding pure species result; this is denoted by $\Delta\hat{d}$. The plots for $\Delta\hat{d}$ in various cases described above are shown in Figs (11.10a) and (11.10b) for the stiff and floppy particles, respectively. For the present parameter set, the degree of segregation from the full model (case A) and the model where only heterogeneous collisions are distinct (case B) is almost identical, while that resulting from differences in the migration velocity (case C) is much weaker.

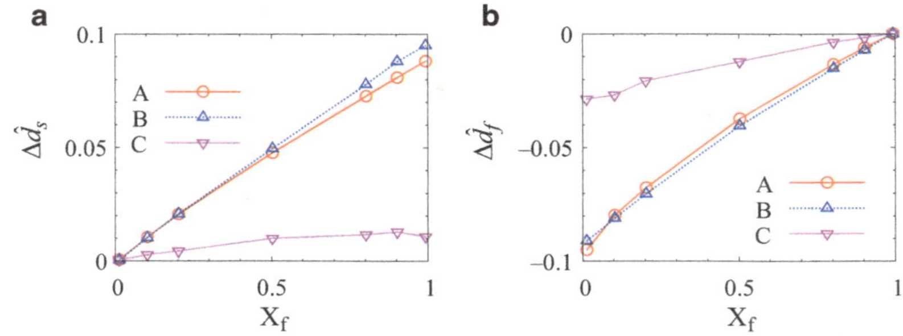

Fig. 11.10 The difference $\Delta\hat{d}$ between $\hat{d}$ of a species in the mixture and the $\hat{d}$ of that species in its pure suspension for (**a**) stiff particles and (**b**) floppy particles in $(\mathrm{Ca}_s, \mathrm{Ca}_f, \phi_v)$=(0.2, 0.5, 0.04) suspension. The cases A–C are described in the text [111]

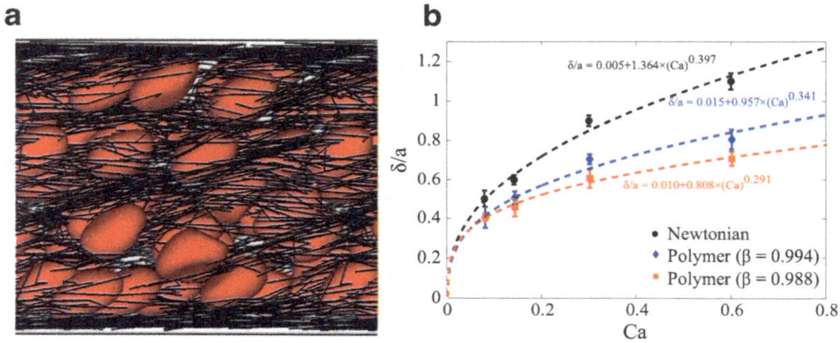

Fig. 11.11 (**a**) Snapshot of a suspension of capsules (Ca = 0.30, ϕ = 0.10) in a polymer solution (β = 0.994, Wi = 20) in a cubic box of size 10a. Polymer molecules are shown as thin black lines. (**b**) Dependence of capsule-depleted layer thickness on Ca for suspensions (phi = 0.10) of capsules in Newtonian and polymer (Weissenberg number Wi = 20) solutions with different values of β in a cubic box of size 10a. Symbols are the simulation results and lines are the fits. The standard deviation is based on results from different initial configurations [54]

Finally, similar studies conducted at higher volume fractions (up to $\phi = 0.2$) show that the effects of heterogeneous collisions and differential migration velocities (cases B and C) are comparable (see [111]). In blood flow in the microcirculation, the average volume fraction is between $\phi_v = 0.1$ and 0.25 and it appears that in this regime, both the migration velocity and the heterogeneous collisions may be playing an important role in the segregation between the various species.

4 Effect of Polymer Additives

The above results illustrate the basic phenomena and mechanisms of cell-free-layer formation and margination. In Sect. 1.4 we reviewed experiments that show that these phenomena are altered when long-chain polymers are added to blood plasma. Pranay et al. [54, 125] have simulated the effects of polymers on the flow of elastic capsules, using a variant of the boundary integral method described above along with simple bead-spring-chain models of polymer molecules [162]. Simulations of an isolated capsule in a polymer solution indicate that wall-induced migration is dramatically attenuated even with very small polymer concentrations. This observation is consistent with numerical simulation results of a single rigid sphere in viscoelastic fluid in Couette flow by D'Avino et al. [163]. They showed that viscoelasticity induces particle motion toward the wall and made a heuristic argument that the viscoelasticity leads to normal stress differences in the solution that are asymmetric in the presence of a wall and this asymmetry leads to the cross-stream migration toward the wall. These results are qualitatively consistent with the observations of Pranay et al. [54], suggesting that the viscoelastic effects that drive the particle toward the wall are competing with the dipole-wall interaction effects that drive a deformable particle away from the wall. Reduction of the migration velocity is expected to lead to thinning of the cell-free layer and this is indeed what is observed in simulations. Figure 11.11a shows a snapshot from a simulation of a suspension of capsules in a polymer solution. Consistent with the expectation from the single particle migration results, as well as experimental results, the addition of polymers attenuates cell-free layer formation. Figure 11.11 shows the cell-free layer thickness as a function of capillary number for a suspension in a Newtonian fluid as well as in polymer solutions with different concentrations (the quantity $1-\beta$ is proportional to polymer concentration). In the simulations the Weissenberg number is held constant at 20 as capillary number increases—in experiments capillary and Weissenberg numbers increase together, since they are both proportional to shear rate. Keeping in mind this difference, we nevertheless see that the results of Fig. 11.11 and 11.3d are qualitatively consistent.

We noted above that experiments indicate that platelet margination is impeded in the presence of polymers, and there is indirect evidence from the tissue scaffold degradation experiments margination of leukocytes is probably disrupted as well. Given the thinning of the cell-free layer in the presence of polymer, one might expect disruption of margination since the RBCs will be present very close the wall, thus knocking the platelets and WBCs back into the bulk flow. This hypothesis awaits systematic experimental and computational study.

5 Conclusions and Outlook

Blood flow is highly complex and exhibits a variety of interesting and physiologically important phenomena. Most of these arise from the particulate nature of blood: examples include the Fåhræus effect, Fåhræus-Lindqvist effect, and margination of leukocytes and platelets. The Fåhræus and Fåhræus-Lindqvist effects are now fairly well understood—these arise due to the deformability of the RBCs, as a result of which they tend to migrate away from vessel walls toward the vessel centers. However, margination, while being known for decades, is still poorly understood.

Some recent progress is nevertheless discernible, primarily due to detailed computer simulations as well as advances in theoretical descriptions of phenomena in flowing suspensions.

Margination occurs primarily due to the size, shape, and deformability differences between the majority component, RBCs, and the leukocytes and platelets, which are present in trace amounts. Simulation efforts have mostly focused on model problems mimicking blood. While these studies are important, they are not well suited to isolate specific mechanisms of margination, as in these studies all of the above three determinants of margination vary simultaneously. Well-controlled studies are necessary where only one of these properties are different at a time. Some recent studies taking this approach have been performed [110, 111], yielding some progress in elucidating the role of deformability on margination. These studies considered a binary suspension in which the two components have the same spherical rest shape and size, but have unequal membrane rigidities. Stiffer particles undergo margination when they are the dilute component in the suspension, in agreement with the margination of leukocytes and platelets, as they are stiffer than the RBCs. Additionally, a simple master equation model for suspension dynamics has been developed that incorporates the two key sources of particle motion in confined deformable particle suspensions: wall-induced migration and hydrodynamic pair collisions. Using this model, it is found that heterogeneous pair collisions, i.e., collisions between a stiff and a floppy particle, contribute significantly to the margination behavior, particularly at low volume fractions; at higher volume fractions, the collisions as well as migration make comparable contributions. In heterogeneous pair collisions, the stiffer particles undergo larger cross-stream displacement than floppy particles, which result in the former getting pushed toward the wall more than the latter. The wall-induced migration also leads to the margination of the stiffer particles, as the stiffer particles have a smaller migration velocity than floppy particles.

There are no well-controlled studies examining the effects of size and shape on margination, though some preliminary results exist for the former [112]. Among these, the effect of shape could be particularly challenging to explore as it is not clear a priori how to prepare systems where the components vary only by shape while keeping their size and deformability the same. Perhaps one could use the volume of the particle as a measure of size and some "characteristic" capillary number as the measure of deformability. Studies of nonspherical particles are also complicated as these particles display a variety of unsteady motions in flow [164], which complicates the interpretation of results on migration and pair collision dynamics. These challenges present exciting opportunities for further research. The outcome of this research is also expected to have practical applications, for example, in the area of optimal drug delivery particle design or in label-free hydrodynamic sorting of multicomponent suspensions in microfluidic devices. This research will also be beneficial for understanding the consequences of various diseases like sickle cell disease and malaria where the mechanical and geometric properties of RBCs are known to be altered—these changes are likely to modify the distribution of cells in the circulation and hence affect various blood functions.

The study of rheological modifiers such as drag-reducing polymers on blood flow is also nascent. Recent simulation studies have shown that the wall-induced migration of RBCs is attenuated in the presence of long-chain polymers [54]. This explains the experimental observations of cell-free-layer thickness reduction with the addition of polymers. However, simulations studies also show an increase in wall shear stress and hence increased dissipation. This result contradicts the in vivo studies showing an increase in cardiac output. Examining the source of this discrepancy and providing a resolution is also expected to be a fertile area for future research. Ultimately, adaptation of blood rheology modifiers for therapies will be greatly facilitated by a microscopic and mechanistic understanding of their effects. The advances described in this chapter provide steps toward this understanding.

Acknowledgment The authors' research on issues related to blood flow has been supported by NSF, grants CBET-0852976 (funded under the American Recovery and Reinvestment Act of 2009) and CBET-1132579. The authors are grateful to Prof. Wilbur Lam and to Kushal Sinha, Pratik Pranay, and Rafael Henriquez Rivera for helpful discussions.

References

1. T. Audesirk, G. Audesirk, *Biology: Life on Earth*, 4th edn. (Prentice-Hall, Upper Saddle River, NJ, 1996)
2. Y.C. Fung, B.W. Zweifach, Annu. Rev. Fluid Mech. **3**, 189 (1971)
3. M. Puig-De-Morales-Marinkovic, K.T. Turner, J.P. Butler, J.J. Fredberg, S. Suresh, Am. J. Physiol. Cell Physiol. **293**(2), C597 (2007)
4. A. Kumar, M.D. Graham, Soft Matter **8**, 10536 (2012)
5. D. Boal, *Mechanics of the Cell*, 2nd edn. (Cambridge University Press, Cambridge, 2012)
6. B. Alberts, A. Johnson, J. Lewis, M. Raff, K. Roberts, P. Walter, *Molecular Biology of the Cell*, 5th edn. (Garland Science, New York, 2007)
7. M.J. Hickey, P. Kubes, Nat. Rev. Immunol. **9**(5), 364 (2009)
8. S.P. Jackson, Blood **109**(12), 5087 (2007)
9. J.A. Leopold, J. Loscalzo, in *Platelets, Thrombosis and the Vessel Wall*, ed. by M.C. Berndt (Harwood Academic Publishers, Amsterdam, 2000)
10. D.M. Wootton, D.N. Ku, Annu. Rev. Biomed. Eng. **1**, 299 (1999)
11. B. Nieswandt, B. Aktas, A. Moers, U.J.H. Sachs, J. Thrombosis Haemostasis **3**(8), 1725 (2005)
12. R. Tran, D.R. Myers, J. Ciciliano, E.L. Trybus Hardy, Y. Sakurai, B. Ahn, Y. Qiu, R.G. Mannino, M.E. Fay, W.A. Lam, J. Cell. Mol. Med. **17**(5), 579 (2013)
13. R.D. Guy, A.L. Fogelson, J.P. Keener, Math. Med. **24**(1), 111 (2007)
14. J. Cho, D.F. Mosher, J. Thrombosis and Haemostasis **4**(7), 1461 (2006)
15. S.W. Schneider, S. Nuschele, A. Wixforth, C. Gorzelanny, A. Alexander-Katz, R.R. Netz, M.F. Schneider, Proc. Natl. Acad. Sci. USA **104**(19), 7899 (2007)
16. J.E. Sadler, Annu. Rev. Biochem. **67**(1), 395 (1998)
17. B. Savage, E. Saldivar, Z. Ruggeri, Cell **84**(2), 289 (1996)
18. T.A. Springer, J. Thrombosis Haemostasis **9**, 130 (2011)
19. C.M. Ward, M.C. Berndt, in *Platelets, Thrombosis and the Vessel Wall*, ed. by M.C. Berndt (Harwood Academic Publishers, Amsterdam, 2000)
20. C. Siedlecki, B. Lestini, K. KottkeMarchant, S. Eppell, D. Wilson, R. Marchant, Blood **88**(8), 2939 (1996)

21. Y.C. Fung, *Biodynamics: Circulation* (Springer, New York, 1996)
22. H.H. Lipowsky, Biorheology **50**, 3 (2013)
23. E.W. Merrill, Physiol. Rev. **49**(4), 863 (1969)
24. M. Brust, C. Schaefer, R. Doerr, L. Pan, M. Garcia, P.E. Arratia, C. Wagner, Phys. Rev. Lett. **110**(7), 078305 (2013)
25. S. Chien, Science **168**(3934), 977 (1970)
26. G.B. Thurston, Biophys. J. **12**(9), 1205 (1972)
27. W.B. Russel, D.A. Saville, W.R. Schowalter, *Colloidal Dispersions* (Cambridge University Press, Cambridge, 1989)
28. B. Neu, H.J. Meiselman, J. Biophys. **83**(5), 2482 (2002)
29. M. Rampling, H. Meiselman, B. Neu, O. Baskurt, Biorheology **41**(2), 91 (2004)
30. R. Fåhræus, T. Lindqvist, Am. J. Physiol. **96**, 562 (1931)
31. A. Pries, D. Neuhaus, P. Gaehtgens, Am. J. Physiol. **263**(6), H1770 (1992)
32. J.H. Barbee, G.R. Cokelet, Microvasc. Res. **3**(1), 6 (1971)
33. G.A. Truskey, F. Yuan, D.F. Katz, *Transport Phenomena in Biological Systems* (Pearson Prentice Hall, 2004)
34. R. Skalak, N. Ozkaya, T.C. Skalak, Annu. Rev. Fluid Mech. **21**, 167 (1989)
35. S. Kim, P.K. Ong, O. Yalcin, M. Intaglietta, P.C. Johnson, Biorheology **46**(3), 181 (2009)
36. A.R. Pries, T.W. Secomb, P. Gaehtgens, Cardiovasc. Res. **32**(4), 654 (1996)
37. K. Svanes, B.W. Zweifach, Microvasc. Res. **1**(2), 210 (1968)
38. Y.C. Fung, Microvasc. Res. **5**(1), 34 (1973)
39. V. Doyeux, T. Podgorski, S. Peponas, M. Ismail, G. Coupier, J. Fluid Mech. **674**, 359 (2011)
40. S.D. Hudson, Phys. Fluids **15**(5), 1106 (2003)
41. S.K. Doddi, P. Bagchi, Int. J. Multiphase Flow **34**(10), 966 (2008)
42. L.G. Leal, Annu. Rev. Fluid Mech. **12**, 435 (1980)
43. G. Danker, P. Vlahovska, C. Misbah, Phys. Rev. Lett. **102**(14), 148102 (2009)
44. Y.L. Chen, M.D. Graham, J.J. de Pablo, K. Jo, D.C. Schwartz, Macromolecules **38**(15), 6680 (2005)
45. J.R. Smart, D.T. Leighton, Phys. Fluids A **3**(1), 21 (1991)
46. M.D. Graham, Annu. Rev. Fluid Mech. **43**(1), 273 (2011)
47. R.M. Jendrejack, D.C. Schwartz, J.J. de Pablo, M.D. Graham, J. Chem. Phys. **120**(5), 2513 (2004)
48. H. Ma, M.D. Graham, Phys. Fluids **17**(8), 083103 (2005)
49. T. Fischer, C. Haest, M. Stöhr-Liesen, H. Schmid-Schönbein, R. Skalak, Biophys. J. **34**(3), 409 (1981)
50. S. Henon, G. Lenormand, A. Richert, F. Gallet, Biophys. J. **76**(2), 1145 (1999)
51. R. Hochmuth, R. Waugh, Annu. Rev. Physiol. **49**(1), 209 (1987)
52. J. Li, M. Dao, C. Lim, S. Suresh, Biophys. J. **88**(5), 3707 (2005)
53. D. Barthes-Biesel, A. Diaz, E. Dhenin, J Fluid Mech. **460**, 211 (2002)
54. P. Pranay, R.G. Henriquez Rivera, M.D. Graham, Phys. Fluids **24**(6), 061902 (2012)
55. G. Segre, A. Silberberg, J Fluid Mech. **14**(1), 136 (2005)
56. G. Segre, A. Silberberg, J Fluid Mech. **14**(1), 115 (2005)
57. B.P. Ho, L.G. Leal, J. Fluid Mech. **65**(02), 365 (1974)
58. J.A. Schonberg, E.J. Hinch, J Fluid Mech **203**, 517 (1989)
59. D. Di Carlo, Lab Chip **9**(21), 3038 (2009)
60. E.J. Lim, T.J. Ober, J.F. Edd, G.H. McKinley, M. Toner, Lab Chip **12**(12), 2199 (2012)
61. J.C. Firrell, H.H. Lipowsky, Am. J. Physiol.-Heart C. **256**(6), H1667 (1989)
62. G.J. Tangelder, H.C. Teirlinck, D.W. Slaaf, R.S. Reneman, Am. J. Physiol.-Heart C. **248**(3), H318 (1985)
63. A.S. Popel, P.C. Johnson, Annu. Rev. Fluid Mech. **37**, 43 (2005)
64. A.W. Browne, L. Ramasamy, T.P. Cripe, C.H. Ahn, Lab Chip **11**(14), 2440 (2011)
65. A. Jain, L.L. Munn, Lab Chip **11**(17), 2941 (2011)
66. H.W. Hou, A.A.S. Bhagat, A.G.L. Chong, P. Mao, K.S.W. Tan, J. Han, C.T. Lim, Lab Chip **10**(19), 2605 (2010)

67. S.C. Hur, N.K. Henderson-MacLennan, E.R.B. McCabe, D. Di Carlo, Lab Chip **11**(5), 912 (2011)
68. M.J. Pearson, H.H. Lipowsky, Am. J. Physiol.-Heart C. **279**(4), H1460 (2000)
69. H.L. Goldsmith, S. Spain, Microvasc. Res. **27**(2), 204 (1984)
70. K.B. Abbitt, G.B. Nash, Am. J. Physiol.-Heart C. **285**(1), H229 (2003)
71. A. Jain, L.L. Munn, PLoS ONE **4**, e7104 (2009)
72. D. Fedosov, J. Fornleitner, G. Gompper, Phys. Rev. Lett. **108**(2) (2012)
73. C. Sun, C. Migliorini, L.L. Munn, Biophys. J. **85**(1), 208 (2003)
74. J.B. Freund, Phys. Fluids **19**(2), 023301 (2007)
75. B. Woldhuis, G.J. Tangelder, D.W. Slaaf, R.S. Reneman, Am. J. Physiol.-Heart C. **262**(4), H1217 (1992)
76. P.A. Aarts, S.A.V.D. Broek, G.W. Prins, G.D. Kuiken, J.J. Sixma, R.M. Heethaar, Arterioscl. Throm. Vas. **8**(6), 819 (1988)
77. W. Tilles, E.C. Eckstein, Microvasc. Res. **33**, 211 (1987)
78. C. Yeh, E.C. Eckstein, J. Biophys. **66**(5), 1706 (1994)
79. L. Crowl, A.L. Fogelson, J. Fluid Mech. **676**, 348 (2011)
80. H. Zhao, E. Shaqfeh, Phys. Rev. E **83**(6) (2011)
81. T. Al Momani, H.S. Udaykumar, J.S. Marshall, K.B. Chandran, Ann. Biomed. Eng. **36**(6), 905 (2008)
82. R.B. Huang, S. Mocherla, M.J. Heslinga, P. Charoenphol, O. Eniola-Adefeso, Mol. Membr. Biol. **27**, 312 (2010)
83. F. Gentile, C. Chiappini, D. Fine, R.C. Bhavane, M.S. Peluccio, M. Cheng, X. Liu, M. Ferrari, P. Decuzzi, J. Biomech. **41**(10), 2312 (2008)
84. K. Loomis, K. McNeeley, R.V. Bellamkonda, Soft Matter **7**(3), 839 (2010)
85. H. Cabral, Y. Matsumoto, K. Mizuno, Q. Chen, M. Murakami, M. Kimura, Y. Terada, M.R. Kano, K. Miyazono, M. Uesaka, et al., Nat. Nanotechnol. **6**(12), 815 (2011)
86. R.E. Serda, B. Godin, E. Blanco, C. Chiappini, M. Ferrari, Biochim. Biophys. Acta **1810**(3), 317 (2011)
87. P. Decuzzi, B. Godin, T. Tanaka, S. Lee, C. Chiappini, X. Liu, M. Ferrari, J. Controlled Release **141**(3), 320 (2010)
88. P. Charoenphol, R.B. Huang, O. Eniola-Adefeso, Biomaterials **31**(6), 1392 (2010)
89. J. Tan, A. Thomas, Y. Liu, Soft Matter **8**, 1934 (2012)
90. F.I. Faruqui, M.D. Otten, P.I. Polimeni, Circulation **75**(3), 627 (1987)
91. C.A. Macias, M.V. Kameneva, J.J. Tenhunen, J.C. Puyana, M.P. Fink, Shock **22**(2), 151 (2004)
92. P.J. Marascalco, H.C. Blair, A. Nieponice, L.J. Robinson, M.V. Kameneva, ASAIO J. **55**(5), 503 (2009)
93. M.V. Kameneva, Z.J.J. Wu, A. Uraysh, B. Repko, K.N. Litwak, T.R. Billiar, M.P. Fink, R.L. Simmons, B. Griffith, H. Borovetz, Biorheology **41**(1), 53 (2004)
94. P.S. Virk, AIChE J. **21**(4), 625 (1975)
95. H.L. Greene, R.F. Mostardi, R.F. Nokes, Polym. Eng. Sci. **20**(7), 499 (1980)
96. N. Antonova, Z. Lazarov, Clin. Hemorheol. Micro. **30**(3–4), 381 (2004)
97. J.J. Pacella, M.V. Kameneva, M. Csikari, E. Lu, F.S. Villanueva, Eur. Heart J. **27**(19), 2362 (2006)
98. T. Sakai, B.M. Repko, B.P. Griffith, J.H. Waters, M.V. Kameneva, Brit. J. Anaesth. **98**(1), 23 (2006)
99. J.J. Pacella, M.V. Kameneva, F.S. Villanueva, Biorheology **46**(5), 365 (2009)
100. X. Chen, D. Zha, J. Xiu, Y. Liao, K. Cui, H. Lin, Z. Jian, F. Hu, X. Huang, B. Zhou, Q. Huang, J. Bin, Y. Liu, Int. J. Cardiol. **147**(1), 112 (2011)
101. A. Cotoia, M.V. Kameneva, P.J. Marascalco, M.P. Fink, R.L. Delude, Shock **31**(3), 258 (2009)
102. M.V. Kameneva, M.S. Polyakova, E.V. Fedoseeva, Fluid Dyn. **25**(6), 956 (1991)
103. P.C. Sousa, P.M. Coelho, M.S.N. Oliveira, M.A. Alves, J. Non-Newton. Fluid Mech. **166** (17–18), 1033 (2011)

104. G. Schettler, H. Schmid-Schonbein, H. Morl, H. Diehm (eds.), *Fluid Dynamics as a localizing Factor for Atherosclerosis* (Springer, Berlin, 1983)
105. J.N. Marhefka, R. Zhao, Z.J. Wu, S.S. Velankar, J.F. Antaki, M.V. Kameneva, Biorheology **46**(4), 281 (2009)
106. R. Zhao, J. Marhefka, J. Antaki, M. Kameneva, Biorheology **47**(3), 193 (2010)
107. A.L. Fogelson, R.D. Guy, Comput. Method. Appl. M. **197**(25–28), 2087 (2008)
108. H. Zhao, E.S.G. Shaqfeh, V. Narsimhan, Phys. Fluids **24**(1), 011902 (2012)
109. E.C. Eckstein, F. Belgacem, Biophys. J. **60**, 53 (1991)
110. A. Kumar, M.D. Graham, Phys. Rev. E **84**(6), 066316 (2011)
111. A. Kumar, M.D. Graham, Phys. Rev. Lett. **109**, 108102 (2012)
112. A. Kumar, R.G. Henriquez Rivera, M.D. Graham, J. Fluid Mech. **738**, 423 (2014)
113. H.S. Lew, Y.C. Fung, Biophys. J. **10**(1), 80 (1970)
114. C. Pozrikidis, *Boundary Integral and Singularity Methods for Linearized Viscous Flow* (Cambridge University Press, Cambridge 1992)
115. A. Kumar, M.D. Graham, J. Comput. Phys. **231**, 6682 (2012)
116. M. Loewenberg, E. Hinch, J. Fluid Mech. **321**(8), 395 (1996)
117. S. Ramanujan, C. Pozrikidis, J. Fluid Mech. **361**, 117 (1998)
118. A.Z. Zinchenko, R.H. Davis, J. Fluid Mech. **455**, 21 (2002)
119. H. Zhao, A.H.G. Isfahani, L.N. Olson, J.B. Freund, J. Comput. Phys. **229**, 3726 (2010)
120. G. Ghigliotti, T. Biben, C. Misbah, J. Fluid Mech. **653**(1), 489 (2010)
121. C. Peskin, Acta Numerica **11**, 479 (2002)
122. D.M. McQueen, C.S. Peskin, ACM SIGGRAPH Computer Graphics **34**(1), 56 (2000)
123. G. Tryggvason, B. Bunner, A. Esmaeeli, D. Juric, N. Al-Rawahi, W. Tauber, J. Han, S. Nas, Y.J. Jan, J. Comput. Phys. **169**(2), 708 (2001)
124. P. Bagchi, Biophys. J. **92**, 1858 (2007)
125. P. Pranay, S.G. Anekal, J.P. Hernandez-Ortiz, M.D. Graham, Phys. Fluids **22**, 123103 (2010)
126. S. Chen, G.D. Doolen, Annu. Rev. Fluid Mech. **30**(1), 329 (1998)
127. A. Ladd, R. Verberg, J. Stat. Phys. **104**(5–6), 1191 (2001)
128. P. Hoogerbrugge, J. Koelman, EPL (Europhys. Lett.) **19**(3), 155 (1992)
129. P.B. Warren, Curr. Opin. Colloid Interface Sci. **3**(6), 620 (1998)
130. R.M. MacMeccan, J.R. Clausen, G.P. Neitzel, C.K. Aidun, J. Fluid Mech. **618**, 13 (2009)
131. I.V. Pivkin, G.E. Karniadakis, Phys. Rev. Lett. **101**(11), 118105 (2008)
132. S. Kim, S.J. Karrila, *Microhydrodynamics: Principles and Selected Applications* (Dover Publicatons, Newyork, 2005)
133. G. Muldowney, J.J.L. Higdon, J. Fluid Mech. **298**, 167 (1995)
134. S.K. Veerapaneni, A. Rahimian, G. Biros, D. Zorin, J. Comput. Phys. **230**, 5610 (2011)
135. M. Deserno, C. Holm, J. Chem. Phys. **109**, 7678 (1998)
136. L. Greengard, V. Rokhlin, J. Comput. Phys. **73**, 325 (1987)
137. H. Hasimoto, J. Fluid Mech. **5**, 317 (1959)
138. A. Rahimian, S.K. Veerapaneni, G. Biros, J. Comput. Phys. **229**, 6466 (2010)
139. J.P. Hernandez-Ortiz, J.J. de Pablo, M.D. Graham, Phys. Rev. Lett. **98**, 140602 (2007)
140. N. Mohandas, E. Evans, Annu. Rev. Bioph. Biom. **23**(1), 787 (1994)
141. N. Mohandas, P.G. Gallagher, Blood **112**(10), 3939 (2008)
142. P. Dimitrakopoulos, Phys. Rev. E **85**(4), 041917 (2012)
143. C. Pozrikidis, J. Comput. Phys. **169**(2), 250 (2001)
144. D.A. Fedosov, B. Caswell, G.E. Karniadakis, Biophys. J. **98**(10), 2215 (2010)
145. T.W. Secomb, *Mechanics of Red Blood Cells and Blood Flow in Narrow Tubes* (Chapman and Hall/CRC, London, 2011), pp. 1–19
146. D. Barthes-Biesel, C. R. Physique **10**(8), 764 (2009)
147. E. Lac, D. Barthes-Biesel, Phys. Fluids **17**(7), 072105 (2005)
148. Z. Peng, A. Mashayekh, Q. Zhu, J. Fluid Mech. **742**, 96 (2014)
149. D. Cordasco, A. Yazdani, P. Bagchi, Phys. Fluids **26**(4), 041902 (2014)
150. T.M. Fischer, Biophys. J **86**(5), 3304 (2004)
151. A. Viallat, M. Abkarian, Int. Jnl. Lab. Hem. **36**(3), 237 (2014)

152. R. Skalak, A. Tozeren, R.P. Zarda, S. Chien, Biophys. J. **13**, 245 (1973)
153. R. Waugh, E. Evans, Biophys. J. **26**(1), 115 (1979)
154. M. Zurita-Gotor, J. Blawzdziewicz, E. Wajnryb, Phys. Rev. Lett. **108**(6), 68301 (2012)
155. V. Narsimhan, H. Zhao, E.S.G. Shaqfeh, Phys. Fluids **25**(6), 061901 (2013)
156. F.R. Da Cunha, E.J. Hinch, J. Fluid Mech. **309**(1), 211 (1996)
157. G.A. Bird, *Molecular Gas Dynamics and the Direct Simulation of Gas Flows* (Oxford University Press, Oxford, 1994)
158. M.S. Ivanov, S.F. Gimelshein, Annu. Rev. Fluid Mech. **30**(1), 469 (1998)
159. G.A. Bird, Annu. Rev. Fluid Mech. **10**(1), 11 (1978)
160. G. Drazer, J. Koplik, B. Khusid, A. Acrivos, J. Fluid Mech. **460**, 307 (2002)
161. K. Koura, Phys. Fluids **29**, 3509 (1986)
162. R.B. Bird, C.F. Curtiss, R.C. Armstrong, O. Hassager, vol. 2, *Dynamics of Polymeric Liquids* (Wiley-Interscience, New York, 1987)
163. G. D'Avino, P. Maffettone, F. Greco, M. Hulsen, J. Non-Newton. Fluid Mech. **165**(9–10), 466 (2010)
164. C. Dupont, A.V. Salsac, D. Barthès-Biesel, J. Fluid Mech. **721**, 180 (2013)

Index

S.E. Spagnolie (ed.), *Complex Fluids in Biological Systems*, Biological and Medical Physics, Biomedical Engineering,
DOI 10.1007/978-1-4939-2065-5

CPSIA information can be obtained at www.ICGtesting.com
Printed in the USA
BVOW10*0818191214

380170BV00001B/3/P

9 781493 920648